Violations of Hyperscaling in Phase Transitions and Critical Phenomena—in Memory of Prof. Ralph Kenna

Violations of Hyperscaling in Phase Transitions and Critical Phenomena—in Memory of Prof. Ralph Kenna

Guest Editors

Bertrand Berche
Yurij Holovatch

MDPI

Basel • Beijing • Wuhan • Barcelona • Belgrade • Novi Sad • Cluj • Manchester

Guest Editors

Bertrand Berche
Laboratoire de Physique et
Chimie Théoriques
Université de Lorraine
Vandœuvre-les-Nancy
France

Yurij Holovatch
Yukhnovskii Institute for
Condensed Matter Physics
National Academy of
Sciences of Ukraine
Lviv
Ukraine

Editorial Office
MDPI AG
Grosspeteranlage 5
4052 Basel, Switzerland

This is a reprint of the Special Issue, published open access by the journal *Entropy* (ISSN 1099-4300), freely accessible at: https://www.mdpi.com/si/entropy/UK478MJKE7.

For citation purposes, cite each article independently as indicated on the article page online and as indicated below:

Lastname, A.A.; Lastname, B.B. Article Title. *Journal Name* **Year**, *Volume Number*, Page Range.

ISBN 978-3-7258-5147-8 (Hbk)
ISBN 978-3-7258-5148-5 (PDF)
https://doi.org/10.3390/books978-3-7258-5148-5

Cover image courtesy of Bertrand Berche

Contents

About the Editors

Bertrand Berche

Bertrand Berche is Professor of Theoretical Physics at the University of Lorraine, France, where he has been teaching and conducting research for over thirty years. After earning his master's degree at Henri Poincaré University in Nancy, he completed a PhD in Statistical Physics in 1991 under the mentorship of Loïc Turban, followed by a habilitation in 1997. His academic career, beginning as a Lecturer in 1991 and having been a Professor since 1998, has been marked by a sustained commitment to both education and research. Passionate about collaboration and international exchange, he has taught and worked with colleagues in Europe, Latin America, and Africa, with particularly enriching experiences in Mauritania, Venezuela, Ukraine, Ecuador, Brazil, and Cameroon. His research interests span diverse topics in theoretical physics, and, in particular, in statistical physics.

Yurij Holovatch

Yurij Holovatch is a professor and chief researcher at the Yukhnovskii Institute for Condensed Matter Physics of the National Academy of Sciences of Ukraine. Yurij studies complex systems and complex networks (ordering, stability, and spreading), phase transitions and critical phenomena in structurally disordered magnets, and scaling of macromolecules (conformational properties of complex polymers). Yurij is a full member of the National academy of sciences of Ukraine. He was awarded the Davydov prize for studies in theoretical and biological physics (2020). Since 2018, he has held the honorary title of visiting professor at Coventry University; and since 2024 he has been an external faculty member at the Complexity Science Hub in Vienna.

entropy MDPI

Editorial

Violations of Hyperscaling in Phase Transitions and Critical Phenomena—In Memory of Prof. Ralph Kenna

Bertrand Berche [1,2,*] and Yurij Holovatch [2,3,4,5,*]

1 Laboratoire de Physique et Chimie Théoriques, Université de Lorraine, BP 70239, CEDEX, 54506 Vandœuvre-les-Nancy, France
2 $\mathbb{L}^4$ Collaboration & Doctoral College for the Statistical Physics of Complex Systems, Leipzig-Lorraine-Lviv-Coventry, 79011 Lviv, Ukraine
3 Yukhnovskii Institute for Condensed Matter Physics, National Academy of Sciences of Ukraine, 79011 Lviv, Ukraine
4 Centre for Fluid and Complex Systems, Coventry University, Coventry CV1 5FB, UK
5 Complexity Science Hub, 1030 Vienna, Austria
* Correspondence: bertrand.berche@univ-lorraine.fr (B.B.); hol@icmp.lviv.ua (Y.H.)

Ralph Kenna (27 August 1964–26 October 2023)

This Special Issue was initiated by Prof. Ralph Kenna, who at the time was a member of the Entropy Editorial Board. The chosen topic lay at the heart of his scientific endeavors, and we are deeply indebted to him for his significant contributions to this field. At the suggestion of the Entropy Editorial Board, we undertook the task of bringing this project to completion. We have done our utmost to honor Ralph's vision, and we respectfully dedicate this Special Issue to his memory.

Ralph Kenna was born in Athlone (Ireland). He was a theoretical physicist and had very diverse centres of interest, such as statistical physics, complex systems, and Irish mythology. After a B.A. in Theoretical Physics (1985) and an M.Sc. (1988) from Trinity College Dublin, he completed his PhD at the University of Graz under Professor Christian

https://doi.org/10.3390/e27080810

Lang in 1993. Ralph then moved to the University of Liverpool from 1994 to 1997 and to Trinity College Dublin from 1997 to 1999. In 2002, he was hired at Coventry University, where he founded the Applied Mathematics Research Centre, joining, in 2018, the Centre for Fluid and Complex Systems. In 2016, he co-founded the $\mathbb{L}^4$ Collaboration and Doctoral College for the Statistical Physics of Complex Systems, joining the Universities of Coventry, Leipzig, Lorraine, and the Institute for Condensed Matter Physics of the National Academy of Sciences of Ukraine in Lviv (ICMP). He was a Fellow of the Institute of Mathematics and its Applications, Member of the Institute of Physics, and an Advisory Board Member of the Middle European Cooperation on Statistical Physics. In 2019, for his important scientific contributions as well as for his initiative in different forms of collaboration with Ukraine and his engagement in the preparation of young scientists, he was conferred the title of Doctor Honoris Causa of the ICMP. Ralph Kenna passed away on 26 October 2023, and is buried in the St. Kieran's Cemetery at Coosan, close to his home town, Athlone. Below, we will mention only some of his scientific results. A more complete description of his life and work can be found in [1].

In statistical physics, Ralph Kenna was a renowned expert in the study of critical phenomena and phase transitions via the analysis of the zeros of the partition function, an approach introduced by Lee and Yang in the 1950s [2,3] and further elaborated by Fisher in 1960s [4].

Together with his collaborators, Ralph Kenna was renowned for developing scaling relations involving logarithmic corrections [5,6], particularly at the upper critical dimension, then extending this to higher dimensions in 2012 [7], following the pioneering work of Michael Fisher on dangerous irrelevant variables [8,9]. This led to the introduction of the new pseudo-critical exponent (koppa) and its logarithmic counterpart (koppa-hat) to govern the finite-size scaling (FSS) of the correlation length and a new form for FSS, called QFSS, to replace standard prescription above the upper critical dimension [10–14]. Spin systems on scale-free networks also display logarithmic corrections that obey the scaling relations developed by Kenna [15].

Ralph Kenna has always been attracted by multidisciplinary subjects and has not hesitated during his career to tackle societal or human sciences problems. In 2010, Kenna and Berche quantified the notion of critical mass of academic research groups [16,17]. Using data from the UK's Research Assessment Exercise 2008 and the French counterpart (then AERES), they tracked how research group quality depends on the size of the group. They found that quality rises linearly with group size up to a point, which they later identified as akin to the Dunbar number in anthropology [18]. Subsequently, together with his collaborators, he used scientometrics to predict the outcome of the UK's Research Excellence Framework 2014. They established that correlations between metrics and peer review are poor [19].

In the field of comparative mythology, Ralph Kenna was a pioneer in applying complex network analysis to the study of Irish and other mythological traditions. His first paper on the topic [20] was downloaded over 30,000 times in 10 years and resulted in considerable media coverage in the international press. Other major works include investigations into the Sagas of Icelanders [21]. Kenna's team found that whether the sagas are historically accurate or not, the properties of the social worlds they record are similar to those of real social networks. The Viking Age in Ireland as portrayed in Cogadh Gaedhel re Gallaibh was next tackled by his group [22]. They developed a measure to place hostility on a spectrum between civil war and international conflict. Their findings quantified and supported the traditional view of the Viking age in Ireland as one of international conflict and challenged recent revisionist claims. Kenna and co-workers also studied Ukrainian mythology. They compared the Kyiv bylyny cycle to other prominent European epics [23].

In the last months of his life, Ralph Kenna struggled with a serious, incurable illness. Continuing his scientific work was also part of this struggle. Among his various projects was the preparation of a Special Issue of the journal *Entropy*. Below is a brief description of the Special Issue scope written by Prof. Kenna:

> *Universality is an emergent phenomenon, at least partially explained by the renormalization group. Because of universality, simplified theoretical models can deliver critical behaviour of real complex systems by trimming back to essentials such as dimensionality, symmetry group, and range of interaction. Universality classes of theoretical models and real systems are characterised by critical exponents that are linked through scaling relations between them. The scaling relations that involve dimensionality are referred to as hyperscaling. Due to the success of mean-field theory in highly connected systems, irrespective of the dimensionality of the systems, dimension-dependent hyperscaling is often said to fail there. That tenet was challenged recently with the introduction of new insights to the renormalization group aimed to rescue hyperscaling in high dimensions.*
>
> *This Special Issue focuses on high-dimensional and other highly connected systems where hyperscaling is traditionally said to fail. We are interested in robustly supported explorations of how hyperscaling can or cannot be re-instated. The intent of this Special Issue is to capture "state of the art" research in high dimensions and high connectivity and as such we welcome new results and reviews of the highest standard. We are also interested in interdisciplinary applications.*

In the process of working on this Special Issue, its scope has been somewhat expanded. The articles offered to readers now contain, along with issues related to scaling above the upper critical dimensionality, other problems of the theory of phase transitions and critical phenomena [24–32]. We consider such an expansion of the subject appropriate, because the consideration of such problems places the question of hyperscaling violation and the peculiarities of scaling above the upper critical dimensionality in a broader context. Moreover, such an expansion of the subject enabled some of the authors who were closely associated with Ralph in their activities to participate in this Special Issue.

The first group of contributions addresses the general topic of finite-size scaling at or above the upper critical dimension—an area now commonly referred to as Q-finite-size scaling—a term coined by Ralph himself.

The opening article is a masterful review by Adelhardt, Koziol, Langheld, and Schmidt [24]. Long-range interactions play a crucial role in a wide variety of quantum systems, yet they remain notoriously challenging to handle from a theoretical standpoint. In this work, the authors provide a comprehensive overview of recent developments in the study of quantum magnets with long-range interactions. They discuss in detail the application of perturbative continuous unitary transformations and stochastic series expansion quantum Monte Carlo techniques to large finite systems. Finite-size scaling methods are then employed to extract the physical properties of the corresponding infinite systems. The review presents a thorough summary of the resulting quantum-critical behavior, including critical exponents across various models, and explores how long-range interactions can be tuned to explore quantum phase transitions beyond the upper critical dimension.

The second contribution is from Moueddene, Donoso, and Berche [25], and it pays tribute to Ralph's favorite area of research, since this paper revisits the scaling relations among the so-called hatted critical exponents, originally introduced by Ralph Kenna, Des Johnston, and Wolfhard Janke. The authors propose alternative derivations for several of these relations. Notably, in the case of the scaling relation involving the behavior of the correlation function, they suggest a corrected form, having identified what they believe to be an error in the original formulation by Kenna and his collaborators.

The contribution by Young [26] deals with finite-size scaling above the upper critical dimension and, in particular, considers boundary-condition-dependent effects that arise there. The case with free-boundary conditions was investigated in detail by the author and is also discussed here, as it is still subject to current—possibly conflicting—investigations. Some numerical results are presented in addition to theoretical arguments.

Next, we turn to contributions in this volume that, while not related to Q-scaling, address broader issues in statistical mechanics. The paper by Folk and Holovatch [27] is dedicated to Ralph Kenna's memory, but for technical reasons, it was not included in the online publication of the Special issue. In this article, the authors report on a part of the work on a larger project, also carried out with Ralph's participation, which consisted of preparing the bilingual commented publication of Ernst Ising's doctoral thesis [33]. The project has been completed this year [34]. Today, the Ising model stands as an archetype for describing collective ordering phenomena, renowned both within physics and across disciplines. Less widely known, however, is that Ernst Ising's doctoral thesis—defended a century ago in 1924—contained more than the solution to what we now call the classical one-dimensional Ising model. Several additional problems, as well as the methods employed to solve them, are examined in this historical note. In particular, the authors highlight the combinatorial technique Ising used to compute the partition function of a chain of elementary magnets. In the thermodynamic limit, this approach yields a partition function expressed in terms of the roots of a specific polynomial. Remarkably, these "Ising roots" obtained via combinatorics coincide with the eigenvalues of the transfer matrix—a formalism only introduced later. The paper also discusses a three-state generalization of the two-state model proposed in Ising's thesis but omitted from his well-known 1925 publication [35]. This extended model may be viewed as a precursor to the now-prevalent class of models featuring multicomponent order parameters.

Oliveira, Alves, Alves, Lima, and Plascak [28] investigate the Biswas–Chatterjee–Sen model of opinion dynamics on three-dimensional Solomon networks using extensive Monte Carlo simulations. Employing finite-size scaling analysis, they provide clear evidence that the model exhibits a second-order phase transition. The authors determine the critical exponents associated with the order parameter, its associated susceptibility, and the correlation length. Their results demonstrate that the model in three dimensions belongs to a universality class distinct from those found in the same model on one- and two-dimensional Solomon networks, as well as from the conventional Ising model defined on these networks.

In their contribution, Iglói and Lin [29] employ a real-space block renormalization group approach to investigate the critical behavior of the random transverse-field Ising spin chain with multispin interactions. They first recover the well-established results for the standard model with nearest-neighbor (two-spin) couplings. Extending their analysis to the case with three-spin interactions, they determine the location of the critical point and show that the associated phase transition is governed by an infinite-disorder fixed point. Notably, the typical correlation–length critical exponent appears to differ from that of the conventional random transverse-field Ising chain, indicating that the model with multispin couplings belongs to a distinct infinite-disorder universality class.

Gessert, Weigel, and Janke [30] analyze the zeros of the partition function in the complex temperature plane (Fisher zeros) and in the complex external field plane (Lee–Yang zeros) for a frustrated Ising model on the honeycomb lattice, featuring competing nearest-neighbor ($J_1 > 0$) and next-nearest-neighbor ($J_2 < 0$) interactions. They investigate the finite-size scaling behavior of the leading zeros and compare their findings with results from a more traditional scaling analysis, based on the logarithmic derivative of the magnetization and the magnetic susceptibility. Although both approaches are subject to significant corrections to scaling due to frustration, the nature of these corrections differs markedly, especially as the interaction ratio $R = J_2/J_1$ is varied. The analysis of the partition function zeros thus emerges as a valuable complement to conventional FSS methods. In particular, the scaling behavior of the zeros provides compelling evidence that the system remains within the Ising universality class down to $R = -0.22$, a regime where conclusions based on traditional FSS become less definitive. This demonstrates the utility of the zeros approach as a powerful additional tool for charting the phase diagrams of models strongly affected by corrections to scaling.

Henkel [31] investigates out-of-equilibrium dynamics in the one-dimensional Glauber–Ising model following a quench to zero temperature. Focusing on single-time and two-time correlation functions on a periodic chain of finite length N, he employs an analytical continuation method. In addition to providing further confirmation of finite-size scaling in non-equilibrium settings, this approach enables a detailed examination of the scaling behavior of the plateau height $C_\infty^{(2)}$, the asymptotic value approached by the two-time autocorrelator deep in the finite-size regime.

Burda and Johnston [32] exploit the correspondence between the partition function of an adsorbing Dyck walk model and the normalization of the Asymmetric Simple Exclusion Process (ASEP) to determine the thermodynamic limit of the locus of ASEP normalization zeros via a conformal mapping approach. They further discuss the equivalence of this method with an electrostatic analogy, which can also be used to derive the zero locus—both for the ASEP and for the related random allocation model. This dual perspective highlights the versatility and robustness of analytical techniques in determining the singularity structure of nonequilibrium systems.

With this collection of articles, their authors, together with the editors of the *Entropy* journal, express their deep respect for the memory of Ralph Kenna—an outstanding physicist and a wonderful person. Having conceived such a collection, he was unable to take part in its editing himself. However, the articles that you will read serve as a continuation of his ideas, and we believe that there will be many more such works.

Funding: This research received no external funding.

Conflicts of Interest: The authors declare no conflicts of interest.

References

1. Berche, B.; Folk, R.; Holovatch, Y.; Mryglod, O. Statistical Physics and Emergent Behaviour: From Spin Models to Cultural Complexity. In Memory of Ralph Kenna (27.08.1964 – 26.10.2023). *Condens. Matter Phys.* **2024**, *27*, 30101. [CrossRef]
2. Yang, C.N.; Lee, T.D. Statistical Theory of Equations of State and Phase Transitions. I. Theory of Condensation. *Phys. Rev.* **1952**, *87*, 404–409. [CrossRef]
3. Lee, T.D.; Yang, C.N. Statistical Theory of Equations of State and Phase Transitions. II. Lattice Gas and Ising Model. *Phys. Rev.* **1952**, *87*, 410–419. [CrossRef]
4. Fisher, M. The nature of critical points. In *Lectures in Theoretical Physics: Proceedings, Boulder Summer Institute for Theoretical Physics*; Barut, A., Brittin, W., Eds.; Gordon and Breach: London, UK, 1968; p. 1.
5. Kenna, R.; Johnston, D.A.; Janke, W. Scaling Relations for Logarithmic Corrections. *Phys. Rev. Lett.* **2006**, *96*, 115701. [CrossRef] [PubMed]
6. Kenna, R.; Johnston, D.A.; Janke, W. Self-Consistent Scaling Theory for Logarithmic-Correction Exponents. *Phys. Rev. Lett.* **2006**, *97*, 155702. [CrossRef] [PubMed]
7. Berche, B.; Kenna, R.; Walter, J.C. Hyperscaling above the upper critical dimension. *Nucl. Phys. B* **2012**, *865*, 115–132. [CrossRef]
8. Privman, V.; Fisher, M.E. Finite-size effects at first-order transitions. *J. Stat. Phys.* **1983**, *33*, 385–417. [CrossRef]
9. Fisher, M.E. Scaling, universality and renormalization group theory. In *Critical Phenomena*; Hahne, F.J.W., Ed.; Springer: Berlin/Heidelberg, Germany, 1983; pp. 1–139.
10. Kenna, R.; Berche, B. Scaling and Finite-Size Scaling above the Upper Critical Dimension. In *Order, Disorder and Criticality*; Holovatch, Y., Ed.; World Scientific: Singapore, 2015; Volume IV, Chapter 1, pp. 1–54. [CrossRef]

11. Flores-Sola, E.; Berche, B.; Kenna, R.; Weigel, M. Finite-size scaling above the upper critical dimension in Ising models with long-range interactions. *Eur. Phys. J. B* **2015**, *88*, 28. [CrossRef]
12. Kenna, R.; Berche, B. Universal finite-size scaling for percolation theory in high dimensions. *J. Phys. A Math. Theor.* **2017**, *50*, 235001. [CrossRef]
13. Berche, B.; Ellis, T.; Holovatch, Y.; Kenna, R. Phase transitions above the upper critical dimension. *Scipost Phys. Lect. Notes* **2022**, *60*. [CrossRef]
14. Kenna, R.; Berche, B. The Enigmatic Exponent 'koppa' and the Story of Finite- Size Scaling Above the Upper Critical Dimension. In *Order, Disorder and Criticality*; Holovatch, Y., Ed.; World Scientific: Singapore, 2024; Volume VIII, Chapter 1, pp. 1–66. [CrossRef]
15. Palchykov, V.; von Ferber, C.; Folk, R.; Holovatch, Y.; Kenna, R. Critical phenomena on scale-free networks: Logarithmic corrections and scaling functions. *Phys. Rev. E* **2010**, *82*, 011145. [CrossRef]
16. Kenna, R.; Berche, B. The extensive nature of group quality. *Europhys. Lett.* **2010**, *90*, 58002. [CrossRef]
17. Kenna, R.; Berche, B. Critical masses for academic research groups and consequences for higher education research policy and management. *High. Educ. Manag. Policy* **2012**, *23*, 1–21. [CrossRef]
18. Dunbar, R. Neocortex size as a constraint on group size in primates. *J. Hum. Evol.* **1992**, *22*, 469–493. [CrossRef]
19. Mryglod, O.; Kenna, R.; Holovatch, Y.; Berche, B. Predicting results of the research excellence framework using departmental *h*-index: Revisited. *Scientometrics* **2015**, *104*, 1013–1017. [CrossRef]
20. MacCarron, P.; Kenna, R. Universal properties of mythological networks. *Europhys. Lett.* **2012**, *99*, 28002. [CrossRef]
21. MacCarron, P.; Kenna, R. Network analysis of the Íslendinga sögur—The Sagas of Icelanders. *Eur. Phys. J. B* **2013**, *86*, 407. [CrossRef]
22. Yose, J.; Kenna, R.; MacCarron, M.; MacCarron, P. Network analysis of the Viking Age in Ireland as portrayed in *Cogadh Gaedhel Re Gall*. *R. Soc. Open Sci.* **2018**, *5*, 171024. [CrossRef]
23. Sarkanych, P.; Fedorak, N.; Holovatch, Y.; MacCarron, P.; Yose, J.; Kenna, R. Network analysis of the Kyiv bylyny cycle–east Slavic epic narratives. *Adv. Complex Syst.* **2022**, *25*, 2240007. [CrossRef]
24. Adelhardt, P.; Koziol, J.A.; Langheld, A.; Schmidt, K.P. Monte Carlo Based Techniques for Quantum Magnets with Long-Range Interactions. *Entropy* **2024**, *26*, 401. [CrossRef]
25. Moueddene, L.; Donoso, A.; Berche, B. Ralph Kenna's Scaling Relations in Critical Phenomena. *Entropy* **2024**, *26*, 221. [CrossRef] [PubMed]
26. Young, A.P. Violations of Hyperscaling in Finite-Size Scaling above the Upper Critical Dimension. *Entropy* **2024**, *26*, 509. [CrossRef]
27. Folk, R.; Holovatch, Y. Ising's Roots and the Transfer-Matrix Eigenvalues. *Entropy* **2024**, *26*, 459. [CrossRef]
28. Oliveira, G.S.; Alves, T.A.; Alves, G.A.; Lima, F.W.; Plascak, J.A. Biswas–Chatterjee–Sen Model on Solomon Networks with Two Three-Dimensional Lattices. *Entropy* **2024**, *26*, 587. [CrossRef]
29. Iglói, F.; Lin, Y.C. Random Quantum Ising Model with Three-Spin Couplings. *Entropy* **2024**, *26*, 709. [CrossRef]
30. Gessert, D.; Weigel, M.; Janke, W. Partition Function Zeros of the Frustrated J1–J2 Ising Model on the Honeycomb Lattice. *Entropy* **2024**, *26*, 919. [CrossRef]
31. Henkel, M. Finite-Size Scaling in the Ageing Dynamics of the 1D Glauber–Ising Model. *Entropy* **2025**, *27*, 139. [CrossRef]
32. Burda, Z.; Johnston, D.A. Partition Function Zeros of Paths and Normalization Zeros of ASEPS. *Entropy* **2025**, *27*, 183. [CrossRef] [PubMed]
33. Ising, E. Beitrag zur Theorie des Ferro- und Paramagnetismus. Ph.D. Thesis, Mathematisch-Naturwissenschaftlichen Fakultät der Hamburgischen Universität, Hamburg, Germany, 1924. Available online: https://icmp.lviv.ua/ising/books/isingthesis.pdf (accessed on 18 July 2025).
34. Folk, R.; Berche, B.; Kenna, R.; Holovatch, Y. *Ernst Ising's Doctoral Thesis One Hundred Years On*; World Scientific: Singapore, 2025; pp. i–xi, 1–202 (to appear).
35. Ising, E. Beitrag zur Theorie des Ferromagnetismus. *Z. Für Physik* **1925**, *31*, 253–258. (A translation to English by Jane Ising, Tom Cummings, Ulrich Harsch is found at Bibliotheca Augustana). Available online: https://www.hs-augsburg.de/~harsch/anglica/Chronology/20thC/Ising/isi_fm00.html (accessed on 18 July 2025). [CrossRef]

entropy

MDPI

Review

Monte Carlo Based Techniques for Quantum Magnets with Long-Range Interactions

Patrick Adelhardt †, Jan A. Koziol †, Anja Langheld † and Kai P. Schmidt *

Department of Physics, Friedrich-Alexander Universität Erlangen-Nürnberg (FAU), 91058 Erlangen, Germany; patrick.adelhardt@fau.de (P.A.); jan.koziol@fau.de (J.A.K.); anja.langheld@fau.de (A.L.)
* Correspondence: kai.phillip.schmidt@fau.de
† These authors contributed equally to this work.

Abstract: Long-range interactions are relevant for a large variety of quantum systems in quantum optics and condensed matter physics. In particular, the control of quantum–optical platforms promises to gain deep insights into quantum-critical properties induced by the long-range nature of interactions. From a theoretical perspective, long-range interactions are notoriously complicated to treat. Here, we give an overview of recent advancements to investigate quantum magnets with long-range interactions focusing on two techniques based on Monte Carlo integration. First, the method of perturbative continuous unitary transformations where classical Monte Carlo integration is applied within the embedding scheme of white graphs. This linked-cluster expansion allows extracting high-order series expansions of energies and observables in the thermodynamic limit. Second, stochastic series expansion quantum Monte Carlo integration enables calculations on large finite systems. Finite-size scaling can then be used to determine the physical properties of the infinite system. In recent years, both techniques have been applied successfully to one- and two-dimensional quantum magnets involving long-range Ising, XY, and Heisenberg interactions on various bipartite and non-bipartite lattices. Here, we summarise the obtained quantum-critical properties including critical exponents for all these systems in a coherent way. Further, we review how long-range interactions are used to study quantum phase transitions above the upper critical dimension and the scaling techniques to extract these quantum critical properties from the numerical calculations.

Keywords: quantum spin systems; long-range interactions; Ising interactions; XY interactions; Heisenberg interactions; Monte Carlo; series expansion; perturbative continuous unitary transformation; stochastic series expansion; quantum phase transitions; critical exponents; quantum simulation

1. Introduction

Since the advent of the theoretical description of classical and quantum phase transitions (QPTs), long-range interactions between degrees of freedom challenged the established concepts and propelled the development of new ideas in the field [1–5]. It is remarkable that, only a few years after the introduction of the renormalisation group (RG) theory by K.G. Wilson in 1971 as a tool to study phase transitions and as an explanation for universality classes [6–11], it was used to investigate ordering phase transitions with long-range interactions. These studies found that the criticality depends on the decay strength of the interaction [1–3]. It then took two decades to develop numerical Monte Carlo (MC) tools capable of simulating basic magnetic long-range models with thermal phase transitions following the behaviour predicted by the RG theory [12,13]. The results of these simulations sparked a renewed interest in finite-size scaling above the upper critical dimension [12,14–19] since "hyperscaling is violated" [13] for long-range interactions that decay slowly enough. In this regime, the treatment of dangerous irrelevant variables (DIVs) in the scaling forms is required to extract critical exponents from finite systems.

Meanwhile, a similar historic development took place regarding the study of QPTs under the influence of long-range interactions. By virtue of pioneering RG studies [20,21], the numerical investigation of long-range interacting magnetic systems has been triggered [22–38]. In particular, Monte Carlo-based techniques became a popular tool to gain quantitative insight into these long-range interacting quantum magnets [22,25,26,29–40]. On the one hand, this includes high-order series expansion techniques, where classical Monte Carlo integration is applied for the graph embedding scheme, allowing extracting energies and observables in the thermodynamic limit [25,29]. On the other hand, there is stochastic series expansion quantum Monte Carlo [39], which enables calculations on large finite systems. To determine the physical properties of the infinite system, finite-size scaling is performed with the results of these computations. Inspired by the recent developments for classical phase transitions [15–19,41], a theory for finite-size scaling above the upper critical dimension for QPTs was introduced [32,34].

When investigating algebraically decaying long-range interactions $\sim r^{-(d+\sigma)}$ with the distance r and the dimension d of the system, there are two distinct regimes: one for $\sigma \leq 0$ (strong long-range interaction) and another one for $\sigma > 0$ (weak long-range interaction) [5,42–45]. In the case of strong long-range interactions, common definitions of internal energy and entropy in the thermodynamic limit are not applicable and standard thermodynamics breaks down [5,42–45]. We will not focus on this regime in this review. For details specific to strong long-range interactions, we refer to other review articles such as Refs. [5,42–45]. For the sake of this work, we restrict the discussion to weak long-range interaction or competing antiferromagnetic strong long-range interactions, for which an extensive ground-state energy can be defined without rescaling of the coupling constant [5].

The interest in quantum magnets with long-range interactions is further fuelled by the relevance of these models in state-of-the-art quantum–optical platforms [5,46–87]. To realise long-range interacting quantum lattice models with a tunable algebraic decay exponent, one can use trapped ions, which are coupled off-resonantly to motional degrees of freedom [5,81–85,88]. Another possibility is to couple trapped neutral atoms to photonic modes of a cavity [5,86,87]. Alternatively, one can realise long-range interactions decaying with a fixed algebraic decay exponent of six or three using Rydberg atom quantum simulators [46–55] or ultracold dipolar quantum atomic or molecular gases in optical lattices [56–73]. Note that, in many of the above-listed cases, it is possible to map the long-range interacting atomic degrees of freedom onto quantum spin models [5,52,89]. Therefore, they can be exploited as analogue quantum simulators for long-range interacting quantum magnets, and the relevance of the theoretical concepts transcends the boundary between the fields.

From the perspective of condensed matter physics, there are multiple materials with relevant long-range interactions [90–106]. The compound $LiHoF_4$ in an external field realises an Ising magnet in a transverse magnetic field [102–105]. A recent experiment with the two-dimensional Heisenberg ferromagnet Fe_3GeTe_2 demonstrates that phase transitions and continuous symmetry breaking can be implemented by circumventing the Hohenberg–Mermin–Wagner theorem with long-range interactions [106]. This material is in the recently discovered material class of 2D magnetic van der Waals systems [107,108]. Further, dipolar interactions play a crucial role in the spin ice state in the frustrated magnetic pyrochlore materials $Ho_2Ti_2O_7$ and $Dy_2Ti_2O_7$ [90–101].

In this review, we are interested in physical systems described by quantum spin models, where the magnetic degrees of freedom are located on the sites of a lattice. We concentrate on the following three paradigmatic types of magnetic interactions between lattice sites: first, Ising interactions, where the magnetic interaction is oriented only in the direction of one quantisation axis; second, XY interactions with a $U(1)$-symmetric magnetic interaction invariant under planar rotations; and third, Heisenberg interactions with a $SU(2)$-symmetric magnetic interaction invariant under rotations in 3D spin space. In the microscopic models of interest, a competition between magnetic ordering and trivial product states, external fields, or quasi-long-range order leads to QPTs.

In this context, the primary research pursuit revolves around how the properties of the QPT depend on the long-range interaction. The upper critical dimension of a QPT in magnetic models with non-competing algebraically decaying long-range interactions is known to depend on the decay exponent of the interaction for a small enough exponent, and decreases as the decay exponent decreases [20,21]. If the dimension of a system is equal to or exceeds the upper critical dimension, the QPT displays mean-field critical behaviour. At the same time, standard finite-size scaling, as well as standard hyperscaling relations are no longer applicable. Therefore, these systems are primary workhorse models to study finite-size scaling above the upper critical dimension. In this case, the numerical simulation of these systems is crucial in order to gauge novel theoretical developments. Further, QPTs in systems with competing long-range interactions do not tend to depend on the long-range nature of the interaction [23–26,29,30,32]. In several cases, long-range interactions then lead to the emergence of ground states and QPTs, which are not present in the corresponding short-range interacting models [27,30,54,55,109–112].

In this review, we are mainly interested in the description and discussion of two Monte Carlo-based numerical techniques, which were successfully used to study the low-energy physics of long-range interacting quantum magnets, in particular with respect to the quantitative investigation of QPTs [22,25,29–32,34–38,40]. The success of Monte Carlo techniques in this field is due to the occurrence of high-dimensional sums and integrals that commonly arise in the formulation of many-particle statistics. In contrast to many deterministic integration techniques, for which the standard error scales exponentially with the dimension of the underlying integral, the standard error of an integral calculated with Monte Carlo integration does not scale with the dimension of the underlying integral. We further chose to review this topic due to our personal involvement with the application and development of these methods [25,29–32,34,35]. On the one hand, we explain in detail how classical Monte Carlo integration can enhance the capabilities of linked-cluster expansions (LCEs) with the pCUT+MC approach (a combination of the perturbative unitary transform approach (pCUT) and MC embedding). On the other hand, we describe how stochastic series expansion (SSE) quantum Monte Carlo (QMC) integration is used to directly sample the thermodynamic properties of suitable long-range quantum magnets on finite systems.

This review is structured as follows. In Section 2, we review the basic concept of a QPT in a condensed way, focusing on the details relevant for this review. We define the quantum-critical exponents and the relations between them in Section 2.1. Here, we also have the first encounter with the generalised hyperscaling relation, which is also valid above the upper critical dimension where conventional hyperscaling breaks down. As the SSE QMC method discussed in this review is a finite-system simulation, we discuss the conventional finite-size scaling below the upper critical dimension in Section 2.2 and the peculiarities of finite-size scaling above the upper critical dimension in Section 2.3. In Section 3, we summarise the basic concepts of Markov chain Monte Carlo integration: Monte Carlo sampling, Markovian random walks, stationary distributions, the detailed balance condition, and the Metropolis–Hastings algorithm. We continue by introducing the series-expansion Monte Carlo embedding method pCUT+MC in Section 4. We start with the basic concepts of a graph expansion in Section 4.1 and introduce the perturbative method of our choice, the perturbative continuous unitary transformation method, in Section 4.2. We introduce the theoretical concepts for setting up a linked-cluster expansion as a full graph decomposition in Section 4.3 and, subsequently, discuss how to practically calculate perturbative contributions in Sections 4.4 and 4.5. We prepare the discussion of the white graph decomposition in Section 4.6 with an interlude on the relevant graph theory in Sections 4.6.1 and 4.6.2 and the important concept of white graphs in Section 4.6.3. Further, in Section 4.7, we discuss the embedding problem for the white graph contributions. Starting from the nearest-neighbour embedding problem in Section 4.7.1, we generalise it to the long-range case in Section 4.7.2 and then introduce a classical Monte Carlo algorithm to calculate the resulting high-dimensional sums in Section 4.7.3. This is followed by some technical aspects on series extrapolations in Section 4.8 and a summary of the entire workflow in Section 4.9. In the next section, the topic changes towards the review of the SSE QMC method, which is an approach to simulate thermodynamic properties of suitable quantum many-body systems on finite systems at a finite temperature. First, we discuss the general concepts of the method in Section 5. We review the algorithm to simulate arbitrary transverse-field Ising models introduced by A. Sandvik [39] in Section 5.1. We then review an algorithm used to simulate non-frustrated Heisenberg models in Section 5.2. After the introduction to the algorithms, we summarise techniques on how to measure common observables in the SSE QMC scheme in Section 5.3. Since the SSE QMC method is a finite-temperature method, we discuss how to rigorously use this scheme to perform simulations at effective zero temperature in Section 5.4. We conclude this section with a brief summary of path integral Monte Carlo techniques used for systems with long-range interactions (see Section 5.5). To maintain the balance between algorithmic aspects and their physical relevance, we summarise several theoretical and numerical results for quantum phase transitions in basics long-range interacting quantum spin models, for which the discussed Monte Carlo-based techniques provided significant results. First, we discuss long-range interacting transverse-field Ising models in Section 6. For ferromagnetic interactions, this model displays three regimes of universality: a long-range mean-field regime for slowly decaying long-range interactions, an intermediate long-range non-trivial regime, and a regime of short-range universality for strong decaying long-range interactions. We discuss the theoretical origins of this behaviour in Section 6.1.1 and numerical results for quantum critical exponents in Section 6.1.2. Since this model is a prime example to study scaling above the upper critical dimension in the long-range mean-field regime, we emphasise these aspects in Section 6.1.3. Further, we discuss the antiferromagnetic long-range transverse-field Ising model on bipartite lattices in Section 6.2 and on non-bipartite lattices in Section 6.3. The next obvious step is to change the symmetry of the magnetic interactions. Therefore, we turn to long-range interacting XY models in Section 7 and Heisenberg models in Section 8. We discuss the long-range interacting transverse-field XY chain in Section 7 starting with the $U(1)$-symmetric isotropic case in Section 7.1, followed by the anisotropic case for ferromagnetic (see Section 7.2) and antiferromagnetic (see Section 7.3) interactions, which display similar behaviour to the long-range transverse-field Ising model on the chain discussed in Section 6. We conclude the discussion of the results with unfrustrated long-range Heisenberg models in Section 8. We focus on the staggered antiferromagnetic long-range Heisenberg square lattice bilayer model in Section 8.1 followed by long-range Heisenberg ladders in Section 8.2 and the long-range Heisenberg chain in Section 8.3. We conclude in Section 9 with a brief summary and with some comments on the next possible steps in the field.

2. Quantum Phase Transitions

This review is part of the Special Issue with the topic "Violations of Hyperscaling in Phase Transitions and Critical Phenomena". In this work, we summarise investigations of low-dimensional quantum magnets with long-range interactions targeting, in particular, quantum phase transitions (QPTs) above the upper critical dimension, where the naive hyperscaling relation is no longer applicable. In this section, we recapitulate the relevant aspects of QPTs needed to discuss the results of the Monte Carlo-based numerical approaches. First, we give a general introduction to QPTs. After that, we discuss in detail the definition of critical exponents and the relations among them in Section 2.1, as well as the scaling below (see Section 2.2) and above (see Section 2.3) the upper critical dimension.

Any non-analytic point of the ground-state energy of an infinite quantum system as a function of a tuning parameter λ is identified with a QPT [113]. This tuning parameter can, for instance, be a magnetic field or pressure, but not the temperature. Quantum phase transitions are a concept of zero temperature as there are no thermal fluctuations and all excited states are suppressed infinitely strong such that the system remains in its ground state. There are two scenarios for how a non-analytic point in the ground-state energy can emerge [113]: First is an actual (sharp) level crossing between the ground-state energy and another energy level. Second, the non-analytic point can be considered as a limiting case of an avoided level crossing. Historically, phase transitions are classified by the lowest order derivative of the free energy that is discontinuous [113,114]. Therefore, a first-order phase transition is discontinuous in the order parameter (first derivative) and a second-order phase transition is discontinuous in the response functions (second derivative). Since, in second-order phase transitions, the order parameter is still continuous across the phase transition, we use the term "continuous phase transition" as an equivalent for "second-order phase transition".

In this review, we are interested in second-order QPTs, which fall into the second scenario. At a second-order QPT, the relevant elementary excitations condense into a novel ground state, while the characteristic length and time scales diverge. Apart from topological QPTs involving a long-range entangled topological phase, such continuous transitions are described by the concept of spontaneous symmetry breaking. On one side of the QPT, the ground state obeys a symmetry of the Hamiltonian, while on the other side, this symmetry is broken in the ground state and a ground-state degeneracy arises.

Following the idea of the quantum-to-classical mapping [113,115], d-dimensional quantum systems can be mapped in the vicinity of a second-order QPT to models of statistical mechanics with a classical (thermal) second-order phase transition in $d+1$ dimensions. In many cases, the models obtained from a quantum-to-classical mapping are rather artificial [113]. However, such mappings often allow categorising QPTs in terms of universality classes and associated critical exponents by the non-analytic behaviour of the classical counterparts [10,11,113,116,117]. The mapping further illustrates that the renormalisation group (RG) theory is also applicable to describe QPTs [10,11,113,116,117].

In the RG theory, each QPT belongs to a non-trivial fixed point of the RG transformation [10,11], whereas a trivial fixed point would, for instance, be a fully ordered state with maximal correlation or a fully disordered state with no correlation at all. Critical exponents are connected to the RG flow in the immediate vicinity of these non-trivial fixed points [10,11,113]. The concept of universality classes arises from the fact that different microscopic Hamiltonians can have a quantum critical point that is attracted by the same non-trivial fixed point under successive application of the RG transformation [10,11]. Due to this, the QPTs in these models have the same critical exponents.

Another remarkable result of the RG theory is the scaling of observables in the vicinity of phase transitions. Historically, the theory of scaling at phase transitions was heuristically introduced before the RG approach [118–124]. The latter provided the theoretical foundation for the scaling hypothesis [6,7]. The main statement of the scaling theory is that the non-analytic contributions to the free energy and correlation functions are mathematically described by generalised homogeneous functions (GHFs) [124]. A function with n variables $f(x_1, x_2, \ldots, x_n)$ is called a GHF, if there exist $a_1, a_2, \ldots, a_n \in \mathbb{R}$ with at least one being non-zero and $a_f \in \mathbb{R}$ such that, for $b \in \mathbb{R}_+$,

$$f(b^{a_1} x_1, b^{a_2} x_2, \ldots, b^{a_n} x_n) = b^{a_f} f(x_1, x_2, \ldots, x_n) . \tag{1}$$

The exponents $a_1, a_2, \ldots, a_n$ are the scaling powers of the variables, and a_f is the scaling power of the function f itself. An in-depth summary of the mathematical properties of GHFs can be found in Appendix B. The most important properties of GHFs are that their derivatives, Legendre transforms, and Fourier transforms are also GHFs. As we will outline in Section 2.2, the theory of finite-size scaling is formulated in terms of GHFs and relates the non-analytic behaviour at QPTs in the thermodynamic limit with the scaling of observables for different finite system sizes. In this, the variables x_i are related to physical parameters like the temperature T, control parameter λ, symmetry-breaking field H, and also irrelevant, more abstract parameters that parameterise the microscopic details of the model like the lattice spacing. Later in this section, we will define irrelevant variables in the context of the RG and GHFs.

Another aspect relevant for this work is that quantum fluctuations are the driving factor with QPTs [113]. In general, fluctuations are more important in low dimensions [117]. The universality class of QPTs for a certain symmetry breaking depends on the dimensionality of the system.

An important aspect regarding this review is the so-called upper critical dimension d_{uc}. The upper critical dimension is defined as a dimensional boundary such that, for systems with dimension $d \geq d_{\mathrm{uc}}$, the critical exponents are those obtained from mean-field considerations. The upper critical dimension is of particular importance for QPTs in systems with non-competing long-range interactions. For sufficiently small decay exponents of an algebraically decaying long-range interaction $1/r^{d+\sigma}$, the upper critical dimension starts to decrease as a function of the decay exponent σ [20,32,34]. In the limiting case of a completely flat decay ($\sigma = -d$) of the long-range interaction resulting in an all-to-all coupling, the model is intrinsically of the mean-field type and mean-field considerations become exact. For a certain value of the decay exponent, the upper critical dimension becomes equal to the fixed spatial dimension, and for decay exponents below this value, the dimension of the system is above the upper critical dimension of the transition [20,32,34]. This makes long-range interacting systems an ideal test bed for studying phase transitions above the upper critical dimension in low-dimensional systems. In particular, long-range interactions can make the upper critical dimension accessible in real-world experiments as the upper critical dimension of short-range models is usually not below three.

Although phase transitions above the upper critical dimension display mean-field criticality, they are still a matter worth studying, since naive scaling theory describing the behaviour of finite systems close to a phase transition (see Section 2.2) is no longer applicable [16,19,125]. Moreover, the naive versions of some relations between critical exponents, as discussed in Section 2.1, do not hold any longer [15,16]. The reason for this issue are the dangerous irrelevant variables (DIVs) in the RG framework [126–128]. During the application of the RG transformation, the original Hamiltonian is successively mapped to other Hamiltonians, which can have infinitely many couplings. All these couplings, in principle, enter the GHFs. In practice, all but a finite number of these couplings are set to zero since their scaling powers are negative, which means they flow to zero under renormalisation. These couplings are, therefore, called irrelevant. This approach of setting irrelevant couplings to zero can be used to derive the finite-size scaling behaviour as described in Section 2.2. However, above the upper critical dimension, this approach breaks down because it is only possible to set irrelevant variables to zero if the GHF does not have a singularity in this limit [126]. Above the upper critical dimension, such singularities in irrelevant parameters exist, which makes them DIVs [127]. We explain the effect of DIVs on scaling in Section 2.3.

2.1. Critical Exponents in the Thermodynamic Limit

As outlined above, a second-order QPT comes with a singularity in the free energy density. In fact, also, other observables experience singular behaviour at the critical point in the form of power-law singularities. For instance, the order parameter m as a function of the control parameter λ behaves as

$$m(\lambda \to \lambda_c^-) \sim |\lambda - \lambda_c|^{\beta} \tag{2}$$

in the ordered phase. Without loss of generality, the system is taken to be in the ordered phase for $\lambda < \lambda_c$ and the notation $\lambda \to \lambda_c^-$ means that λ is approaching λ_c from below, i.e., it is approaching in the ordered phase. In the disordered phase $\lambda > \lambda_c$, the order parameter by definition vanishes such that $m(\lambda \to \lambda_c^+) = 0$. The observables with their respective power law singular behaviour, which is characterised by the critical exponents $\alpha, \beta, \gamma, \delta, \eta, \nu$, and z, are summarised in Table 1 together with how they are commonly defined in terms of the free energy density f, the symmetry-breaking field H, which couples to the order parameter, and the reduced control parameter $r = (\lambda - \lambda_c)/\lambda_c$.

Table 1. Definitions of critical exponents by means of the singularities of thermodynamic quantities for a magnetic phase transition. The free energy density is denoted by f. Note that the control parameter susceptibility associated with the critical exponent α coincides with the heat capacity only for thermal phase transitions, where $r = (T - T_c)/T_c$, while for QPTs, the meaning depends on the control parameter triggering the phase transition [129].

Observable	Definition	Crit. Exp.	Singularity
Characteristic length ξ	via $G(\vec{r})$	ν	$\xi(r \to 0, H = 0) \sim \|r\|^{-\nu}$
Energy gap Δ Charact. time scale ξ_τ	via $G(\vec{r}, \omega)$ $\xi_\tau \sim \Delta^{-1}$	$z\nu$	$\Delta(r \to 0) \sim \|r\|^{z\nu}$ $\xi_\tau(r \to 0) \sim \xi^z \sim \|r\|^{-z\nu}$
Order parameter m	$m = \frac{\partial f}{\partial H}\Big\|_{H=0}$	β δ	$m(r \to 0^-, H = 0) \sim \|r\|^{\beta}$ $m(r = 0, H \to 0) \sim \|H\|^{1/\delta}$
Order-parameter susceptibility χ	$\chi = \frac{\partial m}{\partial H}\Big\|_{H=0}$	γ	$\chi(r \to 0, H = 0) \sim \|r\|^{-\gamma}$
Control-parameter susceptibility χ_r	$\chi_r = \frac{\partial^2 f}{\partial r^2}$	α	$\chi_r(r \to 0, H = 0) \sim \|r\|^{-\alpha}$
Correlation function $G(\vec{r})$	$\frac{\partial \langle m(\vec{r}) \rangle}{\partial H_{\vec{r}=0}}\Big\|_{H_0=0}$	η	$G(\vec{r} \to \infty, r = 0, H = 0)$ $\sim \frac{1}{\|\vec{r}\|^{d-2+\eta}}$

One usually defines reduced parameters like r that vanish at the critical point not only to shorten the notation, but also to express the power-law singularities independent of the microscopic details of the specific model one is looking at. While the value of λ_c depends on these details, the power-law singularities are empirically known to not depend on the microscopic details, but only on more general properties like the dimensionality, the symmetry that is being broken, and, with particular emphasis due to the focus of this review, on the range of the interaction. It is, therefore, common to classify continuous phase transitions in terms of universality classes. These universality classes share the same set of critical exponents. In terms of the RG, this behaviour is understood as distinct critical points of microscopically different models flowing to the same renormalisation group fixed point, which determines the criticality of the system [6,7,10]. Prominent examples for universality classes of magnets are the 2D and 3D Ising ($\mathbb{Z}_2$ symmetry), 3D XY ($O(2)$ symmetry), and 3D Heisenberg ($O(3)$ symmetry) universality classes [113]. It is important to mention that the dimension in the classifications is referring to classical and not quantum systems, and they should not be confused with each other. In fact, the universality class of a short-range interacting non-frustrated quantum Ising model of dimension d lies in the universality of the classical $d+1$-dimensional Ising model.

There are only a few dimensions for which a separate universality class is defined for the different models. For lower dimensions, the fluctuations are too strong in order for a spontaneous symmetry breaking to occur. In the case of the classical Ising model, there is no phase transition for 1D, while for the classical XY and Heisenberg models with continuous symmetries, there is not even a phase transitions for 2D due to the Hohenberg–Mermin–Wagner (HWM) theorem [130,131]. This dimensional boundary is referred to as lower critical dimension d_{lc}. The lower critical dimension is the highest dimension for which no transition occurs, i.e., $d_{\mathrm{lc}} = 1$ for the Ising model and $d_{\mathrm{lc}} = 2$ for the XY and Heisenberg model. For higher dimensions $d \geq 4$, the critical exponents of the mentioned models do not depend on the dimensionality any longer, and they take on the mean-field critical exponents in all dimensions. The underlying reason is that, with increasing dimensions, the local fluctuations become smaller due to the higher connectivity of the system [132]. This has been also exploited in diagrammatic and series expansions in $1/d$ [133–135]. This dimensional boundary, at which the criticality becomes the mean-field one, is called upper critical dimension d_{uc}. Usually, the upper critical dimension is too large to realise a system above its upper critical dimension in the real world. However, long-rang interactions can increase the connectivity of a system in a similar sense as the dimensionality. A sufficiently long-range interaction can, therefore, lower the upper critical dimension to a value that is accessible in experiments.

Finally, it is worth mentioning that the critical exponents are not independent of each other, but obey certain relations [129], namely

$$2-\alpha = (d+z)\nu\,, \tag{3}$$
$$2-\alpha = 2\beta+\gamma\,, \tag{4}$$
$$\gamma = \beta(\delta-1)\,, \tag{5}$$
$$\gamma = (2-\eta)\nu\,. \tag{6}$$

The first relation in Equation (3) is the so-called hyperscaling relation, whose classical analogue (without z) was introduced by Widom [10,136]. The Essam–Fisher relation in Equation (4) [137,138] is reminiscent of a similar inequality, which was proven rigorously by Rushbrooke using thermodynamic stability arguments. Equation (5) is called the Widom relation. The last relation in Equation (6) is the Fisher scaling relation, which can be derived using the fluctuation–dissipation theorem [10,129,138]. Those relations were originally obtained from scaling assumptions of observables close to the critical point, which were only later derived rigorously when the RG formalism introduced to critical phenomena [10,129]. Due to these relations, it is sufficient to calculate three, albeit not arbitrary, exponents to obtain the full set of critical exponents.

The hyperscaling relation Equation (3) is the only relation containing the dimension of the system and is, therefore, often said to break down above the upper critical dimension, where one expects the same mean-field critical exponents independent of the dimension d [10]. It, therefore, deserves special focus in this review since the long-range models discussed will be above the upper critical dimension in certain parameter regimes. Personally, we would not agree that the hyperscaling relation breaks down above the upper critical dimension, but we would rather call Equation (3) a special case of a more general hyperscaling relation:

$$2-\alpha = \left(\frac{d}{\text{ϙ}}+z\right)\nu, \tag{7}$$

with the pseudo-critical exponent ϙ ("koppa") [34]:

$$\text{ϙ} = \begin{cases} 1 & \text{for } d \le d_{\text{uc}} \\ \frac{d}{d_{\text{uc}}} & \text{for } d > d_{\text{uc}}\,. \end{cases} \tag{8}$$

Below the upper critical dimension, the general hyperscaling relation, therefore, relaxes to Equation (3). Above the upper critical dimension, the relation becomes

$$2-\alpha = (d_{\text{uc}}+z)\nu, \tag{9}$$

which is independent of the dimension of the system. For the derivation of this generalised version of the hyperscaling relation for QPTs, see Section 2.3 or Ref. [34]. The derivation of the classical counterpart can be found in Ref. [15] and is reviewed in Ref. [41].

2.2. Finite-Size Scaling below the Upper Critical Dimension

Even though the singular behaviour of observables at the critical point is only present in the thermodynamic limit, it is possible to study the criticality of an infinite system by investigating their finite counterparts. In finite systems, the power-law singularities of the infinite system are rounded and shifted with respect to the critical point, e.g., the susceptibility with its characteristic divergence at the critical point λ_c is deformed to a broadened peak of finite height. The peak's position $r_L = (\lambda_L - \lambda_c)/\lambda_c$ is shifted with respect to the critical point $r=0$. A possible definition of a pseudo-critical point of a finite system is the peak position λ_L. As the characteristic length scale of fluctuations ξ diverges at the critical point, the finite system approaching the critical point will at some point begin to "feel" its finite extent and the observables start to deviate from the ones in the thermodynamic limit. As ξ diverges with the exponent ν like $\xi \sim |r|^{-\nu}$ at the critical point, the extent of rounding in a finite system is related to the value of ν. Similarly, the peak magnitude of finite-size observables at the pseudo-critical point will depend on how strong the singularity in the thermodynamic limit is, which means it depends on the respective critical exponents α, β, γ, and δ. The shifting, rounding, and varying peak magnitude are shown for the susceptibility of the long-range transverse-field Ising model in Figure 1. This dependence of observables in finite systems on the criticality of the infinite system is the basis of finite-size scaling.

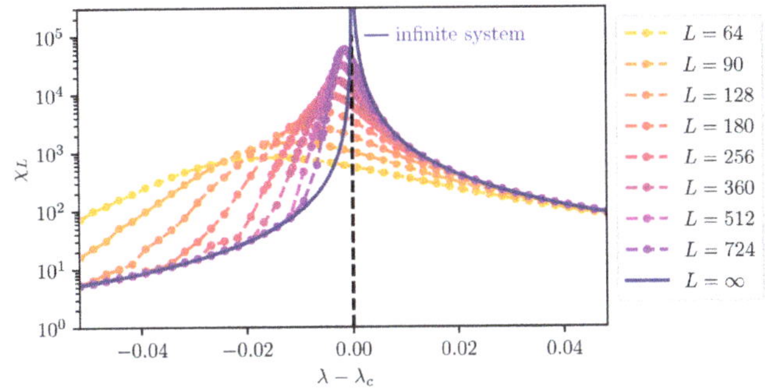

Figure 1. Susceptibility χ_L of the long-range transverse-field Ising chain for different linear system sizes from $L = 64$ to $L = 724$. The smaller the system, the farther away from the critical point the susceptibility starts to deviate from the thermodynamic limit and the farther the peak position shifts away from the critical point marked by the black dotted line.

In a more mathematical sense, the relation between critical exponents and finite-size observables has its origin in the renormalisation group (RG) flow close to the corresponding RG fixed point that determines the criticality [139]. Close to this fixed point, one can linearise the RG flow so that the free energy density and the characteristic length ξ become generalised homogeneous functions (GHFs) in their parameters [124,128,129,140,141]. For a thorough discussion of the mathematical properties of GHFs, we refer to Ref. [124] and Appendix B. This means that the free energy density f and characteristic length scale ξ as functions of the couplings r, H, T, u and the inverse system length L^{-1} obey the relations:

$$f(r, H, T, L^{-1}, u) = b^{-(d+z)} f(b^{y_r} r, b^{y_H} H, b^z T, bL^{-1}, b^{y_u} u) \tag{10}$$

$$\xi(r, H, T, L^{-1}, u) = b\xi(b^{y_r} r, b^{y_H} H, b^z T, bL^{-1}, b^{y_u} u) \tag{11}$$

with the respective scaling dimensions $y_r, y_H, z > 0$, $y_L = 1$, and $y_u < 0$ governing the linearised RG flow with spatial rescaling factor $b > 1$ around the RG fixed point, at which all couplings vanish by definition. All of those couplings are relevant except for u, which denotes the leading irrelevant coupling [10,113]. Relevant couplings are related to real-world parameters that can be used to tune our system away from the critical point like the temperature T, a symmetry-breaking field H, or simply the control parameter r. The irrelevant couplings u do not per se vanish at the critical point like the relevant ones do. However, they flow to zero under the RG transformation and are commonly set to zero in the scaling laws:

$$f(r, H, T, L^{-1}) = b^{-(d+z)} f(b^{y_r} r, b^{y_H} H, b^z T, bL^{-1}) \tag{12}$$

$$\xi(r, H, T, L^{-1}) = b\xi(b^{y_r} r, b^{y_H} H, b^z T, bL^{-1}) \tag{13}$$

by assuming analyticity in these parameters. The generalised homogeneity of thermodynamic observables can be derived from the one of the free energy density f. For example, the generalised homogeneity of the magnetisation:

$$m(r, H, T, L^{-1}) = b^{-(d+z)+y_H} m(b^{y_r} r, b^{y_H} H, b^z T, bL^{-1}) \tag{14}$$

can be derived by taking the derivative of f with respect to the symmetry-breaking field H.

By investigating the singularity of $\xi(r) = \xi(r, H = 0, T = 0, L^{-1} = 0)$ in r via Equation (13), one can show that the scaling power y_r of the control parameter r is related to the critical exponent ν by $y_r = 1/\nu$ [113]. For this, one fixes the value $b^{y_r} r$ of the first argument to ± 1 in the right-hand side of Equation (13) by setting $b = |r|^{-1/y_r}$ such that

$$\xi(r) = |r|^{-1/y_r} \xi(\pm 1) \sim |r|^{-\nu}. \tag{15}$$

Analogously, further relations between the scaling powers and other critical exponents can be derived by looking at the singular behaviour of the respective observables in the corresponding parameters. Overall, one further obtains

$$\alpha = -\frac{d+z-2y_r}{y_r}, \quad \beta = \frac{d+z-y_H}{y_r}, \quad \delta = \frac{y_H}{d+z-y_H}, \quad \gamma = -\frac{d+z-2y_H}{y_r}. \tag{16}$$

From these equations, one can already tell that the critical exponents are not independent of each other. In fact, the scaling relations $2 - \alpha = (d+z)\nu$, $2 - \alpha = 2\beta + \gamma$ and $\gamma = \beta(\delta - 1)$ (see Equations (3)–(5)) can be derived from Equation (16) and $y_r^{-1} = \nu$. By expressing the RG scaling powers y_x in terms of critical exponents, the homogeneity law for an observable $\mathcal{O}$ with a bulk divergence $\mathcal{O}(r, 0, 0, 0) \sim |r|^{\omega}$ is given by

$$\mathcal{O}(r, H, T, L^{-1}) = b^{-\omega y_r} \mathcal{O}(b^{y_r} r, b^{y_H} H, b^z T, bL^{-1}) \tag{17}$$

$$= b^{-\omega/\nu} \mathcal{O}(b^{1/\nu} r, b^{(\beta+\gamma)/\nu} H, b^z T, bL^{-1}). \tag{18}$$

In order to investigate the dependence on the linear system size L, the last argument in the homogeneity law is fixed to $bL^{-1} = 1$ by inserting $b = L$. This readily gives the finite-size scaling form

$$\mathcal{O}_L(r, H, T) = L^{-\omega/\nu} \Psi(L^{1/\nu} r, L^{(\beta+\gamma)/\nu} H, L^z T) \tag{19}$$

with Ψ being the universal scaling function of the observables $\mathcal{O}$. The scaling function Ψ itself does not depend on L any longer, but in order to compare different linear system sizes, one has to rescale its arguments. To extract the critical exponents from finite systems, the observable $\mathcal{O}_L(r, H, T)$ is measured for different system sizes L and parameters (r, H, T) close to the critical point $(r, H, T) = (0, 0, 0)$. The L-dependence according to Equation (19) is then fit with the critical exponents ω, ν, $\beta + \gamma$, and z, as well as the critical point λ_c, which is hidden in the definition of r, as free parameters. It is advisable to fix two of the three parameters r, H, T to their critical values in order to minimise the amount of free parameters in the fit. For example, with $H = T = 0$ and only $r \neq 0$, one can extract the two critical exponents ω and ν alongside λ_c. For further details on a fitting procedure, we refer to Ref. [32]. If one knows the critical point, one can also set $(r, H, T) = (0, 0, 0)$ and look at the L-dependent scaling $\mathcal{O}_L \sim L^{-\omega/\nu}$ directly at the critical point to extract the exponent ratio ω/ν. There are many more possible approaches to extract critical exponents from the FSS law in Equation (19) [142–144]. For relatively small system sizes, it might be required to take corrections to scaling into account [143,144].

2.3. Finite-Size Scaling above the Upper Critical Dimension

In the derivation of finite-size scaling below the upper critical dimension, it was assumed that the free energy density is an analytic function in the leading irrelevant coupling u and, therefore, one can set it to zero. However, this is not the case above the upper critical dimension any longer, and the free energy density f is singular at $u = 0$. Due to this singular behaviour u is referred to as a dangerous irrelevant variable (DIV).

As a consequence, one has to take the scaling of u close to the RG fixed point into account. This is achieved by absorbing the scaling of f in u for small u into the scaling of the other variables [128]:

$$f(r, H, T, L^{-1}, u) = u^{p_{(d+z)}} \overline{f}(u^{p_r} r, u^{p_H} H, u^{p_T} T, u^{p_L} L^{-1}), \tag{20}$$

up to a global power $p_{(d+z)}$ of u. This leads to a modification of the scaling powers in the homogeneity law for the free energy density [128]:

$$f(r, H, T, L^{-1}) = b^{-(d+z)^*} f(b^{y_r^*} r, b^{y_H^*} H, b^{z^*} T, b^{y_L^*} L^{-1}) \tag{21}$$

$$= L^{-(d+z)^*/y_L^*} \mathcal{F}(L^{y_r^*/y_L^*} r, L^{y_H^*/y_L^*} H, L^{z^*/y_L^*} T) \tag{22}$$

with the modified scaling powers [34,128]:

$$(d+z)^* = (d+z) - p_{(d+z)} y_u, \qquad \begin{array}{ll} y_r^* = y_r + p_r y_u, & y_H^* = y_H + p_H y_u, \\ z^* = z + p_z y_u, & y_L^* = 1 + p_L y_u. \end{array} \tag{23}$$

In the classical case [128], y_L^* was commonly set to 1 by choice. This is justified because the scaling powers of a GHF are only determined up to a common non-zero factor [124]. However, for the quantum case [34], this was kept general as it has no impact on the FSS.

As the predictions from the Gaussian field theory and mean field differed for the critical exponents α, β, and δ, but not for the "correlation" critical exponents ν, z, η, and γ [145], the correlation sector was thought not to be affected by DIVs at first [128,142,145]. Later, the Q-FSS, another approach to FSS above the upper critical dimension, pioneered by Ralph Kenna and his colleagues, was developed for classical [15,19,125], as well as for quantum systems [34], which explicitly allowed the correlation sector to also be affected by the DIV. In analogy to the free energy density, the homogeneity law of the characteristic length scale is then also modified to

$$\xi(r, H, T, L^{-1}) = b^{-y_\xi^*} \xi(b^{y_r^*} r, b^{y_H^*} H, b^{z^*} T, b^{y_L^*} L^{-1}) \tag{24}$$

$$= L^{\text{ϙ}} \Xi(L^{y_r^*/y_L^*} r, L^{y_H^*/y_L^*} H, L^{z^*/y_L^*} T) \tag{25}$$

with $y_\xi^* = -1 - p_\xi y_u = -y_r^*/y_r$ in order to reproduce the correct bulk singularity $\xi \sim |r|^{-\nu}$. A new pseudo-critical exponent ϙ ("koppa"):

$$\text{ϙ} = -\frac{y_\xi^*}{y_L^*} = \frac{y_r^*}{y_r y_L^*} = \nu \frac{y_r^*}{y_L^*} \tag{26}$$

is introduced. This exponent describes the scaling of the characteristic length scale with the linear system size. This non-linear scaling of ξ with L is one of the key differences to the previous treatments above the upper critical dimension in Ref. [128].

Analogous to the case below the upper critical dimension, the modified scaling powers y_x^* can be related to the critical exponents:

$$\alpha = -\frac{(d+z)^* - 2y_r^*}{y_r^*}, \tag{27}$$

$$\beta = \frac{(d+z)^* - y_H^*}{y_r^*}, \tag{28}$$

$$\delta = \frac{y_H^*}{(d+z)^* - y_H^*}, \tag{29}$$

$$\gamma = -\frac{(d+z)^* - 2y_H^*}{y_r^*}. \tag{30}$$

By using the mean-field critical exponents for the $\mathcal{O}(n)$ quantum rotor model, one obtains restrictions for the ratios of modified scaling powers:

$$y_r^* = \frac{(d+z)^*}{2}, \qquad y_H^* = \frac{3(d+z)^*}{4} \tag{31}$$

Furthermore, one can link the bulk scaling powers y_r^*, y_H^*, and $(d+z)^*$ to the scaling power y_L^* of the inverse linear system size [34]:

$$(d+z)^* = y_L^* d + \frac{y_r^*}{y_r} z, \tag{32}$$

by looking at the scaling of the susceptibility in a finite system [34,128]. This relation is crucial for deriving an FSS form above the upper critical dimension as the modified scaling power y_L^*, or rather, its relation to the other scaling powers determines the scaling with the linear system size L. For details on the derivation, we refer to Ref. [34]. We want to stress again that the scaling powers of GHFs are only determined up to a common non-zero factor [124]. Therefore, it is evident that one can only determine the ratios of the modified scaling powers, but not their absolute value. The absolute values are subject to choice. Different choices were discussed in Ref. [34], but these choices rather correspond to taking on different perspectives and have no impact on the FSS nor the physics.

From Equation (32) together with Equations (26) and (27), a generalised hyperscaling relation:

$$2 - \alpha = \left(\frac{d}{\varrho} + z\right)\nu, \tag{33}$$

can be derived. This also determines the pseudo-critical exponent:

$$\varrho = \frac{d}{d_{\mathrm{uc}}} \qquad \text{for } d > d_{\mathrm{uc}}\,. \tag{34}$$

Finally, we can express the modified scaling powers in the FSS law for an observable $\mathcal{O}$ with power-law singularity $\mathcal{O}(r,0,0,0) \sim |r|^{\omega}$:

$$\mathcal{O}(r,H,T,L^{-1}) = b^{-\omega y_r^*}\mathcal{O}(b^{y_r^*} r, b^{(\beta+\gamma)y_r^*} H, b^{z^*} T, b^{y_L^*} L^{-1}) \tag{35}$$

$$= L^{-\omega\varrho/\nu}\Psi(L^{\varrho/\nu} r, L^{(\beta+\gamma)\varrho/\nu} H, L^{z^*/y_L^*} T)\,. \tag{36}$$

For $\varrho = 1$ below the upper critical dimension, Equation (36) relaxes to the standard FSS law Equation (19). The scaling in temperature has not yet been studied for finite quantum systems above the upper critical dimension. However, in Ref. [34], it was conjectured that $z^* = y_r^*/y_r z$ based on Equation (32), which is also in agreement with z being the dynamical critical exponent that determines the space–time anisotropy $\xi_\tau \sim \xi^z$, as we will shortly see. This means that the finite-size gap scales as $\Delta_L \sim L^{-z^*/y_L^*} \sim L^{-\varrho z}$ with the system size [34]. Of particular interest is the scaling of the characteristic length scale above the upper critical dimension, for which the modified scaling law Equation (36) also holds with $\omega = -\nu$. Hence, the characteristic length scale in dependence of the control parameter r scales like

$$\xi_L(r) = L^{\varrho}\Xi(L^{\varrho/\nu} r) \tag{37}$$

with the scaling function Ξ. Directly at the critical point $r = 0$, this leads to $\xi_L \sim L^{\varrho}$. Comparing this with the scaling of the inverse finite size gap $\Delta_L^{-1} \sim \xi_{L,\tau} \sim \xi_L^z$ verifies that z still determines the space–time anisotropy. Prior to the Q-FSS [15,17], the characteristic length scale was thought to be bound by the linear system size L [128]. However, this was shown not to be the case by measurements of the characteristic length scale for the classical five-dimensional Ising model [14] and for long-range transverse-field Ising models [34].

For the latter, the data collapse of the correlation length according to Equation (37) is shown in Figure 2 as an example.

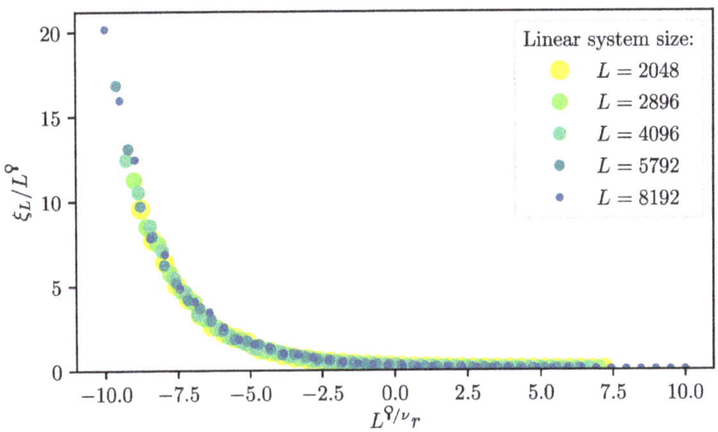

Figure 2. Rescaled correlation length according to Equation (37) for a model above the upper critical dimension with $ϙ = 20/9 \approx 2.\bar{2}$ (to be specific, the long-range transverse-field Ising chain with $\sigma = 0.3$). The control parameter $r \sim h - h_c$ is proportional to the transverse field. The collapse of the data around the critical point $r = 0$ verifies the scaling Equation (37) and, therefore, demonstrates that ξ_L is indeed—in contrast to the prior belief—not bound by the linear system size, but $\xi_L \sim L^{ϙ}$.

3. Monte Carlo Integration

In this section, we provide a brief introduction to Monte Carlo integration (MCI). We focus on the aspects of Markov chain MCI as the basis to formulate the white graph Monte Carlo embedding scheme of the pCUT+MC method in Section 4 and the stochastic series expansion (SSE) quantum Monte Carlo (QMC) algorithm in Section 5 in a self-contained fashion. MCI is the foundation for countless numerical applications, which require the integration over high-dimensional integration spaces. As this review has a focus on "Monte Carlo-based techniques for quantum magnets with long-range interactions", we forward readers with a deeper interest in the fundamental aspects of MCI and Markov chains to Refs. [146–148].

MCI summarises numerical techniques to find estimators for integrals of functions $f : \mathcal{C} \to \mathbb{R}$ over an integration space $\mathcal{C}$ using random numbers. The underlying idea behind MCI is to estimate the integral, or the sum in the case of discrete variables, of the function f over the configuration space by an expectation value:

$$I = \int_{\mathcal{C}} d\omega\, f(\omega) = \int_{\mathcal{C}} d\omega\, \frac{P(\omega)}{P(\omega)} f(\omega) = \int_{\mathcal{C}} d\omega\, P(\omega)\tilde{f}(\omega) = \lim_{S\to\infty} \frac{1}{S} \sum_{i=1}^{i=S} \tilde{f}(\omega_i) \tag{38}$$

with $\omega_i \in \mathcal{C}$ sampled according to a probability density function $P : \mathcal{C} \to \mathbb{R}_{\geq 0}$ (PDF) and the function $\tilde{f}(\omega) = f(\omega)/P(\omega)$ reweighted by the PDF. A famous direct application of this idea is the calculation of the number "pi", which is discussed in great detail in Ref. [148].

In this review, MCI is used for the embedding of white graphs on a lattice to evaluate high orders of a perturbative series expansion or to calculate thermodynamic observables using the SSE framework. In both cases, non-normalised relative weights $\pi(\omega)$ within a configuration space $\mathcal{C}$ arise, which are used for the sampling of the PDF P:

$$P(\omega) = \frac{\pi(\omega)}{\int_{\mathcal{C}} d\omega\, \pi(\omega)}\,, \tag{39}$$

being oftentimes not directly accessible. In the context of statistical physics, $\pi(\omega)$ is often chosen to be the relative Boltzmann weight $e^{-\beta E(\omega)}$ of each configuration ω. While this relative Boltzmann weight is accessible as long as $E(\omega)$ is known, the full partition function to normalise the weights is in general not.

In order to efficiently sample the configuration space C according to the relative weights, the methods in this review use a Markovian random walk to generate $\{\omega_1, \ldots, \omega_m\}$. Let ω_n be the random state of a random walk at a discrete step n. The state ω_{n+1} at the next step is randomly determined according to the conditional probabilities $T(\omega \to \omega')$ (transition probabilities). These transition probabilities are normalised by

$$\int_C d\omega' \, T(\omega \to \omega') = 1 \,. \tag{40}$$

Markovian random walks obey the Markov property, which means the random walk is memory-free and the transition probability for multiple steps factorises into a product over all time steps:

$$T(\omega^{(0)} \to \omega^{(1)} \to \cdots \to \omega^{(m-1)} \to \omega^{(m)}) = \prod_{i=0}^{m-1} T(\omega^{(i)} \to \omega^{(i+1)}) \tag{41}$$

with $\omega^{(0)}$ the start configuration. We require the Markovian random walk to fulfil the following conditions: First, the random walk should have a certain PDF $P(\omega)$ defined by the weights $\pi(\omega)$ in Equation (39) as a stationary distribution. By definition, $P(\omega)$ is a stationary distribution of the Markov chain if it satisfies the global balance condition:

$$\int_C d\omega \, P(\omega) T(\omega \to \omega') = P(\omega') \,. \tag{42}$$

Second, we require the random walk to be irreducible, which means that the transition graph must be connected and every configuration $\omega \in C$ can be reached from any configuration $\omega' \in C$ in a finite number of steps. This property is necessary for the uniqueness of the stationary distribution [147]. Lastly, we require the random walk to be aperiodic (see Ref. [147] for a rigorous definition). Together with the irreducibility condition, this ensures convergence to the stationary distribution [147].

There are several possibilities to design a Markov chain with a desired stationary distribution [146–151]. Commonly, the Markov chain is constructed to be reversible. This means that it satisfies the detailed balance condition:

$$P(\omega) T(\omega \to \omega') = P(\omega') T(\omega' \to \omega) \,, \tag{43}$$

which is a stronger condition for the stationarity of P than the global balance condition in Equation (42). One popular choice for the transition probabilities $T(\omega \to \omega')$ that satisfies the detailed balance condition is given by the Metropolis–Hastings algorithm. Most applications of MCI reviewed in this work are based on the Metropolis–Hastings algorithm [149,150]. In this approach, the transition probabilities $T(\omega \to \omega')$ are decomposed into propositions $\tilde{T}(\omega \to \omega')$ and acceptance probabilities $p_{\text{acc}}(\omega \to \omega')$ as follows:

$$T(\omega \to \omega') = \tilde{T}(\omega \to \omega') p_{\text{acc}}(\omega \to \omega') \,, \tag{44}$$

The probabilities to propose a move $\tilde{T}(\omega \to \omega')$ can be any random walk satisfying the irreducibility and aperiodicity condition. By inserting the decomposition of the transition probabilities Equation (44) into the detailed balance condition Equation (43), one obtains for the acceptance probabilities:

$$\frac{p_{\text{acc}}(\omega \to \omega')}{p_{\text{acc}}(\omega' \to \omega)} = \frac{P(\omega')\tilde{T}(\omega' \to \omega)}{P(\omega)\tilde{T}(\omega \to \omega')} = \frac{\pi(\omega')\tilde{T}(\omega' \to \omega)}{\pi(\omega)\tilde{T}(\omega \to \omega')} \,, \tag{45}$$

where, in the last step, the idea that the unknown normalisation factors (see Equation (39)) of the PDF cancel was used. The condition in Equation (45) is fulfilled by the Metropolis–Hastings acceptance probabilities [150]:

$$p_{\rm acc}(\omega \to \omega') = \min\left(1, \frac{\pi(\omega')\tilde{T}(\omega' \to \omega)}{\pi(\omega)\tilde{T}(\omega \to \omega')}\right). \tag{46}$$

For the special case, for which the proposition probabilities are symmetric $\tilde{T}(\omega \to \omega') = \tilde{T}(\omega' \to \omega)$, Equation (46) reduces to the Metropolis acceptance probabilities:

$$p_{\rm acc}(\omega \to \omega') = \min\left(1, \frac{\pi(\omega')}{\pi(\omega)}\right). \tag{47}$$

As an example, we regard a classical thermodynamic system with Boltzmann weights $e^{-\beta E(\omega)}$ given by the energies of configurations and the inverse temperature to give an intuitive interpretation of the Metropolis acceptance probabilities in Equation (47). The proposition to move from a configuration ω to a configuration ω' with a smaller energy $E(\omega') < E(\omega)$ is always accepted independent of the temperature. On the other hand, the proposition to move to a configuration ω' with a larger energy than ω is only accepted with a probability depending on the ratio of the Boltzmann weights. If the temperature is higher, it is more likely to move to states with a larger energy. This reflects the physics of the system in the algorithm, focusing on the low-energy states at low temperatures and going to the maximum entropy state at large temperatures.

4. Series-Expansion Monte Carlo Embedding

In this section, we provide a self-contained and comprehensive overview of linked-cluster expansions for long-range interacting systems [25,29–31,34,35] using white graphs [152] in combination with Monte Carlo integration for the graph embedding. First, we introduce linked-cluster expansions (LCEs) and discuss perturbative continuous unitary transformations (pCUTs) [153,154] as a suitable high-order series expansion method. We then establish an adequate formalism for setting up LCEs and discuss the calculation of suitable physical quantities in practice. With the help of white graphs, we can employ LCEs for models with long-range interactions and use the resulting contributions in a Monte Carlo algorithm to deal with the embedding problem posed by long-range interactions. This approach we dub pCUT + MC.

4.1. *Motivation and Basic Concepts*

The goal of all our efforts is to calculate physical observables in the thermodynamic limit, i.e., for an infinite lattice $\mathcal{L}$, using high-order series expansions. The starting point of every perturbative problem is

$$\mathcal{H} = \mathcal{H}_0 + \lambda\mathcal{V}, \tag{48}$$

where the Hamiltonian describing the full physical problem $\mathcal{H}$ can be split up into an unperturbed part $\mathcal{H}_0$, which is readily diagonalisable, and a perturbation $\mathcal{V}$ associated with a perturbation parameter λ, which is small compared to the energy scales of $\mathcal{H}_0$. We aim to obtain a power series up to a maximal reachable order $\mathfrak{o}_{\rm max}$ as an approximation of a desired physical quantity:

$$f(\lambda) \approx p_0 + p_1\lambda + p_2\lambda^2 + \dots p_{\mathfrak{o}_{\rm max}}\lambda^{\mathfrak{o}_{\rm max}}, \tag{49}$$

where the coefficients p_i are to be determined by the series expansion. We want to use the information contained in the power series to infer the properties of the approximated function $f(\lambda)$ [155]. The cost of determining the coefficients is associated with an exponential growth in complexity with increasing order [155]. Hence, calculations are performed with the help of a computer programme. Obviously, the computer cannot deal with an infinitely large lattice. Instead, we must look at finite cut-outs consisting of a finite set of lattice sites that are connected by bonds (or links) symbolising the interactions of the Hamiltonian on the lattice. We term these cut-outs *clusters*. If two clusters A and B do not share a common site or conterminously do not have a link that connects any site of A and B with each other ($A \cap B = \emptyset$), then the cluster $C = A \cup B$ is called a *disconnected cluster*. Otherwise, if no such partition into disconnected clusters A and B exists ($A \cap B \neq \emptyset$), the cluster C is called *connected*. We can define quantum operators $\mathcal{M}$ (e.g., a physical observable) on these clusters just as on the infinite lattice.

There are essentially two ways of performing high-order series expansions. The first one is the naive approach of taking a single finite cluster $C \subset \mathcal{L}$ [153,154,156–158] and designing it such that the contribution of $\mathcal{M}(C)$ coincides with the contributions on the infinite lattice $\mathcal{M}(\mathcal{L}) = \mathcal{M}(C)$ up to the considered order in the perturbation parameter. The cluster needs to be chosen large enough such that the perturbative calculations up to the considered order are not affected by the boundaries of the cluster. Another way of performing calculations is to construct the operator contribution on a cluster—coinciding with the infinite lattice contributions up to a given order—by decomposing it into all possible contributions on smaller clusters [155,159–170]. Now, the contributions on many, but smaller clusters must be determined and added up to obtain the contribution on the infinite lattice:

$$\mathcal{M}(\mathcal{L}) = \mathcal{M}(C) = \sum_{C' \subset C} \mathcal{M}(C') \,. \tag{50}$$

In contrast to the previous approach, we willingly accept boundary effects for the many subclusters. Such a cluster decomposition is known to be computationally more efficient because it suffices to calculate the contributions on the Hilbert space of the smaller clusters reducing the overhead of non-contributing processes, and it also suffices to perform the calculations only on a few clusters as many give identical contributions due to symmetries. This can significantly reduce the overhead.

However, there are subtleties about the validity of performing calculations on finite clusters, e.g., when setting up a linked-cluster expansion (linked-cluster means only connected clusters contribute), the operator $\mathcal{M}$ must satisfy a certain property, namely cluster additivity. The quantity $\mathcal{M}$ is called *cluster additive* if and only if the contribution on disconnected clusters $A \cup B$ solely comes from the contributions on its constituent connected clusters A and B. This means we can simply add the contributions of A and B from the smaller connected clusters to obtain the one for the disconnected cluster, i.e.,

$$\mathcal{M}(A \cup B) = \mathcal{M}(A) + \mathcal{M}(B), \tag{51}$$

as illustrated schematically in Figure 3.

Figure 3. Schematic illustration of cluster additivity. The contribution of a disconnected cluster $A \cup B$ (clusters within dashed circle) made up of individual connected clusters A and B (yellow areas) is the sum of its individual parts.

We can also understand cluster additivity in the language of perturbation theory where acting on bonds in every order forms a cluster of perturbatively active bonds. If such a cluster is connected, we call these processes *linked*. So, cluster additivity simultaneously means that only linked processes will contribute. Cluster additivity is at the heart of the *linked-cluster theorem*, which states that only linked processes will contribute to the overall contribution in the thermodynamic limit. To set up a linked-cluster expansion, we want to exploit cluster additivity and the linked-cluster theorem so that we can "simply" add up the contributions from individual connected clusters to obtain the desired power series in the thermodynamic limit.

An example of a cluster additive quantity is the ground-state energy E_0. Imagine we want to calculate the ground-state energy of a non-degenerate ground-state subspace, then cluster additivity is naturally fulfilled:

$$E_0(A \cup B) = E_0(A) + E_0(B), \tag{52}$$

and we can calculate the ground-state energy on $A \cup B$ from its individual parts A and B. However, cluster additivity is not satisfied in general. We can construct a counterexample by considering the first excited state with energy E_1. For example, consider the first excitation above the ferromagnetic ground state of the transverse-field Ising model in the low-field limit that is a single spin flip dressed with quantum fluctuations induced by the transverse field [113]. We usually refer to such excitations as *quasiparticles* (qp). Here, we cannot add the contributions on clusters A and B to obtain the excitation energy on cluster $A \cup B$:

$$E_1(A \cup B) \neq E_1(A) + E_1(B)\,. \tag{53}$$

How to set up a linked-cluster expansion for intensive properties is not obvious, and it seemed out of reach after the introduction of linked-cluster expansions in the 1980s [159–161]. Only several years later, it was noticed by Gelfand [162] that additivity can be restored for excited states when properly subtracting the ground-state energy. This approach was later generalised to multiparticle excitations [154,156,164,165] and observables [154,163].

In the following, we first introduce a perturbation theory method that maps the original problem in Equation (48) to an effective one. We will show that the derived effective Hamiltonian and observables satisfy cluster additivity. In the subsequent section, we make use of the property and show how we can set up a linked-cluster expansion for energies of excited states and observables by properly subtracting contributions from lower energy states.

4.2. Perturbation Method: Perturbative Continuous Unitary Transformations

The first step towards setting up a linked-cluster expansion is to find a perturbation method that satisfies cluster additivity, which is generically not given and a non-trivial task [171]. Here, we use perturbative continuous unitary transformations (pCUTs) [153,154] that transform the original Hamiltonian perturbatively order by order into a quasiparticle-conserving Hamiltonian, reducing the original many-body problem to an effective few-body problem. We start discussing how to solve the flow equation to obtain the pCUT method and show afterwards how the Hamiltonian decomposes into additive parts that can be used for a linked-cluster expansion.

We strive to solve the usual problem of perturbation theory of Equation (48). The unperturbed part $\mathcal{H}_0$ can be easily diagonalised exactly with a spectrum that has to be equidistant and bounded from below. Additionally, the perturbation $\mathcal{V}$ must be a sum of operators T_n:

$$\mathcal{V} = \sum_{n=-N}^{N} T_n\,, \tag{54}$$

containing all processes changing the energy by n quanta and—if properly rescaled—corresponding to the same number of quasiparticles n. The goal of the pCUT method is to find an optimal basis in which the many-body problem of the original Hamiltonian reduces to an effective few-body problem. For that, we introduce a unitary transformation depending on a continuous flow parameter ℓ and define

$$\mathcal{H}(\ell) = U^{\dagger}(\ell)\mathcal{H}U(\ell)\,. \tag{55}$$

In the limiting case $\ell = 0$, we require $\mathcal{H}(0) = \mathcal{H}$ to recover the original Hamiltonian, and for $\ell = \infty$, we require $\lim_{\ell\to\infty} \mathcal{H}(\ell) = \mathcal{H}_{\text{eff}}$ so that the unitary transformation maps the original to the desired effective Hamiltonian. We can rewrite the unitary transformation as

$$U(\ell) = \mathcal{T}_\ell \exp\left(-\int_0^\ell \eta(\ell')\mathrm{d}\ell'\right), \tag{56}$$

where η is the anti-hermitian generator generating the unitary transformation and $\mathcal{T}_\ell$ the ordering operator for the flow parameter. Taking the derivatives of Equation (55) and Equation (56) in ℓ, we eventually arrive at the flow equation:

$$\frac{\mathrm{d}\mathcal{H}(\ell)}{\mathrm{d}\ell} = [\eta(\ell), \mathcal{H}(\ell)]\,. \tag{57}$$

Flow equations have been studied for quite some time in mathematics and physics with a variety of applications [172–180]. It was Knetter and Uhrig [153] who proposed a perturbative ansatz for the generator of continuous unitary transformations along the lines of Mielke [180], introducing the quasiparticle generator (also known as the "MKU generator") for the pCUT method:

$$\eta^{\text{qp}}(\ell)_{i,j} := \text{sgn}(\mathcal{H}_{0\,i,i} - \mathcal{H}_{0\,j,j})\mathcal{H}_{i,j}(\ell)\,, \tag{58}$$

where the indices i, j refer to blocks of the Hamiltonian labelling the quasiparticle number. Diagonal blocks $\mathcal{H}_{i,i}$ contain all processes conserving the number of quasiparticles i, while off-diagonal blocks $\mathcal{H}_{i,j}$ contain all processes changing the quasiparticle number from i to j. The reasoning behind the ansatz can be explained by looking at $\text{sgn}(\mathcal{H}_{0\,i,i} - \mathcal{H}_{0\,j,j})$, where processes $i \to j$ in $\mathcal{H}_{i,j}$ are assigned the opposite sign of the inverse processes $j \to i$, and therefore, the idea is to "rotate away" off-diagonal blocks by the unitary transformation during the flow of ℓ, while processes that do not change the quasiparticle number are not transformed away due to $\text{sgn}(0) = 0$, but get renormalised during the flow. Consequently, in the limit $\ell \to \infty$, we obtain an effective Hamiltonian $\mathcal{H}_{\text{eff}}$ that is block diagonal in n. This idea is depicted in Figure 4.

Next, we make a perturbative ansatz for the Hamiltonian during the flow:

$$\mathcal{H}(\ell) = \mathcal{H}_0 + \sum_{\substack{k=1 \\ \sum_j n_j = k}}^{\infty} \lambda_1^{n_1} \ldots \lambda_{N_\lambda}^{n_{N_\lambda}} \sum_{\dim(\boldsymbol{m})=k} F(\ell;\boldsymbol{m})T(\boldsymbol{m})\,, \tag{59}$$

with the notation

$$\boldsymbol{m} = (m_1, m_2, m_3, \ldots, m_k)\,, \tag{60}$$
$$m_i \in \{0, \pm 1, \pm 2, \ldots, \pm N\}\,, \tag{61}$$
$$\dim(\boldsymbol{m}) = k\,, \tag{62}$$
$$T(\boldsymbol{m}) = T_{m_1} T_{m_2} T_{m_3} \ldots T_{m_k}\,, \tag{63}$$

with $F(\ell;\boldsymbol{m})$ being undetermined real functions. We introduce N_λ distinct expansion parameters instead of just a single λ to keep the notation as general as possible because in Section 4.6.3, about white graphs, we will need multiple expansion parameters to encode additional information. Inserting Equations (58) and (59) into the flow equation (57), we can solve the equation perturbatively order-by-order as we obtain a recursive set of differential equations for $F(\ell;\boldsymbol{m})$.

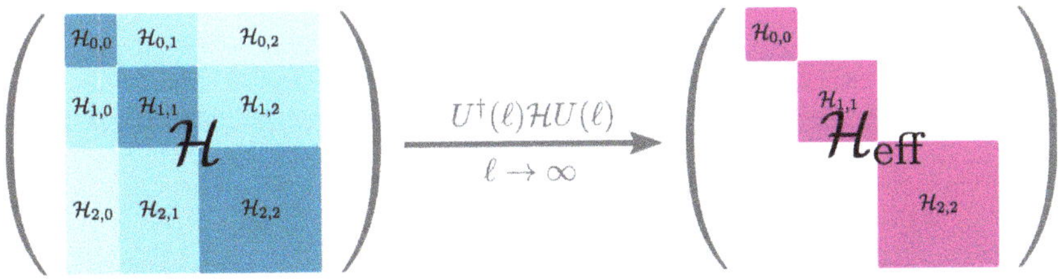

Figure 4. Illustration of the method of perturbative continuous unitary transformations (pCUTs) transforming the original Hamiltonian on the left to a block-diagonal quasiparticle-conserving effective Hamiltonian on the right. The desired effective Hamiltonian is given in the limit $\ell \to \infty$ of the flow parameter ℓ of the continuous unitary transformation $\mathcal{H}(\ell) = U^\dagger(\ell)\mathcal{H}(0)U(\ell)$. While the different quasiparticle sectors interact with each other by the off-diagonal blocks in the original Hamiltonian, the off-diagonal blocks are zero in the effective Hamiltonian as they are "rotated away" during the flow.

To recover the original Hamiltonian $\mathcal{H}$, we have to demand the correct initial conditions $F(0;\boldsymbol{m}) = 1$ for $|\boldsymbol{m}| = 1$ and $F(0;\boldsymbol{m}) = 0$ for $|\boldsymbol{m}| > 1$. We can solve the differential equations (cf. Ref. [153]) exactly for $\ell \to \infty$, yielding

$$\mathcal{H}_{\text{eff}} = \mathcal{H}_0 + \sum_{\substack{k=1 \\ \sum_j n_j = k}}^{\infty} \lambda_1^{n_1} \dots \lambda_{N_\lambda}^{n_{N_\lambda}} \sum_{\substack{\dim(\boldsymbol{m})=k, \\ M(\boldsymbol{m})=0}} C(\boldsymbol{m})T(\boldsymbol{m}), \tag{64}$$

with $F(\infty;\boldsymbol{m}) = C(\boldsymbol{m}) \in \mathbb{Q}$ being exact rational coefficients and the restriction $M(\boldsymbol{m}) = \sum_{i=1}^{k} m_i = 0$ making the products $T(\boldsymbol{m})$ quasiparticle-conserving [153]. Hence, the commutator of the effective Hamiltonian with the unperturbed diagonal part of the original Hamiltonian vanishes ($[\mathcal{H}_{\text{eff}}, \mathcal{H}_0] = 0$). Note that, so far, the effective Hamiltonian (64) is model independent. It only depends on the overall structure of Equation (54). The generic form of the Hamiltonian comes at the cost of an additional normal ordering usually by applying the Hamiltonian to a cluster. Of course, it could also be performed explicitly by using the hard-core bosonic commutation relations, but the former approach can be handled much easier by a computer programme. Yet, we achieved our goal of obtaining a block-diagonal Hamiltonian:

$$\mathcal{H}_{\text{eff}} = \bigoplus_{n=0} \mathcal{H}_{\text{eff}\,n}\,, \tag{65}$$

where $\mathcal{H}_{\text{eff}\,n}$ is the effective irreducible Hamiltonian of n quasiparticle processes (see also Section 4.3). Let us emphasise again that this block-diagonal structure allows us to solve the n quasiparticle blocks individually, which significantly reduces the complexity of the original many-body problem to an effective one that is block-diagonal in the quasiparticle number n.

If we want to calculate an effective observable, we can make an ansatz along the same lines [154]. We insert the perturbative ansatz:

$$\mathcal{O}(\ell) = \sum_{\substack{k=1 \\ \sum_j n_j = k}}^{\infty} \lambda_1^{n_1} \dots \lambda_{N_\lambda}^{n_{N_\lambda}} \sum_{i=1}^{k+1} \sum_{\dim(\boldsymbol{m})=k} G(\ell;\boldsymbol{m};i)\mathcal{O}(\boldsymbol{m};i) \tag{66}$$

with undetermined functions $G(\ell;\boldsymbol{m};i)$. The operator product is defined as

$$\mathcal{O}(\boldsymbol{m};i) = T_{m_1} \dots T_{m_{i-1}} \mathcal{O} T_{m_{i+1}} \dots T_{m_k}\,. \tag{67}$$

Inserting exactly the same generator (58) and the ansatz for the observable in Equation (66) instead of the Hamiltonian into the flow Equation (57), we arrive at

$$\mathcal{O}_{\text{eff}} = \sum_{\substack{k=1 \\ \sum_j n_j = k}}^{\infty} \lambda_1^{n_1} \dots \lambda_{N_\lambda}^{n_{N_\lambda}} \sum_{i=1}^{k+1} \sum_{\dim(\boldsymbol{m})=k} \tilde{C}(\boldsymbol{m};i)\mathcal{O}(\boldsymbol{m};i) \tag{68}$$

with $\tilde{C}(\boldsymbol{m};i) = G(\infty;\boldsymbol{m};i) \in \mathbb{Q}$ by solving the resulting set of differential equations for $\ell \to \infty$ [154]. Note that the last sum does not contain a restriction $M(\boldsymbol{m}) = 0$, and therefore—in contrast to the effective Hamiltonian—effective observables are not (necessarily) quasiparticle-conserving.

We have just derived the effective form of the Hamiltonian and observables in the pCUT method that have a very generic form depending only on the structure of the perturbation of Equation (54). As already stated, the model dependence of our approach comes into play when performing a linked-cluster expansion by applying the effective Hamiltonian or observable to finite clusters. But how do we know if the effective quantities are cluster additive? We follow the argumentation of Refs. [181,182] by looking at the original Hamiltonian (48) that trivially satisfies cluster additivity as long as all bonds represent a non-vanishing term in $\mathcal{V}$ between sites ("bond equals interaction"). Thus, the Hamiltonian on a disconnected cluster $A \cup B$:

$$\mathcal{H}|_{A\cup B} = \mathcal{H}|_A + \mathcal{H}|_B \tag{69}$$

is cluster additive because $\mathcal{H}|_A$ and $\mathcal{H}|_B$ are non-interacting. Here, we denote the restriction of the Hamiltonian $\mathcal{H}$ to a cluster C as $\mathcal{H}|_C$. We can further insert this property into the flow equation:

$$\begin{aligned} \frac{\mathrm{d}\mathcal{H}(\ell)|_{A\cup B}}{\mathrm{d}\ell} &= [\eta(\ell)|_{A\cup B}, \mathcal{H}(\ell)|_{A\cup B}] \\ \frac{\mathrm{d}\mathcal{H}(\ell)|_A}{\mathrm{d}\ell} + \frac{\mathrm{d}\mathcal{H}(\ell)|_B}{\mathrm{d}\ell} &= [\eta(\ell)|_A + \eta(\ell)|_B, \mathcal{H}(\ell)|_A + \mathcal{H}(\ell)|_B] \\ &= [\eta(\ell)|_A, \mathcal{H}(\ell)|_A] + [\eta(\ell)|_B, \mathcal{H}(\ell)|_B]\,. \end{aligned} \tag{70}$$

Here, we used the property of $\mathcal{H}(0)$ that it commutes on disconnected clusters and the fact that the Hamiltonian is continuously transformed during the flow starting from $\ell = 0$. Therefore, the derivative and the commutator can be split up, acting on each cluster individually and preserving cluster additivity during the flow. Consequently, the effective Hamiltonian:

$$\mathcal{H}_{\text{eff}}|_{A\cup B} = \mathcal{H}_{\text{eff}}|_A + \mathcal{H}_{\text{eff}}|_B \tag{71}$$

in the limit $\ell \to \infty$ is cluster-additive as well. The same proof holds for effective observables. Another more physical argument is that the effective pCUT Hamiltonian can be written as a sum of nested commutators of T-operators [152,167]. For instance, considering the perturbation $\mathcal{V} = T_{-1} + T_0 + T_{+1}$, the effective Hamiltonian looks like

$$\mathcal{H}_{\text{eff}} = \mathcal{H}_0 + \lambda T_0 + \lambda^2[T_{+1}, T_{-1}] + \frac{\lambda^3}{2}([[T_{+1}, T_0], T_{-1}] + [T_{+1}, [T_0, T_{-1}]]) + \dots. \tag{72}$$

Splitting up $T_n = \sum_l \tau_{n,l}$ into local operators acting on bonds l, the nested commutators vanish for processes that are not linked. Hence, the linked-cluster theorem is fulfilled, and the effective Hamiltonian is cluster-additive. To emphasise the linked-cluster property, the generic effective Hamiltonian is often written as

$$\mathcal{H}_{\text{eff}} = \mathcal{H}_0 + \sum_{\substack{k=1 \\ \sum_j n_j = k}}^{\infty} \lambda_1^{n_1} \dots \lambda_{N_\lambda}^{n_{N_\lambda}} \sum_{\substack{\dim(\boldsymbol{m})=k \\ M(\boldsymbol{m})=0}} \sum_{\substack{C \\ |\mathcal{E}_C| \le k}} C(\boldsymbol{m}) \sum_{\substack{l_1,\dots,l_k \\ \cup_{i=1}^k l_i = C}} \tau_{m_1,l_1} \dots \tau_{m_k,l_k}\,, \tag{73}$$

where the sum over C runs over all possible connected clusters with maximal k bonds ($k \ge |\mathcal{E}_C|$) [152]. The notation $\mathcal{E}_C$ will be clarified in Section 4.6.2, where graphs are formally introduced. In this context, it is simply the set of bonds of a connected cluster C and $|\mathcal{E}_C|$ the number of bonds in this set. The condition $\bigcup_{i=1}^{k} l_i = C$ arising from the linked-cluster theorem ensures that the cluster consisting of active links and sites during a process must match with the bonds and sites of the connected cluster C. For observables, the generalised condition $\bigcup_{i=1}^{k} l_i \cup x = C$ holds, where the index x can either refer to a site (local observable) or a link (non-local observable), and we have

$$\mathcal{O}_{\text{eff}} = \sum_{\substack{k=1 \\ \sum_j n_j = k}}^{\infty} \lambda_1^{n_1} \ldots \lambda_{N_\lambda}^{n_{N_\lambda}} \sum_{i=1}^{k+1} \sum_{\dim(\boldsymbol{m})=k} \sum_{\substack{C \\ |\mathcal{E}_C| \leq k}} \tilde{C}(\boldsymbol{m}; i) \times \sum_{\substack{l_1,\ldots,l_k \\ \bigcup_{i=1}^{k} l_i \cup x = C}} \tau_{m_1,l_1} \ldots \tau_{m_{i-1},l_{i-1}} \mathcal{O}_x \tau_{m_i,l_i} \ldots \tau_{m_k,l_k} \,. \tag{74}$$

Although we showed that the effective Hamiltonian and observables are cluster-additive and, therefore, fulfil the linked-cluster theorem, to set up a linked-cluster expansion, there are important subtleties remaining when we restrict the effective Hamiltonian and observables to the quasiparticle basis, which we need to address before we can discuss how to perform the calculations in practice.

4.3. Unravelling Cluster Additivity

In this subsection, we need to clarify how we can use the cluster-additive property of the effective pCUT Hamiltonian and observables to set up a linked-cluster expansion not only for the ground-state energy, but also for the energies of excited states. Many aspects of this section are based on the original work of Ref. [154], in which a general formalism was developed for how to derive suitable quantities for the calculation of multiparticle excitations and observables. We further develop this formalism by inferring the concept of cluster additivity for the quasiparticle basis, introducing the notion of particle additivity. The term "additivity" in this context was recently introduced by Ref. [171].

We start by recalling that the effective Hamiltonian is block-diagonal, and we can write the Hamiltonian operator as a sum of irreducible operators of n quasiparticle processes:

$$\mathcal{H}_{\text{eff}} = \sum_{n=0} \mathcal{H}_{\text{eff},n} \,. \tag{75}$$

We can express the n quasiparticle processes in second quantisation in terms of local (hard-core) bosonic operators $b_i^\dagger$ creating and b_i annihilating a quasiparticle at site i. When considering quantum magnets like we do in this review, a hard-core repulsion comes into play allowing only a single quasiparticle at a given site [154]. For instance, in the ferromagnetic ground-state of the 2D Ising model, an elementary excitation is given by a single spin flip, which can be interpreted as a quasiparticle excitation [113]. Obviously, at most one excitation on the same site is allowed. Different particle flavours τ can also be accounted for by incorporating an additional index τ of the operator $b_{i,\tau}^{(\dagger)}$. To keep the notation simple, we will drop this additional index in the following. The irreducible operators in the second quantisation and normal-ordered form then read

$$\begin{aligned}
\mathcal{H}_{\text{eff}\,0} &= \epsilon_0 \mathbb{1}\,, \\
\mathcal{H}_{\text{eff}\,1} &= \sum_i \sum_j t_{i;j} b_j^\dagger b_i \,, \\
\mathcal{H}_{\text{eff}\,2} &= \sum_{i_1,i_2} \sum_{j_1,j_2} t_{i_1,i_2;j_1,j_2} b_{j_2}^\dagger b_{j_1}^\dagger b_{i_2} b_{i_1} \,, \\
&\vdots \\
\mathcal{H}_{\text{eff}\,n} &= \sum_{i_1,\ldots,i_n} \sum_{j_1,\ldots,j_n} t_{i_1,\ldots,i_n;j_1,\ldots,j_n} b_{j_n}^\dagger \ldots b_{j_1}^\dagger b_{i_n} \ldots b_{i_1} \,.
\end{aligned} \tag{76}$$

Written in normal order, the meaning of these processes is directly clear when acting on states in the quasiparticle basis. The prefactors $t_{i_1\dots,i_n;j_1,\dots,j_n}$ are to be determined by applying the effective pCUT Hamiltonian (64) to an appropriately designed cluster C. Let us consider the quasiparticle basis on a connected cluster C:

$$\{|0\rangle_C, |1\rangle_C, |2\rangle_C, \dots, |n\rangle_C\} = \{|0\rangle_C, |1; i\rangle_C, |2; i_1, i_2\rangle_C, \dots, |n; i_1, i_2, \dots, i_n\rangle_C\}, \quad (77)$$

where the number n specifies the number of particle excitations and the indices i_j denote the positions of the n (local) excitations. The effective pCUT Hamiltonian is quasiparticle conserving, so let us restrict it to N particle states, like when evaluating its matrix elements in this basis. If we evaluate by acting on a state with fewer particles than particles involved in the process, then the irreducible operator annihilates more particles than there are and the contribution is zero, that is $\mathcal{H}_n|^N = 0$ for $N < n$. When we determine the action of $\mathcal{H}_{\text{eff}\,n}|^n$ for $N = n$, this allows us to determine all prefactors $t_{i_1\dots,i_n;j_1,\dots,j_n}$ defining the action of $\mathcal{H}_{\text{eff}\,n}$ on the entire unrestricted Hilbert space. The second quantisation presents a natural generalisation of the Hamiltonian restricted to a finite number of particles to an arbitrary number of particles [154]. We can construct $\mathcal{H}_n|^N$ for $N > n$ from $\mathcal{H}_n|^n$ since the latter completely defines the action $\mathcal{H}_{\text{eff}\,n}$ on the entire Hilbert space.

Although everything seems fine so far to set up a linked-cluster expansions, let us tell you that it is not. We finished the motivation in Section 4.1 with the statement that we cannot simply add up contributions for energies of excited states (cf. Equation (53)). The reason is that, although we showed that $\mathcal{H}_{\text{eff}}$ is cluster additive, the irreducible operators restricted to the N particle basis $\mathcal{H}_{\text{eff}\,n}|^N$ are in fact not. To grasp a better understanding of the abstract concept of cluster additivity and why setting up a linked cluster expansion for higher particle channels usually fails, let us consider the following basis on a disconnected cluster $A \cup B$:

$$\begin{aligned}\{|0\rangle_{A\cup B}, |1\rangle_{A\cup B}, |2\rangle_{A\cup B}, \dots\} = \{&|0\rangle_A \otimes |0\rangle_B, |0\rangle_A \otimes |1\rangle_B, |1\rangle_A \otimes |0\rangle_B, \\ &|0\rangle_A \otimes |2\rangle_B, |1\rangle_A \otimes |1\rangle_B, |2\rangle_A \otimes |0\rangle_B, \dots\},\end{aligned} \quad (78)$$

where $|n\rangle_C$ represents all possible n-particle states living on a cluster C. While there is only one way to decompose the zero-particle states $|0\rangle_{A\cup B}$ on the disconnected cluster $A \cup B$, one-particle states $|1\rangle_{A\cup B}$ decompose into two sets of states with a particle on cluster A ($|1\rangle_A \otimes |0\rangle_B$) and a particle on B ($|0\rangle_A \otimes |1\rangle_B$). For two-particle excitations $|2\rangle_{A\cup B}$, there are three possibilities to distribute the particles. In general, N-particle states have the form $|k\rangle_A \otimes |N-k\rangle_B$ with $k \in \{0, 1, \dots, N\}$, and there are $N+1$ possibilities to decompose the states.

When restricting Equation (51) to these N-quasiparticle states, a cluster-additive Hamiltonian must decompose as

$$\mathcal{H}_{\text{eff}}^{\text{cl. add.}}|_{A\cup B}^N = \bigoplus_{k=0}^{N} \left(\mathcal{H}_{\text{eff}}|_A^k + \mathcal{H}_{\text{eff}}|_B^{N-k} \right), \quad (79)$$

where we introduce the notation $|_C^n$ restricting the Hamiltonian to all n-quasiparticle states on a cluster C. The direct sum in Equation (79) is introduced to emphasise that, for a cluster additive Hamiltonian, there must not be any particle processes between the two disconnected clusters A and B. The Hilbert space on the disconnected cluster $A \cup B$ can be seen as the natural extension of the Hilbert spaces on cluster A and B, and we can define the operators on the clusters A and B in terms of the Hilbert space on $A \cup B$ as a tensor product:

$$\begin{aligned}\mathcal{H}_{\text{eff}}|_A^k &:= \mathcal{H}_{\text{eff}}|^k \otimes \mathbb{1}|^{N-k}, \\ \mathcal{H}_{\text{eff}}|_B^k &:= \mathbb{1}|^{N-k} \otimes \mathcal{H}_{\text{eff}}|^k,\end{aligned} \quad (80)$$

where operators on the left of $\otimes$ are defined on the Hilbert space of A and the operators to the right on the Hilbert space of B. The issue with Equation (79) is that, when we restrict the particle basis to N on the disconnected cluster $A \cup B$, there are contributions from lower particles channels coming from the $N+1$ possibilities to distribute the N particles on the two clusters. For example, if we look at the one-particle space:

$$\mathcal{H}_{\text{eff}}^{\text{cl. add.}}|^{1}_{A\cup B} = \left(\mathcal{H}_{\text{eff}}|^{1}_{A} + \mathcal{H}_{\text{eff}}|^{0}_{B}\right) \oplus \left(\mathcal{H}_{\text{eff}}|^{0}_{A} + \mathcal{H}_{\text{eff}}|^{1}_{B}\right), \tag{81}$$

we see that, in addition to the one-particle contributions, we obtain additional zero-particle contributions. The left part of the direct sum stems from acting on $|1\rangle_A \otimes |0\rangle_B$ and the right side from acting on $|0\rangle_A \otimes |1\rangle_B$. There are always two possibilities to distribute the one-particle excitation on the two clusters where the other cluster is always unoccupied, which gives an additional zero-particle contribution. This fact is schematically illustrated in Figure 5a.

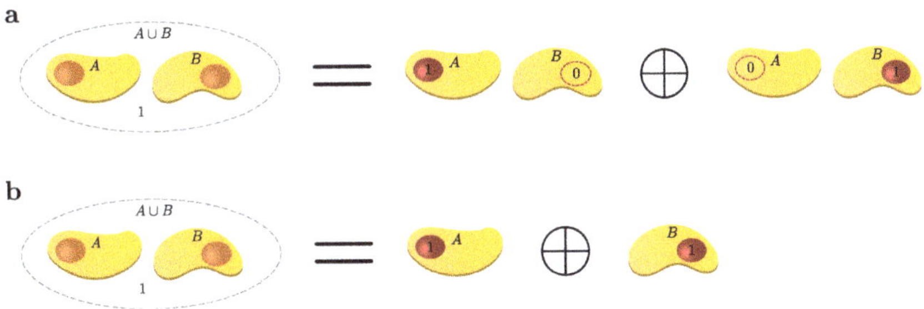

Figure 5. Schematic illustration of cluster additivity (**a**) and particle additivity in the one-particle basis (**b**) on a disconnected cluster $A \cup B$ (grey area) consisting of individual connected clusters A and B (yellow areas). (**a**) Cluster additivity in the one-particle basis translates to one particle being on cluster A and zero on B and vice versa. To calculate the contribution on the disconnected cluster $A \cup B$, both contributions need to be considered including the cases of zero occupancy. (**b**) Particle additivity is fulfilled when the one-particle contribution on a disconnected cluster $A \cup B$ is simply the sum of one-particle contributions on the connected clusters A and B.

This is not the desired behaviour for a linked-cluster expansion. We would like the general notion of Equation (51) to directly translate to the particle-restricted basis, which is illustrated in Figure 5b. In other words, our goal is to find a notion of (cluster) additivity in the restricted particle basis such that we can simply add up the N-particle contributions of individual clusters without caring about lower particle channels. We define *particle additivity* as

$$\mathcal{H}_{\text{eff}}^{\text{add.}}|^{N}_{A\cup B} := \mathcal{H}_{\text{eff}}|^{N}_{A} \oplus \mathcal{H}_{\text{eff}}|_{B}{}^{N}, \tag{82}$$

where we demand the other contributions from Equation (79) to vanish. The crucial thing to notice is that irreducible operators (76) in the particle basis:

$$\mathcal{H}_{\text{eff}\,N}|^{N}_{A\cup B} \equiv \mathcal{H}_{\text{eff}}^{\text{add}}|^{N}_{A\cup B} \tag{83}$$

are, in fact, particle-additive [154,171]. In the following, we will show that this is indeed the case. First, we remember that, for the ground-state energy, we can trivially add up the contributions. Starting from the definition for cluster additivity (79), we have

$$\begin{aligned} \mathcal{H}_{\text{eff}}^{\text{cl. add}}|^{0}_{A\cup B} &= \mathcal{H}_{\text{eff}}|^{0}_{A} + \mathcal{H}_{\text{eff}}|^{0}_{B} \\ &\equiv \mathcal{H}_{\text{eff}\,0}|^{0}_{A} + \mathcal{H}_{\text{eff}\,0}|^{0}_{B} \equiv \mathcal{H}_{\text{eff}}^{\text{add}}|^{0}_{A\cup B}\,. \end{aligned} \tag{84}$$

Second, from restricting the decomposition of Equation (75) to the $N = 1$ particle channel, we can express the irreducible one-particle operator as

$$\mathcal{H}_{\text{eff}\,1}|^1_{A\cup B} = \mathcal{H}^{\text{cl. add.}}_{\text{eff}}|^1_{A\cup B} - \mathcal{H}_{\text{eff}\,0}|^1_{A\cup B}\,. \tag{85}$$

We recall that, by calculating $\mathcal{H}_{\text{eff}\,0}|^0$, we can automatically derive $\mathcal{H}_{\text{eff}\,0}$ on the entire Hilbert space, which subsequently defines $\mathcal{H}_{\text{eff}\,0}|^1$. Therefore, by inserting the definition for cluster additivity (79), we obtain

$$\begin{aligned}
\mathcal{H}^{\text{cl. add.}}_{\text{eff}}|^1_{A\cup B} - \mathcal{H}_{\text{eff}\,0}|^1_{A\cup B} =& \left[\left(\mathcal{H}_{\text{eff}}|^1_A + \mathcal{H}_{\text{eff}}|^0_B\right) \oplus \left(\mathcal{H}_{\text{eff}}|^0_A + \mathcal{H}_{\text{eff}}|^1_B\right)\right] \\
&- \left[\left(\mathcal{H}_{\text{eff}\,0}|^1_A + \mathcal{H}_{\text{eff}\,0}|^0_B\right) \oplus \left(\mathcal{H}_{\text{eff}\,0}|^0_A + \mathcal{H}_{\text{eff}\,0}|^1_B\right)\right] \\
=& \left(\mathcal{H}_{\text{eff}}|^1_A - \mathcal{H}_{\text{eff}\,0}|^1_A\right) \oplus \left(\mathcal{H}_{\text{eff}}|^1_B - \mathcal{H}_{\text{eff}\,0}|^1_B\right) \\
=& \mathcal{H}_{\text{eff}\,1}|^1_A \oplus \mathcal{H}_{\text{eff}\,1}|^1_B \\
\equiv& \mathcal{H}^{\text{add.}}_{\text{eff}}|^1_{A\cup B}\,,
\end{aligned} \tag{86}$$

where we used the definition of Equation (82) in the last line. Hence, we have proven Equation (83) for $N < 2$. The above proof can be readily extended to $N \geq 2$.

We achieved our goal of finding a notion of cluster additivity in the particle basis, which we termed particle additivity. We can determine the desired particle-additive quantities by using the subtraction scheme:

$$\mathcal{H}^{\text{add.}}_{\text{eff}}|^N \equiv \mathcal{H}_{\text{eff}\,N}|^N = \mathcal{H}_{\text{eff}}|^N - \sum_{n=0}^{N-1} \mathcal{H}_{\text{eff}\,n}|^N, \tag{87}$$

which comes from Equation (75) by restricting it to an N-particle basis. This is an inductive scheme starting from $N = 0$ calculating the irreducible additive quantity $\mathcal{H}_{\text{eff}\,0}|^0$. This result can be used to calculate the subsequent irreducible additive quantity $\mathcal{H}_{\text{eff}\,1}|^1$ for $N = 1$. Then, for $N = 2$, we use the results from $N = 0$ and $N = 1$ to calculate $\mathcal{H}_{\text{eff}\,2}|^2$, and so on. Again, it is important that $\mathcal{H}_{\text{eff}\,n}|^n$ completely defines the operator $\mathcal{H}_{\text{eff}\,n}$ and, therefore, any $\mathcal{H}_{\text{eff}\,n}|^m$.

When considering effective observables, the particle number is no longer conserved and more types of processes are allowed. We need to generalise Equation (75) for effective observables by introducing an additional sum over d that is the change in the quasiparticle number. An effective observable, thus, decomposes as

$$\mathcal{O}_{\text{eff}} = \sum_{n=0} \sum_{d\geq -n} \mathcal{O}_{\text{eff}\,n,d}\,, \tag{88}$$

where $\mathcal{O}_{\text{eff}\,n,d}$ are irreducible contributions [154]. When writing them in second quantisation, we have

$$\mathcal{O}_{\text{eff}\,n,d} = \sum_{i_1,\ldots,i_n} \sum_{j_1,\ldots,j_{n+d}} \tilde{t}_{i_1,\ldots,i_n;j_1,\ldots,j_{n+d}} b^\dagger_{j_{n+d}} \ldots b^\dagger_{j_1} b_{i_n} \ldots b_{i_1}\,. \tag{89}$$

We can directly see that the d quasiparticles are created because there are d additional creation operators. When d is negative, d quasiparticles are annihilated. We can infer a notion for cluster additivity and particle additivity along the same lines:

$$\mathcal{O}^{\text{cl. add.}}_{\text{eff}}|^{N\to N+d}_{A\cup B} = \bigoplus_{k=0}^{N} \left(\mathcal{O}_{\text{eff}}|^{k\to k+d}_A + \mathcal{O}_{\text{eff}}|^{N-k\to N-k+d}_B\right), \tag{90}$$

$$\mathcal{O}^{\text{add.}}_{\text{eff}}|^{N\to N+d}_{A\cup B} = \mathcal{O}_{\text{eff}}|^{N\to N+d}_A \oplus \mathcal{O}_{\text{eff}}|^{N\to N+d}_B\,. \tag{91}$$

To determine the particle additive parts, we can use an analogue subtraction scheme as described in Equation (87), which can be denoted as

$$\mathcal{O}_{\mathrm{eff}\,N,d}\big|^{N\to N+d} = \mathcal{O}_{\mathrm{eff}}\big|^{N\to N+d} - \sum_{n=0}^{N-1} \mathcal{O}_{\mathrm{eff}\,n,d}\big|^{N\to N+d} . \tag{92}$$

If we want to calculate $\mathcal{O}_{\mathrm{eff}\,N,d}|^{N\to N+d}$, then we have to inductively apply Equation (92).

There are several things to be noted at this point. First, not all perturbation theory methods satisfy cluster additivity (79), and in this case, we cannot write operators as a direct sum any longer. There will be quasiparticle processes between one and the other cluster changing the number of particles on each cluster [155,162]. This is sketched in Figure 6 by the presence of an additional term.

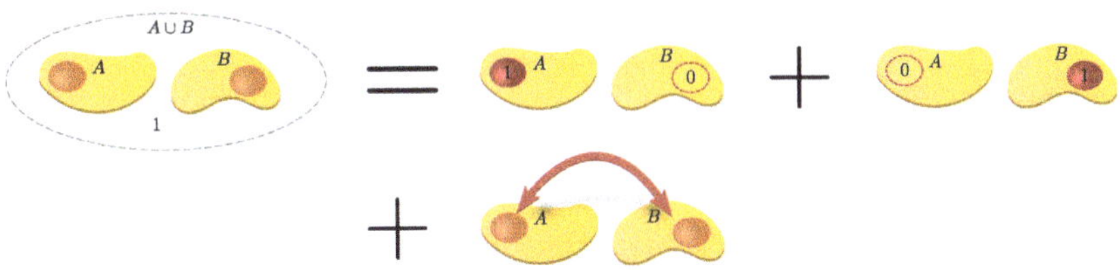

Figure 6. For operators that are not cluster additive, the contribution on the disconnected cluster $A \cup B$ originates not only from the sum of the contributions where a single particle is on A or B, but also from contributions where the particle can hop between the two connected clusters.

When falsely performing a linked-cluster expansion, it can be noticed immediately that the approach breaks down. A symptom of non-cluster additivity is the presence of contributions of lower orders than expected from the number of edges of the graph [155]. When calculating reduced contributions in a linked-cluster expansion, we subtract only contributions of connected subgraphs, which leaves non-zero contributions of disconnected clusters when the perturbation theory method is not cluster-additive. However, there are notable exceptions when a linked-cluster expansion for energies of excited states is still correct even though the perturbation theory method is not cluster-additive [163–165,183–186]. This is only possible when the considered excitation does not couple with a lower particle channel, i.e., lower lying states are described by a distinct set of quantum numbers lying in another symmetry sector [155,165,187]. For instance, consider the elementary quasiparticle excitation in a high-field expansion for the transverse-field Ising model (TFIM), then the structure of the perturbation is $T_{-2} + T_0 + T_2$, and therefore, the first excited state does not couple directly with the ground state (there is no $T_{\pm 1}$, which is due to symmetry). If one wants to draw a comparison, we can think of this as being similar to the case when calculating the excitations in the density matrix renormalisation group. It is no problem to target an excited state if it is in a different symmetry sector than the ground state, but if it is in the same symmetry sector described by the same set of quantum numbers, then the Hamiltonian needs to be modified to project out the ground state [188].

Recently, a minimal transformation to an effective Hamiltonian was discovered that preserves cluster additivity. This method, called "projective cluster-additive transformation" [171], can be used analogously and is even more efficient for the calculation of high-order perturbative series. In this review, however, we stick to the well-established pCUT method.

4.4. Calculating Cluster Contributions

At this point, we may ask how to evaluate physical quantities on finite clusters in practice. To evaluate these quantities, we must evaluate them in the quasiparticle basis. In general, when setting up a cluster expansion, may it be non-linked or linked, it is important to subtract the contributions from all possible subclusters to prevent over-counting. Mathematically, for a quantity $\mathcal{M}$ ($\mathcal{H}_{\text{eff}}$ or $\mathcal{O}_{\text{eff}}$), this can be written as

$$\overline{\mathcal{M}}|_{C}^{\cdots} = \mathcal{M}|_{C}^{\cdots} - \sum_{C' \subset C} \overline{\mathcal{M}}|_{C'}^{\cdots}, \tag{93}$$

where the sum runs over all real subclusters C' in cluster C, and we call the resulting quantity $\overline{\mathcal{M}}|_{C}^{\cdots}$ *reduced*. Starting from the smallest possible cluster (e.g., a single bond between two sites), this formula can be inductively applied to determine the reduced quantity on increasingly big clusters. An essential observation to make is that reduced operators vanish on disconnected clusters by construction if the operator $\mathcal{M}$ is additive since we subtract all contributions from individual subclusters. As the linked-cluster theorem applies, we can set up a cluster expansion:

$$\overline{\mathcal{M}}(\mathcal{L})|_{C}^{\cdots} = \sum_{C \subset \mathcal{L}} \overline{\mathcal{M}}(C)|_{C}^{\cdots} \tag{94}$$

of connected clusters, but we need to consider reduced quantities $\overline{\mathcal{M}}$ to prevent over-counting. For a light notation, we will drop the bar in the sections below as we will only consider reduced contributions on graphs anyway.

Now, we are ready to look at the problem from a more practical point of view. From the previous subsection, we know how cluster additivity translates into the particle basis and how to construct particle-additive parts, namely the irreducible quasiparticle contributions. We decompose the effective Hamiltonian into its irreducible contributions by explicitly calculating:

$$\mathcal{H}_{\text{eff}\,0}|^{0} = \mathcal{H}_{\text{eff}}|^{0} \tag{95}$$
$$\mathcal{H}_{\text{eff}\,1}|^{1} = \mathcal{H}_{\text{eff}}|^{1} - \mathcal{H}_{\text{eff}\,0}|^{1}, \tag{96}$$
$$\mathcal{H}_{\text{eff}\,2}|^{2} = \mathcal{H}_{\text{eff}}|^{2} - \mathcal{H}_{\text{eff}\,1}|^{2} - \mathcal{H}_{\text{eff}\,0}|^{2} \tag{97}$$
$$\vdots \tag{98}$$
$$\mathcal{H}_{\text{eff}\,N}|^{N} = \mathcal{H}_{\text{eff}}|^{N} - \sum_{n=0}^{N-1} \mathcal{H}_{\text{eff}\,n}|^{N}. \tag{99}$$

Again, consider the effective Hamiltonian in the second quantisation made up of hard-core bosonic operators $b_i^{(\dagger)}$ annihilating (creating) quasiparticles and the quasiparticle counting operator $n_i = b_i^\dagger b_i$ occurring in the unperturbed Hamiltonian $\mathcal{H}_0$. We also consider a connected cluster C, and we denote n quasiparticle states on this cluster as $|n; i_1, \ldots, i_n\rangle_C$ with the quasiparticles on the sites i_1 to i_n. Note that, for multiple quasiparticle flavours or multiple sites within a lattice unit cell, this notation can be generalised to $|n; i_1, \ldots, i_n, \tau_1, \ldots, \tau_n\rangle_C$ by introducing additional indices τ_i. To lighten the notation, in the following, we stick to the former case. Let us consider the three lowest particle channels:

$n = 0$ We can directly calculate the ground-state energy $E_0(C)$ on a cluster C as it is already additive:

$$E_0(C) = \langle 0|\mathcal{H}_{\text{eff}}|0\rangle_C, \tag{100}$$

as can be seen from Equation (95).

$n = 1$ To calculate the irreducible amplitudes $t_{i;j}^{(1)}(C)$ associated with the hopping process $b_j^\dagger b_i$ in $\mathcal{H}_{\text{eff}\,1}$, we need to subtract the zero-particle channel, as can be seen from Equation (96). However, we only need to subtract the ground-state energy if the hopping process is local, $b_i^\dagger b_i$, since the ground-state energy only contributes to diagonal processes. Thus, we calculate

$$\begin{aligned} t_{i;j}^{(1)}(C) &= \langle 1;j|\mathcal{H}_{\text{eff}}|1;i\rangle_C && \text{if } i \neq j\,, \\ t_{i;i}^{(1)}(C) &= \langle 1;i|\mathcal{H}_{\text{eff}}|1;i\rangle_C - E_0(C) && \text{else}\,. \end{aligned} \tag{101}$$

$n = 2$ In the two-particle case, we have to distinguish between three processes: pair hoppings ($t_{i,j;k,l}^{(2)}(C)\, b_l^\dagger b_k^\dagger b_j b_i$ with four distinct indices), correlated hoppings ($t_{i,j;i,k}^{(2)}(C)\, b_k^\dagger b_j\, n_i$), and density–density interactions ($t_{i,j;i,j}^{(2)}(C)\, n_j n_i$). The free quasiparticle hopping is already irreducible, and nothing has to be done, but for the correlated hopping contribution, we have to subtract the free one-particle hopping. In the case of the two-particle density–density interactions, we need to subtract the local one-particle hoppings, as well as the ground-state energy, as this process is diagonal (cf. Equation (97)). Therefore, we calculate

$$\begin{aligned} t_{i,j;k,l}^{(2)}(C) &= \langle 2;k,l|\mathcal{H}_{\text{eff}}|2;i,j\rangle_C && \text{if } i \neq j \neq k \neq l\,, \\ t_{i,j;i,k}^{(2)}(C) &= \langle 2;i,k|\mathcal{H}_{\text{eff}}|2;i,j\rangle_C - t_{j;k}^{(1)}(C) && \text{if } i \neq j \neq k\,, \\ t_{i,j;i,j}^{(2)}(C) &= \langle 2;i,j|\mathcal{H}_{\text{eff}}|2;i,j\rangle_C - t_{i;i}^{(1)}(C) - t_{j;j}^{(1)}(C) - E_0(C) && \text{if } i \neq j\,. \end{aligned} \tag{102}$$

An analogous procedure can be applied for effective observables. Here, we need to determine the irreducible contributions for a fixed d. The subtraction scheme is given by

$$\mathcal{O}_{\text{eff}\,0,d}|^{0\to d} = \mathcal{O}_{\text{eff}}|^{0\to d} \tag{103}$$

$$\mathcal{O}_{\text{eff}\,1,d}|^{1\to 1+d} = \mathcal{O}_{\text{eff}}|^{1\to 1+d} - \mathcal{O}_{\text{eff}\,0,d}|^{1\to 1+d} \tag{104}$$

$$\mathcal{O}_{\text{eff}\,2,d}|^{2\to 2+d} = \mathcal{O}_{\text{eff}}|^{2\to 2+d} - \mathcal{O}_{\text{eff}\,1,d}|^{2\to 2+d} - \mathcal{O}_{\text{eff}\,0,d}|^{2\to 2+d} \tag{105}$$

$$\vdots \tag{106}$$

$$\mathcal{O}_{\text{eff}\,N,d}|^{N\to N+d} = \mathcal{O}_{\text{eff}}|^{N\to N+d} - \sum_{n=0}^{N-1} \mathcal{O}_{\text{eff}\,n,d}|^{N\to N+d}\,. \tag{107}$$

For $d = 0$, we recover exactly the same subtraction procedure as before. It is straightforward to generalise this procedure for $d \neq 0$. Let us specifically consider the example we will encounter in the next section, when calculating correlations for the spectral weight. The effective observable is applied to the unperturbed ground state $|0\rangle$. Hence, there are only contributions out of the ground state ($N = 0$), and the effective observables decomposes into

$$\mathcal{O}_{\text{eff}} = \mathcal{O}_{\text{eff}\,0,0} + \mathcal{O}_{\text{eff}\,0,1} + \mathcal{O}_{\text{eff}\,0,2} + \ldots\,. \tag{108}$$

Since only $N = 0$ processes contribute, nothing needs to be subtracted and the effective observable $\mathcal{O}_{\text{eff}}|^{0\to d} = \mathcal{O}_{\text{eff}\,0,d}|^{0\to d}$ is irreducible and already particle-additive.

Although, for these types of calculations, the lowest orders can be analytically determined by hand, the calculations usually become cumbersome quickly and must be evaluated using a computer programme to push to higher perturbative orders (in most cases, a maximal order ranging from 8 to 20 is achievable). Such a programme reads the information of the cluster (site and bond information), the "bra" and "ket" states, the coefficient lists $C(\boldsymbol{m})$ or $\tilde{C}(\boldsymbol{m};i)$ from Equations (64) and (68) up to the desired order, the structure of the Hamiltonian $\mathcal{H}$, the local τ-operators from the perturbation $\mathcal{V}$, and if necessary, the observable. The input states, as well as the τ-operators should be efficiently implemented in the eigenbasis of $\mathcal{H}_0$ bitwise encoding the information as known, for instance, from exact diagonalisation. If possible, the calculation should be performed with rational coefficients for the exact representation of the perturbative series up to a desired order. The routine of the programme is then to iterate through the operator sequences from the coefficient list C or $\tilde{C}$ and to consecutively apply the τ-operators by systematically iterating over all bonds of the cluster and calculating the action of the operator, saving the intermediate states for the action of the next operator in the sequence. As intermediate states are superpositions of basis states, they are saved in associative containers:

$$|\psi\rangle = \sum_j c_j |j\rangle \,, \tag{109}$$

where $|j\rangle$ is the bit representation of a basis state. The key of the associative container is the basis state $|j\rangle$ and the associated value, the prefactor c_j. The bitwise representation of the basis states $|j\rangle$, as well as the τ-operators allow for a fast access and modification during the calculation.

So far, we have introduced the pCUT method for calculating the perturbative contributions on clusters. We demonstrated that the resulting effective quasiparticle-conserving Hamiltonian is cluster-additive and showed how to extract particle-additive irreducible contributions. We introduced a subtraction scheme for the effective Hamiltonian and observables that can be easily applied to calculate additive quantities to set up a linked-cluster expansion. Finally, we briefly discussed an efficient implementation of the pCUT method.

4.5. *Energy Spectrum and Observables*

Having established the basic theoretical framework for the pCUT method, we want to give a short overview of the physical quantities that are most frequently calculated with this approach. To this end, we assume that we consider a single suitably designed cluster for the calculations instead of setting up an LCE as a full graph decomposition. We will see how we can calculate the desired quantities without thinking about the abstract concepts necessary for linked-cluster expansions, and with the insights from this section, it will be easier to recognise what we are aiming at. Here, we first consider the energy spectrum of the Hamiltonian. We derive both the control-parameter susceptibility as the second derivative of the ground-state energy, as well as the elementary excitation gap from the effective one-quasiparticle Hamiltonian. Second, we consider observables. In the pCUT approach, often, spectral weights are calculated, which are of great importance for inelastic neutron scattering experiments.

4.5.1. Ground-State Energy and Elementary Excitation Gap

Following the above-described recursive scheme, we start with the zero-quasiparticle channel assuming a non-degenerate unperturbed ground state, which is the situation in all applications discussed in this review. The ground-state energy can be directly calculated from the cluster as in Equation (100). We consider a suitably designed cluster that is large enough to accommodate all fluctuations of a given maximal order $\mathfrak{o}_{\max}$ and has periodic boundary conditions to correctly account for translational invariance. We calculate $E_0 = \langle 0|\mathcal{H}_{\text{eff}}|0\rangle$ and obtain a high-order perturbative series:

$$E_0 = \sum_{\mathfrak{o}=0}^{\mathfrak{o}_{\max}} p_{\mathfrak{o}} \lambda^{\mathfrak{o}} \tag{110}$$

in the expansion parameter λ, which is valid in the thermodynamic limit up to the given maximal order $\mathfrak{o}_{\max}$. We can extract the ground-state energy per site by dividing E_0 by the number of sites of the cluster $\epsilon_0 = E_0/N$. By taking the second derivative, we obtain the control-parameter susceptibility:

$$\chi = -\frac{\mathrm{d}^2\epsilon_0}{\mathrm{d}\lambda^2} \,. \tag{111}$$

We are usually interested in the quantum-critical properties of the model and, therefore, analyse the behaviour about the quantum-critical point λ_c. The control-parameter susceptibility shows the diverging power-law behaviour:

$$\chi \propto |\lambda - \lambda_c|^{-\alpha} \tag{112}$$

with the associated critical exponent α, as we know from Table 1.

Turning to the one-quasiparticle channel, we calculate the hopping amplitudes following Equation (101). Here, we use open boundary conditions, again with a cluster large enough to accommodate all fluctuations contributing to the hopping process. Note that, in our notation, we denote $t_{i;j}$ for hopping amplitudes on a graph or cluster level and the character a for processes in the thermodynamic limit. As for the ground-state energy, we can directly infer the contribution in the thermodynamic limit if the contributing fluctuations do not feel finite-size effects, and thus, we use $a(\boldsymbol{j}-\boldsymbol{i})$ in the following. Further, we can generalise our notation to multiple particle types or larger unit cells containing multiple sites, as mentioned earlier, by introducing additional indices ξ, τ. We denote a hopping from unit cell $\boldsymbol{i}$ to $\boldsymbol{j}$ by $\boldsymbol{\delta} = \boldsymbol{j} - \boldsymbol{i}$ and within the unit cells from ξ to τ. We calculate $a_{\xi,\tau}(\boldsymbol{\delta}) = \langle 1; \boldsymbol{j}, \tau | \mathcal{H}_{\text{eff}} | 1; \boldsymbol{j} - \boldsymbol{\delta}, \xi \rangle$ by fixing $\boldsymbol{j}$, due to translational symmetry. The effective one-quasiparticle Hamiltonian in the second quantisation then is

$$\mathcal{H}_{\text{eff}}^{\text{1qp}} := \mathcal{H}_{\text{eff}}|^1 = \mathcal{H}_{\text{eff}\,0}|^1 + \mathcal{H}_{\text{eff}\,1}|^1 = \epsilon_0 N + \sum_{\boldsymbol{j},\boldsymbol{\delta},\xi,\tau} a_{\xi,\tau}(\boldsymbol{\delta})\, b^\dagger_{\boldsymbol{j},\tau} b_{\boldsymbol{j}-\boldsymbol{\delta},\xi}. \tag{113}$$

Applying the Fourier transform for a discrete lattice:

$$b_{\boldsymbol{j},\tau} = \frac{1}{\sqrt{N}} \sum_{\boldsymbol{k}} b_{\boldsymbol{j},\tau} \exp(\mathrm{i}\boldsymbol{k}\boldsymbol{j}) \qquad b^\dagger_{\boldsymbol{j},\tau} = \frac{1}{\sqrt{N}} \sum_{\boldsymbol{k}} b^\dagger_{\boldsymbol{j},\tau} \exp(-\mathrm{i}\boldsymbol{k}\boldsymbol{j}), \tag{114}$$

we can diagonalise the resulting Hamiltonian in momentum space:

$$\begin{aligned} \mathcal{F}\left(\mathcal{H}_{\text{eff}}^{\text{1qp}}\right) &= \epsilon_0 N + \sum_{\boldsymbol{k}} \boldsymbol{b}^\dagger_{\boldsymbol{k}} \Omega(\boldsymbol{k}) \boldsymbol{b}_{\boldsymbol{k}} = \sum_{\boldsymbol{k}} \Omega_{m,n}(\boldsymbol{k}) b^\dagger_{\boldsymbol{k},m} b_{\boldsymbol{k},n} \\ &= \sum_{\boldsymbol{k}} \omega_\nu(\boldsymbol{k}) \beta^\dagger_{\boldsymbol{k},\nu} \beta_{\boldsymbol{k},\nu}, \end{aligned} \tag{115}$$

introducing the operators $\beta^{(\dagger)}_{\boldsymbol{k},\nu}(\boldsymbol{k})$ that diagonalise the matrix $\Omega_{m,n}$. The eigenenergies $\omega_\nu(\boldsymbol{k})$ are the associated bands of the one-quasiparticle dispersion. In case of a trivial unit cell or a single particle flavour, the dispersion matrix becomes a scalar, and we can directly express the single-banded dispersion as

$$\omega(\boldsymbol{k}) = a(\boldsymbol{0}) + 2\sum_{\boldsymbol{\delta}} a(\boldsymbol{\delta}) \cos(\boldsymbol{k}\boldsymbol{\delta}), \tag{116}$$

where the sum over $\boldsymbol{\delta}$ is restricted to symmetry representatives and we assumed real hopping amplitudes $a(\boldsymbol{\delta})$. We determine the hopping amplitudes $a(\boldsymbol{\delta})$ as a perturbative series by performing the calculations on the properly designed cluster. Note that, even for a single cluster, we need to perform a subtraction scheme because, in order to obtain the irreducible contribution $a(\boldsymbol{0})$ explicitly, we need to subtract the ground-state energy from $\langle \boldsymbol{i} | \mathcal{H}_{\text{eff}} | \boldsymbol{i} \rangle$. When we determine the elementary excitation gap at the minimum of the lowest band:

$$\Delta = \min_{\boldsymbol{k},\nu} \omega_\nu(\boldsymbol{k}) = \sum_{o=0}^{o_{\max}} p_o \lambda^o, \tag{117}$$

we can as well extract the gap directly as such a series. The gap closing shows a power-law behaviour:

$$\Delta \propto |\lambda - \lambda_c|^{z\nu} \tag{118}$$

about the critical point with the critical exponent $z\nu$.

4.5.2. Spectral Properties

Neutron scattering is a powerful method resolving spatial and dynamical structures in condensed matter physics since thermal neutrons have a de Broglie wavelength of similar length scale as interatomic distances and their energy is of the same order of magnitude as typical excitations [189]. By measuring the change in momentum and kinetic energy in inelastic neutron scattering experiments determining the dynamic response, not only static properties like magnetic order can be resolved, but also dynamic properties like spin–spin correlations [190]. The dynamic response $S_{\alpha\beta}(\boldsymbol{k},\omega)$ can be determined as it is proportional to the cross-section:

$$\frac{\mathrm{d}^2\sigma}{\mathrm{d}\Omega\mathrm{d}\omega} \propto \sum_{\alpha,\beta} \mathcal{S}_{\alpha,\beta}(\boldsymbol{k},\omega) \tag{119}$$

of inelastic neutron scattering [189,190]. We follow the derivations in Refs. [155,191,192] and start with the definition of the *dynamic response*:

$$\mathcal{S}_{\alpha,\beta}(\boldsymbol{k},\omega) = \frac{1}{2\pi N}\sum_{\boldsymbol{i},\boldsymbol{j}}\int_{-\infty}^{\infty}\mathrm{d}t\exp\{\mathrm{i}[\omega t - \boldsymbol{k}(\boldsymbol{j}-\boldsymbol{i})]\}\langle S_{\boldsymbol{j}}^{\alpha}(t)S_{\boldsymbol{i}}^{\beta}(0)\rangle_T\,, \tag{120}$$

which is the space and time Fourier transform of the spin correlation function $\langle S^{\alpha}(t)S^{\beta}(0)\rangle_T$ with $\alpha,\beta\in\{x,y,z,+,-\}$ and $\langle\cdot\rangle_T$ referring to the thermal expectation value. In the limit of vanishing temperature $T=0$, the expectation value simplifies to $\langle\cdot\rangle = \langle\psi_0|\cdot|\psi_0\rangle$ with $|\psi_0\rangle$ being the ground state. Then, we call $\mathcal{S}_{\alpha,\beta}(\boldsymbol{k},\omega)$ the *dynamic structure factor*. We introduce a complete set of energy eigenstates $\{|\psi_\Lambda\rangle\}$, where Λ denotes a set of quantum numbers, for instance $n,\boldsymbol{k}$, where n is the number of quasiparticles and $\boldsymbol{k}$ the lattice momentum. Writing the dynamic structure factor in terms of these energy eigenstates, this yields the dynamic structure factor in the spectral form as a sum:

$$\mathcal{S}_{\alpha,\beta}(\boldsymbol{k},\omega) = \sum_{\Lambda}\mathcal{S}^{\Lambda}_{\alpha,\beta}(\boldsymbol{k},\omega)\,, \tag{121}$$

where $\mathcal{S}^{\Lambda}_{\alpha,\beta}(\boldsymbol{k},\omega)$ is called the *exclusive structure factor* or *spectral weight* associated with the quantum numbers Λ. We insert $\mathbb{1} = \sum_{\Lambda}|\psi_\Lambda\rangle\langle\psi_\Lambda|$ into the correlation function and switch to the Heisenberg picture, where $S_{\boldsymbol{j}}^{\alpha}(t) = e^{\mathrm{i}\mathcal{H}t}S_{\boldsymbol{j}}^{\alpha}(0)e^{-\mathrm{i}\mathcal{H}t}$. The spectral weight then reads

$$\begin{aligned}\mathcal{S}^{\Lambda}_{\alpha,\beta}(\boldsymbol{k},\omega) = \frac{1}{2\pi N}\sum_{\boldsymbol{i},\boldsymbol{j}}\int_{-\infty}^{\infty}\mathrm{d}t\exp[\mathrm{i}(\omega - E_\Lambda + E_0)t]\exp[\mathrm{i}\boldsymbol{k}(\boldsymbol{j}-\boldsymbol{i})]\\ \times\langle\psi_0|S_{\boldsymbol{j}}^{\alpha}(0)|\psi_\Lambda\rangle\langle\psi_\Lambda|S_{\boldsymbol{i}}^{\beta}(0)|\psi_0\rangle\,,\end{aligned} \tag{122}$$

which, after integrating over time t, yields

$$\mathcal{S}^{\Lambda}_{\alpha,\beta}(\boldsymbol{k},\omega) = \delta(\omega - E_\Lambda + E_0)\frac{1}{N}\sum_{\boldsymbol{i},\boldsymbol{j}}\langle\psi_0|S_{\boldsymbol{j}}^{\alpha}(0)|\psi_\Lambda\rangle\langle\psi_\Lambda|S_{\boldsymbol{i}}^{\beta}(0)|\psi_0\rangle\exp[\mathrm{i}\boldsymbol{k}(\boldsymbol{j}-\boldsymbol{i})]\,. \tag{123}$$

If we consider the case $(S_{\boldsymbol{i}}^{\alpha})^{\dagger} = S_{\boldsymbol{i}}^{\beta}$ or some observable with $(\mathcal{O}_{\boldsymbol{i}})^{\dagger} = \mathcal{O}_{\boldsymbol{i}}$, which could be a linear combination of spin operators, then the above expression further simplifies to

$$\mathcal{S}^{\Lambda}(\boldsymbol{k},\omega) = \delta(\omega - E_\Lambda + E_0)\frac{1}{N}\left|\sum_{\boldsymbol{i}}\langle\psi_\Lambda|\mathcal{O}_{\boldsymbol{i}}|\psi_0\rangle e^{-\mathrm{i}\boldsymbol{k}\boldsymbol{i}}\right|^2\,. \tag{124}$$

By the above equations, we can identify the spectral weights $\mathcal{S}^{\Lambda}(\boldsymbol{k})$ as

$$\mathcal{S}^{\Lambda}(\boldsymbol{k}) = \frac{1}{N}\left|\sum_{i}\langle\psi_{\Lambda}|\mathcal{O}_{i}|\psi_{0}\rangle e^{-\mathrm{i}\boldsymbol{k}\boldsymbol{i}}\right|^{2} . \tag{125}$$

These quantities are usually visualised as a heat map where the dispersion $\omega(\boldsymbol{k})$ is plotted against the momentum $\boldsymbol{k}$ and the value $\mathcal{S}^{\Lambda}(\boldsymbol{k},\omega)$, that is the intensity of the scattering signal associated with $|\psi_{\Lambda}\rangle$, is colour-coded.

In the pCUT approach, we want to reformulate the observable in terms of an effective one $\mathcal{O}_{\text{eff}}$. Here, we want to restrict ourselves to the one-quasiparticle spectral weight. In Ref. [191], you can also find a formulation for the two-quasiparticle case. For the 1qp spectral weight, Λ are the quantum numbers defining one-quasiparticle states, and we denote $S^{\Lambda}(\boldsymbol{k}) \equiv S_{\tau}^{1\text{qp}}(\boldsymbol{k})$ in the following. Since the pCUT method is a perturbative approach, we want to reformulate the problem in the language of $\mathcal{H}_0$ states:

$$|\psi_0\rangle = U|0\rangle , \tag{126}$$

$$|\psi_{1\text{qp}}\rangle = U|1;\boldsymbol{k},\tau\rangle , \tag{127}$$

where we introduce the momentum states $|1;\boldsymbol{k},\tau\rangle$ with additional index τ denoting a quantum number like a flavour of the excitation or denoting a site in a unit cell. The momentum states are defined via the Fourier transform:

$$|1;\boldsymbol{k},\tau\rangle = \frac{1}{\sqrt{N}}\sum_{\boldsymbol{j}}\exp(\mathrm{i}\boldsymbol{k}\boldsymbol{j})|1;\boldsymbol{j},\tau\rangle . \tag{128}$$

Inserting these identities, we obtain

$$\begin{aligned}\mathcal{S}_{\tau}^{1\text{qp}}(\boldsymbol{k}) &= \frac{1}{N}\left|\sum_{\boldsymbol{i}}\langle\psi_{1\text{qp}}|\mathcal{O}_{\boldsymbol{i}}|\psi_0\rangle\exp(-\mathrm{i}\boldsymbol{k}\boldsymbol{i})\right|^{2} = \frac{1}{N}\left|\sum_{\boldsymbol{i}}\langle 1;\boldsymbol{k},\tau|U^{\dagger}\mathcal{O}_{\boldsymbol{i}}U|0\rangle\exp(-\mathrm{i}\boldsymbol{k}\boldsymbol{i})\right|^{2} \\ &= \frac{1}{N^2}\left|\sum_{\boldsymbol{i},\boldsymbol{j}}\langle 1;\boldsymbol{j},\tau|\mathcal{O}_{\text{eff},\boldsymbol{i}}|0\rangle\exp[\mathrm{i}\boldsymbol{k}(\boldsymbol{j}-\boldsymbol{i})]\right|^{2} ,\end{aligned} \tag{129}$$

where we defined $\mathcal{O}_{\text{eff},\boldsymbol{i}} = U^{\dagger}\mathcal{O}_{\boldsymbol{i}}U$. For a problem with translational invariance, we can fix the site $\boldsymbol{i}$ of the observable and introduce $\boldsymbol{\delta} = \boldsymbol{j}-\boldsymbol{i}$, which yields

$$\mathcal{S}_{\tau}^{1\text{qp}}(\boldsymbol{k}) = \frac{1}{N^2}\left|\sum_{\boldsymbol{i},\boldsymbol{\delta}}\langle 1;\boldsymbol{i}+\boldsymbol{\delta},\tau|\mathcal{O}_{\text{eff},\boldsymbol{i}}|0\rangle\exp(\mathrm{i}\boldsymbol{k}\boldsymbol{\delta})\right|^{2} = \left|\sum_{\boldsymbol{\delta}}\tilde{a}_{\tau}(\boldsymbol{\delta})\exp(\mathrm{i}\boldsymbol{k}\boldsymbol{\delta})\right|^{2} , \tag{130}$$

where we used $\tilde{a}_{\tau}(\boldsymbol{\delta}) = \langle 1;\boldsymbol{i}+\boldsymbol{\delta},\tau|\mathcal{O}_{\text{eff},\boldsymbol{i}}|0\rangle$. Note that the form of the irreducible contribution of the effective observable is $\mathcal{O}_{\text{eff}\,0,1} = \sum_{\boldsymbol{\delta}}\tilde{a}_{\tau}(\boldsymbol{\delta})b^{\dagger}_{\boldsymbol{\delta},\tau}$, exactly as denoted above in Equation (108). As long as the processes $\tilde{a}(\boldsymbol{\delta})$ and $\tilde{a}(-\boldsymbol{\delta})$ are equivalent by means of the model symmetries, we can use $\tilde{a}(\boldsymbol{\delta}) = \tilde{a}(-\boldsymbol{\delta})$ and further simplify $\mathcal{S}_{\tau}^{1\text{qp}}(\boldsymbol{k})$ to

$$\mathcal{S}_{\tau}^{1\text{qp}}(\boldsymbol{k}) = \left|\tilde{a}_{\tau}(\boldsymbol{0}) + 2\sum_{\boldsymbol{\delta}}\tilde{a}_{\tau}(\boldsymbol{\delta})\cos(\boldsymbol{k}\boldsymbol{\delta})\right|^{2} \equiv |s_{\tau}(\boldsymbol{k})|^{2} . \tag{131}$$

So, in the pCUT method, we determine $\tilde{a}_{\tau}(\boldsymbol{\delta})$ as a perturbative series with which we can determine $\mathcal{S}_{\tau}^{1\text{qp}}(\boldsymbol{k})$. Note that this observable is already an irreducible contribution according to Equation (92), and no subtraction scheme needs to be performed. Because we consider the contribution $\mathcal{O}_{\text{eff}}|^{0\to 1}$ of the observable, the initial particle number is $n = 0$, and nothing needs to be subtracted. To extract the quantum-critical behaviour, we need to evaluate the expression at the critical momentum $\boldsymbol{k}_c$ as

$$\mathcal{S}_{\tau}^{1\text{qp}}(k_c) = \sum_{o=0}^{o_{\max}} p_o \lambda^o, \tag{132}$$

yielding a perturbative series for this quantity as well. It shows a diverging critical behaviour that goes as

$$\mathcal{S}_{\tau}^{1\text{qp}}(\boldsymbol{k}_c) \propto |\lambda - \lambda_c|^{-(2-z-\eta)\nu} \tag{133}$$

with the associated critical exponent $(2 - z - \eta)\nu$. After determining the three quantities χ, Δ, and $\mathcal{S}_{\tau}^{1\text{qp}}(k_c)$ described above, we can use the extrapolation techniques described in Section 4.8 to extract estimates for the critical point λ_c and the associated critical exponents. However, we continue with the description of a linked-cluster expansion as a full graph decomposition for long-range interacting systems. The next step on the way is to formally introduce graphs, discuss their generation, and the associated concept of white graphs.

4.6. White Graph Decomposition

In this section, we first give a brief introduction to graph theory, which forms the basis for understanding how linked-cluster expansions as a graph decomposition work. Then, we discuss how to generate all topologically distinct graphs up to a certain number of edges corresponding to the maximal number of active edges at a given perturbative order. We conclude this section by explaining the concept of white graphs where edge colours are ignored as a topological attribute in the classification of graphs. White graphs are essential for tackling long-range interactions, and therefore, every graph decomposition in this review is, in fact, a white graph decomposition.

4.6.1. Graph Theory

So far, we already defined clusters as a cut-out of the infinite lattice with a finite set of sites and a set of bonds connecting those sites. More generally, only considering the topology of clusters without restricting them to the geometry of lattices, we can define a *graph* as a tuple $\mathcal{G} = (\mathcal{V}_\mathcal{G}, \mathcal{E}_\mathcal{G})$ consisting of a (finite) set of vertices $\mathcal{V}_\mathcal{G}$ and a (finite) set of edges $\mathcal{E}_\mathcal{G}$. An edge $e \in \mathcal{E}_\mathcal{G}$ consists of a pair of vertices $\{\mu, \nu\}$, and these vertices $\mu, \nu \in \mathcal{V}_\mathcal{G}$ are called *adjacent*. The *degree* of a vertex is the number of edges connecting it to other vertices of the graph. In the following, we only consider *undirected*, *simple*, and *connected* graphs, which means there are neither directed edges, multiple edges between two vertices, nor loops (no edge that is joining a vertex to itself), and there always exists a path of edges connecting any two vertices of a graph [155,193,194]. As an example, we depict a graph $\mathcal{G} = (\mathcal{V}_\mathcal{G}, \mathcal{E}_\mathcal{G})$ with $\mathcal{V}_\mathcal{G} = \{0,1,2,3,4,5\}$ and $\mathcal{E}_\mathcal{G} = \{\{0,1\},\{0,2\},\{0,4\},\{0,5\},\{1,3\},\{2,3\}\}$ in Figure 7.

A *subgraph* $\mathcal{G}'$ of a graph $\mathcal{G}$ (we write $\mathcal{G}' \subset \mathcal{G}$) is defined as a subset of $\mathcal{V}'_\mathcal{G} \subset \mathcal{V}_\mathcal{G}$ and $\mathcal{E}'_\mathcal{G} \subset \mathcal{E}_\mathcal{G}$ [193–195]. We call $\mathcal{G}'$ a proper subgraph if $\mathcal{V}'_\mathcal{G}$ and $\mathcal{E}'_\mathcal{G}$ are proper subsets of $\mathcal{V}_\mathcal{G}$ and $\mathcal{E}_\mathcal{G}$.

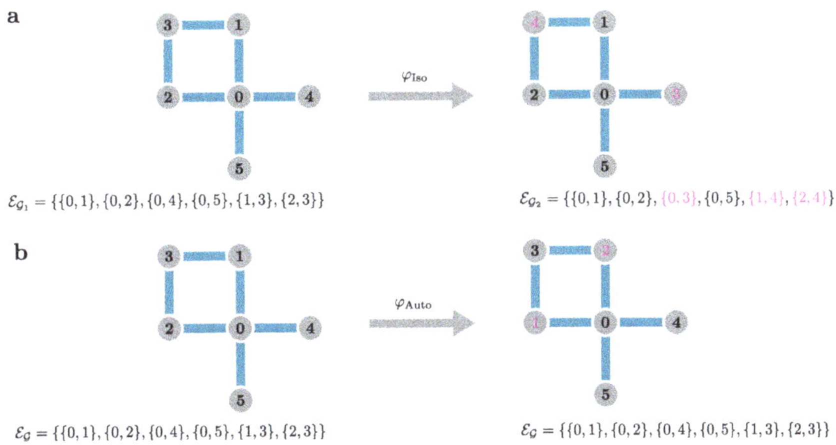

Figure 7. Illustration of a graph isomorphism $\varphi_{\rm Iso}$ and automorphism $\varphi_{\rm Auto}$ map for an example graph. (**a**) The mapping $\varphi_{\rm Iso}(3) = 4$, $\varphi_{\rm Iso}(4) = 3$ and identity for the remaining vertices is a graph isomorphism preserving the adjacency of vertices. If such an isomorphism exists between two graphs, they are topologically equivalent. (**b**) Under a graph automorphism, the edge set $\mathcal{E}_{\mathcal{G}}$ remains invariant, i.e., the graph is mapped onto itself. Here, it is exemplified for the mapping $\varphi_{\rm Auto}(1) = 2$, $\varphi_{\rm Auto}(2) = 1$ and identity for the remaining vertices. A graph automorphism is, therefore, a special case of a graph isomorphism, which leaves the edge set invariant.

To set up a full graph decomposition, it is essential to define how to distinguish different graphs. If there exists an isomorphism between two graphs, we call them *topologically equivalent*; otherwise, they are *topologically distinct*. A *graph isomorphism* $\mathrm{Iso}(\mathcal{G}_1, \mathcal{G}_2)$ is a bijective map $\varphi_{\rm Iso}$ between the vertex sets of two graphs, such that

$$\begin{aligned} &\varphi_{\rm Iso} : \mathcal{V}_{\mathcal{G}_1} \to \mathcal{V}_{\mathcal{G}_2} \\ &\{\mu, \nu\} \in \mathcal{E}_{\mathcal{G}_1} \Leftrightarrow \{\varphi_{\rm Iso}(\mu), \varphi_{\rm Iso}(\nu)\} \in \mathcal{E}_{\mathcal{G}_2}\,. \end{aligned} \tag{134}$$

So, an isomorphism preserves adjacency and non-adjacency, i.e., μ and ν are adjacent if and only if the vertices $\varphi(\mu)$ and $\varphi(\nu)$ are adjacent for any vertices $\mu, \nu \in \mathcal{V}_{\mathcal{G}}$ [193,194,196]. A special case are *graph automorphisms* $\mathrm{Auto}(\mathcal{G})$, which are maps of a graph on itself, so we have

$$\begin{aligned} &\varphi_{\rm Auto} : \mathcal{V}_{\mathcal{G}} \to \mathcal{V}_{\mathcal{G}} \\ &\{\mu, \nu\} \in \mathcal{E}_{\mathcal{G}} \Leftrightarrow \{\varphi_{\rm Auto}(\mu), \varphi_{\rm Auto}(\nu)\} \in \mathcal{E}_{\mathcal{G}}\,. \end{aligned} \tag{135}$$

In other words, a graph automorphism is a permutation of the vertex set preserving adjacency [194,196]. The number of graph automorphisms $|\mathrm{Auto}(\mathcal{G})|$ of a given graph gives the number of its symmetries, and we call it the *symmetry number* $s_{\mathcal{G}} = |\mathrm{Auto}(\mathcal{G})|$ of a graph [155,195]. Examples of a graph isomorphism and automorphism are depicted in Figure 7. Further, if $\mathcal{G}_1 \subset \mathcal{G}_2$, the mapping (134) is injective instead of bijective such that

$$\begin{aligned} &\varphi_{\rm Mono} : \mathcal{V}_{\mathcal{G}_1} \to \mathcal{V}_{\mathcal{G}_2} \\ &\{\mu, \nu\} \in \mathcal{E}_{\mathcal{G}_1} \Rightarrow \{\varphi_{\rm Mono}(\mu), \varphi_{\rm Mono}(\nu)\} \in \mathcal{E}_{\mathcal{G}_2}, \end{aligned} \tag{136}$$

and we call it a *subgraph isomorphism* or *monomorphism* $\mathrm{Mono}(\mathcal{G}_1, \mathcal{G}_2)$ [195,197,198].

An example of a monomorphism is depicted in Figure 8a. Monomorphisms will later become of utmost importance as they give the number of embeddings of the subgraph onto a graph, which is the infinite lattice.

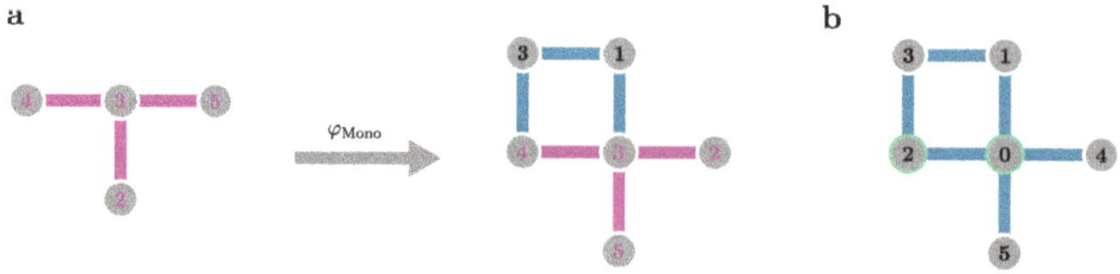

Figure 8. Depiction of a graph monomorphism φ_{Mono} and a graph with colour attributes. (**a**) An example showing a graph monomorphism. The smaller graph on the left is mapped onto the bigger graph on the right, which is the same graph as in Figure 7. Explicitly, the mapping φ_{Mono} is given by $\varphi_{\text{Mono}}(2) = 4$, $\varphi_{\text{Mono}}(3) = 0$, $\varphi_{\text{Mono}}(4) = 2$, and $\varphi_{\text{Mono}}(5) = 5$. (**b**) A coloured graph with the "green" colour attribute assigned to vertices 0 and 2 ($\mathcal{A}_{\mathcal{V}} = \{\{0, \text{green}\}, \{2, \text{green}\}\}$). If mappings are applied to coloured graphs, the colour set must be left invariant, i.e., vertices with a colour must be mapped onto each other and vertices with no colour as well.

To account for hopping processes during the embedding of graphs, we need to assign additional attributes to graphs like colouring their vertices [195]. Then, a coloured graph $\mathcal{G}_c$ is a tuple $(\mathcal{V}_\mathcal{G}, \mathcal{E}_\mathcal{G}, \mathcal{A}_\mathcal{V})$, where $\mathcal{A}_\mathcal{V}$ is a set of pairs $\{\mu, a\}$ with $\mu \in \mathcal{V}$ and a the colour attribute. In Figure 8b, an example is depicted with $\mathcal{A}_\mathcal{V} = \{\{0, \text{green}\}, \{2, \text{green}\}\}$. We can extend the above definitions for isomorphisms and automorphisms and say that they must preserve the vertex colour, i.e., only vertices of the same colour can be mapped onto each other. Of course, this reduces the cardinality $|\text{Auto}(\mathcal{G}_c)|$ and, therefore, the symmetry number $s_{\mathcal{G}_c}$ associated with the coloured graph. We can later exploit the colour information for matching hopping vertices of the graph with the actual hopping of the quasiparticle on the lattice. Note that, also, colouring edges of graphs with attributes $\mathcal{A}_\mathcal{E}$ is useful to distinguish different types of interactions of a Hamiltonian on the graph level and very similar properties as for coloured vertices hold. As stated above, two graphs are topologically distinct if there does not exist a graph isomorphism between the two graphs. Thus, for coloured graphs, the colour attribute serves as another topological attribute. The importance of this will become apparent later in Section 4.6.3. For a more elaborate overview of graph theory in the context of linked-cluster expansions, we recommend Ref. [195].

4.6.2. Graph Generation

For the graph generation of (undirected) simple, connected graphs, we need to define an ordering between all graph isomorphs such that it is possible to pick a unique representative of all graph isomorphs that is called *canonical representative* [199]. One challenge lays in efficiently generating graphs as the number of connected graphs grows exponentially with the number of edges and the idea behind every algorithm must be to restrict the number of possible graph isomorph candidates when generating new edge sets by permutation. A well known algorithm is McKay's algorithm [199,200], which exploits the degree information of the vertices, sets up a search tree for vertices with the same degree, and uses the ordering to check if the canonical representative of the current graph already exists. We recommend using open-source libraries for the graph generation and calculation of additional symmetries. For instance, there is "*nauty*" [201] and "*networkX*" [202] or the "*Boost Graph Library*" [203].

There are various conventions for how to write the graph information to a file. One could simply save the site and edge set as lists or save its *adjacency matrix*, where the rows and columns refer to the sites and a non-zero entry in the corresponding matrix element marks the existence of an edge between the sites [155,193,194]. Here, we suggest to simply use a *bond list*. Each entry in the list contains the edge information of the graph, denoted as n_e, μ, ν, where $e \in \mathcal{E}_\mathcal{G}$ with $\mu, \nu \in \mathcal{V}_\mathcal{G}$ adjacent to e and n_e is just a number associated with the edge e. The number n_e can be interpreted as a specific expansion parameter corresponding to this bond. In the simplest case, there is just one expansion parameter and therefore a single number n_e for all edges. Assigning multiple expansion parameters becomes especially important in the next Section 4.6.3. Usually, the symmetry number $s_{\mathcal{G}}$ of a graph is calculated on the fly when generating a set of graphs and should be saved as well. In our workflow, we save the generated graphs into bond lists and create a list containing all connected graphs (e.g., the file names) along with their symmetry number $s_{\mathcal{G}}$. These types of lists suffice for the calculation of the ground-state energy. When calculating 1qp irreducible processes for the dispersion or for the spectral weight observable, we can think of these processes breaking the graph symmetry. Therefore, after the graph generation, we consider all possible processes on a graph and assign colour attributes to the start and end vertices. Due to the symmetry of the processes, we assign the same colour for hoppings and distinct colours for processes of the spectral weight (When calculating the hopping processes of the effective 1qp Hamiltonian from vertex μ to ν, it has the symmetry $t_{\mu;\nu} = t_{\nu;\mu}$ with the inverse process giving the same contributions as long as the prefactors are real due to the Hermiticity of the Hamiltonian. Thus, we can use the same colours for start and end vertex. The 1qp process for the spectral weight $\tilde{t}_{\mu;\nu}$ is distinct because a quasiparticle is created at vertex μ and then subsequently hops to vertex ν. Therefore, $\tilde{t}_{\mu;\nu}$ and $\tilde{t}_{\nu;\mu}$ are, in general, not equivalent processes $\tilde{t}_{\mu;\nu} \neq \tilde{t}_{\nu;\mu}$, and we have to use distinct colours for the start and end vertices. But, as mentioned in Section 4.5.2, the processes can become equivalent by means of general model symmetries, for instance due to underlying graph symmetries). We calculate the symmetry number $s_{\mathcal{G}_c}$ associated with the coloured graph. In the end, we create a list, where each entry contains the graph, its symmetry number $s_{\mathcal{G}}$, a representative process (start and end vertex), and the associated symmetry number of the coloured graph $s_{\mathcal{G}_c}$, counting the number of processes that give the same contributions as the representative process due to symmetries.

After generating the graphs and the lists, we can employ perturbation theory calculations on the graph level, viewing graphs as abstract clusters with vertices as lattice sites connected by bonds. A programme as described above reads in the graph and process information and repeatedly performs the pCUT calculations on every graph. The resulting graph contributions must be added up in such a way that the information of the lattice geometry is restored by weighting the graph contributions with embedding factors. That is, how many ways a graph can be fit onto a lattice apart from translational invariance (cf. Ref. [155]). For the conventional embedding of contributions from models with just nearest-neighbour interactions, the number of graphs can be reduced as the lattice geometry puts a restriction on the graphs. For example, graphs containing cycles of odd length cannot be embedded on the square lattice. In contrast, for long-range interactions, no such restriction exists, and therefore, the lattice can be seen as a single fully connected graph where every site interacts with each other. Hence, we have to generate every simple connected graph up to a given number of edges as all graphs contribute. Before we turn to the embedding problem in more detail, we will first deal with the challenges long-range interactions pose and address the problem by using white graphs.

4.6.3. White Graphs for Long-Range Interactions

In many cases in perturbation theory, there may be more types of expansion parameters, e.g., due to the geometry of the lattice, as can be seen in the n-leg Heisenberg ladder [152]:

$$\mathcal{H} = \mathcal{H}_0 + \lambda_{\perp}\mathcal{V} + \lambda_{\parallel}\mathcal{V} \tag{137}$$

with an expansion parameter $\lambda_{\perp}$ associated with the rungs and another one $\lambda_{\parallel}$ associated with the legs. The interaction $\mathcal{V}$ is given by XY interactions, and the series expansion is performed about the Ising limit $\mathcal{H}_0$. Setting up a graph decomposition for this model, the canonical approach would be to associate each expansion parameter with a distinct edge colour, blue for $\lambda_{\perp}$ and purple for $\lambda_{\parallel}$.

In Figure 9a (left), all graphs with two edges and two colours (one for $\lambda_{\perp}$ and one for $\lambda_{\parallel}$) are depicted on the left. It is necessary to incorporate the edge colour information as the graphs can only be embedded correctly on the infinite lattice when the edge colour matches the bond type of the lattice. Another common type of perturbation problem is where the perturbation splits into different types of interactions, although associated with the same perturbation parameter. For instance, the problem can look like

$$\begin{aligned}\mathcal{H} &= \mathcal{H}_0 + \lambda\mathcal{V}\\ &= \mathcal{H}_0 + \lambda(\mathcal{V}_x + \mathcal{V}_x + \mathcal{V}_z)\,,\end{aligned} \tag{138}$$

which is essentially the form of the Hamiltonian when performing a high-field series expansion [204] for Kitaev's honeycomb model [205], where each site has one x-, y-, and z-type Ising interaction bond to one of its neighbours on the honeycomb lattice, respectively. Here, we associate the three different interaction types with three types of edge colours, blue, green, and purple for x-, y-, and z-bonds. See Figure 9b (left) for an illustration of all three graphs with two edges and three colours (there are only three possibilities because colours must alternate due to the constraints posed by Kitaev's honeycomb model). In both cases, the edge colour is an additional topological attribute of the graph, leading to exponentially more graphs with the number of colours, which becomes relevant when pushing to high orders in perturbation.

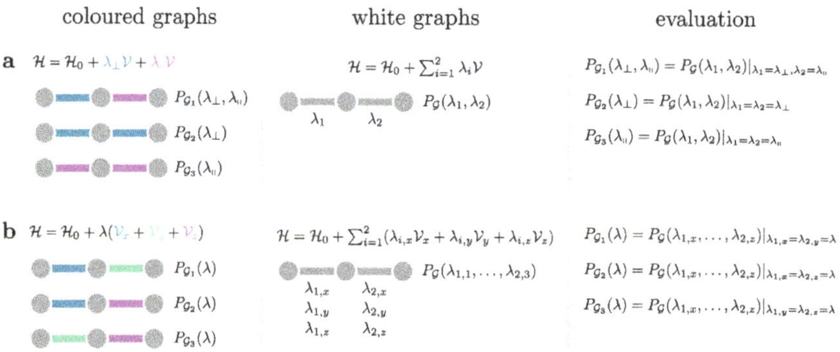

Figure 9. In contrast to the conventional approach using coloured graphs (left), where different expansion parameters or different interaction types are associated with an edge colour, for white graphs (centre) the edge colour is ignored in the topological classification of graphs. Instead additional information is tracked, e.g., by associating each link with abstract expansion parameters and only substituting these abstract contribution during the embedding procedure, reintroducing the correct colour information (right), hence the name white graphs. (**a**) For the problem of Equation (137) on linear graphs with two edges, there are three distinct graphs as the expansion parameters $\lambda_\perp$ and $\lambda_{\shortparallel}$ are associated with individual edge colours (left), but there is only one white graph, as we associated one abstract expansion parameter for each edge (centre). When substituting the abstract expansion parameters with the physical one, reintroducing the correct colour, we can recover the polynomial contributions of the conventional approach (right) (cf. Ref. [152]). (**b**) For the problem of Equation (138) also on linear graphs with two edges, there are also three topologically distinct graphs (left), but for the white graph contribution, we have to introduce multiple abstract expansion parameters for each edge, due to the three flavours $f \in \{x, y, z\}$ (centre). The substitution works analogously to recover the polynomial contribution form the conventional approach (right). Parameters that are not explicitly set are set to zero (cf. Ref. [204]).

In the case of long-range interactions, such an approach becomes unfeasible. A Hamiltonian with long-range interactions is of the form

$$\begin{aligned}\mathcal{H} &= \mathcal{H}_0 + \lambda\sum_{\delta}\frac{1}{|\delta|^{d+\sigma}}\mathcal{V}\\ &= \mathcal{H}_0 + \sum_{\delta}\lambda_\delta\mathcal{V},\end{aligned} \tag{139}$$

where δ is the distance between interacting sites, d the dimension of the lattice, and σ the long-range decay exponent and $\lambda_\delta = \lambda|\delta|^{-(d+\sigma)}$. Applying the conventional approach from above, we would associate each of the infinitely many perturbations V_δ with its own edge colour. The only obvious way to resolve this problem would be to truncate the sum over δ only considering very small distances. Instead, the use of *white graphs* [152] can be a solution to problems of this kind [25,29]. The idea is to change our view of how to tackle these Hamiltonians. We ignore the edge colours of the graph—significantly reducing the number of graphs for a given order—and instead, encode the colour information in the expansion parameters on the graph level in a more abstract way. This is performed by associating each edge e of the graph $\mathcal{G}$ with a different "abstract" perturbation parameter λ_e such that we can track how often τ-operators acted on each edge e, yielding a multivariable polynomial:

$$P_{\mathcal{G}}(\{\lambda_e\}) = \sum_m v_m(\mathcal{G}) M_{\mathcal{G},m} = \sum_m v_m(\mathcal{G}) \prod_{e\in\mathcal{E}_{\mathcal{G}}} \lambda_e^{n_{e,m}}\,, \tag{140}$$

with the sum over all monomial contributions. The individual monomial contributions consist of a prefactor $v_m(\mathcal{G})$, and its monomial dependency $M_{\mathcal{G},m}$. $M_{\mathcal{G},m}$ comprises a product of expansion parameters λ_e associated with edge e and their respective integer powers $n_{e,m} \geq 1$ tracking how often each bond was active during the calculation. Let us emphasise that we simply wrote the white graph contribution for ϵ_0, $t_{\mu;\nu}$, or $\tilde{t}_{\mu;\nu}$ explicitly as a polynomial $P_{\mathcal{G}}(\{\lambda_e\})$. It is only later during the embedding of these abstract generalised contributions when the proper link colour is reintroduced by substituting the expansion parameters by the actual expansion parameters for each realisation on the actual lattice. This is the origin of the name "white graphs" because, during the calculation on the graph, the colour of the links is unspecified (but encoded in the multivariable polynomial), and only during the embedding, the colour of the edges is reintroduced. In Figure 9, we illustrate the white graph concept in the middle. On the right, we depict how to recover the polynomial of the conventional graph contribution from the abstract white graph contribution for the models in Equations (137) and (138) by correctly substituting the abstract expansion parameters. The main difference between the model (138) compared to (137) is that, for the different interaction types, multiple parameters are associated with each edge. To account for three types of interaction flavours, we have to consider three expansion parameters $\lambda_{e,f}$ with $f \in \{x, y, z\}$ per edge e.

For long-range interactions, we can straightforwardly apply the exact same white graph scheme. We substitute the abstract expansion parameter with the correct algebraically decaying interaction strength depending on the distance between the interacting sites on the lattice [25,29]. For this, we have to use the substitution:

$$\lambda_e \mapsto \frac{1}{|\delta|^{(d+\sigma)}} \tag{141}$$

with $\delta = \boldsymbol{i}_\nu - \boldsymbol{i}_\mu$ and $e = \{\mu, \nu\} \in \mathcal{E}_{\mathcal{G}}$. It is also possible to incorporate multiple interaction types, but the substitution for the different interaction types must then be performed before the embedding, as was done in Refs. [31,35,206].

So far, we explained how to resolve the problem of infinitely many perturbation parameters by introducing white graphs. We managed to reduce the number of expansion parameters from infinity to the number of edges, i.e., the order of the perturbation. Yet, the polynomial in Equation (140) still grows exponentially with the number of expansion parameters. Following the description in Ref. [152], we can further mitigate this issue by an efficient representation of white graph contributions. The abstract concept is to track the relevant information in a quantity $M(\boldsymbol{n})$ as a generalised monomial with the property $M(\boldsymbol{n}_1 + \boldsymbol{n}_2) = M(\boldsymbol{n}_1)M(\boldsymbol{n}_2)$ and use $\boldsymbol{n}_i$ as an abstract parameter that encodes the tracked information. In Equation (140), the monomial quantity $M(\boldsymbol{n})$ is just $M_{\mathcal{G},m}$, tracking how often each edge was active during the calculation by associating each edge with its own abstract expansion parameter. We can generalise the expression of Equation (109) for states comprising additional information:

$$|\psi\rangle = \sum_{i,j} c_{i,j} M(\boldsymbol{n}_i)|j\rangle = \sum_j |j\rangle \left(\sum_i c_{i,j} M(\boldsymbol{n}_i) \right) . \tag{142}$$

Instead of a simple superposition of states $|j\rangle$ with associated prefactor c_j as in Equation (109), we have an additional superposition over all monomials $M(\boldsymbol{n}_i)$ comprising the information of all the distinct processes encoded in $M(\boldsymbol{n}_i)$ leading to the state $|j\rangle$. We can make use of this factorisation property by using nested containers. The key of the outer container is the basis state $|j\rangle$, and the value contains an inner container with the monomial $M(\boldsymbol{n}_i)$ as the key and the prefactor $c_{i,j}$ as the value. Thus, the action of a τ-operator on a state $|j\rangle$ can be calculated independent of the action on the monomial $M(\boldsymbol{n}_i)$. For a flat container, we would have to calculate the action on the same state $|j\rangle$ multiple times. To further improve the efficiency of the calculation, we can directly encode into the programme which edge of a graph was active in a bitwise representation. Using the information of the number of edges of a graph and tracking the applied perturbative order during the calculation, we can neglect subcluster contributions on the fly and reduce the computational overhead even further. Therefore, we can directly calculate the reduced contribution on the graph without the need for explicitly subtracting subcluster contributions.

Wrapping things up, we explained how the use of white graphs can be applied to models with long-range interactions resolving the issue of infinitely many graphs or expansion parameters at any finite perturbative order. Instead of treating each perturbation of the long-range interaction as another topological attribute in the classification of graphs, we associate an abstract expansion parameter with each edge of a graph, and only during the embedding on the lattice, we substitute these expansion parameters with the actual bond information of long-range interacting sites. An efficient representation of white graphs can further help to reduce the computational overhead.

4.7. Monte Carlo Embedding of White Graphs

We discussed ways to set up a linked-cluster expansion, either by designing a single appropriately large cluster hosting all relevant fluctuations up to a given perturbative order or by setting up a linked-cluster expansion as a full graph decomposition, where the latter is the more efficient way to perform high-order series expansions. Of course, a quantity $\mathcal{M}$ must be cluster-additive in the first place. To decompose $\mathcal{M}(\mathcal{L})$ on the lattice $\mathcal{L}$ into many smaller subcluster contributions, we can add up the contributions:

$$\mathcal{M}(\mathcal{L}) = \sum_{C \subset \mathcal{L}} \mathcal{M}(C) , \tag{143}$$

where $\mathcal{M}(C)$ are reduced contributions on a cluster C to prevent overcounting. It is not necessary to calculate the contribution of every possible cluster since many clusters have the same contribution. It suffices to calculate the contribution of a representative cluster, only containing the relevant topological information, that is a graph defined by its vertex set $\mathcal{V}_{\mathcal{G}}$ and its edge set $\mathcal{E}_{\mathcal{G}}$. We can see a graph $\mathcal{G}$ as representing an equivalence class $[\mathcal{G}]$, whose elements are all clusters C realising all possible embeddings of a graph on the lattice [195]. Figuratively, all elements of the equivalence class are related by translational and rotational symmetry or are different geometric realisations on the lattice. We can now split the sum over all clusters into one sum over all possible graphs on the lattice and another one over all elements in the equivalence class:

$$\mathcal{M}(\mathcal{L}) = \sum_{C \subset \mathcal{L}} \mathcal{M}(C) = \sum_{\mathcal{G} \subset \mathcal{L}} \sum_{C \in [\mathcal{G}]} \mathcal{M}(C) , \tag{144}$$

where it suffices to calculate $\mathcal{M}(C)$ once for all $C \in [\mathcal{G}]$ [195]. We are left counting the number of elements C in the equivalence class such that we can write

$$\mathcal{M}(\mathcal{L}) = \sum_{\mathcal{G} \subset \mathcal{L}} \mathcal{W}(\mathcal{G}, \mathcal{L}) \mathcal{M}(\mathcal{G}) \,, \tag{145}$$

where the embedding factor $\mathcal{W}(\mathcal{G}, \mathcal{L})$ is simply a number counting the number of embeddings [195]. The important point in our line of thought is that we can calculate the quantity $\mathcal{M}(\mathcal{G})$ only once on the graph level and multiply the contribution with a weight that is the embedding factor $\mathcal{W}$ and sum up the resulting contribution for all graphs to obtain the desired quantity $\mathcal{M}(\mathcal{L})$ in the thermodynamic limit. We are essentially left with determining the embedding factors after calculating the graph contributions $\mathcal{M}(\mathcal{G})$. Depending on the topology of the graph, the number of possible embeddings is different and, therefore, also the embedding factor. When calculating one quasiparticle processes for the 1qp dispersion or the 1qp spectral weight, it is important that we account for the graph vertices' colour attributes for the definition of the equivalence class. If conventional graph contributions are considered, we directly obtain the correct contribution $M(\mathcal{G})$. If there are multiple physical parameters or different interaction flavours, the graph edges have to be colour matched with the bonds of the lattice. If white graphs are used, we do not have to match any colours and this can be ignored; however, the white graph contributions have to be evaluated appropriately substituting the abstract expansion parameters with the correct physical parameter for each embedding. Regarding continuative reading on the embedding, we want to point out the standard literature for linked-cluster expansions [152,155,187], as well as Ref. [195]. Nonetheless, we show in the following how the embedding procedure works for models with nearest-neighbour interactions specifically on the example of the ground-state energy and one-quasiparticle processes yielding the 1qp dispersion of the Hamiltonian. Likewise, we derive the 1qp spectral weight. With these findings, we can turn to the embedding problem of long-range interacting models and eventually describe the Monte Carlo algorithm as a solution to this problem.

4.7.1. Conventional Nearest-Neighbour Embedding

Ground-state energy: We start with the simplest case of calculating the ground-state energy on the infinite lattice $\mathcal{L}$. We know that $\mathcal{H}_{\text{eff}}$ is cluster additive and that the ground-state energy is an extensive quantity, so we can directly calculate

$$E_0 = \sum_{\mathcal{G} \subset \mathcal{L}} \mathcal{W}(\mathcal{G}, \mathcal{L}) E_0(\mathcal{G}) \,. \tag{146}$$

In other words, this means we have to multiply the ground-state energy contributions on the graphs with the correct embedding factor and add up the resulting weighted contributions for every graph. For the definition of the embedding factor, we follow the formalism introduced by Ref. [195] and write

$$\mathcal{W}(\mathcal{G}, \mathcal{L}) = \frac{|\text{Mono}(\mathcal{G}, \mathcal{L})|}{|\text{Auto}(\mathcal{G})|} = \frac{1}{s_{\mathcal{G}}} |\text{Mono}(\mathcal{G}, \mathcal{L})| \,. \tag{147}$$

The embedding factor $\mathcal{W}$ is the number of subgraph isomorphisms (monomorphisms) of the graph $\mathcal{G}$ on the lattice $\mathcal{L}$ divided by the number of graph automorphisms, i.e., the symmetry number of the graph. It is necessary to divide by the symmetry number $s_{\mathcal{G}}$ because the number of monomorphisms is in general not equal to the number of subgraphs because there may be multiple monomorphisms that map to the identical subgraph [195]. To properly account for this, we have to divide by the number of automorphisms.

In Figure 10, you see an example for the embedding problem. We can recognise the fact that there are multiple monomorphisms by the presence of arrows illustrating the ambiguity of mapping onto a specific subgraph embedding. Further, we always have to consider reduced graph contributions subtracting all subgraph contributions as

$$E_0(\mathcal{G}) = \langle 0|\mathcal{H}_{\text{eff}}|0\rangle - \sum_{\mathcal{G}'\subset\mathcal{G}} \mathcal{W}(\mathcal{G}',\mathcal{G})\, E_0(\mathcal{G}')\,. \tag{148}$$

Here, $\mathcal{W}(\mathcal{G}',\mathcal{G})$ is the embedding factor for subgraphs $\mathcal{G}'$ with respect to the considered graph $\mathcal{G}$. Note again that, with an efficient white-graph implementation, reduced contributions are calculated on the fly without the need for explicit subtraction. We state the subtraction scheme for completeness. Going back to the embedding problem for the infinite lattice $\mathcal{L}$, the embedding factor will be extensive as the ground-state energy is extensive as well. Usually, intensive quantities such as the energy per site $\epsilon_0 = E_0/N$ are calculated, where N is the number of sites, and an arbitrary edge of the graph is fixed on the lattice. This then gives

$$\epsilon_0 = \sum_{\mathcal{G}\subset\mathcal{L}} w(\mathcal{G},\mathcal{L})E_0(\mathcal{G}), \tag{149}$$

where the normalised embedding factor is

$$w(\mathcal{G},\mathcal{L}) = \frac{\mathcal{W}(\mathcal{G},\mathcal{L})}{N} = \frac{q}{s_{\mathcal{G}}}|\text{Mono}(\mathcal{G}_c,\mathcal{L}_c)|, \tag{150}$$

denoted with a lower case w, where q is the coordination number of the lattice (the number of neighbours of any site) [155,195]. One can think about fixing sites as colouring two adjacent vertices on the graph and two adjacent sites on the lattice and considering the monomorphism with respect to the additional colour attributes (for instance, $\mathcal{A}_{\mathcal{V}} = \{\{\mu,\text{yellow}\},\{\nu,\text{blue}\}\}$) of the coloured graph $\mathcal{G}_c$ and lattice $\mathcal{L}_c$. In Figure 10, we depict the number of embeddings for a given graph on the square lattice and show how the embedding factor is calculated for this example. Note that it would be equally valid to just fix a single site. Then, one would not have to account for the coordination number q and the number of monomorphism would be larger by a factor of q, making it computationally more expensive. In the end, we obtain the ground-state energy (per site) as a high-order perturbative series in an expansion parameter λ up to a computationally feasible maximal order $\mathfrak{o}_{\max}$ as in Equation (110).

1qp dispersion: We now turn to the one-quasiparticle channel and calculate the hopping amplitudes. For a quasiparticle hopping δ on the lattice with additional hopping within the unit cell from ξ to τ or a quasiparticle changing its flavour from ξ to τ, we need to calculate

$$a_{\xi,\tau}(\delta) = \langle 1;\boldsymbol{i}+\delta,\tau|\mathcal{H}_{\text{eff}}|1;\boldsymbol{i},\xi\rangle = \sum_{\mathcal{G}\subset\mathcal{L}}\sum_{(\mu,\nu)\in\mathcal{P}} w(\mathcal{G}_c,\mathcal{L}_c)\, t_{\mu;\nu}(\mathcal{G})\,, \tag{151}$$

where we fix the initial vertex μ and end vertex ν of the quasiparticle process on the graph to the initial site $\boldsymbol{i}$ and end site $\boldsymbol{i}+\delta$ in real space. We can choose an arbitrary site $\boldsymbol{i}$ without loss of generality due to the translational symmetry of the models considered. (If we consider larger unit cells, we also need to fix the sites ξ, τ witin the unit cells $\boldsymbol{i}$ and $\boldsymbol{i}+\delta$.) The fixing of sites can be formally achieved by assigning colours to the initial and end sites on the graph and on the lattice. The second sum goes over all representative hopping processes $\mathcal{P}$, as described above in Section 4.6.2. The embedding factor here is similar to Equation (150) and reads

$$w(\mathcal{G}_c,\mathcal{L}_c) = \frac{s_{\mathcal{G}_c}}{s_{\mathcal{G}}}|\text{Mono}(\mathcal{G}_c,\mathcal{L}_c)|\,, \tag{152}$$

where the colour attribute comes from fixing the sites to the hopping. We account for the reduced symmetry of coloured graphs stemming from the representative hopping by multiplying by the symmetry number of the coloured graph $s_{\mathcal{G}_c}$.

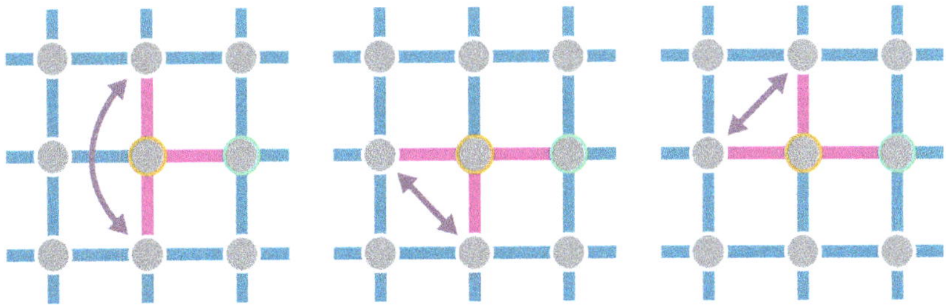

Figure 10. The graph with three edges given in Figure 8 is embedded on the infinite lattice. We consider a coloured graph with additional colour attributes (yellow and green) for two specific vertices. The reason behind colouring the vertices might by to simply fix the graph due to translational and rotational symmetry of the lattice as it is done when calculating the ground-state energy or due to the presence of a one-quasiparticle process from one coloured site to the other. For the ground-state energy contribution, the embedding factor would be $w(\mathcal{G}_c, \mathcal{L}_c) = \frac{q}{s_{\mathcal{G}}}|\mathrm{Mono}(\mathcal{G}_c, \mathcal{L}_c)| = \frac{4}{6} \times 6 = 4$. There are six possible embeddings (monomorphisms) on the infinite lattice, when the coloured vertices on the graph are correctly mapped to the coloured sites on the lattice. There are only three geometrically distinct embeddings, but the number of monomorphisms is two-times bigger due to an ambiguity of mapping the subgraph on the graph as illustrated by the arrows.

The embedding factor is then calculated with respect to the coloured graph $\mathcal{G}_c$ and lattice $\mathcal{L}_c$. Again, we need to be careful as we have to consider reduced and additive contributions, i.e., we have to determine

$$t_{\mu;\nu}(\mathcal{G}) = \langle 1;\nu|\mathcal{H}_{\mathrm{eff}}|1;\mu\rangle - \delta_{\mu,\nu}E_0(\mathcal{G}) - \sum_{\mathcal{G}'\subset\mathcal{G}}\sum_{(\mu,\nu)\in\mathcal{P}} \mathcal{W}(\mathcal{G}'_c,\mathcal{G}_c)\, t_{\mu;\nu}(\mathcal{G}')\,, \tag{153}$$

for each contribution on a graph $\mathcal{G}$. After having determined the ground-state energy per site ϵ_0 and the hopping amplitudes $a_{\xi,\tau}(\delta)$ on the lattice $\mathcal{L}$, we can derive the effective one-quasiparticle Hamiltonian (113). As we have seen in Section 4.5, we can readily derive the one-quasiparticle gap as a series in the perturbation parameter λ as in Equation (117).

1qp spectral weight: Lastly, we can do the same for one-quasiparticle spectral weights. We calculate the process amplitudes:

$$\tilde{a}_\tau(\delta) = \langle 1;\delta,\tau|\mathcal{O}_{\mathrm{eff},i}|0\rangle = \sum_{\mathcal{G}\subset\mathcal{L}}\sum_{(\mu,\nu)\in\mathcal{P}} w(\mathcal{G}_c,\mathcal{L}_c)\,\tilde{t}_{\mu;\nu}(\mathcal{G})\,. \tag{154}$$

Note that a process creating a quasiparticle at μ that subsequently hops to ν with the contribution $\tilde{t}_{\mu;\nu}(\mathcal{G})$ is in general distinct to the inverse process, i.e., $\tilde{t}_{\mu;\nu}(\mathcal{G}) \neq \tilde{t}_{\nu;\mu}(\mathcal{G})$. On the other hand, the 1qp hopping processes of the Hamiltonian fulfil $t_{\mu;\nu}(\mathcal{G}) = t_{\nu;\mu}(\mathcal{G})$ as long as the hopping amplitudes are real. While for hopping processes of the Hamiltonian, we can use the same colour (for start and end vertex) as a topological attribute, here, we must use two different colours, leading to a smaller symmetry number of the coloured graphs. On the graph level, only subgraph contributions must be subtracted:

$$\tilde{t}_{\mu;\nu}(\mathcal{G}) = \langle 1;\nu|\mathcal{O}_{\mathrm{eff},\mu}|0\rangle - \sum_{\mathcal{G}'\subset\mathcal{G}}\sum_{(\mu,\nu)\in\mathcal{P}} \mathcal{W}(\mathcal{G}'_c,\mathcal{G}_c)\, t_{\mu;\nu}(\mathcal{G}')\,. \tag{155}$$

With the contributions $\tilde{a}_\tau(\delta)$, the spectral weight can be determined with Equation (131), and evaluating this quantity for example at the critical momentum $\boldsymbol{k}_c$, we again obtain a series (132) in the perturbation parameter λ.

For the conventional embedding problem, we can also consider several generalisations, which we want to briefly mention. First, we could consider quasiparticle processes between different particle types. In a 1qp process, a particle flavour could change from ξ to τ and the graph contribution would be denoted by including the additional indices $t_{\mu,\xi;\nu,\tau}$ and

$\tilde{t}_{\mu;\nu,\tau}$. The rest of the formalism is identical. Second, we may want to consider different interaction types like in Equation (138) or more expansion parameter like in Equation (137). In such a case, when considering coloured graphs, we also have to consider their coloured edges, which must be matched with the appropriate bonds on the lattice. Then, the coloured graphs $\mathcal{G}_c$ are given by the tuple $(\mathcal{V}_\mathcal{G}, \mathcal{E}_\mathcal{G}, \mathcal{A}_\mathcal{E}, \mathcal{A}_\mathcal{V})$, where $\mathcal{A}_\mathcal{E}$ are the edge colour attributes and $\mathcal{A}_\mathcal{V}$ the vertex colour attributes. Now, every embedding factor w needs to be determined using coloured graphs with respect to $\mathcal{A}_\mathcal{E}$ and $\mathcal{A}_\mathcal{V}$. When using white graphs, the additional edge colour can be ignored and the embedding is as before; however, we need to appropriately substitute the abstract expansion parameters with the actual physical expansion parameters.

By now, we have everything together to calculate the ground-state energy, the 1qp dispersion, and the 1qp spectral weight in the thermodynamic limit from a linked-cluster expansion set up as a full graph decomposition. Therefore, we need to calculate the embedding factor, multiply them with the associated graph contribution, and add up the resulting weighted contributions. The embedding factor can be determined using available graph libraries like the *Boost Graph Library* [203] as only automorphisms and monomorphisms (with colours) need to be determined. For long-range interactions, however, we cannot just simply calculate the embedding factor because, for every graph, there are infinitely many possible embeddings, even when accounting for translational invariance.

4.7.2. Embedding for Models with Long-Range Interactions

In this section, we restrict ourselves to a single physical perturbation parameter. If we had more than one perturbation parameter, this would introduce an additional functional dependency, which we want to avoid as we perform the Monte Carlo summation. A convenient way around this is to sample the parameters and perform the Monte Carlo summation for each parameter ratio like it was performed in [31,35,206] for anisotropic XY interactions, distinct ladder interactions, and for XXZ interactions. In the following, we restrict ourselves to a single quasiparticle flavour due to simplicity and a trivial unit cell, as we have not generalised the algorithm to larger unit cells yet. The starting point to describe the embedding problem for long-range interactions is the embedding formulas given in the previous section in Equations (149), (151), and (154) for the ground-state energy per site, 1qp hopping processes, and the 1qp spectral weight, respectively, which we have to rewrite and adapt for long-range interactions.

Ground-state energy: Starting with the embedding of the ground-state energy per site (149), we can write

$$
\begin{aligned}
\epsilon_0 &= \sum_{\mathcal{G}} w(\mathcal{G},\mathcal{L}) E_0(\mathcal{G}) \\
&= \sum_{\mathfrak{o}=2}^{\mathfrak{o}_{\max}} \sum_{\mathcal{G}} w(\mathcal{G},\mathcal{L}) E_0^{(\mathfrak{o})}(\mathcal{G}) \\
&= \sum_{\mathfrak{o}=2}^{\mathfrak{o}_{\max}} \sum_{n_s=2}^{\mathfrak{o}} \sum_{\substack{\mathcal{G} \\ |\mathcal{V}_\mathcal{G}|=n_s}} w(\mathcal{G},\mathcal{L}) E_0^{(\mathfrak{o})}(\mathcal{G}) \,.
\end{aligned}
\tag{156}
$$

We have replaced $\mathcal{G} \subset \mathcal{L}$ with $\mathcal{G}$ in the sum to emphasise that the graphs are not restricted to the nearest-neighbour lattice geometry any longer due to the long-range interactions, making the lattice a fully connected graph. From the first to second line, we decomposed the ground-state energy contribution from graphs into contributions of individual orders $E_0(\mathcal{G}) = \sum_{\mathfrak{o}=2}^{\mathfrak{o}_{\max}} E_0^{(\mathfrak{o})}$. Note that the minimum order of the ground-state energy is $\mathfrak{o} = 2$. From the second to third line, we introduced a second sum over vertices n_s and restricted the sum over all graphs to a sum over graphs with a fixed number of vertices n_s. Next, we can reformulate the embedding factor in Equation (147) as

$$w(\mathcal{G},\mathcal{L}) = \sum_{c\in\mathcal{C}} \frac{1}{s_{\mathcal{G}}}\,, \tag{157}$$

where we replaced the expression for the number of monomorphisms (divided by N) with a sum over all possible configurations $\mathcal{C}$. A configuration is nothing else than the current embedding of graphs. As we will see later, we can calculate the contributions of multiple graphs simultaneously as the Monte Carlo sum only depends on the number of sites (we use "site" also as a synonym for "vertex position"). When using the word configuration, we think about it as the current set of vertex positions on the lattice. The sum over all configurations comprises individual sums for each vertex over all lattice sites excluding configurations where vertices overlap, as shown in the next subsection in Equation (180). One may falsely conclude that a sum over all configurations should result in the un-normalised embedding factor $\mathcal{W}$, but as we will see in the following, by substituting the abstract expansion parameters with the physical long-range interactions, only the relative distance between sites is relevant for the contribution in the summand, irrespective of the absolute position of the sites on the lattice. We can also see the reason behind splitting the graph set into sets with graphs of a fixed number of vertices because we can now group all graph contributions with a given number of sites into a single integrand because for long-range interactions there are no constraints on the embeddings (except for overlaps) and the integrand only depends on the (relative) position of the vertices.. Further, the white graph contributions $E_0^{(o)}(\mathcal{G})$ still need to be replaced with the correct colour, i.e., the general expansion parameters need to be substituted with the algebraically decaying long-range interaction depending on the current configuration. In reality, the contribution $E_0^{(o)}(\mathcal{G},c)$ depends on the current configuration c. Thus, replacing the expression with an explicit sum is necessary as the contribution for each configuration is different and $w(\mathcal{G},\mathcal{L})$ cannot just be a number. The substitution must look like

$$E_0^{(o)}(\mathcal{G},c) := E_0^{(o)}\left(\mathcal{G};\{\lambda_e \mapsto |\delta|^{-(d+\sigma)}\}\right) = \sum_m v_m(\mathcal{G}) \prod_{e\in\mathcal{E}_{\mathcal{G}}} \frac{1}{|\delta|^{-n_{e,m}(d+\sigma)}}\,, \tag{158}$$

where the index m of the sum runs over all monomials of the contribution $E_0^{(o)}(\mathcal{G},c)$ and the product is over all edges $e = \{\mu,\nu\} \in \mathcal{E}_{\mathcal{G}}$ of the graph $\mathcal{G}$ with the adjacent vertices $\mu,\nu \in \mathcal{V}_{\mathcal{G}}$. The power law in the product arises from substituting the expansion parameters $\lambda_e \mapsto |\delta|^{-\alpha}$ on the edges e with the appropriate algebraically decaying long-range interaction of the current embedding (cf. Equation (140)). The adjacent vertices μ,ν are embedded on the lattice sites $\boldsymbol{i}_\mu$ and $\boldsymbol{i}_\nu$ with the distance $\boldsymbol{\delta} = \boldsymbol{i}_\nu - \boldsymbol{i}_\mu$, and the multiplicity $n_{e,m} \in \mathbb{N}$ comes from the power n_e of the associated expansion parameter λ_e. This way, we can reduce the many expansion parameters from the white graph contribution to a single physical perturbation parameter λ, and by reordering the expression (156) of the ground-state energy, we have

$$\epsilon_0 = \sum_{o=2}^{o_{\max}} \sum_{n_s=2}^{o} \sum_{c\in\mathcal{C}} \sum_{\substack{\mathcal{G}\\ |\mathcal{V}_{\mathcal{G}}|=n_s}} \frac{1}{s_{\mathcal{G}}} E_0^{(o)}(\mathcal{G},c)\lambda^o\,. \tag{159}$$

We can define

$$f_{n_s}^{(o)}(c) := \sum_{\substack{\mathcal{G}\\ |\mathcal{V}_{\mathcal{G}}|=n_s}} \frac{1}{s_{\mathcal{G}}} E_0^{(o)}(\mathcal{G},c)\,, \tag{160}$$

$$S[f_{n_s}^{(o)}] := \sum_{c\in\mathcal{C}} f_{n_s}^{(o)}(c)\,, \tag{161}$$

where $f_{n_s}^{(o)}(c)$ is the integrand function and $S[\cdot]$ denotes the associated sum over all possible embeddings on the lattice that will be evaluated using a classical Monte Carlo algorithm. Since Monte Carlo runs are usually performed for a batch of different seeds, we introduce an additional sum over seeds averaging the Monte Carlo runs, which we denote as

$$\overline{S}[\cdot] = \frac{1}{N_{\text{seeds}}} \sum_{s=1}^{N_{\text{seeds}}} S[\cdot] \,. \tag{162}$$

Eventually, this yields the expression:

$$\epsilon_0 = \sum_{o=2}^{o_{\max}} \sum_{n_s=2}^{o} \overline{S}[f_{n_s}^{(o)}] \lambda^o \,. \tag{163}$$

To express the ground-state energy ϵ_0 as a perturbative series, we write

$$\epsilon_0 = p_0 + \sum_{o=2}^{o_{\max}} p_o \lambda^o \quad \text{with} \quad p_i = \sum_{n_s=2}^{o} \overline{S}[f_{n_s}^{(o)}] \,, \tag{164}$$

where we have to sum up the contributions of multiple Monte Carlo runs to obtain the series prefactors p_i for $i > 0$. The zeroth prefactor p_0 is simply given by the ground-state energy of $\mathcal{H}_0$. As a result of the Monte Carlo summation, these prefactors p_i carry an uncertainty that can be estimated from the standard deviation obtained by averaging over the results for all seeds.

1qp dispersion: We now turn to extracting the one-quasiparticle dispersion. In Equation (116), we have seen that the dispersion can be analytically determined in terms of the hopping amplitudes of Equation (151) up to some perturbative order. For nearest-neighbour interactions, it is an analytic function in $\boldsymbol{k}$; however, for long-range interactions, there are infinitely many hoppings possible at any order, so we can neither explicitly determine the hopping amplitudes $a(\boldsymbol{\delta})$, nor is it possible to have the dispersion as an analytical function in $\boldsymbol{k}$. This would introduce a functional dependence in the integrand that we cannot sample. Instead, we will have to evaluate the dispersion for certain values $\boldsymbol{k}^\star$ to obtain an explicit series in the perturbation parameter λ. Evaluating Equation (116) at $\boldsymbol{k} = \boldsymbol{k}^\star$ and inserting Equation (151), we can write

$$\begin{aligned} \omega(\boldsymbol{k} = \boldsymbol{k}^\star) &= a(\boldsymbol{0}) + 2\sum_{\boldsymbol{\delta}} a(\boldsymbol{\delta}) \cos(\boldsymbol{k}^\star \boldsymbol{\delta}) \\ &= \sum_{\boldsymbol{\delta}} \Xi(\boldsymbol{k}^\star, \boldsymbol{\delta}) a(\boldsymbol{\delta}) = \sum_{\boldsymbol{\delta}} \Xi(\boldsymbol{k}^\star, \boldsymbol{\delta}) \sum_{\mathcal{G}} \sum_{(\mu,\nu)\in\mathcal{P}} w(\mathcal{G}_c, \mathcal{L}_c) t_{\mu;\nu}(\mathcal{G}) \,, \end{aligned} \tag{165}$$

where we rewrote the sum over $\boldsymbol{\delta}$ by introducing the function:

$$\Xi(\boldsymbol{k}^\star, \boldsymbol{\delta}) = \begin{cases} 1 & \boldsymbol{\delta} = \boldsymbol{0}, \\ 2\cos(\boldsymbol{k}^\star \boldsymbol{\delta}) & \boldsymbol{\delta} \neq \boldsymbol{0}. \end{cases} \tag{166}$$

Again, we can split $t_{\mu;\nu}$ into the contributions of individual orders $t_{\mu;\nu}^{(o)}$ and split the graph set into subsets of a fixed number of sites $n_s = |\mathcal{V}_\mathcal{G}|$, yielding

$$\omega(\boldsymbol{k}^\star) = \sum_{o=1}^{o_{\max}} \sum_{n_s=2}^{o+1} \sum_{\substack{\mathcal{G} \\ |\mathcal{V}_\mathcal{G}|=n_s}} \sum_{(\mu,\nu)\in\mathcal{P}} \sum_{\boldsymbol{\delta}} \Xi(\boldsymbol{k}^\star, \boldsymbol{\delta}) w(\mathcal{G}_c, \mathcal{L}_c) t_{\mu;\nu}^{(o)}(\mathcal{G}) \,. \tag{167}$$

Note here that the second sum goes until $o + 1$, while for the ground-state energy, it runs only until o. The maximal number of sites at a given order is $o + 1$ because graphs with o edges can maximally have $o + 1$ sites. For the ground-state energy fluctuations, every quasiparticle that is created has to be annihilated again, so at order $\mathfrak{o}$, a process can only touch maximal $\mathfrak{o}-1$ edges, which restricts the sum to $\mathfrak{o}$ sites.

Now, we argue that we can drop the sum over δ by thinking differently about this embedding problem for the dispersion. The information of the start and end vertex of the hopping process is encoded into vertex colours, and when finding the subgraph monomorphisms for the embedding on the infinite lattice $\mathcal{L}$, the colours of the vertices must match the coloured sites on the lattice, i.e., the hopping vertices are fixed to the hopping sites on the lattice. Since the long-range interactions allow any hopping—i.e., of any distance—at any order, it is not useful to think in this picture. Instead, we should think about the embedding problem analogous to the one for the ground-state energy, where no such hopping constraint exists and the embedding factor $\mathcal{W}$ is simply proportional to a sum over all configurations. This is valid as we let the sum over all lattice sites and account for constraints on δ by multiplying with the symmetry number of the coloured graph $s_{\mathcal{G}_c}$. The relevant hopping information of the vertices, which was previously fixed by coloured vertices, is anyway encoded into the cosine terms. Hence, we can make the substitution:

$$\sum_{\delta} \Xi(\boldsymbol{k}^\star,\delta) w(\mathcal{G}_c,\mathcal{L}_c) = \sum_{c\in C} \frac{s_{\mathcal{G}_c}}{s_{\mathcal{G}}} \cos(\boldsymbol{k}^\star\delta)\,, \tag{168}$$

where we account for the reduced symmetry of the graph due to the hopping by multiplying with the symmetry number of the coloured graph $s_{\mathcal{G}_c}$. As before, we need to substitute the general white graph contribution with the actual algebraically decaying long-range interactions of the current embedding:

$$t^{(\mathfrak{o})}_{\mu;\nu}(\mathcal{G},c) := t^{(\mathfrak{o})}_{\mu;\nu}(\mathcal{G};\{\lambda_e \mapsto |\delta|^{-\alpha}\}) = \sum_m v_m(\mathcal{G}) \prod_{e\in\mathcal{E}_{\mathcal{G}}} \frac{1}{|\delta|^{-n_{e,m}\alpha}}\,. \tag{169}$$

Inserting Equation (168) into Equation (167), we end up with the expression:

$$\omega(\boldsymbol{k}^\star) = \sum_{\mathfrak{o}=1}^{\mathfrak{o}_{\max}} \sum_{n_s=2}^{\mathfrak{o}+1} \sum_{c\in C} \sum_{\substack{\mathcal{G}\\ |\mathcal{V}_{\mathcal{G}}|=n_s}} \sum_{(\mu,\nu)\in\mathcal{P}} \frac{s_{\mathcal{G}_c}}{s_{\mathcal{G}}} t^{(\mathfrak{o})}_{\mu;\nu}(\mathcal{G},c)\cos(\boldsymbol{k}^\star\delta)\lambda^{\mathfrak{o}}\,. \tag{170}$$

For a lighter notation, we again define the integrand function and the Monte Carlo sum:

$$f^{(\mathfrak{o})}_{n_s,\boldsymbol{k}^\star}(c) := \sum_{\substack{\mathcal{G}\\ |\mathcal{V}_{\mathcal{G}}|=n_s}} \sum_{(\mu,\nu)\in\mathcal{P}} \frac{s_{\mathcal{G}_c}}{s_{\mathcal{G}}} t^{(\mathfrak{o})}_{\mu;\nu}(\mathcal{G},c)\cos(\boldsymbol{k}^\star\delta)\,, \tag{171}$$

$$S[f^{(\mathfrak{o})}_{n_s,\boldsymbol{k}^\star}] := \sum_{c\in C} f^{(\mathfrak{o})}_{n_s,\boldsymbol{k}^\star}(c)\,. \tag{172}$$

Introducing an average over a batch of seeds for the MC sum $\overline{S}[\cdot]$, we obtain

$$\omega(\boldsymbol{k}^\star) = \sum_{\mathfrak{o}=1}^{\mathfrak{o}_{\max}} \sum_{n_s=2}^{\mathfrak{o}+1} \overline{S}[f^{(\mathfrak{o})}_{n_s,\boldsymbol{k}^\star}]\lambda^{\mathfrak{o}}\,. \tag{173}$$

The perturbative series of the dispersion evaluated at $\boldsymbol{k}=\boldsymbol{k}^\star$ can then be expressed as

$$\omega(\boldsymbol{k}^\star) = p_0 + \sum_{\mathfrak{o}=1}^{\mathfrak{o}_{\max}} p_{\mathfrak{o}}\lambda^{\mathfrak{o}} \quad \text{with} \quad p_i = \sum_{n_s=2}^{\mathfrak{o}+1} \overline{S}[f^{(\mathfrak{o})}_{n_s,\boldsymbol{k}^\star}]\,. \tag{174}$$

The sum prefactors p_i for $i>0$ can be determined by summing up the individual contributions from the Monte Carlo runs and the prefactor p_0 is given by the energy gap of $\mathcal{H}_0$.

1qp spectral weight: Lastly, the evaluation for the spectral weight observable is analogous to the 1qp dispersion. The integrand and Monte Carlo sum are defined as

$$f^{(o)}_{n_s,\boldsymbol{k}^\star}(c) := \sum_{\substack{\mathcal{G} \\ |\mathcal{V}_\mathcal{G}|=n_s}} \sum_{(\mu,\nu)\in\mathcal{P}} \frac{s_{\mathcal{G}_c}}{s_\mathcal{G}} \tilde{t}^{(o)}_{\mu;\nu}(\mathcal{G},c)\cos(\boldsymbol{k}^\star\boldsymbol{\delta}) \,, \tag{175}$$

$$S[f^{(o)}_{n_s,\boldsymbol{k}^\star}] := \sum_{c\in C} f^{(o)}_{n_s,\boldsymbol{k}^\star}(c) \,, \tag{176}$$

which we use to calculate

$$s(\boldsymbol{k}^\star) = p_0 + \sum_{o=1}^{o_{\max}} p_o\lambda^o \quad \text{with} \quad p_i = \sum_{n_s=2}^{o+1} \overline{S}[f^{(o)}_{n_s,\boldsymbol{k}^\star}] \,. \tag{177}$$

We determine $s(\boldsymbol{k}^\star)$ by again calculating the series prefactors p_i for $i > 0$ and then determine the 1qp spectral weight with $\mathcal{S}^{1\text{qp}}(\boldsymbol{k}^\star) = |s(\boldsymbol{k}^\star)|^2$.

It should be noted that we have to perform

$$n = \frac{o_{\max}(o_{\max}-1)}{2} \times N_{\text{seeds}} \tag{178}$$

Monte Carlo runs for a series of order $o_{\max}$ with N_{seeds}. This means the number of runs grows quadratically with the maximal order $o_{\max}$.

So far, we have derived the necessary formalism for how to express the embedding problem of models with long-range interactions, but we have not talked about how to evaluate the Monte Carlo sums $S[\cdot]$ for the integrand functions f. In the next section, we investigate how to evaluate such sums by introducing a suitable Monte Carlo algorithm.

4.7.3. Monte Carlo Algorithm for the Long-Range Embedding Problem

We are left with evaluating the Monte Carlo sum $S[\cdot]$, which runs over all configurations C of graphs. The embeddings on the lattice depend only on the number of vertices of a graph $\mathcal{G}$, and there is no constraint by the edge set as in the nearest-neighbour case because every site of the lattice interacts with any other site, making it a fully connected graph. The only restriction for models with long-range interactions is that the vertices of the graph are not allowed to overlap as they do not self-interact and an overlap would imply infinite interaction strength resulting from the term $|\boldsymbol{\delta}|^{-(d+\sigma)}$. In conclusion, we can write the sum over all possible configurations as number-of-sites-many sums over all possible lattice positions:

$$S[f^{(o)}_{n_s}] = \sum_{c\in C} f^{(o)}_{n_s}(c) = \sum_{\boldsymbol{i}_1}{}' \cdots \sum_{\boldsymbol{i}_{n_s}}{}' f^{(o)}_{n_s}(\boldsymbol{i}_1,\ldots,\boldsymbol{i}_{n_s}) \,, \tag{179}$$

$$S[f^{(o)}_{n_s,\boldsymbol{k}^\star}] = \sum_{c\in C} f^{(o)}_{n_s,\boldsymbol{k}^\star}(c) = \sum_{\boldsymbol{i}_1}{}' \cdots \sum_{\boldsymbol{i}_{n_s}}{}' f^{(o)}_{n_s,\boldsymbol{k}^\star}(\boldsymbol{i}_1,\ldots,\boldsymbol{i}_{n_s}) \,, \tag{180}$$

where the primed sum $\sum_{\boldsymbol{i}}{}'$ over vertex position $\boldsymbol{i}$ is the short notation for excluding overlaps with any other vertex position. For example, for the ground-state contribution with three sites on a one-dimensional chain, this sum would look like

$$S[f^{(o)}_{n_s}] = \sum_{k=-\infty}^{\infty} \sum_{\substack{j=-\infty \\ j\neq k}}^{\infty} \sum_{\substack{i=-\infty \\ i\neq j \\ i\neq k}}^{\infty} f^{(o)}_{n_s}(i,j,k) \,. \tag{181}$$

Due to the overlap constraint, these are nested high-dimensional sums over the integrand functions $f^{(o)}_{n_s}$, which are in general hard to solve. The dimensionality of the MC problem is given by $d_{\text{sum}} = n_s \cdot d$ because higher dimensions d of the system introduce additional sums for each component. If we wanted to evaluate the Monte Carlo sum in two dimensions for contributions with eight sites, which already occurs in eighth-order perturbation theory, we would have to determine the integral value of 16 nested sums over the integrand function $f_8^{(8)}$. This makes it clear that the evaluation of such sums becomes challenging very quickly. In the following, we use the short notation f_{n_s}, interchangeable for both the ground-state energy integrand $f_{n_s}^{(o)}$ and the 1qp process integrands $f_{n_s,\boldsymbol{k}^\star}^{(o)}$.

The first approach to tackle the problem of evaluating these sums using conventional numerical integration techniques was pioneered by S. Fey and K.P. Schmidt already in 2016 [25]. They managed to successfully determine the phase diagram and critical exponents from the closing of the gap of the long-range transverse-field Ising model on a one-dimensional chain with ferro- and antiferromagnetic interactions. While they were successful with their approach over a large range of decay exponents in one dimension, the extraction of the critical properties for small decay exponents was challenging. The two-dimensional problem was out of reach with this approach as the number of nested sums doubles and the sums converge significantly more slowly. Here, Monte Carlo integration came into play as it is known to be a powerful integration technique for high-dimensional problems where conventional integration fails. The reason behind the slow convergence of such high-dimensional sums is often that the configuration space where the integrand mainly contributes to the integral is significantly smaller than the entire configuration space. In 2019, Fey et al. [29] introduced a Markov chain Monte Carlo algorithm to efficiently sample the relevant configuration space. They were able to determine the quantum-critical properties of the long-range TFIM on two-dimensional lattices to even higher precision than previously for the one-dimensional chain, extending the accessible range of decay exponents without having to forfeit perturbative orders in higher dimensions.

In the following, we describe the Markov chain Monte Carlo algorithm introduced by Ref. [29] to evaluate the high-dimensional nested sums. To sample the relevant configuration space efficiently, we use importance sampling with respect to some convenient probability weight $\pi(c)$ with respect to a configuration c and the associated partition function $Z = \sum_c \pi(c)$. We can insert an identity into Equation (179) or Equation (180) and rewrite it as

$$S[f_{n_s}] = \sum_{c\in C} \underbrace{\frac{\pi(c)}{Z}\frac{Z}{\pi(c)}}_{=1} f_{n_s}(c) = Z\left\langle \frac{f_{n_s}(c)}{\pi(c)} \right\rangle_\pi , \tag{182}$$

where $\pi(c)/Z$ can be interpreted as the probability of being in configuration c. The integrand now reads as the contribution f_{n_s} (we dropped the order o and momentum $\boldsymbol{k}^\star$ as indices to lighten the notation) from configuration c multiplied by its probability, which allows us to write the sum as the expectation value $\langle\cdot\rangle_\pi$ of $f_{n_s}(c)/\pi(c)$ with respect to the weight π. We later call this sum "target sum". Since the partition function Z is not known a priori, we also introduce a "reference sum":

$$S\left[f_{n_s}^{\mathrm{ref}}\right] = \sum_{c\in C} f_{n_s}^{\mathrm{ref}}(c) = Z\left\langle \frac{f_{n_s}^{\mathrm{ref}}(c)}{\pi(c)} \right\rangle_\pi , \tag{183}$$

over a reference function $f_{n_s}^{\mathrm{ref}}$. We require this sum to be analytically solvable to avoid introducing an additional source of error. We denote its analytical expression as $S_{n_s}^{\mathrm{ref}}$. The reference function $f_{n_s}^{\mathrm{ref}}$ should behave similarly to the integrand function of interest f_{n_s}. This means that the reference sum and target sum should have considerable contributions in the same area of the configuration space and their asymptotic behaviour should be similar as well. Although we could make, in principle, an arbitrary choice for the reference function, the latter properties guarantee to lead to good convergence. In one dimension, we choose the reference integrand as

$$f_{n_s}^{\mathrm{ref}}(c) = \prod_{n=1}^{n_s-1} \frac{1}{|i_{n+1} - i_n|^\rho} = \prod_{n=1}^{n_s-1} \frac{1}{|\delta_n|^\rho} \tag{184}$$

with $\delta_n = i_{n+1} - i_n$ and the reference integrand exponent ρ, which is a free simulation parameter. We can solve the reference sum as follows:

$$
\begin{aligned}
S\left[f_{n_s}^{\text{ref}}\right] &= \sum_{c\in C} \prod_{n=1}^{n_s-1} \frac{1}{|i_{n+1} - i_n|^\rho} \\
&= \sum_{i_1}{}' \cdots \sum_{i_N}{}' \frac{1}{|i_2 - i_1|^\rho} \cdots \frac{1}{|i_{n_s} - i_{n_s-1}|^\rho} \\
&= n_s! \sum_{i_1<i_2} \cdots \sum_{i_{n_s-1}<i_{n_s}} \frac{1}{|i_2 - i_1|^\rho} \cdots \frac{1}{|i_{n_s} - i_{n_s-1}|^\rho} \\
&= n_s! \sum_{i_1<i_2} \cdots \sum_{i_{n_s-1}<i_{n_s}} \frac{1}{|\delta_1|^\rho} \cdots \frac{1}{|\delta_{n_s-1}|^\rho} \\
&= n_s! \prod_{n=1}^{n_s-1} \sum_{\delta=1}^{\infty} \frac{1}{\delta^\rho} = n_s! \zeta(\rho)^{n_s-1} ,
\end{aligned}
\tag{185}
$$

where $\zeta(\rho)$ is the Riemann ζ function and we accounted for $n_s!$ possibilities to randomly embed the vertices by ordering the indices of the sums. One major difference between the reference and the target sum is that, in the target sum, many different graph contributions contribute. In fact, the reference sum above is exactly the contribution of order $\mathfrak{o} = n_s - 1$ of a chain graph with n_s vertices and the contribution from the associated target sum is the same up to a linear factor. In higher dimensions, we cannot choose a contribution proportional to the one of a chain graph any longer since it cannot be solved analytically. Instead, we make simplifications to the reference sum and require that the reference sum is still good enough to capture the same properties as the target sum. We choose to decouple the dimensions in the reference integrand:

$$
f_{n_s}^{\text{ref}}(c) = \prod_{n=1}^{d} \prod_{\nu=1}^{n_s-1} \frac{1}{(1 + |i_{\nu+1,n} - i_{\nu,n}|)^\rho} = \prod_{n=1}^{d} \prod_{\nu=1}^{n_s-1} \frac{1}{(1 + |\delta_{\nu,n}|)^\rho}
\tag{186}
$$

and explicitly allow overlaps in the reference sum, such that it can be solved analytically as follows:

$$
\begin{aligned}
S\left[f_{n_s}^{\text{ref}}\right] &= \sum_{i_{1,1}=-\infty}^{\infty} \cdots \sum_{i_{n_s,d}=-\infty}^{\infty} \frac{1}{(1 + |i_{2,1} - i_{1,1}|)^\rho} \cdots \frac{1}{(1 + |i_{n_s,d} - i_{n_s-1,d}|)^\rho} \\
&= \prod_{n=1}^{d(n_s-1)} \sum_{\delta=-\infty}^{\infty} \frac{1}{(1 + |\delta|)^\rho} = \prod_{n=1}^{d(n_s-1)} \left(2\sum_{\delta=0}^{\infty} \frac{1}{(1+\delta)^\rho} - 1 \right) \\
&= (2\zeta(\rho) - 1)^{d(n_s-1)} .
\end{aligned}
\tag{187}
$$

Although the exponent ρ can be chosen freely, we want to achieve similar asymptotic behaviour as the target integrand; therefore, we choose $\rho = 1 + \sigma$ in one dimension as the reference sum exactly behaves like the target integrand of a chain graph. In two dimensions, we made some simplifications to the reference sum, and we have to adopt the parameter as $\rho = (d + \sigma)/2$ for $\sigma < 5$, $\rho = 3$ for $5 \leq \sigma < 7$, and $\rho = 3.5$ for $\sigma \geq 7$. This is by no means a rigorous choice, but it empirically proved to produce good convergence [29,207]. Solving Equation (183) for Z and inserting it into Equation (182), we obtain

$$
S[f_{n_s}] = \frac{\left\langle \frac{f_{n_s}(c)}{\pi(c)} \right\rangle_\pi}{\left\langle \frac{f_{n_s}^{\text{ref}}(c)}{\pi(c)} \right\rangle_\pi} S_{n_s}^{\text{ref}} .
\tag{188}
$$

We use the analytic expression of Equation (185) in 1D or Equation (187) in 2D for the reference sum $S_{n_s}^{\text{ref}}$. We got rid of the partition function Z and now can use this expression in our Monte Carlo run to determine the sum $S[\cdot]$ using the analytic expression of the reference sum $S_{n_s}^{\text{ref}}$ while tracking the running averages in the numerator and denominator expressions.

We are left with just one missing ingredient, which is the choice of the probability weight $\pi(c)$. For our choice to be a probability weight, it must fulfil $\pi(c) \geq 0$, and we want the weight to be the largest if both the reference and the target integrand contribute the most. An obvious choice may be the quadratic mean:

$$\pi(c) = \sqrt{\left(f_{n_s}^{\text{ref}}(c)\right)^2 + \left(f_{n_s}(c)\right)^2}\,, \tag{189}$$

which is always ≥ 0 and rotationally invariant in $f_{n_s}^{\text{ref}}$ and f_{n_s}. However, we also want both quantities to contribute equally to the probability weight on average over all configurations. As the contributions of the target and reference sum may differ significantly, we introduce a factor for rescaling:

$$R = \left|\frac{S\left[f_{n_s}^{\text{ref}}\right]}{S[f_{n_s}]}\right|, \tag{190}$$

which can be estimated in an in-advance calibration run. We then use an adjusted probability weight:

$$\pi(c) = \sqrt{\left(f_{n_s}^{\text{ref}}(c)\right)^2 + R^2\left(f_{n_s}(c)\right)^2} \tag{191}$$

for the actual Monte Carlo run. The weight needs to be evaluated at every Monte Carlo step to track the running averages of the numerator and denominator in Equation (188).

Now, we have everything together to describe the actual Monte Carlo algorithm. We employ a Markov chain Monte Carlo algorithm, where we need to sample the configuration space $\mathcal{C}$ according to the probability weight π in Equation (191). Each configuration with a non-zero weight must be in principle accessible by the Markov chain. On the one hand, we propose a high acceptance rate of the Monte Carlo steps to sample the large configuration space efficiently, not staying in the same configuration for too long. On the other hand, we want to sample closely confined configurations with rather small distances between vertices more often than configurations that are farther apart (configurations with large distances between vertices) such that we capture the asymptotic behaviour of the model. The interaction strength decays algebraically with the distance between vertices, leading to smaller contributions for configurations in which sites are far apart. What we call a confined configuration, therefore, depends on the decay exponent of the long-range interaction σ. In the algorithm, we have the free exponent parameter ρ (for the reference sum) and γ (for probability distributions), which can be changed to tweak this behaviour, but are usually chosen similar or equal to $d + \sigma$. In two dimensions, we had to adapt the values of ρ to obtain a similar asymptotic behaviour for the reference and target sum due the approximations we made. An optimal choice of these parameters ensures fast convergence of the Monte Carlo sum. The current embedding configuration should be represented as an array container where the entries are the positions of the graph vertices. In one dimension, entries are simple integers, while in higher dimensions, the position needs to be represented as a mathematical vector. For small decay exponents σ, very large distances can occur between vertices from time to time, which need to be squared when calculating the absolute distance. This can lead to an integer overflow, and therefore, the use of a 128-bit integer may be considered. Further, we define functions in the programme for the target integrand f_{n_s} and for the reference integrand $f_{n_s}^{\text{ref}}$, where the current configuration is passed as a parameter and the function returns the contribution from the integrand evaluated for the current configuration.

Turning back to the sampling scheme, the idea of the Markov chain is to interpret the vertex positions on the lattices as random walkers. We randomly select a graph vertex and then draw a new position from a probability distribution such that the move fulfils the detailed balance condition. In each Monte Carlo step, we perform the following two moves:

Shift move: This Monte Carlo move is implemented to introduce confined random fluctuations to the current configuration independent of the strength of the algebraically decaying long-range interactions. It is especially important for larger decay exponents σ when the configurations are much more likely to be confined. First, we randomly select a vertex $n_{\text{sel}} \in \{1, \ldots, n_s\}$ drawn from a discrete uniform distribution with $p^{\text{sel}} = 1/n_s$. Second, for the fluctuation, we draw a shift value $d^{\text{prop}} \in \{-n_s, \ldots, n_s\}$ from a discrete uniform distribution $p^{\text{shift}} = 1/(2n_s + 1)$. In one dimension, we have to draw a single time, and in higher dimensions, we draw repeatedly for each component. Subsequently, we add the shift to the position of the selected vertex and propose the position:

$$i^{\text{prop}} = i_{n_{\text{sel}}} + d^{\text{prop}} \,. \tag{192}$$

We might have proposed a position that is already occupied by another vertex, so we have to check for overlaps. In one dimension, we reset the proposed position to the original one if there is an overlap, while in higher dimensions, we explicitly allow overlaps. As we remember from above, this distinction is also present in the reference sums in one dimension in Equation (185) compared to higher dimensions in Equation (187). If an overlap occurs in dimensions higher than one, then the target summand is explicitly set to zero such that these configurations cannot contribute (otherwise, the sum would become infinity). Then, we calculate the Metropolis acceptance probability:

$$\begin{aligned} p_{\text{acc}}^{\text{shift}} &= \min\left(1, \frac{\pi(c^{\text{prop}})}{\pi(c^{\text{curr}})}\right) \\ &= \min\left(1, \frac{\sqrt{(f_{n_s}^{\text{ref}}(c^{\text{prop}}))^2 + R^2(f_{n_s}(c^{\text{prop}}))^2}}{\sqrt{(f_{n_s}^{\text{ref}}(c^{\text{curr}}))^2 + R^2(f_{n_s}(c^{\text{curr}}))^2}}\right), \end{aligned} \tag{193}$$

by determining the probability weights π of the current and the proposed configuration. The result of the target and reference function calls should be saved into variables to prevent redundant and expensive function calls at each Monte Carlo step. Note that the transition weights $\tilde{T}(c^{\text{curr}} \to c^{\text{prop}}) = p^{\text{sel}} \times p^{\text{shift}}$ cancel out as we draw only from uniform distributions. Lastly, the minimum function is implemented by drawing a random number $y \in [0,1)$, and we accept the proposed move if $y < p_{\text{acc}}^{\text{shift}}$ and update the current configuration. An example of such a shift move is depicted in Figure 11a.

Rift move: In contrast to the previous move, which should introduce fluctuations to the configuration independent of the current one and independent of the long-range interaction strength, "rift moves" are introduced to better capture the correct asymptotic behaviour induced by the algebraically decaying interactions. The moves are able to propose very large distances between vertices, but are also able to do the opposite, closing the "rift" between vertices when the configuration is split into essentially two clusters. At first, we select a site $n_{\text{sel}} \in \{1, \ldots, n_s - 1\}$ from the vertex set with discrete uniform probability $p^{\text{sel}} = 1/(n_s - 1)$, explicitly excluding the last site. In one dimension, we can order the vertex set such that the first vertex is the one with the smallest positional value and the last the one with the largest value, so we order by $i_n < i_m$, where n, m are vertex indices and i_n, i_m the associated sites on the lattice. The same ordering was also performed when we solved the reference sum in Equation (185). In higher dimensions, a similar ordering comes at a much higher computational cost, so we stick to the vertex numbering given by the array indices, i.e., the order is $n < m$. Here, it is also important that the vertex labelling of the reference sum coincides with the labelling of the chain graph. To capture the physical asymptotics of the system, we draw random values from a ζ-function distribution. In one dimension, we draw from

$$p^{\text{rift}}(r^{\text{prop}}) = \frac{(r^{\text{prop}})^{-\gamma}}{\zeta(\gamma)}, \tag{194}$$

yielding a power-law distribution with $r^{\text{prop}} > 0$ with the free exponent parameter γ. We choose $\gamma = d + \sigma$ for obvious reasons. The distance to the next vertex is given by $r^{\text{curr}} = i_{n_{\text{sel}}+1} - i_{n_{\text{sel}}}$, and r^{prop} is the proposed distance drawn from the ζ distribution. Since we ordered by the position and only selected sites in $\{1, \ldots, n_s - 1\}$, it is sufficient to draw positive values only. We shift all indices $i_n > i_{n_{\text{sel}}}$ according to

$$i_n^{\text{prop}} = i_n + (r^{\text{prop}} - r^{\text{curr}}) \,. \tag{195}$$

In higher dimensions, we have no such ordering and, therefore, extend such a distribution to negative values (we refer to it as a "double-sided" ζ-function distribution) and draw random values from

$$p^{\text{rift}}(r^{\text{prop}}) = \frac{(1 + |r^{\text{prop}}|)^{-\gamma}}{2\zeta(\gamma) - 1} \tag{196}$$

for each component. Note that the additional one is introduced to prevent divergence when sites overlap. After drawing the new distance r^{prop}, we shift all vertices componentwise with $n > n_{\text{sel}}$ according to

$$i_{n>n_{\text{sel}}}^{\text{prop}} = i_{n>n_{\text{sel}}} + (r^{\text{prop}} - r^{\text{curr}}) \,. \tag{197}$$

The underlying idea is that, if there is a large distance between two vertices $i_{n_{\text{sel}}}$ and $i_{n_{\text{sel}}+1}$, we can close the "rift" of the entire configuration instead of introducing a new one between $i_{n_{\text{sel}}+1}$ and $i_{n_{\text{sel}}+2}$. The transition weights for this move are given by

$$\tilde{T}(c^{\text{curr}} \to c^{\text{prop}}) = p^{\text{sel}} \times p(r^{\text{new}}) \,, \tag{198}$$
$$\tilde{T}(c^{\text{prop}} \to c^{\text{curr}}) = p^{\text{sel}} \times p(r^{\text{curr}}) \,. \tag{199}$$

With these, we can calculate the Metropolis–Hastings acceptance probability in one dimension:

$$\begin{aligned} p_{\text{acc}}^{\text{rift}} &= \min\left(1, \frac{\pi(c^{\text{prop}})}{\pi(c^{\text{curr}})} \frac{p(c^{\text{curr}})}{p(c^{\text{prop}})}\right) \\ &= \min\left(1, \frac{\sqrt{(f_{n_s}^{\text{ref}}(c^{\text{prop}}))^2 + R^2(f_{n_s}(c^{\text{prop}}))^2}}{\sqrt{(f_{n_s}^{\text{ref}}(c^{\text{curr}}))^2 + R^2(f_{n_s}(c^{\text{curr}}))^2}} \frac{(r^{\text{prop}})^{\gamma}}{(r^{\text{curr}})^{\gamma}}\right) \end{aligned} \tag{200}$$

and, likewise, in higher dimensions:

$$p_{\text{acc}}^{\text{rift}} = \min\left(1, \frac{\sqrt{(f_{n_s}^{\text{ref}}(c^{\text{prop}}))^2 + R^2(f_{n_s}(c^{\text{prop}}))^2}}{\sqrt{(f_{n_s}^{\text{ref}}(c^{\text{curr}}))^2 + R^2(f_{n_s}(c^{\text{curr}}))^2}} \frac{\prod_{n=1}^{d}(1 + |r_n^{\text{prop}}|)^{\gamma}}{\prod_{n=1}^{d}(1 + |r_n^{\text{curr}}|)^{\gamma}}\right) . \tag{201}$$

As above, we randomly draw $y \in [0, 1)$, accept if $y < p_{\text{acc}}^{\text{rift}}$, and update the current configuration if the proposed configuration is accepted. In Figure 11b, you can find a typical rift move illustrated.

To implement the Monte Carlo algorithm, we just have to introduce a loop, where for each loop iteration we perform a Monte Carlo step consisting of those two moves.

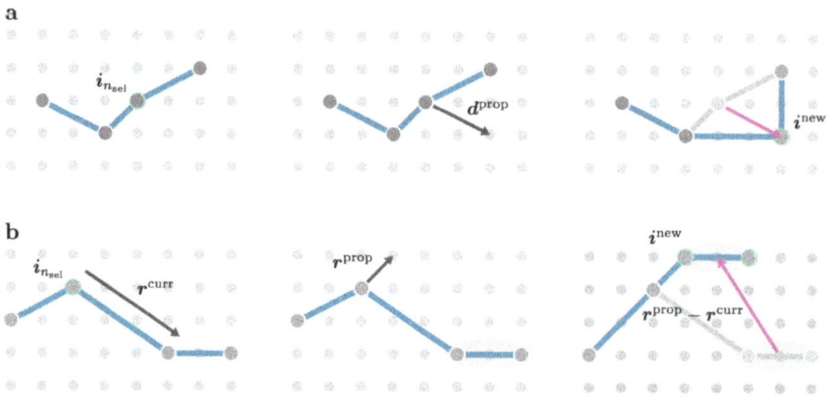

Figure 11. Exemplary Monte Carlo moves for a linear graph on a two-dimensional square lattice. (**a**) During a shift move, a vertex $n_{sel} \in \{1, \ldots, n_s\}$ is selected randomly from a uniform distribution. Then, a shift vector $\boldsymbol{d}^{prop}$ is drawn (uniformly for each component), which moves the selected site to $\boldsymbol{i}^{prop} = \boldsymbol{i}_{n_{sel}} + \boldsymbol{d}^{prop}$ if accepted. (**b**) For rift moves, a vertex is selected from $n_{sel} \in \{1, \ldots, n_s - 1\}$. Instead of drawing from uniform distributions, rift moves account for the correct asymptotic behaviour of the system by drawing a new distance to the next vertex from a ζ-function distribution (from a normal ζ function in one dimension and from a double-sided ζ function in higher dimension for each component). If accepted, the vertices $n > n_{sel}$ are shifted to the new position $\boldsymbol{i}^{prop}_{n>n_{sel}} = \boldsymbol{i}_{n>n_{sel}} + (\boldsymbol{r}^{prop} - \boldsymbol{r}^{curr})$.

To ensure that fluctuations to the current embedding introduce enough new configurations (shift moves) while making sure that the Monte Carlo algorithm does not spend too much time in physically unlikely configurations, the shift move probability was set to $p^{shift} = 0.7$ and the rift move probability to $p^{rift} = 0.3$ [29,207] (The "single-site rift move" was later completely replaced with the "multi-site" rift move presented in this review). This choice proved to produce good convergence behaviour over a large range of decay exponents and different models. After performing a move—accepted or not—we update the estimate for the target and reference sum in the numerator and denominator in Equation (188), respectively, and keep track of the statistics like the variance of target and reference sum. To determine the value $S[f_{n_s}]$ of the Monte Carlo run, the ratio of both quantities has to be multiplied by the analytical value of the reference sum as in Equation (188).

We can test the convergence of the MC algorithm by considering the 1qp gap $\Delta = \omega(k = 0)$ of the one-dimensional ferromagnetic long-range transverse-field Ising model (LRTFIM). We benchmark the second-order series coefficient p_2. According to Equation (174), the coefficient from the MC approach is $p_2^{MC} = \overline{S}[f^{(2)}_{2,0}] + \overline{S}[f^{(2)}_{3,0}]$ and the exact value is given by $p_2^{exact} = 2(\zeta(2(1+\sigma)) - \zeta(1+\sigma)^2)$ (see Ref. [25]).

In Figure 12, we show the MC error:

$$\varepsilon_{MC} = \left| \frac{p_2^{exact} - p_2^{MC}}{p_2^{exact}} \right| \tag{202}$$

as a function of the number of steps N_{steps} for a hundred MC runs with a distinct seed for each. As can be seen from the figure, the MC error goes to zero with $N_{steps}^{-1/2}$, as generally expected for MC simulations.

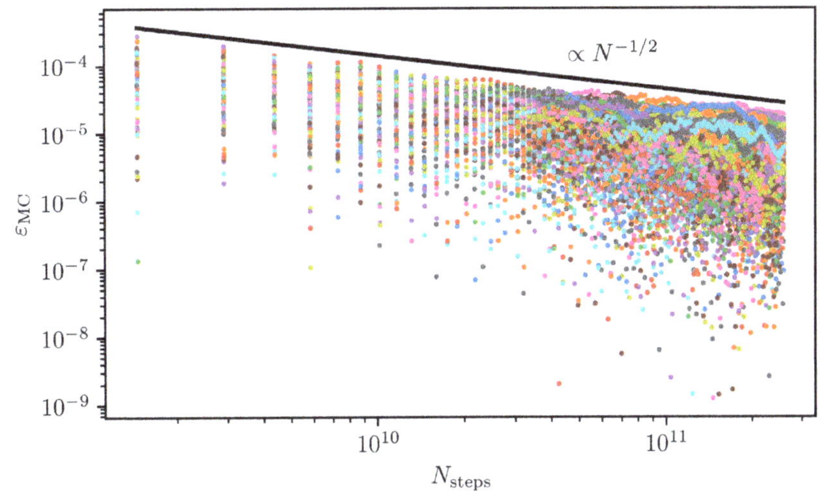

Figure 12. Monte Carlo error $\varepsilon_{\mathrm{MC}} = |p_2^{\mathrm{exact}} - p_2^{\mathrm{MC}}|/|p_2^{\mathrm{exact}}|$ (coloured data points) on a log–log scale as a function of the number of steps N_{steps}. The plot shows data from a hundred 12-hour-long MC runs with a distinct seed each. The MC error $\varepsilon_{\mathrm{MC}}$ goes to zero following a $N_{\mathrm{steps}}^{-1/2}$ convergence behaviour (indicated by the black line), as generally expected from an MC algorithm.

Further improvements to the algorithm can be made by recentring the configuration to the origin since the graphs on the lattice may drift away due to the random walk. Also, for integer powers occurring in the target integrand from powers of the expansion parameters, it may be better to use plain multiplication than using generic integer power functions. Most importantly, lookup tables should be used for the non-integer powers stemming from the algebraically decaying interaction strengths for distances $|\boldsymbol{i}_1 - \boldsymbol{i}_2| < \delta_{\mathrm{max}}$ within a cutoff δ_{max} that occur most commonly during the MC run. The integrand function also depends on $\boldsymbol{k}$ as cosine terms are present. It is useful to compile the code for a desired $\boldsymbol{k}$ value, so lookup tables can be defined at compile time due to the periodicity of the cosine. For instance, in the simplest case $k = 0$, the cosine is always 1 and for $k = \pi$ the cosine has only values of 1 and -1.

Let us emphasise that, for the pCUT+MC approach, we need to perform individual runs for every perturbative order, for every possible number of sites at a given order, for different values of momentum $\boldsymbol{k}$ if necessary (e.g., for a dispersion), and for different seeds so that an average value can be calculated. Typical runs are performed with $N_{\mathrm{seeds}} = 5$–20 seeds for 6–24 hours. For instance, for the long-range transverse-field Ising model, we were able to extract perturbative series up to order 10 for the 1qp spectral weight, order 11 for the 1qp gap, and order 13 for the ground-state energy. Much higher orders in perturbation are likely not feasible with the current implementation as the number of Monte Carlo runs scales quadratically with the order. In the future, to further improve the efficiency of the approach, one may come up with additional Monte Carlo moves for the Markov chain that can change the number of sites. As a result, the algorithm would only scale linearly with the perturbative order, but potentially at the cost of slower convergence. Maybe a compromise would be a favourable option, where the lower orders that converge faster are computed using an algorithm with the additional move and higher orders with the algorithm presented above.

Another important issue worth mentioning is the fact that, so far, the above algorithm can only be applied to lattices with a trivial single-site unit cell. To generalise the algorithm to arbitrary lattice geometries in the future, we would need to introduce another Monte Carlo move. We would keep the moves introduced above for moving the vertices along the underlying Bravais lattice. A new move then changes the position within the unit cell. However, new subtleties emerge from introducing a larger unit cell, as we would have to fix two vertices within the unit cells for hopping processes, while the remaining vertices can be moved freely. A larger unit cell also means that we have to calculate the entire matrix of the dispersion, as can be seen in Equation (115). Entries of the matrix are in general complex-valued, which needs to be accounted for as well. Lastly, the behaviour of the Monte Carlo algorithm for the system is altered due to the additional Monte Carlo move, which may impact the convergence. So, the choice of the reference sum should probably be adapted as well to achieve the desired convergence. See also the discussion in Ref. [207].

4.8. Series Extrapolation

We are interested in extracting the quantum-critical properties from the perturbative series obtained from the Monte Carlo embedding. DlogPadé extrapolations are an established and powerful method that allows us to extrapolate high-order series even beyond the radius of convergence and determine the quantum-critical point and associated critical exponents. A more elaborate description of DlogPadés and its application to critical phenomena can be found in Refs. [208,209].

We have given a high-order perturbative series of a physical quantity $\kappa(\lambda)$ in the perturbation parameter λ. See Section 4.5 for typical quantities of interest. A Padé extrapolant is defined as

$$P[L,M]_\kappa = \frac{P_L(\lambda)}{Q_M(\lambda)} = \frac{p_0 + p_1\lambda + \cdots + p_L\lambda^L}{1 + q_1\lambda + \cdots + q_M\lambda^M}, \tag{203}$$

where $p_i, q_i \in \mathbb{R}$ and the degrees L, M of the numerator polynomial $P_L(\lambda)$ and denominator polynomial $Q_M(\lambda)$ are restricted to $\mathfrak{o}_{\max} = L + M$, where $\mathfrak{o}_{\max}$ is the maximal perturbative order. The coefficients p_i and q_i are fixed by a set of linear equations by cross-multiplying Equation (203) with $Q_M(\lambda)$ and requiring $\mathfrak{o}_{\max} = L + M$, i.e., that all higher order terms must vanish on the left side of the equation [208]. We introduce

$$P[L,M]_\mathcal{D} = \frac{\mathrm{d}}{\mathrm{d}\lambda}\ln(\kappa) \tag{204}$$

as the Padé extrapolant of the logarithmic derivative of κ that must satisfy $\mathfrak{o}_{\max} - 1 = L + M$, as we lose one perturbative order due to differentiation. The DlogPadé extrapolant of κ can now be defined as

$$\mathrm{d}P[L,M]_\kappa = \exp\left(\int_0^\lambda P[L,M]_\mathcal{D}\,\mathrm{d}\lambda'\right). \tag{205}$$

Given that the quantity of interest κ shows a second-order phase transition with a dominant power law $\kappa \sim |\lambda - \lambda_c|^{-\theta}$ about the critical point λ_c, we can extract the critical point λ_c and the associated critical exponent θ with DlogPadé extrapolants (Although we can extract the critical properties of a second-order quantum phase transition, we are blind to first-order phase transitions as the series expansion of a single physical quantity cannot capture level crossings of the analysed quantity at the critical point λ_c). We can determine estimates for the critical point λ_c by analysing the poles of the extrapolant $P[L,M]_\mathcal{D}$. We have to identify the physical pole whose position determines λ_c and exclude spurious extrapolants that have non-physical poles in the complex plane close to the real line with $\lambda < \lambda_c$. If λ_c is known, we can define biased DlogPadés by

$$P[L,M]_{\theta^\star} = (\lambda_c - \lambda)\frac{\mathrm{d}}{\mathrm{d}\lambda}\ln(\kappa). \tag{206}$$

Here, defective extrapolants have to be removed as well by excluding all extrapolants that have poles in the vicinity of $\lambda < \lambda_c$. We can extract estimates for the critical exponent θ by calculating the residua. For an unbiased estimate, we calculate

$$\theta = \mathrm{Res}\, P[L,M]_{\mathcal{D}}|_{\lambda=\lambda_c} = \left.\frac{P_L(\lambda)}{\frac{\mathrm{d}}{\mathrm{d}\lambda}Q_M(\lambda)}\right|_{\lambda=\lambda_c}, \tag{207}$$

and for the biased estimate, we calculate

$$\theta^\star = \mathrm{Res}\, P[L,M]_{\theta^\star}|_{\lambda=\lambda_c} = \left.\frac{P_L(\lambda)}{Q_M(\lambda)}\right|_{\lambda=\lambda_c}. \tag{208}$$

Biased DlogPadé extrapolants were also used in the past trying to extract the exponent of multiplicative logarithmic corrections at the upper critical dimension [25,29,31,35]. At the upper critical dimensions, there are corrections to the dominant power-law behaviour of the form

$$\kappa \sim |\lambda - \lambda_c|^{-\theta}(\ln(\lambda - \lambda_c))^{p_\theta}\,. \tag{209}$$

We can bias the critical point λ_c and the critical exponent θ to the known mean-field value and define the extrapolant:

$$P[L,M]_{p_\theta^\star} = -\ln(1 - \lambda/\lambda_c)[(\lambda_c - \lambda)\mathcal{D}(\lambda) - \theta]\,, \tag{210}$$

so we can determine the estimate $p_\theta^\star$ by calculating the residuum of the Padé extrapolant.

We take the series of mean coefficients p_i for the quantities given by Equations (164), (174), and (177) and calculate a set of extrapolants to obtain reliable estimates for the critical properties of interest. We calculate estimates that satisfy $L + M = \mathfrak{o} \leq \mathfrak{o}_{\max}$. We exclude all defective extrapolants as briefly described above and arrange the remaining DlogPadés into families with $L - M = \mathrm{const.}$ Usually, we only allow $|L - M| \leq 3$ or $|L - M| \leq 2$ since more diagonal families are expected to converge faster to the real physical value as a function of the perturbative order $\mathfrak{o} = L + M$ [208]. We further take the mean over the highest order extrapolants of each family that has at least two or three extrapolants to obtain an estimate for the critical point and exponents. The uncertainty obtained from averaging over the extrapolants is by no means a rigorous error, but is, rather, a "subjective" measure for the uncertainty obtained from systematically analysing extrapolants [209]. From experience, the estimated critical values are relatively stable to a small uncertainty in the series, and the uncertainty obtained from averaging over the highest order extrapolants is comparatively large. Therefore, the series of mean coefficients can simply be used for extrapolating the observables as performed in all previous publications [29–31,34,35,206]. A more rigorous approach can also be applied by repeatedly applying the extrapolation scheme to the series of individual seeds and averaging the critical values afterwards.

4.9. *Workflow of Series Expansion Monte Carlo Embedding*

In the previous subsections, we comprehensively described the pCUT+MC method for models with long-range interactions. The approach in its entirety is quite involved. Hence, to conclude this section, we want to give a short overview of the individual steps necessary and the workflow associated.

The workflow is sketched in Figure 13.

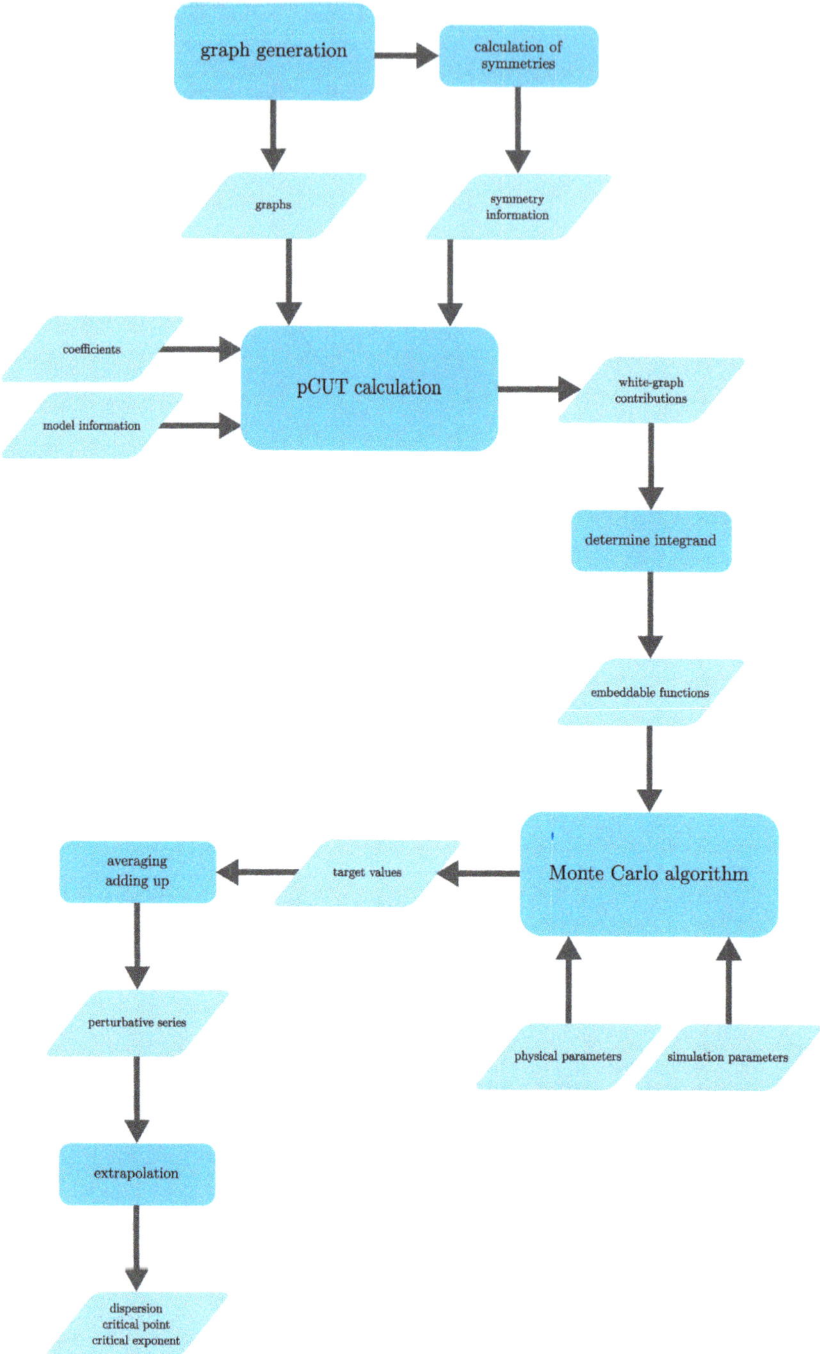

Figure 13. The workflow of the presented pCUT+MC method consists of several steps. There are three major steps. First, there is the graph generation, which only has to be performed once. Second is the calculation of the graph contributions with the pCUT method, and third is the Monte Carlo algorithm to embed the contributions on the lattice to determine the perturbative series of the quantity of interest in the thermodynamic limit.

The approach starts with the generation of graphs with a software programme like "*nauty*" [201] and saving the graphs to bond files. The symmetry number $s_{\mathcal{G}}$ of the graph can be obtained on the fly as it is usually a by-product of searching for further graph isomorphs. For the 1qp dispersion or spectral weight, we want to calculate quasiparticle processes on these graphs; however, we do not want to calculate every process on a graph since many are related by symmetries and give the exact same contribution. Therefore, we calculate additional symmetry numbers $s_{\mathcal{G}_c}$ of coloured graphs. In a programme, we iterate over all possible vertex pairs and colours of the associated vertices (two colours for the 1qp spectral weight and one colour for 1qp processes of the Hamiltonian), choose a representative process, and save the associated vertices and the symmetry number of the coloured graph to a list. While, for the calculation of the ground-state energy, a list containing the graph names suffices, for particle processes, we want to have a list that contains the graph name, symmetry number, start and end vertex of the process, and a symmetry number associated with the process counting the number of equivalent processes. On the upside, it suffices to generate these lists and graphs once.

The second step is to iterate over the list entries, read the graph as a cluster, read the associated input states, and apply the pCUT method. Of course, such a programme must also read the pCUT coefficients $C(\boldsymbol{m})$ or $\tilde{C}(\boldsymbol{m};i)$ for observables and the model information, i.e., the τ-operators. The programme iterates over the different operator sequences, through the operators from each sequence, systematically iterates over all edges of a graph, and applies the τ-operators. We additionally associate each edge of a graph with its own expansion parameter, such that we obtain the white graph contributions.

Having calculated the white graph contributions over every representative process for every graph, we need to convert the white graph contributions to an integrand function—also embeddable function—which is a callable function in our Monte Carlo algorithm. We have a programme that substitutes the expansion parameter with the algebraic decay expression ($\lambda_e \mapsto |\delta|^{-d+\sigma}$). This must be an expression in the programming language of choice, such that the Monte Carlo programme can call this function. Further, we multiply the white graph contribution with the symmetry factor $s_{\mathcal{G}_c}/s_{\mathcal{G}}$ for 1qp processes and with $1/s_{\mathcal{G}}$ for the ground-state energy to properly account for the symmetries during the embedding. Also, all white graph contributions of a given order and number of vertices are grouped into one such function. The functions are saved in a file and included in the Monte Carlo algorithm code. If the white graph contributions are associated with models with different interaction flavours, we must sample for certain parameter values during this step.

The physical parameters are also fixed like the momentum $\boldsymbol{k}$ and the decay exponent $d+\sigma$, and of course, the code must be adapted for different lattice geometries. There are also simulation parameters like the number of seeds, the ratio between shift and rift moves, and the simulation exponents ρ of the reference sum and γ for the rift move ζ-function distribution. The Monte Carlo algorithm is then executed for each embeddable function with a fixed order, number of sites, and seed, yielding the target values of Equation (188).

We average the target values for multiple runs with different seeds, add them up according to Equations (164), (174), and (177), and obtain a perturbative series in λ. Afterwards, we employ DlogPadé extrapolation to extract critical quantities of interest like the critical point λ_c and the associated critical exponent α for the ground-state energy, $z\nu$ for the 1qp gap, or $(2-z-\eta)\nu$ for the 1qp spectral weight. We can also use the extrapolations to construct the dispersion close to the critical point.

5. Stochastic Series Expansion Quantum Monte Carlo

In this section, we discuss the method of stochastic series expansion (SSE) QMC. This class of QMC algorithms is closely related to path integral (PI) QMC and samples configurations according to the Boltzmann distribution of a quantum mechanical Hamiltonian. This sampling is achieved by extending the configuration space in the imaginary-time direction by operator sequences. The objective is to evaluate thermal expectation values for operators at a finite temperature on a finite system.

The canonical partition function of a system with a quantum mechanical Hamiltonian $\mathcal{H}$ can be expressed as

$$Z = \mathrm{Tr}\{\exp(-\beta\mathcal{H})\} = \sum_{|\alpha\rangle}\langle\alpha|\exp(-\beta\mathcal{H})|\alpha\rangle \tag{211}$$

with β being the inverse temperature and the sum over an arbitrary orthonormal basis $\{|\alpha\rangle\}$. The task is to bring Equation (211) into the form of

$$Z = \sum_{\omega\in C}\pi(\omega) \tag{212}$$

with all weights $\pi(\omega)$ required to be non-negative.

Of course, for a Hamiltonian that is traced over in its eigenbasis (or a classical system), Equation (211) is already in the form of Equation (212), and the system can be directly sampled by a Metropolis–Hastings algorithm (see Equation (46)). For a general quantum mechanical problem, we do not have access to the eigenstates of a system and require a reformulation of Equation (211).

The SSE QMC idea resolves this issue in the following way: Given a Hamiltonian $\mathcal{H}$, a computational orthonormal basis $\{|\alpha\rangle\}$ is chosen in which the trace is evaluated. Further, there should exist a decomposition of the Hamiltonian:

$$\mathcal{H} = -\sum_i \mathcal{H}_i \tag{213}$$

into operators $\mathcal{H}_i$. $\mathcal{H}_i$ and $\{|\alpha\rangle\}$ are chosen such that the following two conditions are met:

- No-branching rule:

$$\mathcal{H}_i|\beta\rangle \propto |\gamma\rangle \in \{|\alpha\rangle\} \qquad \forall\,\mathcal{H}_i \quad \forall|\beta\rangle \in \{|\alpha\rangle\}, \tag{214}$$

  ensuring that no superpositions of basis states are created by acting with $\mathcal{H}_i$.
- Non-negative real matrix elements in the computational basis:

$$\langle\beta|\mathcal{H}_i|\gamma\rangle \geq 0 \qquad \forall\,\mathcal{H}_i \quad \forall|\beta\rangle, |\gamma\rangle \in \{|\alpha\rangle\}\,. \tag{215}$$

The second condition is not strictly necessary, but makes sure that no sign problem arises, which would lead to exponentially hard computational complexity. In general, it is not necessarily possible to find a computational basis in which this condition can be fulfilled for all Hamiltonians. However, if the negative matrix elements contribute in such a way that they always occur in pairs and the minus signs cancel, the condition can be relaxed without inducing a sign problem. We will encounter this case for the antiferromagnetic Heisenberg models in Section 5.2.

For operators $\mathcal{H}_i$ that are diagonal in the computational basis, the conditions (214) and (215) never pose a problem as diagonal operators intrinsically obey the no-branching rule and can always be made non-negative by adding a suitable constant to the Hamiltonian. The main difficulty is to find a computational basis in which the off-diagonal matrix elements are non-negative, which is not necessarily possible, as mentioned above. In particular, for fermionic or frustrated systems, negative signs typically occur.

In order to reformulate Equation (211) in the form of Equation (212), a high-temperature expansion for the partition function is performed:

$$Z = \mathrm{Tr}\{\exp(-\beta\mathcal{H})\} = \sum_{|\alpha\rangle}\langle\alpha|\exp(-\beta\mathcal{H})|\alpha\rangle \tag{216}$$

$$= \sum_{|\alpha\rangle}\sum_{n=0}^{\infty}\frac{\beta^n}{n!}\langle\alpha|(-\mathcal{H})^n|\alpha\rangle \tag{217}$$

$$= \sum_{|\alpha\rangle}\sum_{n=0}^{\infty}\frac{\beta^n}{n!}\langle\alpha|(\sum_i\mathcal{H}_i)^n|\alpha\rangle\,. \tag{218}$$

In general, the evaluation of $\langle\alpha|(\sum_i \mathcal{H}_i)^n|\alpha\rangle$ is not feasible. The way the SSE tackles this expression is by expanding the product of sums as

$$\left(\sum_i \mathcal{H}_i\right)^n = \sum_{S_n}\prod_{k=1}^{n}\mathcal{H}_{i(k)} \tag{219}$$

as a sum over all occurring operator sequences S_n resulting from the exponentiation. The additional dimension created by the operator sequence is usually referred to as imaginary time in analogy to path-integral formulations. Inserting Equation (219) into Equation (218), one obtains

$$Z = \sum_{|\alpha\rangle}\sum_{n=0}^{\infty}\sum_{S_n}\frac{\beta^n}{n!}\langle\alpha|\prod_{k=1}^{n}\mathcal{H}_{i(k)}|\alpha\rangle\,. \tag{220}$$

Note that each of the summands in Equation (220) is non-negative by design due to the condition (215) and can be interpreted as the relative weight of a configuration. By comparing Equation (220) with Equation (212), we see that it is of a suitable form for a Markov chain Monte Carlo sampling. We identify the direct product of the set of all basis states $\{|\alpha\rangle\}$ with the set of all sequences as the configuration space:

$$\mathcal{C} = \{|\alpha\rangle\}\times\bigcup_{n=0}^{\infty}\{S_n\}\,. \tag{221}$$

The weight of a configuration is given by

$$\pi(|\alpha\rangle, S_n;\beta) = \frac{\beta^n}{n!}\langle\alpha|\prod_{k=1}^{n}\mathcal{H}_{i(k)}|\alpha\rangle\,. \tag{222}$$

In the next step, we discuss the structure of each configuration consisting of a computational basis state $|\alpha\rangle$ and an operator sequence S_n. Regarding Equation (220), we stress that the action of the product of operators in the sequence onto the basis state is crucial. Due to the no-branching rule (see Equation (214)), the action of $\prod_{k=1}^{n}\mathcal{H}_{i(k)}$ onto the basis state can be interpreted as a discrete propagation of the state α in imaginary time according to the operators in the sequence. We define the state at propagation index $p\in\{0,\ldots,n\}$ to be

$$|\alpha(p)\rangle = \prod_{k=1}^{p}\mathcal{H}_{i(k)}|\alpha\rangle\,. \tag{223}$$

Due to the periodic boundary condition of the trace and the orthogonality of the basis states $\{|\alpha\rangle\}$, only sequences for which $|\alpha(n)\rangle = |\alpha(0)\rangle$ have a non-zero weight (see Equation (222)).

At this point, we can demonstrate why it does not matter if some matrix elements are negative in some instances, e.g., for some antiferromagnetic spin models on bipartite lattices. If a matrix element of an operator is negative, but it is ensured by the Hamiltonian and the periodicity of the trace that there is always an even number of negative matrix elements in a sequence with non-vanishing weight, then the definition of weights as in Equation (222) is nevertheless possible.

In the discussion so far, we considered sequences of all possible lengths. In order to formulate algorithms sampling the configuration space efficiently, a scheme with a fixed sequence length $\mathcal{L}$ can be introduced, in which all sequences with $n < \mathcal{L}$ are padded with identity operators $\mathbb{1}$ to length $\mathcal{L}$ and all sequences $n > \mathcal{L}$ are discarded. The physical justification to discard all sequences above a certain fixed length $\mathcal{L}$ is that they are exponentially suppressed for sufficiently large $\mathcal{L}$. In short, this is the case because the mean operator number $\langle n \rangle$ is proportional to the mean energy $\langle \mathcal{H} \rangle$ and the variance is related to the specific heat in the following way:

$$\langle \mathcal{H} \rangle = -\frac{\langle n \rangle}{\beta} \qquad\qquad C = \langle n^2 \rangle - \langle n \rangle^2 - \langle n \rangle \,. \tag{224}$$

A derivation and discussion of these statements can be found in Refs. [39,210–213]. From Equation (224), we can infer that the infinite sum over all sequence lengths can be truncated at a finite $\mathcal{L} \propto \beta N$. From rearranging Equation (224) and using the extensivity of the mean energy, the mean sequence length scales as $\langle n \rangle \propto \beta N$ proportional to the inverse temperature and system size N. As the mean sequence length has a finite value, the idea is to choose an $\mathcal{L} > \langle n \rangle$ great enough to be able to sample all but a negligible amount of operator sequences. Further, we will argue that the introduction of a large enough cutoff results in an exponentially small and negligible error. In the limit of $\beta \to \infty$, the specific heat has to vanish. Therefore, the variance of the mean sequence length is proportional to $\langle n \rangle$. From this, it is concluded that the weights of sequences vanish exponentially for large enough sequence lengths n [212].

We, therefore, introduce a large enough cutoff for the sequence length $\mathcal{L}$ and consider operator sequences with a fixed length $\mathcal{L}$. The expression for the partition function can then be written as

$$Z \approx \sum_{S_\mathcal{L}} \sum_{|\alpha\rangle} \frac{\beta^n (\mathcal{L}-n)!}{\mathcal{L}!} \langle \alpha | \prod_{k=1}^{\mathcal{L}} \mathcal{H}_{i(k)} | \alpha \rangle \,. \tag{225}$$

The new configuration space includes all sequences of length $\mathcal{L}$ where the shorter sequences are padded by inserting unity operators. The random insertion of unity operators into a sequence of $n < \mathcal{L}$ non-trivial operators results in $\binom{\mathcal{L}}{n} = \frac{\mathcal{L}!}{n!(\mathcal{L}-n)!}$ sequences of length $\mathcal{L}$. The modified prefactor in Equation (225) accounts for this overcounting. Although Equation (225) is, strictly speaking, an approximation to the partition function, we want to note that, in practice, this does not cause a systematic error as $\mathcal{L}$ can be chosen large enough in a dynamical fashion such that, during the finite simulation time, no sequence of length $n > \mathcal{L}$ would occur. Therefore, we will not consider the fixed-length scheme as an approximation below.

To summarise: Following the argumentation discussed in this section, it is possible to bring any partition function into the form of Equation (225), whereby all the summands are non-negative if a suitable decomposition and computational basis can be found. The next step is to implement a Markov chain MC sampling on the configuration space. As the configuration space largely depends on the model, the sampling is performed in a model-dependent way. In Section 5.1, we will introduce an algorithm to sample the (LR)TFIM. In Section 5.2, we will further describe an algorithm to sample unfrustrated (long-range) Heisenberg models. In Section 5.3, the measurement of a variety of operators is discussed. Further, we will discuss how to sample systems at effectively zero temperature in Section 5.4 as this review considers the zero-temperature physics of quantum phase transitions. In Section 5.5, we give a short overview of other MC algorithms for long-range models based on path integrals.

5.1. Algorithm for Arbitrary Transverse-Field Ising Models

In this section, we describe the SSE algorithm to sample arbitrary transverse-field Ising models (TFIMs) of the form

$$H = \sum_{i\neq j} J_{i,j}\sigma_i^z\sigma_j^z - \sum_i h_i\,\sigma_i^x \tag{226}$$

as introduced by A. Sandvik in Ref. [39]. The Pauli matrices σ_i^κ with $\kappa \in \{x,z\}$ describe N spins $1/2$ located on the lattice sites i,j. The transverse-field strength at a lattice site i is $h_i > 0$, and the Ising couplings between sites i and j have the strength $J_{i,j} \in \mathbb{R}$. For $J_{i,j} > 0$, the interaction is antiferromagnetic and it is energetically favourable for the spins to anti-align, while for $J_{i,j} < 0$, the interaction is ferromagnetic and it is energetically favourable for the spins to align.

Choosing the σ^z-eigenbasis $\{|\alpha\rangle\} = \{|\sigma_1^z,\ldots,\sigma_N^z\rangle\}$ for the SSE formulation avoids the sign problem for arbitrary Ising interactions. The Hamiltonian is decomposed using the following operators:

$$\mathcal{H}_{0,0} = \mathbb{1} \tag{227}$$
$$\mathcal{H}_{i,0} = h_i\sigma_i^x \qquad i > 0 \tag{228}$$
$$\mathcal{H}_{i,i} = h_i\mathbb{1} \tag{229}$$
$$\mathcal{H}_{i,j} = |J_{i,j}| - J_{i,j}\sigma_i^z\sigma_j^z \qquad i,j > 0,\ i \neq j\,. \tag{230}$$

We call $\mathcal{H}_{0,0}$ a trivial operator, $\mathcal{H}_{i,0}$ a field operator, $\mathcal{H}_{i,i}$ a constant operator, and $\mathcal{H}_{i,j}$ an Ising operator. The operator $\mathcal{H}_{i,i}$ is associated with the site i, even though it is proportional to $\mathbb{1}$. With these operators, we can rewrite Equation (226) up to an irrelevant constant as

$$\mathcal{H} = -\sum_{i=1}^{N}\sum_{j=0}^{N}\mathcal{H}_{i,j}\,. \tag{231}$$

Note that Equation (231) is a sum over field, constant, and Ising operators. The constant operators are not part of the original Hamiltonian, but will be relevant for algorithmic purposes. The trivial operators (Equation (227)) are not relevant to express the Hamiltonian, but are necessary for the fixed-length sampling scheme. The proposed decomposition fulfils the no-branching property, and there are no negative matrix elements of the operators in the computational basis due to the positive constant $|J_{i,j}|$ that is added to the Ising operators. The possible matrix elements for Ising operators $\mathcal{H}_{i,j} = |J_{i,j}| - J_{i,j}\sigma_i^z\sigma_j^z$ acting on a pair of spins i,j are given by

$$\langle\uparrow\uparrow_{i,j}|\mathcal{H}_{i,j}|\uparrow\uparrow_{i,j}\rangle = \langle\downarrow\downarrow_{i,j}|\mathcal{H}_{i,j}|\downarrow\downarrow_{i,j}\rangle = \begin{cases} 2|J_{i,j}| & \text{for } J_{i,j} < 0 \\ 0 & \text{for } J_{i,j} > 0 \end{cases} \tag{232}$$

$$\langle\uparrow\downarrow_{i,j}|\mathcal{H}_{i,j}|\uparrow\downarrow_{i,j}\rangle = \langle\downarrow\uparrow_{i,j}|\mathcal{H}_{i,j}|\downarrow\uparrow_{i,j}\rangle = \begin{cases} 0 & \text{for } J_{i,j} < 0 \\ 2|J_{i,j}| & \text{for } J_{i,j} > 0\,. \end{cases} \tag{233}$$

This implies that only sequences where (anti)ferromagnetic Ising bonds are placed on (anti)aligned spins have a non-vanishing weight.

The partition function (Equation (225)) in the fixed-length scheme reads

$$Z = \sum_{S_\mathcal{L}}\sum_{|\alpha\rangle}\frac{\beta^n(\mathcal{L}-n)!}{\mathcal{L}!}\prod_{k=1}^{\mathcal{L}}\langle\alpha(k)|\mathcal{H}_{i(k),j(k)}|\alpha(k-1)\rangle\,. \tag{234}$$

The propagated states $|\alpha(p)\rangle$ only change in the imaginary-time propagation if a field operator, the only off-diagonal operator in the chosen basis, acts on it. A configuration only has a non-zero weight if all the Ising operators in the sequence are placed on spins that have the correct alignment. Constant operators are included in the Hamiltonian for algorithmic purposes, which will become important in the off-diagonal update described below.

Before going into the description of the Markov chain to sample the configuration space, it is illustrative to visualise a configuration with non-vanishing weight defined by a state $|\alpha\rangle$ and an operator sequence $S_{\mathcal{L}}$. In Figure 14, an exemplary configuration with non-vanishing weight is illustrated for the one-dimensional TFIM and visualisations of the different operators Equations (228)–(230) in such a configuration are shown.

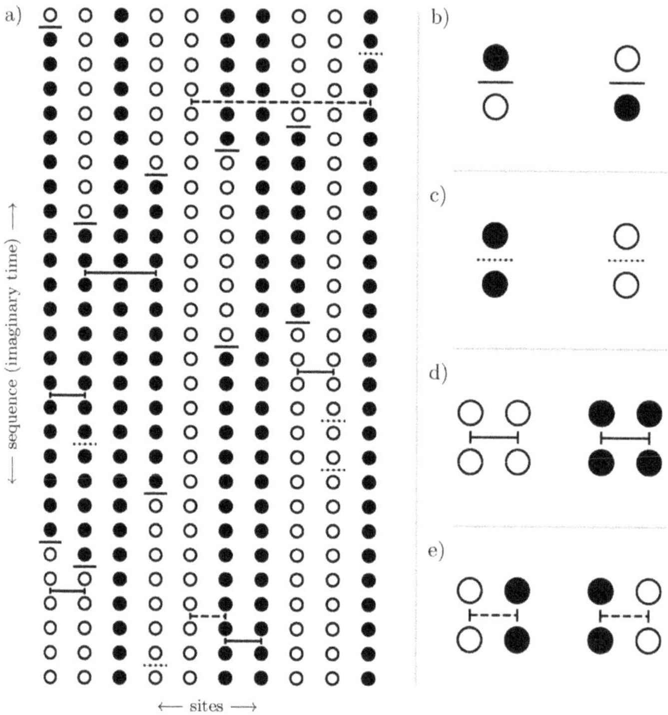

Figure 14. (**a**) An SSE configuration of a transverse-field Ising chain with $N = 10$ sites and a sequence length $\mathcal{L} = 27$. The spatial spin direction goes from left to right, and the imaginary-time dimension goes from bottom to top. The number of trivial operators in the operator sequence $S_{\mathcal{L}}$ is 8. Filled (empty) circles represent spins aligned in the $\sigma_i^z = +1$ (-1) direction. The propagated states $|\alpha(p)\rangle$ correspond to the p-th row from below with the lowest row being state $|\alpha\rangle$. (**b**) A depiction of the possible vertices for field operators. (**c**) A depiction of the possible vertices for constant operators. (**d**) A depiction of the allowed vertices for ferromagnetic Ising operators. Note that a ferromagnetic Ising vertex can only connect sites that are connected by a ferromagnetic bond in the Hamiltonian. (**e**) A depiction of the allowed vertices for antiferromagnetic Ising operators. Note that an antiferromagnetic Ising vertex can only connect sites that are connected by an antiferromagnetic bond in the Hamiltonian.

Before going into the details of the algorithm, we introduce the main concept of the update scheme and the crucial obstacles that are encountered in setting up an efficient algorithm [39]. Each step of the Markov chain sampling of configurations is performed by performing a so-called diagonal update followed by an off-diagonal quantum cluster update [39]. In the diagonal update, one iterates over the sequence and exchanges trivial operators with constant or Ising operators and vice versa while propagating the states along the sequence. In the off-diagonal update, constant operators are exchanged with field operators while preserving the weight of the configuration. The main obstacle that is circumvented with this update procedure is the following: It is a non-trivial and non-local task to insert off-diagonal operators into the sequence without creating a configuration with vanishing weight. The first problem is that one cannot simply insert a single field operator into the sequence as it breaks the periodicity of the propagated states $|\alpha(\mathcal{L})\rangle \neq |\alpha(0)\rangle$, leading to a vanishing weight due to the orthonormality of the computational basis. Therefore, field operators can only occur in an even number for each site to preserve the periodicity in imaginary time. The second issue involves Ising operators placed on a pair of sites i and j. If one places a field operator at one of the sites before and behind the Ising operator in the sequence, this preserves the periodicity in imaginary time, but the matrix element of the Ising operator becomes zero as the spins will be misaligned with respect to the sign of the Ising coupling (see Equation (232)). These issues can be tackled by a non-local off-diagonal update [39], which we will thoroughly discuss after the diagonal update.

In the diagonal update, the number of non-trivial operators n is altered by exchanging trivial operators with constant and Ising operators in the operator sequence and vice versa. This does not change the states $|\alpha(p)\rangle$, including the state $|\alpha\rangle = |\alpha(0)\rangle$. Starting from propagation step $p = 0$ and state $|\alpha(0)\rangle$, one iterates over the sequence step-by-step and conducts the exchange of trivial operators with non-trivial diagonal operators. If a field operator is encountered at the current propagation step, the state is propagated by flipping the respective spin, and the iteration proceeds. If a diagonal operator is encountered, a local update following the Metropolis–Hastings algorithm as described in Section 3 is performed and the total transition probability is made of a proposition probability $\tilde{T}$ and an acceptance probability p_{acc}.

If a trivial operator is encountered at propagation step k, a non-trivial diagonal operator gets proposed with the probability

$$\tilde{T}(\mathcal{H}_{0,0} \to \mathcal{H}_{i,j}) = \frac{M_{i,j}}{C} \tag{235}$$

taking into account the weight $M_{i,j}$ of the proposed operator with the normalising constant C:

$$M_{i,j} = \begin{cases} 2|J_{ij}| & \text{for } i \neq j\,, \\ h_i & \text{for } i = j\,, \end{cases} \tag{236}$$

$$C = \sum_i h_i + 2\sum_{i\neq j} |J_{ij}|\,. \tag{237}$$

The weight is essentially given by the matrix elements of the respective operators $\mathcal{H}_{i,j}$ with the special case that the Ising operators are a priori handled as if they were allowed. At a later stage, it is checked if the spins are correctly aligned at the considered propagation step k and state $|\alpha(k)\rangle$, and the operator gets rejected if this is not the case. This has the benefit that one does not need to check every single bond for correct alignment for the insertion of a single operator, which scales like $\mathcal{O}(N^2)$ for long-range models. This sampling can be performed in a constant time complexity in the number of elements in the distribution by using the so-called walker method of aliases [214]. Instructions on how to set up a walker sampler from a distribution of discrete weights and how to draw from this distribution can be found in Appendix C or in Ref. [214]. Drawing from the discrete distribution of weights, an operator $\mathcal{H}_{i,j}$ gets proposed to be inserted. On the other side, if a non-trivial diagonal operator is encountered, it is always proposed to be replaced by a trivial operator:

$$\tilde{T}(\mathcal{H}_{i,j} \to \mathcal{H}_{0,0}) = 1\,. \tag{238}$$

The acceptance probabilities are then chosen according to the Metropolis–Hastings algorithm in Equation (46). For a constant field operator $\mathcal{H}_{i,i}$, this gives the acceptance probability:

$$p_{\text{acc}}(\mathcal{H}_{0,0} \to \mathcal{H}_{i,i}) = \min\left(1, \frac{\beta M_{i,i}}{\mathcal{L}-n} \frac{C}{M_{i,i}}\right) \tag{239}$$

$$= \min\left(1, \frac{\beta(\sum_i h_i + 2\sum_{i\neq j}|J_{ij}|)}{\mathcal{L}-n}\right). \tag{240}$$

Similarly, for an Ising operator $\mathcal{H}_{i,j}$, one has

$$p_{\text{acc}}(\mathcal{H}_{0,0} \to \mathcal{H}_{i,j}) = \min\left(1, \frac{\beta\langle\alpha(k)|\mathcal{H}_{i,j}|\alpha(k-1)\rangle}{\mathcal{L}-n} \frac{C}{M_{i,j}}\right) \tag{241}$$

$$= \frac{\langle\alpha(k)|\mathcal{H}_{i,j}|\alpha(k-1)\rangle}{M_{i,j}} \min\left(1, \frac{\beta(\sum_i h_i + 2\sum_{i\neq j}|J_{ij}|)}{\mathcal{L}-n}\right) \tag{242}$$

up to a factor $\langle\alpha(k)|\mathcal{H}_{i,j}|\alpha(k-1)\rangle / M_{i,j}$, which is either 1 or 0 depending on if the spins at the current propagation step k are correctly aligned or misaligned. An Ising bond with misaligned spins would lead to a vanishing weight of the configuration and is, therefore, not allowed.

Up to this factor, the acceptance probability is the same independent of the type of non-trivial diagonal operator that is proposed to be inserted as the operator weight $M_{i,j}$ arising in the weight for the newly proposed configuration cancels with the same factor in the proposition probability. This is because we already chose to propose an operator by considering its respective weight $M_{i,j}$. As the acceptance probabilities do not differ, one can, therefore, also first check if one accepts to insert any non-trivial diagonal operator and, only if this is accepted, draw the precise operator to be inserted. Of course, if the chosen operator is an Ising operator, it still has to be checked if the spins are correctly aligned to prevent a non-vanishing weight.

The cancellation of the operator weights $M_{i,j}$ also makes it easier to perform the reverse process and replace a non-trivial diagonal operator with a trivial one in the sense that the acceptance probability for inserting $\mathcal{H}_{0,0}$:

$$p_{\text{acc}}(\mathcal{H}_{i,j} \to \mathcal{H}_{0,0}) = \min\left(1, \frac{\mathcal{L}-(n-1)}{\beta(\sum_i h_i + 2\sum_{i\neq j}|J_{ij}|)}\right) \tag{243}$$

does not depend on the current non-trivial diagonal operator $\mathcal{H}_{i,j}$.

If a proposition gets rejected, the iteration along the sequence continues and the procedure starts again for the next operator. After each diagonal update sweep (iteration over the whole sequence) during the equilibration, trivial operators are appended to the end of the sequence such that $\mathcal{L} > 4n/3$ [39]. This allows dynamically adjusting the sequence length $\mathcal{L}$ for the fixed-length scheme to ensure a sufficiently long sequence.

For an efficient implementation of the off-diagonal quantum cluster update, it is crucial to introduce the concept of operator legs. An operator at position p in the sequence has legs with the numbers $4p, \ldots, 4p+3$. For an Ising operator, these legs are associated with two legs per site, one upper and one lower leg (see Figure 15a).

Figure 15. Assignment of leg numbers to operator legs for an operator at propagation step p. (**a**) Illustration of a two-site operator with four real vertex legs. (**b**) Illustration of a single-site operator with only two real vertex legs and two ghost legs.

For a constant or field operator, only two of the four legs are real vertex legs since these operators act only on a single site (see Figure 15b). The remaining two legs are called ghost legs and are considered solely due to algorithmic reasons as it is numerically beneficial to let every operator have the same number of legs. This allows calculating the propagation index p of an operator from the leg numbers using integer division with $p = \lfloor leg/4 \rfloor$. Trivial operators can be described with four ghost legs or can be ignored entirely in the sequence for the off-diagonal update.

For the chosen representation of the Ising model within the SSE, one can subdivide the configuration into disjoint clusters that extend in space, as well as in imaginary time with Ising operators acting as bridges in real space and with constant and field operators acting as delimiters in imaginary time. These clusters of spins can be flipped by replacing the delimiting constant operators with field operators and vice versa without changing the weight $\pi(|\alpha\rangle, S_{\mathcal{L}}; \beta)$. If the cluster winds around the boundary in imaginary time, the respective spins in state $|\alpha\rangle$ have to be updated as well. In order to flip half of the configuration on average and obtain a good mixing, all clusters are constructed and the probability to flip a specific cluster is chosen to be $1/2$ for each cluster separately.

For the construction of the clusters, the propagated spin states along imaginary time are not needed. The entire problem can be dealt with in the language of vertices with legs. The relevant information for the cluster formation is which legs are connected. This information about the connection between legs of different operators is stored in a doubly linked list. The list is set up in the following way: At the index of a vertex leg i, we store the index j of the leg it is connected to, i.e., $\text{list}[i] = j$ and vice versa $\text{list}[\text{list}[i]] = \text{list}[j] = i$. This segmentation is illustrated for an exemplary configuration in Figure 16 together with the doubly linked leg list for the depicted configuration. The doubly linked list can already be set up during the diagonal update when the sequence is traversed either way. An efficient algorithm to set up the data structure is described in Refs. [213,215].

In comparison to general off-diagonal loop updates [215–217], the formation of clusters in the presented off-diagonal cluster update for the TFIM is fully deterministic. The whole configuration is divided into disjoint clusters, all of which will be built and flipped with probability $1/2$. It is beneficial to already decide whether a cluster is flipped or not before constructing the cluster so the constant and field operators and spin states can already be processed during the construction. A leg that is processed during the construction of a cluster is marked as visited in order to not process the same leg twice. This is also the reason why all the clusters have to be constructed even if they will not be flipped, as otherwise, the same cluster would get constructed starting from another leg later on. Further, we introduce a stack for the legs that were visited during the construction, but whose connections are yet to be processed .

The formation of each cluster starts by choosing a leg that has not yet been visited in the current off-diagonal update. At the beginning of the cluster formation, it is randomly determined if the cluster is flipped or not. If the leg corresponds to an Ising vertex, the cluster branches out to all four legs of the vertex. This means that all four legs of the operator are put on the stack. If the leg corresponds to a constant or field operator, the type of operator is exchanged if the cluster is flipped and only this primal leg is put on the stack. Next, the following logic is repeated until the stack is empty. We pop a leg l from the stack. If the leg has been visited already, we continue with a new leg from the stack. Else, we determine the new leg $l' = \text{list}[l]$ using the doubly linked list. We mark both legs l and l' as visited. If one passes by the periodic boundary in imaginary time while going from leg l to l' and the cluster will be flipped, the corresponding site belonging to the legs in the state $|\alpha\rangle$ has to be flipped. If l' belongs to an Ising operator, we add all legs of the Ising operator that have not been visited yet to the stack. If l' belongs to a constant or field operator, we exchange the operator if the cluster is said to be flipped. This procedure is repeated until the stack is empty. After that, one proceeds with the next cluster starting from a leg that has not yet been visited in the current cluster update. If no such leg is left, the cluster update is finished.

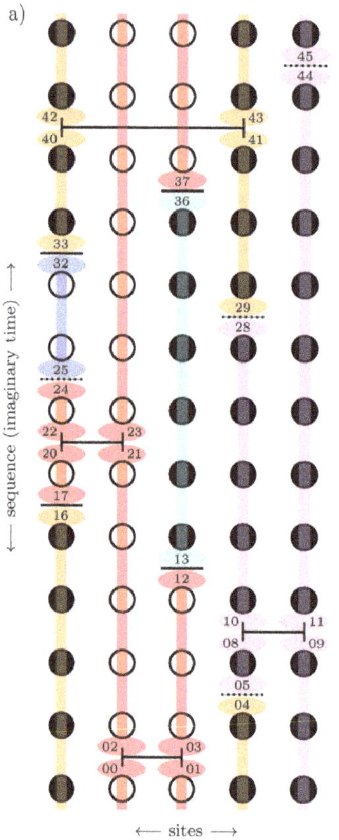

b)

entry	value
00	23
01	37
02	21
03	12
04	43
05	08
06	−1
07	−1
08	05
09	45
10	28
11	44
12	03
13	36
14	−1
15	−1
16	42
17	20
18	−1
19	−1
20	17
21	02
22	24
23	00
24	22
25	32
26	−1
27	−1
28	10
29	41
30	−1
31	−1
32	25
33	40
34	−1
35	−1
36	13
37	01
38	−1
39	−1
40	33
41	29
42	16
43	04
44	11
45	09
46	−1
47	−1

Figure 16. Illustration for the segmentation and construction of the doubly linked list for an exemplary configuration of the ferromagnetic transverse-field Ising model. (**a**) Segmentation of a configuration into disjoint clusters including the numbers of the operator legs. Ghost legs are not depicted. The coloured lines with ellipses at each end depict the operator legs that are linked. Each colour represents one cluster in the off-diagonal update. (**b**) Depiction of a doubly linked list for the configuration shown in (**a**). The left column represents the entry numbers in the list and the right column the corresponding legs to which the entry is connected. The colour represents the clusters in (**a**) to which the connection belongs. Ghost legs are linked to the value −1 and are shaded in grey.

In addition to this cluster update, spins that have no operators acting on them in the sequence can be flipped with a probability of 1/2 (thermal spin flip).

In summary, the off-diagonal update exchanges field $\mathcal{H}_{i,0}$ and constant operators $\mathcal{H}_{i,i}$ with each other and changes the state $|\alpha\rangle$. Combining the diagonal update with the off-diagonal cluster update samples the entire configuration space.

5.2. Algorithm for Unfrustrated Heisenberg Models

In this section, we describe the algorithm to sample arbitrary unfrustrated spin-1/2 Heisenberg models with the SSE framework [213]. By unfrustrated Heisenberg models, we mean Hamiltonians:

$$\mathcal{H} = \sum_{b\in \mathrm{AF}} J_b \vec{S}_{i(b)}\vec{S}_{j(b)} + \sum_{b\in \mathrm{F}} J_b \vec{S}_{i(b)}\vec{S}_{j(b)} \tag{244}$$

written as a sum over three-component interactions between two sites $i(b)$ and $j(b)$ connected by bond b, where each bond is either ferromagnetic (F) with $J_b < 0$ or antiferromagnetic (AF) with $J_b > 0$, with the property that there is no loop of lattice sites connected by bonds that contains an odd number of antiferromagnetic bonds, as this would lead to frustration. The spin operators are a compact notation for a three-component vector of spin operators $\vec{S}_i = (S_i^x, S_i^y, S_i^z)^T$. The coupling strengths J_b can have a priori arbitrary amplitudes. It is crucial to look at unfrustrated Heisenberg models in order to define non-negative weights for configurations as the off-diagonal components of the antiferromagnetic operators have a negative matrix element. In an unfrustrated model, these matrix elements always occur an even number of times in the operator sequence of any valid configuration, which makes it possible to construct a non-negative SSE weight [213].

For the SSE algorithm, the σ^z-eigenbasis $\{|\alpha\rangle\} = \{|\sigma_1^z, \dots, \sigma_N^z\rangle\}$ is chosen, but the σ^x- or σ^y-eigenbasis would work the same way. The Hamiltonian is decomposed into

$$\mathcal{H} = -\sum_{b\in \mathrm{AF}} \left(\mathcal{H}_{1,b}^{\mathrm{AF}} + \mathcal{H}_{2,b}^{\mathrm{AF}} - \frac{|J_b|}{4} \right) - \sum_{b\in \mathrm{F}} \left(\mathcal{H}_{1,b}^{\mathrm{F}} + \mathcal{H}_{2,b}^{\mathrm{F}} - \frac{|J_b|}{4} \right) \tag{245}$$

using the following operators:

$$\mathcal{H}_{1,b}^{\mathrm{F}} = \frac{|J_b|}{4} - J_b S_{i(b)}^z S_{j(b)}^z \tag{246}$$

$$\mathcal{H}_{2,b}^{\mathrm{F}} = -\frac{J_b}{2}(S_{i(b)}^+ S_{j(b)}^- + S_{i(b)}^- S_{j(b)}^+) \tag{247}$$

$$\mathcal{H}_{1,b}^{\mathrm{AF}} = \frac{|J_b|}{4} - J_b S_{i(b)}^z S_{j(b)}^z \tag{248}$$

$$\mathcal{H}_{2,b}^{\mathrm{AF}} = -\frac{J_b}{2}(S_{i(b)}^+ S_{j(b)}^- + S_{i(b)}^- S_{j(b)}^+)\,. \tag{249}$$

The diagonal operators $\mathcal{H}_{1,b}^{\mathrm{F}}$ and $\mathcal{H}_{1,b}^{\mathrm{AF}}$ in Equations (246) and (248) are defined in the same fashion as for the TFIM up to the factor of $1/4$ due to the usage of spin operators $S^\kappa = \sigma^\kappa/2$ instead of Pauli matrices σ^κ with $\kappa \in \{x, y, z\}$. The contribution to the weight of a sequence of these operators is either $|J_b|/2$ if the bond fulfils the (anti)ferromagnetic condition and zero otherwise. Although the expressions for the ferromagnetic and antiferromagnetic bonds look the same, we distinguish between these two bonds to highlight that these objects behave differently within the off-diagonal update.

As $J_b > 0$ for antiferromagnetic bonds, the off-diagonal operators $\mathcal{H}_{2,b}^{\mathrm{AF}}$ do not fulfil the non-negativity of matrix elements (see Equation (215)) in the computational basis. Therefore, they must always appear in an even number of times in the operator sequence to avoid the sign problem. For the ferromagnetic bonds, this restriction is not necessary.

Analogous to the TFIM, the operator sequence additionally contains trivial operators $\mathcal{H}_{0,0} = \mathbb{1}$, which are not part of the Hamiltonian, but are used to pad the sequence to a fixed length $\mathcal{L}$. In contrast to the TFIM, there is no need for further artificial operators like the constant field operators in the TFIM used to limit the cluster in the cluster update. In the case of the Heisenberg model, the non-local off-diagonal update is constructed in the form of loops instead of a cluster with several branches that need to be limited in imaginary time. An exemplary SSE configuration for a Heisenberg chain using the decomposition from above can be seen in Figure 17.

Similar to the sampling of the TFIM in Section 5.1, each step of the Markov chain sampling of configurations is achieved by performing a diagonal update followed by a non-local off-diagonal update. The diagonal update exchanges trivial operators with diagonal operators, while the off-diagonal update exchanges diagonal bond operators with the respective off-diagonal bond operators.

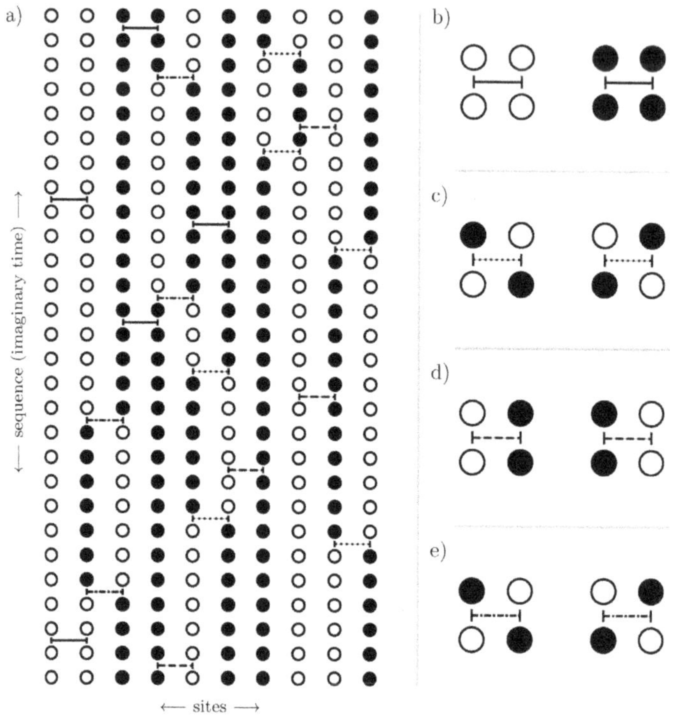

Figure 17. (**a**) An SSE configuration of a Heisenberg chain with ten sites and a sequence length of 27. The number of trivial operators is 5. Filled (empty) circles represent spins in the $\sigma_i^z = +1(-1)$ direction. The propagated states $|\alpha(p)\rangle$ correspond to the p-th row in the configuration. (**b**) A depiction of the allowed vertices for ferromagnetic diagonal operators. (**c**) A depiction of the allowed vertices for ferromagnetic off-diagonal operators. (**d**) A depiction of the allowed vertices for antiferromagnetic diagonal operators. (**e**) A depiction of the allowed vertices for antiferromagnetic off-diagonal operators. Note that (anti)ferromagnetic vertices can only connect sites that are connected by (anti)ferromagnetic bonds in the Hamiltonian.

Due to the structure of the diagonal operators (see Equations (246) and (248)), similar to the TFIM, the diagonal update is performed similarly to the diagonal update of the TFIM described in Section 5.1. Nonetheless, to make the description of the algorithm for unfrustrated Heisenberg models self-contained, we recapitulate the diagonal update and adapt it to the Heisenberg case.

The diagonal update exchanges trivial operators with diagonal operators in the sequence and vice versa. It will, therefore, not change the states $|\alpha(p)\rangle$, including the state $|\alpha\rangle = |\alpha(0)\rangle$. Starting from $|\alpha\rangle$ at $k = 0$, the sequence is again iterated over by k. If the operator at the current position k in the sequence is an off-diagonal operator operator, the state $|\alpha(k-1)\rangle$ is propagated by applying the off-diagonal operator and the iteration proceeds. If one encounters diagonal or trivial operators in the sequence, the Metropolis–Hastings algorithm (see Section 3) is performed and the transition probability to an altered configuration is again split into a proposition probability $\tilde{T}$ and an acceptance probability p_{acc}.

Analogous to the algorithm for the TFIM, if a trivial operator is encountered, a non-trivial diagonal operator gets proposed with the probability:

$$\tilde{T}(\mathcal{H}_{0,0} \to \mathcal{H}_{1,b}^{\text{AF/F}}) = \frac{M_b}{C}, \tag{250}$$

taking into account the weight M_b of the proposed operator and the normalising constant C:

$$M_b = \frac{|J_b|}{2} \qquad\qquad C = \frac{1}{2}\sum_b |J_b|\,. \tag{251}$$

The diagonal update for the Heisenberg model only differs from the one of the TFIM by the detailed probabilities M_b and the normalising constant C. The factor of $1/4$ in the bond weights in comparison to the Ising model comes from using spin operators in contrast to Pauli matrices in the Hamiltonian. This sampling according to the weight M_b can be performed in a constant-time complexity in the number of elements in the distribution by using the walker method of aliases [214]. An introduction on how to set up a walker sampler from a distribution of discrete weights and how to draw from this distribution can be found in Appendix C or in Ref. [214]. Drawing from the discrete distribution of weights, an operator $\mathcal{H}_{1,b}^{\mathrm{AF/F}}$ gets proposed to be inserted. On the other side, if a non-trivial diagonal operator is encountered, it is always proposed to be replaced by a trivial operator:

$$\tilde{T}(\mathcal{H}_{1,b}^{\mathrm{AF/F}} \to \mathcal{H}_{0,0}) = 1, \tag{252}$$

exactly as for the TFIM. The acceptance probabilities are then chosen according to the Metropolis–Hastings algorithm Equation (46). The respective Metropolis–Hastings probabilities with which the proposed replacements at position k in the operator sequence are accepted are given by the ratios of the configuration weights before and after the potential replacement and the ratio of the proposition probabilities:

$$p_{\mathrm{acc}}(\mathcal{H}_{0,0} \to \mathcal{H}_{1,b}^{\mathrm{AF/F}}) = \min\left(1, \frac{\beta\langle\alpha(k)|\mathcal{H}_{1,b}^{\mathrm{AF/F}}|\alpha(k-1)\rangle}{\mathcal{L}-n}\frac{C}{M_b}\right) \tag{253}$$

$$= \frac{\langle\alpha(k)|\mathcal{H}_{1,b}^{\mathrm{AF/F}}|\alpha(k-1)\rangle}{M_b}\min\left(1, \frac{\beta(\sum_b |J_b|)}{2(\mathcal{L}-n)}\right), \tag{254}$$

$$p_{\mathrm{acc}}(\mathcal{H}_{1,b}^{\mathrm{AF/F}} \to \mathcal{H}_{0,0}) = \min\left(1, \frac{2(\mathcal{L}-(n-1))}{\beta(\sum_b |J_b|)}\right). \tag{255}$$

As in the case of the TFIM, the specific matrix element $\langle\alpha(k)|\mathcal{H}_{1,b}^{\mathrm{AF/F}}|\alpha(k-1)\rangle$ for the propagated states $|\alpha(k-1)\rangle$ and $|\alpha(k)\rangle = |\alpha(k-1)\rangle$ has to be considered, which is either M_b, and cancels with this factor in the acceptance probability, or is 0, and the acceptance probability vanishes. Intuitively, this factor simply checks if the spins are correctly aligned or misaligned with respect to the sign of the coupling J_b and only allows the operator to be inserted if the spins $i(b)$ and $j(b)$ of the propagated state $|\alpha(k-1)\rangle$ are correctly aligned. If the proposition is accepted, the replacement of the operator at index k is performed, and one continues with the next propagation index $k+1$ in imaginary time. If the proposition is denied, the current operator remains, and one continues with the next propagation index $k+1$ in imaginary time.

As for the TFIM, a fixed-length scheme is used and the sequence length $\mathcal{L}$ limiting the amount of non-trivial operators n has to be dynamically adjusted. During the thermalisation of the sampling procedure, trivial operators are appended to the end of the sequence after each diagonal update (iteration over the whole sequence) such that $\mathcal{L} > 4n/3$ [39].

In order to efficiently implement an off-diagonal loop update, it is crucial to introduce the concept of operator legs. An operator at position p in the sequence has four legs with the numbers $4p, \dots, 4p+3$, two legs (an upper and a lower leg) for each of the spin sites. Trivial operators can be described with four ghost legs or can be ignored entirely in the sequence for the off-diagonal update.

For the unfrustrated Heisenberg models, the off-diagonal update is performed using a loop update. During the update, loops are constructed in the space–time configuration (see Figure 17 for an example) and the spin states on the loop are flipped along the way, which changes the operator vertex types. Thereby, diagonal operators become off-diagonal operators and vice versa. It is noteworthy that the weight of a configuration is not modified by this since the weights of diagonal and off-diagonal operators are the same in the isotropic Heisenberg model. Therefore, flipping loops with a fixed probability satisfies detailed balance as long as the probability to construct the reverse loop is the same.

For the Heisenberg model, the creation of the loops is deterministic. Whenever a loop enters an operator during the loop construction, there is only one exit leg that creates a valid operator for the bond type (AF or F). Flipping the spins along the loop exchanges an diagonal operator $\mathcal{H}_{1,b}^{\mathrm{AF/F}}$ with its off-diagonal counterpart $\mathcal{H}_{2,b}^{\mathrm{AF/F}}$ and vice versa. The entrance leg and exit leg never belong to the same site. For ferromagnetic operators, the propagation direction of the loop in imaginary time remains the same (see Figure 18). For antiferromagnetic operators, the propagation direction of the loop in imaginary time changes (see Figure 18). When the loop crosses the periodic boundary in imaginary time at a site i, the computational basis state $|\alpha\rangle$ has to be updated by flipping the spin at site i. As each leg only belongs to one loop, the configuration can be split into disjoint loops. Therefore, it is possible to construct all loops during the off-diagonal update and flip each loop independently with a probability of $1/2$.

Combining the diagonal and off-diagonal update allows sampling arbitrary unfrustrated Heisenberg models. This includes antiferromagnetic Heisenberg ladders and bilayer systems with an unfrustrated long-range interaction.

a)

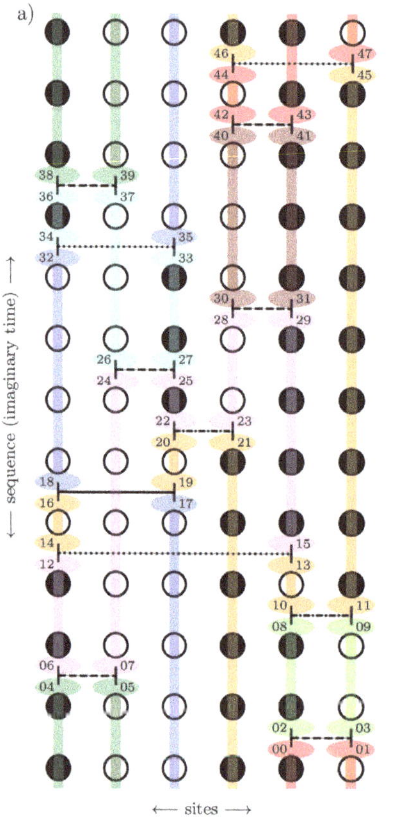

b)

entry	value
00	43
01	47
02	08
03	09
04	38
05	39
06	12
07	24
08	02
09	03
10	13
11	45
12	06
13	10
14	16
15	29
16	14
17	35
18	32
19	20
20	19
21	46
22	25
23	28
24	07
25	22
26	37
27	33
28	23
29	15
30	40
31	41
32	18
33	27
34	36
35	17
36	34
37	26
38	04
39	05
40	30
41	31
42	44
43	00
44	42
45	11
46	21
47	01

Figure 18. Illustration showing the construction of the doubly linked vertex list and the off-diagonal deterministic loop update for unfrustrated Heisenberg models. As an example, a Heisenberg chain with periodic boundary conditions and nearest-neighbour antiferromagnetic and next-nearest-neighbour ferromagnetic interactions is considered. (**a**) Illustration of a configuration including the

numbers of the operator legs belonging to the respective operators. The coloured lines with ellipses at each end depict the operator legs that are linked. Each colour represents one loop in the off-diagonal update. (**b**) Depiction of a doubly linked list for the configuration shown in (**a**). The left column represents the entry numbers in the list and the right column the corresponding legs to which the entry is connected. The colour represents the loops in (**a**) to which the connection belongs.

5.3. Observables

In this section, we introduce some observables that can be easily measured within the SSE formalism introduced. Throughout this section, we do not use the fixed-length scheme, but keep the general form with operator sequences of fluctuating length to keep the notation short and simple. The sampling of observables can be performed analogously in the fixed-length scheme. When implementing the formulas for the observables in the fixed-length scheme, one just has to keep in mind that the sequences used in the formulas are the same ones, but without the padded trivial operators. This means, for instance, that a sum over all propagation steps p becomes a sum over all non-trivial propagation steps in the fixed-length scheme.

Up to this point in Section 5, we have focused on the partition function of a quantum-mechanical Hamiltonian:

$$Z = \sum_{\omega \in \mathcal{C}} \pi(\omega),$$

expressed as a sum over non-negative weights $\pi(\omega)$ of a configuration $\omega \in \mathcal{C}$. However, we are actually not interested in calculating the partition function itself, but rather, the expectation values of observables.

Analogous to the partition function, quantum mechanical expectation values can be expanded into a high-temperature series:

$$\langle A \rangle = \frac{1}{Z} \sum_{\{|\alpha\rangle\}} \sum_{n=0} \sum_{S_n} \frac{(-\beta)^n}{n!} \langle \alpha | A \prod_{k=1}^{n} \mathcal{H}_{l(k)} | \alpha \rangle \,. \tag{256}$$

It is important to realise that configurations with a non-vanishing weight $\pi(\alpha, S_n; \beta)$ constituting the partition function do not necessarily contribute to the expectation value $\langle A \rangle$ with the same weight or a non-vanishing weight at all [213]. In general, only for operators A that are diagonal in the computational basis $\{|\alpha\rangle\}$, the expectation value can be written as

$$\langle A \rangle = \frac{1}{Z} \sum_{\{|\alpha\rangle\}} \sum_{n=0} \sum_{S_n} A(\alpha) \pi(\alpha, S_n; \beta) \tag{257}$$

with $A(\alpha) = \langle \alpha | A | \alpha \rangle$. For instance, the magnetisation in the z-direction or any n-th moment thereof is such a diagonal operator when using the σ^z-eigenbasis $\{|\alpha\rangle\} = \{|\sigma_1^z, \ldots, \sigma_N^z\rangle\}$ as a computational basis. The statistics of the MC estimates for such observables can be improved by realising that [218]

$$\pi(\alpha, S_n; \beta) = \pi(\alpha(p), S_n(p); \beta) \tag{258}$$

with $S_n(p)$ being the sequence obtained from cyclically permuting S_n for p times [218]. This means that the expectation value Equation (257) can be expressed as

$$\langle A \rangle = \frac{1}{Z} \sum_{\{|\alpha\rangle\}} \sum_{n=0} \sum_{S_n} \frac{1}{n} \sum_{p=0}^{n-1} A(\alpha(p)) \pi(\alpha, S_n; \beta), \tag{259}$$

where $A(\alpha)$ is additionally averaged over imaginary time with $A(\alpha(p)) = \langle \alpha(p) | A | \alpha(p) \rangle$.

For off-diagonal operators, one needs to find customised formulas for the expectation value. However, some of these expectation values are accessible in a quite general way. For instance, one can show that the mean energy has a rather simple formula [210,211,218]:

$$\langle \mathcal{H} \rangle = -\frac{\langle n \rangle}{\beta} . \tag{260}$$

Moreover, for the operators $\mathcal{H}_i$ (see Equation (213)), one finds [210,211,218]

$$\langle \mathcal{H}_i \rangle = -\frac{\langle n_i \rangle}{\beta} , \tag{261}$$

where n_i is the amount of operators $\mathcal{H}_i$ occurring in the operator sequence. Similarly, one can derive a formula for the heat capacity [210,211,218]:

$$C = \langle n(n-1) \rangle - \langle n \rangle^2 = \left\langle n^2 \right\rangle - \langle n \rangle^2 - \langle n \rangle . \tag{262}$$

However, for small temperatures, the heat capacity is calculated as the small difference of large numbers.

Linear Response and Correlation Functions

An important class of observables is linear response functions like susceptibilities. The linear response of an observable A to a perturbation $\mathcal{H} \to \mathcal{H} - \lambda B$ tuned by perturbation parameter λ is given by a Kubo integral [212,219]:

$$\chi_{A,B} = \left. \frac{\partial \langle A \rangle}{\partial \lambda} \right|_{\lambda=0} \tag{263}$$

$$= \int_0^\beta \langle A(\tau) B(0) \rangle \mathrm{d}\tau - \beta \langle A \rangle \langle B \rangle . \tag{264}$$

Once again, if A or B are off-diagonal operators, one has to consider case by case and find customised formulas that can be used to sample the specific observables within the SSE formulation. We will focus on the case where A and B are diagonal in the chosen computational basis. Important examples are the magnetic susceptibility or its local version of spin–spin correlation functions averaged over imaginary time, i.e., the correlation function at zero frequency when regarding Equation (264) as a Laplace transformation from imaginary time to frequency space. To make this type of observable accessible to a sampling in the SSE formulation, the imaginary-time correlation function $\langle A(\tau) B(0) \rangle$ is expanded in temperature analogous to the partition function:

$$\langle A(\tau) B(0) \rangle = \frac{1}{Z} \mathrm{Tr}\left(e^{-\beta \mathcal{H}} e^{\tau \mathcal{H}} A e^{-\tau \mathcal{H}} B \right) \tag{265}$$

$$= \frac{1}{Z} \sum_{\{|\alpha\rangle\}} \left\langle \alpha \middle| \sum_{k=0}^{\infty} \left(\frac{(\beta-\tau)^k}{k!} (-\mathcal{H})^k \right) A \sum_{l=0}^{\infty} \left(\frac{\tau^l}{l!} (-\mathcal{H})^l \right) B \middle| \alpha \right\rangle \tag{266}$$

$$= \frac{1}{Z} \sum_{\{|\alpha\rangle\}} \sum_{l,k=0}^{\infty} \frac{(\beta-\tau)^k}{k!} \frac{\tau^l}{l!} \left\langle \alpha \middle| (-\mathcal{H})^k A (-\mathcal{H})^l B \middle| \alpha \right\rangle \tag{267}$$

$$= \frac{1}{Z} \sum_{\{|\alpha\rangle\}} \sum_{l,k=0}^{\infty} \frac{(\beta-\tau)^k}{k!} \frac{\tau^l}{l!} A_l B_0 \left\langle \alpha \middle| (-\mathcal{H})^{l+k} \middle| \alpha \right\rangle , \tag{268}$$

where, in the last step, the operators A and B were replaced by their respective eigenvalues $A_l = \langle \alpha(l) | A | \alpha(l) \rangle$ and $B_0 = \langle \alpha(0) | B | \alpha(0) \rangle$ of the state $|\alpha(p)\rangle$ at the propagation steps $p = l, 0$.

By replacing the sum over k by a sum over $n := l + k$ and inserting the decomposition of the Hamiltonian, this takes the form

$$\langle A(\tau)B(0)\rangle = \frac{1}{Z}\sum_{\{|\alpha\rangle\}}\sum_{n=0}^{\infty}\sum_{l=0}^{n}\sum_{S_n}\frac{(\beta-\tau)^{n-l}}{(n-l)!}\frac{\tau^l}{l!}A_l B_0\left\langle \alpha \left| \prod_{i=0}^{n-1}\mathcal{H}_{a_i,b_i} \right| \alpha \right\rangle \tag{269}$$

$$= \frac{1}{Z}\sum_{\{|\alpha\rangle\}}\sum_{n=0}^{\infty}\sum_{l=0}^{n}\sum_{S_n}\frac{(\beta-\tau)^{n-l}}{(n-l)!}\frac{\tau^l}{l!}A_l B_0\frac{n!}{\beta^n}\pi(\alpha,S_n;\beta) \tag{270}$$

$$= \frac{1}{Z}\sum_{\{|\alpha\rangle\}}\sum_{n=0}^{\infty}\sum_{l=0}^{n}\sum_{S_n}\frac{n!}{(n-l)!l!}\left(1-\frac{\tau}{\beta}\right)^{n-l}\left(\frac{\tau}{\beta}\right)^{l}A_l B_0\pi(\alpha,S_n;\beta) \tag{271}$$

$$= \frac{1}{Z}\sum_{\{|\alpha\rangle\}}\sum_{n=0}^{\infty}\sum_{l=0}^{n}\sum_{S_n}\binom{n}{l}\left(1-\frac{\tau}{\beta}\right)^{n-l}\left(\frac{\tau}{\beta}\right)^{l}\pi(\alpha,S_n;\beta)\frac{1}{n}\sum_{p=0}^{n-1}A_{p+l}B_p, \tag{272}$$

where, in the last step, the average over imaginary time was taken in order to improve the statistics. From Equation (272), the connection between the discrete propagation steps of the SSE and the continuous imaginary time τ becomes apparent [211,212]. An imaginary-time separation τ corresponds to a binomial distribution of separations l of SSE propagation steps:

$$B(l|\tau,n) = \binom{n}{l}\left(1-\frac{\tau}{\beta}\right)^{n-l}\left(\frac{\tau}{\beta}\right)^{l}, \tag{273}$$

which is peaked around $l = n\tau/\beta$ [211]. If one is interested in the spectral properties of the system, one can use this formula for sampling imaginary-time correlation functions $\langle A(\tau)B(0)\rangle$. However, there are more efficient ways to calculate imaginary-time correlation functions by embedding the SSE configuration into continuous imaginary time. We refer to Refs. [212,220] for details, as this is out of the scope of this review.

To get to the linear response function $\chi_{A,B}$ in Equation (264), the imaginary time integral of in Equation (272) has to be calculated. This integral can be analytically calculated:

$$\int_0^\beta \mathrm{d}\tau \left(1-\frac{\tau}{\beta}\right)^{n-l}\left(\frac{\tau}{\beta}\right)^{l} = \beta\int_0^1 \mathrm{d}u\,(1-u)^{n-l}u^l \tag{274}$$

$$= \beta\frac{(n-l)!}{n!}l!\int_0^1 \mathrm{d}u\,(1-u)^n \tag{275}$$

$$= \beta\frac{(n-l)!}{n!}l!\frac{1}{n+1} \tag{276}$$

by performing l partial integrations. Inserting Equation (272) into the formula (264) and performing the imaginary time integral Equation (276) yield

$$\chi_{A,B} = \frac{1}{Z}\sum_{\{|\alpha\rangle\}}\sum_{n=0}^{\infty}\sum_{S_n}\pi(\alpha,S_n;\beta)\frac{\beta}{n(n+1)}\sum_{l=0}^{n}\sum_{p=0}^{n-1}A_{p+l}B_p - \beta\langle A\rangle\langle B\rangle. \tag{277}$$

One can further separate the $l = n$ term while using the periodicity $A_{p+n} = A_p$ in imaginary time and rewrite the sums over l and p as a product of two sums. This eventually yields [210]

$$\chi_{A,B} = \left\langle \frac{\beta}{n(n+1)}\left[\sum_{p=0}^{n-1}A_pB_p + \left\{\sum_{p=0}^{n-1}A_p\right\}\left\{\sum_{p=0}^{n-1}B_p\right\}\right]\right\rangle_{\pi(\alpha,S_n;\beta)} - \beta\langle A\rangle\langle B\rangle. \tag{278}$$

From an algorithmic perspective, the two sums in Equation (278) need to be calculated by traversing the operator sequence. The effort for measuring $\chi_{A,B}$, therefore, scales like the algorithm with complexity $\mathcal{O}(\beta N)$ when the observables A_p, B_p, and A_pB_p are not calculated from scratch at every propagation step p, but gradually updated and averaged while propagating the state through the sequence.

For the special case of the susceptibility, one needs to set $A = m = \sum_i \sigma_i^z$ and $B = M = N \cdot m$. The last term $N\beta\langle m\rangle^2$ in Equation (278) can be dropped for simulations of finite systems due to $\langle m\rangle = 0$. Explicitly, this gives

$$\chi = N\left\langle \frac{\beta}{n(n+1)}\left[\sum_{p=0}^{n-1} m_p m_p + \left\{\sum_{p=0}^{n-1} m_p\right\}\left\{\sum_{p=0}^{n-1} m_p\right\}\right]\right\rangle_{\pi(\alpha,S_n;\beta)} \tag{279}$$

for the susceptibility. Once again, we want to stress that this formula is not formulated in the fixed-length scheme and the average over the propagation index p, therefore, refers to the propagation steps of non-trivial operators. Averaging over all propagation steps in the fixed-length scheme is incorrect.

For the spin–spin correlation functions $G_{i,j}(\omega = 0)$, the operators are set to $A = \sigma_i^z$ and $B = \sigma_j^z$, leading to

$$G_{i,j}(\omega = 0) = \int_0^\beta \left\langle \sigma_i^z(\tau)\sigma_j^z(0)\right\rangle \mathrm{d}\tau \tag{280}$$

$$= \left\langle \frac{\beta}{n(n+1)}\left[\sum_{p=0}^{n-1} \sigma_{i,p}^z\sigma_{j,p}^z + \left\{\sum_{p=0}^{n-1} \sigma_{i,p}^z\right\}\left\{\sum_{p=0}^{n-1} \sigma_{j,p}^z\right\}\right]\right\rangle_{\pi(\alpha,S_n;\beta)}, \tag{281}$$

where it was already used that $\langle\sigma_i^z\rangle = 0$ for any finite system. Similar to the zero-frequency spin–spin correlation function, one can also calculate an equal-time spin–spin correlation function:

$$G_{i,j}(\tau = 0) = \left\langle \sigma_i^z\sigma_j^z\right\rangle = \left\langle \frac{1}{n}\sum_{p=0}^{n-1} \sigma_{i,p}^z\sigma_{j,p}^z\right\rangle_{\pi(\alpha,S_n;\beta)}. \tag{282}$$

Sampling the correlations among all sites i and j at all propagation steps p leads to a complexity of $\mathcal{O}(\beta N^3)$, which is worse than the computational complexity $\mathcal{O}(\beta N)$ of the SSE QMC updates. Even though one could eliminate the average over imaginary time in the case of Equation (282) to reduce the complexity to $\mathcal{O}(N^2)$ at the expense of statistical accuracy, this is not possible for Equation (281) as the sum over imaginary time is intrinsic in the definition of the zero-frequency correlation function. However, by realising that the correlations $\sigma_i^z\sigma_j^z$ between two sites i and j are not altered at the order $\mathcal{O}(\beta N)$-times in imaginary time, but only $\mathcal{O}(\beta)$, one can measure the correlation functions Equations (281) and (282) in $\mathcal{O}(\beta N^2)$ time. This is achieved by traversing imaginary time while memorising the propagation step last$[i]$ of the last preceding spin flip operator for every site separately. When a spin flip occurs at a site j at propagation step p, one needs to update the correlation function:

$$G[i,j] \to G[i,j] + \sigma_{i,p}^z\sigma_{j,p}^z(p - \max(\mathrm{last}[i], \mathrm{last}[j])) \tag{283}$$

for every site i before the local magnetisation is propagated with $\sigma_{j,p+1}^z = -\sigma_{j,p}^z$. One further needs to update last$[j] = p$. In order to avoid that the measurement dominates the simulation for large systems, we only measure the correlations between a fixed site $i = 1$ and $j \in \{1,\ldots,N\}$. This finally reduces the complexity of the measurement to $\mathcal{O}(\beta N)$.

Correlation functions are observables that contain much information about a system's state and, in particular, its order. Moreover, its decay at the critical point with the distance between two spins is connected to the critical exponent η (anomalous dimension). Away from the critical point, the correlation function usually decays exponentially with long-range models being one exception to this. The length scale of this exponential decay is given by the correlation length ξ, which itself is an interesting quantity at a continuous phase transition as its divergence is the reason for systems becoming scale-free at the critical point. As the correlation function does not decay exponentially for long-range system, which we focus on in this review, we will use the more general term "characteristic length scale" to refer to the length scale at which the correlations switch to their long-distance behaviour. In particular, with respect to the Q-FSS for systems above the upper critical dimension, the characteristic length scale is a crucial quantity as its scaling was predicted to be different from the standard FSS with $\xi_L \sim L^{\text{ϙ}}$ instead of $\xi_L \sim L$ (see Section 2.3 or Ref. [34] for details on scaling above the upper critical dimension in quantum systems). This quantity, therefore, played a crucial role in confirming the Q-FSS hypothesis. However, the characteristic length scale is a subtle quantity, which is hard to extract or even define on a finite lattice. For a finite system, the definition of a characteristic length scale is not unique, and there are several definitions for ξ_L, which will converge to ξ_∞ for $L \to \infty$ [221]. For long-range systems, finding a suitable definition for the characteristic length scale is even more difficult, as the correlation function decays algebraically even away from the critical point [222].

Common definitions that are tailored for correlation lengths, which specify the exponential decay of a correlation function at long distances, such as the second moment:

$$\xi_\infty = \sqrt{\frac{1}{2D}\frac{\int |\vec{r}|^2 G(\vec{r})\,\mathrm{d}r}{\int G(\vec{r})\,\mathrm{d}r}} \tag{284}$$

therefore, might yield $\xi_\infty = \infty$ in an infinite system not only at the critical point, but also away from the critical point [223].

One possible definition for long-range quantum systems is [224]

$$\xi_L^{(\mathrm{LR}\omega)} = \frac{1}{q_{\min}}\left[\frac{\tilde{G}_L(0,\omega=0) - \tilde{G}_L(q_{\min},\omega=0)}{\tilde{G}_L(q_{\min},\omega=0)}\right]^{1/\sigma} \tag{285}$$

with $G(q,\omega)$ being the Fourier transform of $G_{i,j}(\omega)$ from real space to momentum space and $q_{\min} = 2\pi/L$ the smallest wavevector fitting on the finite lattice. By inserting the Gaussian propagator $\tilde{G}(q,\omega=0) \sim (b_\sigma q^\sigma + m^2)^{-1}$ for long-range interactions [20,21] into Equation (285), one obtains

$$\begin{aligned}\xi_L^{(\mathrm{LR}\omega)} &= \frac{1}{q_{\min}}\left[\frac{b_\sigma q_{\min}^\sigma + m_L^2}{m_L^2} - 1\right]^{1/\sigma} \\ &= b_\sigma^{1/\sigma} m_L^{-2/\sigma},\end{aligned} \tag{286}$$

so that the momentum dependency cancels. Another definition for the same quantity, but using the equal-time Gaussian propagator $G_L(0,\tau=0) \sim (2\sqrt{\tilde{g}}\sqrt{b_\sigma q^\sigma + m^2})^{-1}$ [212], is

$$\begin{aligned}\xi_L^{(\mathrm{LR}\tau)} &= \frac{1}{q_{\min}}\left[\frac{\tilde{G}_L^2(0,\tau=0) - \tilde{G}_L^2(q_{\min},\tau=0)}{\tilde{G}_L^2(q_{\min},\tau=0)}\right]^{1/\sigma} \\ &= b_\sigma^{1/\sigma} m_L^{-2/\sigma}.\end{aligned} \tag{287}$$

By inserting the scaling of the gap $m_\infty \sim |r|^{z\nu} = |r|^{\sigma\nu/2}$ for the n-component quantum rotor model in the long-range mean-field regime [20,21], which is relevant for most applications discussed below, one obtains the scaling of $\xi^{(LR)}$ in the thermodynamic limit to be

$$\xi_\infty^{(LR)} \sim |r|^{-\nu}, \tag{288}$$

which is the singularity one expects from the characteristic length scale.

5.4. Sampling at Effectively Zero Temperature

The SSE QMC approach can be used to stochastically calculate the thermal average of operators for systems of a finite size at a finite temperature. We discussed in Section 2 that QPTs are a concept defined in the thermodynamic limit at zero temperature. Further, we presented in Sections 2.2 and 2.3 how the critical point, as well as critical exponents can be extracted from certain observables calculated on finite systems at zero temperature. The goal of many works that use SSE QMC sampling is to calculate the expectation value of these observables on finite systems at an effective zero temperature [32,34]. We define the notion of effective zero temperature as a temperature at which, in practice, only the ground state contributes to the thermal average, as all excited states are exponentially suppressed [32,34]. This definition can be applied only to gapped systems as gapless systems at infinitely small, but finite temperature lead to a thermal average over the infinitesimally low-lying excitations. As the SSE QMC approach is sampling finite systems, there cannot be a non-analytic point in the ground-state energy associated with a second-order QPT [113]. Closely related with this statement, the ground state of a finite system cannot spontaneously break the symmetry of the Hamiltonian, as it is the case at a second-order QPT [113]. An intuition to this statement can be built from a perturbative point of view. On a finite system, there is always a finite-order perturbative process coupling the hypothetical ground states of a symmetry-broken phase. Due to this finite-order coupling, there will always be a level repulsion between the states and no true degeneracy occurs. This means that, on a finite system, there is always a finite-size energy gap between the ground state and the excitations. As discussed in Section 2.2 on a finite system, there is a pseudo-critical point close to the parameter values of the QPT in the thermodynamic limit. For a system with gapped phases on both sides of the QPT, we expect the relevant finite-size gap to be the smallest in the vicinity of the pseudo-critical point. The finite-size gap at the pseudo-critical point is expected scales as $L^{-z\varphi}$ with the linear system size L to the power of the dynamical correlation length exponent z [34,113,129,212] and $\varphi = 1$ for a QPT below the upper critical dimension (see Section 2.3). This statement can be directly derived from the finite-size scaling form of the energy gap:

$$\Delta_L(r) = L^{-z\varphi}\Psi_\Delta(L^{\varphi/\nu} r)\,. \tag{289}$$

Note that the scaling dependence for systems above the upper critical dimension has not been confirmed yet by numerical studies. To perform simulations at effective zero temperature, there are two main approaches in the contemporary literature. Firstly, one just scales the simulation temperature with L for different linear system sizes. This is a valid approach for many systems as $z = 1$ is a common value for the critical exponent, indicating a space–time symmetry in the correlations. For long-range interacting systems such as the LRTFIM or unfrustrated long-range Heisenberg models, one has $z \leq 1$, which makes the scaling sufficient, but overly ambitious [21]. An improved version of this naive approach is to scale the simulation temperature with $L^{z\varphi}$ for different linear system sizes. This follows the expected scaling of the finite-size gap. The correct prefactor of the scaling is not known and accounted for in the scaling, and corrections to scaling for small system sizes can lead to errors. A more sophisticated approach to determine a suitable effective zero temperature for a finite system is discussed in Refs. [32,34,225]. This technique was introduced in Ref. [225] as the beta-doubling method. The idea is to study the convergence of observables $\langle \mathcal{O}_L(r,\beta)\rangle$ for a fixed system size and parameter value r at successively doubled inverse temperatures $\beta = 2^n$ of the simulation. The fraction of $\langle \mathcal{O}_L(r,\beta)\rangle / \langle \mathcal{O}_L(r,\beta_{\max})\rangle$ is considered with $\beta_{\max}$ being the largest considered β value. This quantity is used to probe the convergence of the observable $\langle \mathcal{O}_L(r,\beta)\rangle$ towards zero temperature. The temperature convergence can be gauged using a plot, as demonstrated in Figure 19. Note that the value of $\langle \mathcal{O}_L(r,\beta_{\max})\rangle / \langle \mathcal{O}_L(r,\beta_{\max})\rangle$ is always at one. Therefore, as a rule of thumb, one can validate simulation temperatures for which at least the two points prior to the last point also have a value close to one.

Although the beta-doubling scheme has more overhead than a naive scaling with $L^{z_{\text{q}}}$, it provides multiple advantages. First of all, one does not need to know the dynamical critical exponents z beforehand. Secondly, in most cases, it is easy to implement: if one uses either way a simulated annealing scheme by successively cooling the simulation temperature during the thermalisation of the algorithm, the beta-doubling scheme can be easily incorporated in this procedure. Thirdly, the beta doubling provides a direct demonstration that a simulation is sampling zero-temperature properties. It captures intrinsically the relevant scaling $L^{z_{\text{q}}}$, as well as potentially large prefactors and corrections to this scaling. Fourthly, it does not need to be applied to all parameter values r, as it suffices to perform the beta-doubling procedure only for the parameter values where the relevant finite-size gap is expected to be the smallest and use this temperature for the whole parameter range.

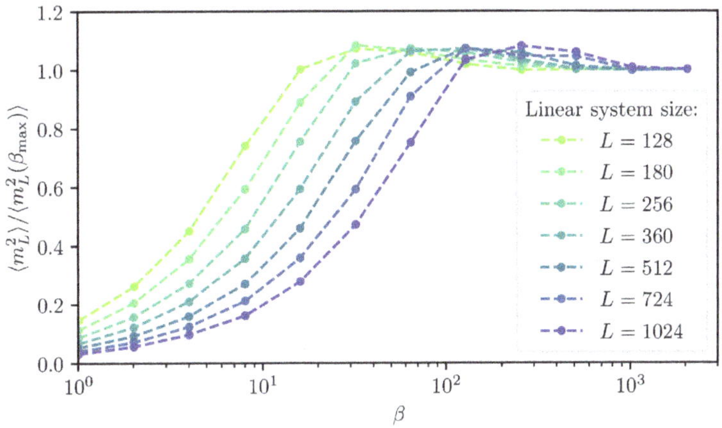

Figure 19. Illustration of the beta-doubling method for the one-dimensional LRTFIM in the short-range regime with decay exponent $\sigma = 2.5$ at a transverse field of $h = 1.25$. The simulation starts at $\beta = 1$ (leftmost points) for every system size. The inverse temperature is then doubled in every beta-doubling step until the maximum $\beta_{\max} = 2048$ is reached. All of the shown magnetisation curves seem to be converging to zero temperature. Larger systems with $L > 1024$ were discarded as they do not appear to be fully converged yet.

In general, the beta-doubling scheme can help to reduce computational cost by performing simulations at a tailored effective zero temperature. If computational time is not a critical factor, in many practical applications, one can use the temperature that is sufficient for the largest system and sample all smaller systems at the same temperature.

5.5. Overview: Path Integral Quantum Monte Carlo Algorithms for Long-Range Models

The QMC methods discussed so far are both based on the SSE formulation [39,213]. In parallel with the development of the SSE-based QMC methods for long-range systems, path integral (PI) QMC methods operating in continuous imaginary time have been introduced for extended Bose–Hubbard models with long-range density–density interactions [215,226–228] (see (290)). The main application of these methods in the context of long-range interactions is the determination and classification of ground states [64,229,230]. Since we focus on reviewing QPTs in long-range interacting systems, which are usually simulated using the SSE, we only skim over the basic concepts relevant for the PI QMC methods. One motivation for the development of these methods was the study of the Mott insulator to superfluid transition in the presence of disorder [231–233]. Further, long-range density–density interactions were added to the algorithms [226]. A great application case for these algorithms is the study of extended Bose–Hubbard models [64,229,230], which are an effective description of ultracold atomic or molecular gases trapped in optical lattice potentials [56,58,61,62,64,68,229,234–238]. For a comprehensive explanation on how to derive effective Bose–Hubbard models for these systems, see Refs. [56,236–238].

A basic Hamiltonian studied with the world-line QMC methods in this context [226–228] reads

$$\mathcal{H} = -\mu \sum_i n_i + \sum_{ij} V_{i,j} n_i n_j - t \sum_{\langle i,j \rangle} \left(a_i^\dagger a_j + a_j^\dagger a_i \right) \tag{290}$$

with (hard-core) bosonic degrees of freedom on the lattice site i described by creation $a_i^\dagger$, annihilation a_i, and particle number n_i operators. The coupling μ describes a chemical potential and $V_{i,j} \leq 0$ the density–density interaction between particles at sites i and j, and the hopping amplitude $t > 0$.

PI QMC methods formulated in imaginary time are based on the PI formulation of the partition function:

$$Z = \mathrm{Tr}\left\{ e^{-\beta\mathcal{H}} \right\} = \lim_{K\to\infty} \mathrm{Tr}\left\{ \left(e^{-\Delta\tau\mathcal{H}} \right)^K \right\} \quad \text{with} \quad \Delta\tau = \beta/K \tag{291}$$

using the Suzuki–Trotter decomposition [115]. The PI formulation is used to extend the configuration space to all imaginary-time trajectories $|\alpha(\tau)\rangle$ of computational basis states [39,226], which are periodically closed in β ($|\alpha(0)\rangle = |\alpha(\beta)\rangle$) and which are connected by an imaginary-time evolution ($\langle\alpha(\tau_2)|e^{-\mathcal{H}(\tau_2-\tau_1)}|\alpha(\tau_1)\rangle$). These methods can be implemented in discrete [239] or continuous [226,240] imaginary time.

In the SSE approach, the configuration space is extended using the operator sequences arising from a high-temperature expansion of the exponential in the partition function. These configurations can be regarded as a series of discretely propagated states $|\alpha(p)\rangle$ with an artificial propagation index p labelling its position in the sequence. This introduction of an additional dimension to obtain a configuration space that can be sampled using MC in the PI and SSE QMC approach hints that both approaches are closely related. Studies of the close relationships between the SSE and PI representations can be found in Refs. [211,241]. Concepts developed in one of the two pictures can be equivalently formulated in the other one. For example, the concept of loops [215,216,242] and directed loops [215,217,242] can be applied in both formulations. The sampling of the Heisenberg model as described in Section 5.2 can be transferred to the PI QMC picture, as it is a direct application of the directed loop idea [217,225].

Since we will focus in this review on quantum-critical properties of magnetic systems, we only briefly summarise the major applications of PI QMC methods for extended (hard-core) Bose–Hubbard models with dipolar density–density interactions [64,229,230]. The PI QMC is used in this field to determine ground-state phase diagrams studying the emergence of solid, supersolid, and superfluid phases [64,229,230]. It has been demonstrated in Ref. [64] that long-range interactions stabilise more solid phases than the chequerboard solid with a filling of $1/2$ present in systems with nearest-neighbour interactions. It has also been shown that long-range interactions lead to the emergence of supersolid ground states [64]. The great benefit of the QMC simulations in comparison to other methods (e.g., mean-field calculations [111]) is the quantitative nature of these methods. Therefore, the numerical study of these models is highly relevant, since experimental progress in cooling and trapping atoms and molecules with dipolar electric or magnetic moments [67,69,72,73,243] enables experiments to realise extended Hubbard models with long-range interactions [68]. We expect the implementation of SSE QMC techniques for this kind of model to be straightforward using the directed loop approach [217,225].

6. Long-Range Transverse-Field Ising Models

In this section, we review the ground-state quantum phase diagrams of the long-range transverse-field Ising model (LRTFIM) with algebraically decaying long-range interactions.

We emphasise how the Monte Carlo-based techniques introduced in this work are a reliable way to obtain critical exponents of quantum phase transitions (QPTs) in this model. The Hamiltonian of the LRTFIM is given by

$$\mathcal{H} = \frac{J}{2}\sum_{i\neq j}\frac{1}{|\vec{r}_i - \vec{r}_j|^{d+\sigma}}\sigma_i^z\sigma_j^z - h\sum_i \sigma_i^x \tag{292}$$

with Pauli matrices σ_i^κ and $\kappa \in \{x,z\}$ describing spins 1/2 located on the lattice sites $\vec{r}_i$. The Ising coupling is tuned by the parameter J. For $J > 0$, the Ising interaction is antiferromagnetic and, for $J < 0$, ferromagnetic. The amplitude of the transverse field is denoted by h. The positive parameter $(d+\sigma)$ governs the algebraic decay of the Ising interaction. Here, d denotes the spatial dimension of the system and σ is a tunable real-valued parameter. The limiting cases of the algebraically decaying long-range interaction are the nearest-neighbour interaction for $\sigma = \infty$ and an all-to-all coupling for $(d+\sigma) = 0$. As a side note, there is literature where $\alpha \equiv d + \sigma$ is used as a parameter to tune the decay of the long-range interaction. For this review, we stay with $d+\sigma$ in order to treat systems with different spatial dimensions d on the same footing.

The Hamiltonian (292) can be treated using the pCUT+MC method as described in Section 4. A generic, perturbative starting point for the LRTFIM is the high-field limit. The transverse field term is regarded as the unperturbed Hamiltonian and the Ising interaction as the perturbation. Using the Matsubara–Matsuda transformation [25,29,89]:

$$\sigma_i^x = 1 - 2b_i^\dagger b_i \qquad \sigma_i^z = b_i^\dagger + b_i \ , \tag{293}$$

the Hamiltonian (292) can be brought into a hard-core bosonic quasi-particle picture:

$$\mathcal{H} = \epsilon_0 N + \sum_i b_i^\dagger b_i + \sum_{i\neq j}\lambda_{i,j}\left(b_i^\dagger b_j + b_i^\dagger b_j^\dagger + b_i b_j^\dagger + b_i b_j\right) . \tag{294}$$

Here, ϵ_0 is a constant, N is the number of sites, and $b_i^{(\dagger)}$ is a hard-core bosonic quasiparticle annihilation (creation) operator. The perturbation parameters are given by

$$\lambda_{i,j} = \frac{J}{4h}\frac{1}{|\vec{r}_i - \vec{r}_j|^{(d+\sigma)}} . \tag{295}$$

In the pCUT language, the perturbation associated with the perturbation parameter λ can be written as $\mathcal{V} = T_{-2} + T_0 + T_2$, where the hopping processes $b_i^\dagger b_j$ are contained in T_0 and the pair creation (annihilation) processes are in T_2 (T_{-2}).

Further, the Hamiltonian (292) is of the form to straightforwardly apply the SSE QMC algorithm reviewed in Section 5.1 (see Equation (226)). We recall that this QMC algorithm is sign-problem-free for arbitrary, potentially frustrated, Ising interactions.

The studied effects of algebraically decaying long-range Ising interactions on the ground-state phase diagram can be categorised into three scenarios:

First, for ferromagnetic interactions, there is a QPT with $\mathbb{Z}_2$ symmetry breaking between a ferromagnetic low-field phase and a symmetric x-polarised high-field phase. The intriguing effect of long-range interaction is the emergence of three regimes with distinct types of universality of the QPT. For large values of the parameter $\sigma > 2 - \eta_{\mathrm{SR}}$, the universality class is the same as in the nearest-neighbour model. For intermediate σ values of $2d/3 < \sigma < 2 - \eta_{\mathrm{SR}}$, the critical exponents change as a function of σ and the criticality can be described by a non-trivial long-range theory. For $\sigma < 2d/3$, the long-range interaction lowers the upper critical dimension below the physical dimension of the model and the transition becomes a long-range mean-field transition. We review the recent studies investigating this model in Section 6.1 with a particular focus on the ϕ^4-theory in Section 6.1.1 and on numerical studies investigating the critical properties in Section 6.1.2. The aspects regarding FSS above the upper critical dimension are summarised in Section 6.1.3.

The second relevant category is antiferromagnetic Ising interactions where the nearest-neighbour couplings form a bipartite lattice. In this case, there is a QPT with $\mathbb{Z}_2$ symmetry breaking between an antiferromagnetic low-field phase and a symmetric x-polarised high-field phase. The current state of literature indicates no change in the universality class of the phase transition in dependence of the long-range interaction. We summarise and review the recent progress in studying this case in Section 6.2.

The last aspect is scenarios where the nearest-neighbour couplings do not form a bipartite lattice. In this case, the long-range interactions may substantially alter the quantum phase diagram of the nearest-neighbour model. We summarise and review the recent progress regarding this case in Section 6.3.

6.1. *Ferromagnetic Long-Range Transverse-Field Ising Models*

The ferromagnetic TFIM with nearest-neighbour interactions in $d \geq 1$ dimensions is a paradigmatic model to display a QPT in a $(d+1)$D-Ising universality class. This transition describes the non-analytic change in the ground-state between a $\mathbb{Z}_2$ symmetry-broken low-field phase and a symmetric x-polarised high-field phase. By adding algebraically decaying long-range interactions to this model, one can study how such interactions can alter non-universal, as well as universal properties of a QPT and which new features can emerge.

In the case of ferromagnetic interactions, the algebraically decaying long-range interaction stabilises the symmetry-broken phase and the QPT shifts to larger h values [25,29,32,34]. An intuitive way to understand this behaviour is that, due to the long-range interaction, more connections are introduced that align the spins in the z-direction, and the energy cost for a single spin flip increases. In the limit $\sigma \to 0$, the critical value of the transverse field diverges as the cost for a single spin flip becomes extensive.

In this review, we focus exclusively on the regime of so-called weak long-range interactions with an algebraic decay exponent $\sigma > 0$ [21]. The research field of strong long-range interactions considers $\sigma \leq 0$ [5,42–45]. For weak long-range interactions, the Hamiltonian (292) is well defined in the thermodynamic limit, and common properties such as the ground-state additivity and thermodynamic quantities are well defined [5]. For strong long-range interactions, the energy of the ferromagnetically aligned state is superextensive, and in order to study the model in the thermodynamic limit, an appropriate rescaling of the Ising coupling with the system size is required [5].

The significance of the ferromagnetic LRTFIM comes from its paradigmatic nature of displaying changes in the universal behaviour due to the long-range interaction [20,21,25,29,32]. It has been known since the advent of the theory of classical phase transitions that algebraically decaying long-range interactions are a potential knob to alter the fixed-point structure of the renormalisation flow and, therefore, change the universal properties (e.g., critical exponents) [1,2,140]. Coming from the limit $\sigma = \infty$ of nearest-neighbour interactions, the QPT of a d-dimensional model remains in the $(d+1)$D-Ising universality class until $\sigma = 2 - \eta_{\rm SR}$ ($\eta_{\rm SR}$ is the anomalous dimension critical exponent of the nearest-neighbour universality class). This means that the critical exponents are constant as a function of σ and the fixed point associated with the QPT remains the one of the nearest-neighbour model. For $\sigma = 2 - \eta_{\rm SR}$, the RG fixed point associated with the QPT changes to the one of a non-trivial long-range interacting theory, and the critical exponents become σ-dependent [21]. The upper critical dimension of this theory with long-range interaction is lowered with respect to the short-range model as $d_{\rm uc} = 3\sigma/2$. If the (fixed) spatial system dimension d is larger than or equal to the σ-dependent upper critical dimension, the universality class describing the QPT enters a long-range mean-field regime. The three different regimes, as well as their boundaries in dependence of the dimensionality of the system are visualised in Figure 20.

In the following, we recapitulate the basic field-theoretical arguments leading to the distinction between three universality regimes (see Section 6.1.1). We will further emphasise the most relevant aspects for the two long-range regimes. In the non-trivial, intermediate regime, the precise values of the critical exponents depend on σ and need to be determined numerically. Therefore, we review the recent numerical studies computing the critical exponents in Section 6.1.2. To conclude the discussion, we outline how to utilise ferromagnetic long-range Ising models to study QPTs above the upper critical dimension (see Section 6.1.3).

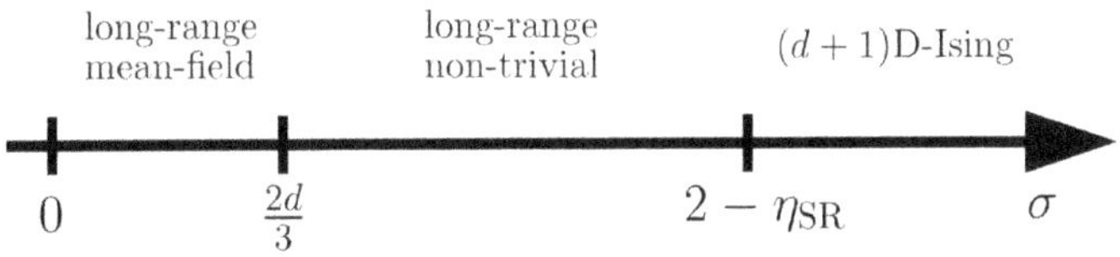

Figure 20. Sketch of the three distinct universality regimes of the QPT in the ferromagnetic LRTFIM. For one- and two-dimensional systems, all three regimes exist, and the boundaries can be obtained using the expressions in the figure. For $d \geq 3$, there is only the long-range mean-field and the nearest-neighbour mean-field regime with a boundary at $\sigma = 2$.

6.1.1. ϕ^4-Theory for Quantum Rotor Models with Long-Range Interactions

Starting with the short-range interacting action (see Equation (A3)) of the n-component quantum rotor model, Dutta et al. [20] introduced an action for long-range interacting rotor models, by adding the algebraic decay between interacting fields as an additional term to the action in real space. Note the the action for the Ising model is equivalent to the one-component (scalar) rotor model [113]. By performing a Fourier transformation of the resulting action, the authors obtained the action:

$$\begin{aligned} S_{\tilde{\phi}} = &\int \frac{d^d q}{(2\pi)^d} \int \frac{d\omega}{2\pi} [\tilde{g}\omega^2 + r + aq^\sigma + bq^2] \tilde{\phi}(q, i\omega) \tilde{\phi}(-q, -i\omega) \\ &+ u \int \frac{d\omega_1}{2\pi} \cdots \frac{d\omega_4}{2\pi} \int \frac{d^d q_1}{(2\pi)^d} \cdots \frac{d^d q_4}{(2\pi)^d} \delta^d(q_1 + \cdots + q_4) \delta(\omega_1 + \cdots + \omega_4) \times \\ &\times [\tilde{\phi}(q_1, i\omega_1) \tilde{\phi}(q_2, i\omega_2)] [\tilde{\phi}(q_3, i\omega_3) \tilde{\phi}(q_4, i\omega_4)] \end{aligned} \tag{296}$$

with $a, b > 0$, σ from the decay exponent $d + \sigma$ of the microscopic model, and r, u coupling constants as in the nearest-neighbour case. The q^σ term, which is new compared to the nearest-neighbour theory, results from the Fourier transformation of the added long-range interacting term:

$$\int \mathrm{d}^d x \int \mathrm{d}^d y \int \mathrm{d}\tau \, \frac{\phi(x, \tau)\phi(y, \tau)}{|x - y|^{d+\sigma}} \tag{297}$$

of the order-parameter field ϕ [40]. As the relevant modes for the QPT are the ones with long wavelengths, it is clear that, for $\sigma \geq 2$, the q^2 term gives the leading contribution in q; the q^σ term can be neglected and, therefore, the system is, in this case, described by the short-range interacting theory. In contrast to this, a different critical behaviour distinct from the short-range case is possible for $\sigma < 2$. With the same argument as above, the q^2 term becomes negligible for $\sigma < 2$. In the following, we will focus on this regime, where the behaviour differs from the nearest-neighbour case. In order to gain insight into the QPT affected by long-range interactions, an investigation of the Gaussian theory is a good starting point to enter the framework of perturbative renormalisation group calculations [20,21,113] like the ϵ-expansion in Ref. [20]. From Equation (296), one can directly read off the propagator of the Gaussian theory as

$$G_0(q, \omega)^{-1} = q^\sigma + \tilde{g}\omega^2 + r \,. \tag{298}$$

From field-theoretical power-counting arguments, Dutta et al. [20] derived several properties of the Gaussian theory. Using the divergence of the mass renormalisation term, the lower critical dimension below which no phase transition is occurring can be derived as

$d_{\mathrm{lc}} = \sigma/2$ [20]. By regarding the Gaussian propagator, one finds directly that $\eta = 2 - \sigma$ as η is defined by the deviation from 2 of the leading power by which the momentum enters the propagator (see Equation (A37)). Note that, in the long-range case, the Gaussian theory has an $\eta \neq 0$. From further power-counting analysis, it is possible to derive more critical exponents from the Gaussian theory [20]:

$$\gamma = 1 \qquad \nu = \frac{1}{\sigma} \qquad z = \frac{\sigma}{2} \qquad \eta = 2 - \sigma. \tag{299}$$

In the limit $\sigma \to 2$, the short-range mean-field exponents are recovered [20]. With the argument that the hyperscaling relation still holds directly at the upper critical dimension, Dutta et al. [20] derived the upper critical dimension by inserting the long-range Gaussian exponents (see Equation (299) and $\alpha = 0$).

$$2 - \alpha = \nu(d + z) \longrightarrow 2 = \frac{1}{\sigma}(d_{\mathrm{uc}} + \frac{\sigma}{2}) \Leftrightarrow d_{\mathrm{uc}} = \frac{3\sigma}{2}. \tag{300}$$

In addition to the investigation of the Gaussian part of the action in Equation (296), one can derive non-trivial exponents below the upper critical dimension by deriving perturbative corrections to the Gaussian exponents in an ϵ-expansion around the upper critical dimension [113]. Dutta et al. [20] established with a one-loop renormalisation group expansion of flow equations that the Gaussian fixed point is stable for $d \geq d_{\mathrm{uc}}$. Therefore, the long-range Gaussian exponents (see Equation (299)) are valid for $\sigma \leq 2D/3$. For $d < d_{\mathrm{uc}}$, the correction indicates a flow to a non-trivial fixed point. The first-order corrections of the ϵ-expansion for ν and γ are provided [20], and the first-order corrections to η and z are argued to be zero [20]. It should be noted that the dynamic correlation exponent $z \neq 1$ for $\sigma < 2$ and, therefore, correlations are not isotropic in the space and imaginary-time direction [20,21]. "For any value of $\sigma < 2$, η sticks to its mean-field value $2 - \sigma$, because the renormalization does not generate new q^{σ} terms" [20]. This also fits well with the claim that $\sigma = 2 - \eta_{\mathrm{SR}}$, because, then, there is no discontinuity at the boundary [1–3,21,244].

A recent study by Defenu et al. [21] with a functional renormalisation group approach investigated the flow of couplings in an effective action with anomalous dimension effects. This analysis showed that the boundary between the short-range and the non-trivial long-range regime does not occur at $\sigma = 2$, but at $\sigma = 2 - \eta_{\mathrm{SR}}$. So, it is shifted towards lower σ by the anomalous dimension of the short-range criticality. This anomalous dimension consideration in [21] is mainly driven by the following argument: From the functional renormalisation group ansatz, the authors derived two flow equations for the coupling of the long-range coupling Z_k (connected to a in Equation (296)) and the short-range coupling $Z_{2,k}$ (connected to b in Equation (296)):

$$\partial_t Z_k = (2 - \sigma - \eta) Z_k \tag{301}$$

$$\partial_t \eta = \frac{\partial_z Z_{2,k}}{Z_{2,k}}. \tag{302}$$

The first flow Equation (301) has fixed points for $\eta = 2 - \sigma$ or $\lim_{k \to 0} Z_k = 0$. The first fixed point results in long-range exponents, whereas the second fixed point means that the long-range coupling becomes irrelevant.

In addition to this anomalous dimension effect, the authors in [21] also calculated critical exponents in the non-trivial intermediate phase with their functional renormalisation group technique. They also compared their results with Monte Carlo simulations of the dissipative non-Ohmic spin chains, which can be mapped to the same behaviour as the long-range interacting quantum Ising model [245], showing reasonable agreement.

6.1.2. Critical Exponents and Critical Points for One- and Two-Dimensional Systems

The criticality of the ferromagnetic LRTFIM has been studied by a wide variety of methods with different strengths and weaknesses. Overall, all methods qualitatively agree with each other and confirm the three expected regimes from the field-theoretical analysis [20,21], namely a long-range mean field, a short-range regime, and an intermediate non-trivial long-range regime, which connects the two limiting regimes with monotonously changing exponents. In the limit $\sigma \to 0$, the value of the critical field h_c diverges as the energy cost for a single spin flip in the low-field phase becomes extensive and the quantum fluctuations introduced by the transverse field fail to compete with the Ising interaction. In the limit $\sigma \to \infty$, the value of h_c converges towards the exact value for the nearest-neighbour transverse-field Ising model [25,29,32,246].

In the case of the LRTFIM on the one-dimensional linear chain, the model has been studied by perturbative continuous unitary transformation with Monte Carlo embedding (pCUT + MC) [29,31,34,247], the functional renormalisation group (FRG) [21], the density matrix renormalisation group (DMRG) [28,248], as well as stochastic series expansion (SSE) [32,34,247,249], path integral (PI) [33] QMC, and QMC with stochastic parameter optimisation (QMC+SPO) [250].

The critical value of the transverse field was calculated with the pCUT method [29,31,34] by estimating the gap closing using DlogPadé extrapolation of the perturbative gap series coming from the high-field phase, by the DMRG using finite-size scaling of the fidelity susceptibility [28], and by SSE QMC calculations using finite-size scaling of the magnetisation [32,249]. The finite-temperature transition points were studied by PI QMC calculations [33] for $\sigma = 0.5$ in the long-range mean field regime and $\sigma = -0.95$ in the strong long-range regime.

Depending on the method, different critical exponents were extracted. Only in the case of pCUT + MC and SSE QMC calculations[34,247], all three regimes were investigated. By extracting three independent critical exponents, the authors were able to provide the full set of critical exponents for each method individually. In the case of the pCUT + MC approach [34,247], the exponents were computed by (biased) DlogPadé extrapolants of high-order series of the gap ($z\nu$), the one-quasiparticle static spectral weight ($(2 - z - \eta)\nu$), and the control-parameter susceptibility (α). In SSE QMC studies [34,247], the exponents were calculated using data collapses of the magnetisation (β/ν and ν) and the order-parameter susceptibility (γ/ν). The DMRG study [28] applied the method of data collapse to the fidelity susceptibility to extract the single exponent ν in the non-trivial long-range regime. Another DMRG study [248] extracted critical exponents by using finite-size scaling for the energy gap (z) and the squared order parameter ($2\beta/\nu$ and ν). Using the FRG [21], the exponents ν and z were calculated, from which one can additionally calculate the exponent α. In the PI QMC study [33], the finite-temperature criticality of the LRTFIM was investigated for $\sigma = 0.5$ in the long-range mean field regime and $\sigma = -0.95$ in the strong long-range regime. The study includes the extraction of an exponent θ_t, which can be related to ν, from the Binder cumulant and classical order-parameter susceptibility via finite-size scaling, as well as the exponent $\gamma\theta_t$ from finite-size scaling of the classical order-parameter susceptibility. In Figure 21, the critical field h_c and the canonical critical exponents are plotted for the discussed studies.

The two-dimensional case of the square lattice is less studied. The critical values of the transverse field were only calculated by the pCUT + MC [29] and SSE QMC method [32,249]. There is so far no study that has extracted the full set of critical exponents altogether. With pCUT [29], the gap exponents $z\nu$ were also extracted by DlogPadé extrapolants of the gap series. In the FRG study [21], the critical exponents z and ν were calculated, and in the SSE QMC study [32], the critical exponents β and ν were extracted from the data collapses of the magnetisation. These results are summarised in Figure 22. Note that we present the results from pCUT + MC calculations from Ref. [29] with one additional order to the maximal order in the perturbation parameter.

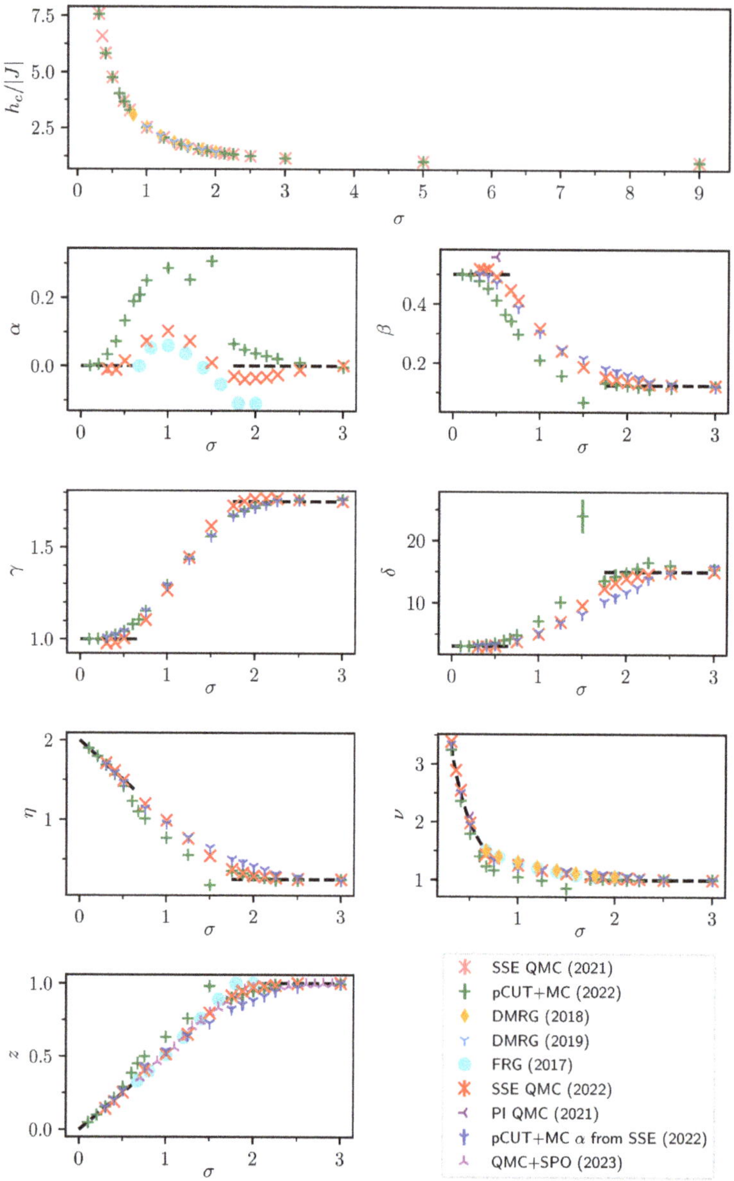

Figure 21. Critical field values and exponents from numerical studies of the ferromagnetic LRTFIM on the linear chain. The panels display $h_c/|J|$ (top), α (second row left), β (second row right), γ (third row left), δ (third row right), η (fourth row left), ν (fourth row right), and z (bottom). The labels refer to the references in the following way: "SSE QMC (2021) [32,249]", "SSE QMC (2022) [34,247]", "pCUT+MC (2022) [34,247]", "pCUT+MC α from SSE (2022) [34,247]", "DMRG (2018) [28]", "DMRG (2019) [248]", "FRG (2017) [21]", "PI QMC (2021) [33]", and "QMC+SPO (2023) [250]". The values for the critical exponents of the transition in the nearest-neighbour model $\alpha = 0$, $\beta = 1/8$, $\gamma = 7/4$, $\delta = 15$, $\eta = 1/4$, $\nu = 1$, and $z = 1$ [246,251,252] and in the long-range mean-field regime $\alpha = 0$, $\beta = 1/2$, $\gamma = 1$, $\delta = 3$, $\eta = 2 - \sigma$, $\nu = 1/\sigma$, and $z = \sigma/2$ [20,21] are given by the dashed lines.

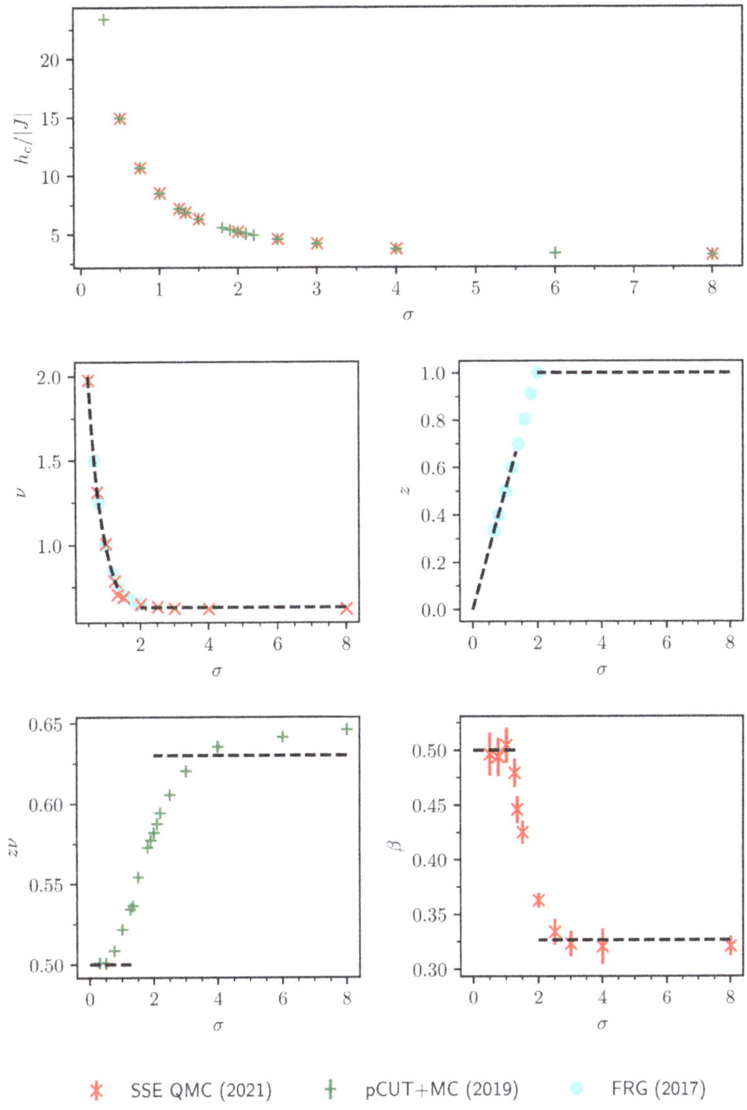

Figure 22. Critical field values and exponents from numerical studies of the ferromagnetic LRTFIM on the two-dimensional square lattice. The panels display $h_c/|J|$ (top), ν (middle left), z (middle right), $z\nu$ (bottom left), and β (bottom right). The data points "SSE QMC (2021)" for h_c/J, ν, and β are from Ref. [32,249]. The data points "pCUT+MC (2019)" for h_c/J and $z\nu$ are from Ref. [29]. The data points "FRG (2017)" for ν and z originate from the functional RG study in Ref. [21]. The values for the critical exponents of the transition in the nearest-neighbour model $\nu = 0.629971(4)$, $\beta = 0.326419(3)$, and $z = 1$ [253] and in the long-range mean-field regime $\nu = 1/\sigma$, $z = \sigma/2$, and $\beta = 0.5$ [20,21] are given by the dashed lines.

Overall, the different methods all perform well even for relatively small σ. For the pCUT and SSE QMC studies, where all three regimes were simulated, the limiting cases of long-range mean-field criticality and nearest-neighbour criticality were correctly reproduced, which underlines the reliability of the presented results in the intermediate regime. The main challenge that the methods are facing is to sharply resolve the regime boundaries due to rounding effects, which are probably due to the limited length scales in the simulation. Additionally, the boundary to the long-range mean field regime is spoiled by logarithmic corrections to scaling at the upper critical dimension. Even in the case of pCUT + MC, which is a method operating in the thermodynamic limit, there is an intrinsic limit of the length scale due to the finite order of the series. In particular, the rounding makes it hard to verify (or falsify) the claim that the boundary between the short-range and non-trivial long-range regime is shifted $2 \to 2 - \eta_{\text{SR}}$. For 1D, the boundary is shifted to $\sigma = 1.75$, which could be resolvable in extensive simulation. However, in 2D, the boundary is only marginally shifted to $\sigma \approx 1.964$, which is probably not resolvable by the reviewed methods. The best boundary resolution is probably given by the SSE QMC study [34]. If one is interested in studying the shift of the boundary, one could, therefore, push the SSE QMC simulations of the one-dimensional chain to larger systems for $\sigma \lessapprox 2$. Another difficulty for the methods operating on finite systems is the breakdown of the common hyperscaling relation and standard FSS above the upper critical dimension, which affects the processing of data in the long-range mean-field regime for $\sigma < 2d/3$. The scaling above the upper critical dimension was already outlined in Section 2.3. However, we will recapitulate the most important points in the following Section 6.1.3, not only because this modified scaling was used in several of the above-mentioned methods to extract critical exponents [32–34], but also because the LRTFIM constitutes a testing ground for quantum Q-FSS scaling and the modified scaling of the correlation length was confirmed in a study of the LRTFIM [34].

6.1.3. Scaling above the Upper Critical Dimension

As discussed in Section 6.1.1, for $\sigma < 2 - \eta_{\text{SR}}$, the action describing the QPT and its universality changes from short range to long range. In this long-range regime, the upper critical dimension $d_{\text{uc}} = 3\sigma/2$ was found to depend on the value of σ [20,21]. This means that the connectivity of the system increases for decreasing σ and a smaller spatial dimension $d < 3$ is already sufficient to reach or even exceed the upper critical dimension. The sigma below which the model is above its upper critical dimension is $\sigma_{\text{uc}} = 2d/3$. In this regime, it is not possible to correctly extract all critical exponents from the standard FSS, and the common hyperscaling relation:

$$2 - \alpha = (d + z)\nu \tag{303}$$

becomes invalid as the critical exponents become independent of the dimension d. The origin of this discrepancy comes from the effect of dangerous irrelevant variables that one has to take into account in the derivation of scaling above the upper critical dimension. When doing so, one obtains a modified hyperscaling relation:

$$2 - \alpha = \left(\frac{d}{\varrho} + z\right)\nu \tag{304}$$

with the exponent $\varrho = \max(1, d/d_{\text{uc}})$. This relation can be used for the conversion of different critical exponents [34,35].

Additionally, the standard FSS scaling form of an observable $\mathcal{O}$ with power law singularity $\mathcal{O}(r, L^{-1} = 0) \sim |r|^{\omega}$ is modified to [34]

$$\mathcal{O}(r, L^{-1}) = L^{-\omega\varrho/\nu}\Psi(L^{\varrho/\nu}r) \tag{305}$$

following the formulation of the Q-FSS for classical [15–19,41] and quantum [34] systems. Finite-size methods that rely on the method of data collapse or other FSS-based techniques have to use this adapted formula to extract all the critical exponents successfully [34]. For a more elaborate description of quantum Q-FSS, please refer to Section 2.3 or to Ref. [34].

On the other side, the LRTFIM can also be used as a testing ground for quantum Q-FSS. One key difference that distinguishes the Q-FSS from standard FSS is that the correlation sector is also affected by the DIV in the case of the Q-FSS. This leads to a modified scaling of the characteristic length scale $\xi_L(r) = L^{\varphi}\Xi(L^{\varphi/\nu}r)$ with the system size instead of $\xi_L(r) \sim L\Xi(L^{1/\nu}r)$, as it is for standard FSS. The characteristic length scale was measured in the SSE QMC study [34] in the long-range mean-field regime, and the exponent φ was extracted by a data collapse with fixed mean field value $\nu = \sigma^{-1}$. The extracted values coming from Ref. [34] are presented in Figure 23 and are clearly in line with the predictions made by the Q-FSS.

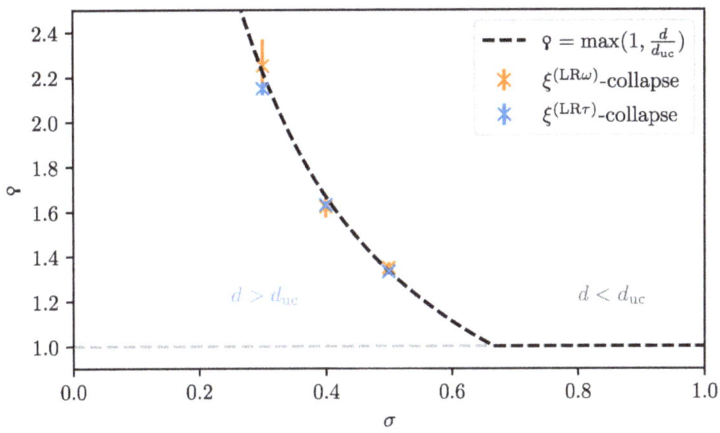

Figure 23. Exponent φ extracted by the data collapse of $\xi^{(LR\omega)}$ and $\xi^{(LR\tau)}$ for different decay exponents of the LRTFIM on the linear chain [34]. The black dashed line depicts the prediction by the Q-FSS $\varphi = \max(1, d/d_{uc})$, while the grey dashed line shows the prediction from standard FSS $\varphi = 1$. In the regime above the upper critical dimension for $\sigma < 2/3$, the predictions start to deviate and the extracted values for φ are clearly in line with the Q-FSS scenario. Figure adapted from Ref. [34].

6.2. *Antiferromagnetic Long-Range Transverse-Field Ising Models on Bipartite Lattices*

In this section, we review results for the antiferromagnetic LRTFIM on lattices where the coupling structure of the nearest-neighbour couplings forms a bipartite lattice. A lattice—or, more generally, a graph—is called bipartite if the sites can be split into two disjoint sets A and B so that there are no edges between sites within each of the two sets A and B. In the case of a lattice, A and B are called bipartite sublattices. Note that, only in the nearest-neighbour limit on bipartite lattices, there is an exact duality between the ferromagnetic and antiferromagnetic TFIM, and the quantum-critical properties coincide. This duality comes from a sublattice rotation of π along the x-axis for the spins in one of the two sublattices. In the nearest-neighbour limit, there is an x-polarised symmetric high-field phase and a $\mathbb{Z}_2$ symmetry-broken low-field phase. The low-field phase is adiabatically connected to the zero-field limit. In the zero-field limit, there are two ground states associated with the states where the spins on the sites in one set of the bipartite sublattices are pointing in one direction while the others point in the other direction. Note that these states can be directly mapped onto the two ferromagnetic ground states by the above-mentioned sublattice rotation. As for the ferromagnetic nearest-neighbour TFIM, the QPT between these two phases is of $(d+1)$D-Ising universality.

The antiferromagnetic long-range interaction beyond the nearest-neighbours induces a hierarchy of competing interactions. Due to the long-range interactions, there is no longer a bipartite coupling structure as the spins cannot be aligned such that the energy of all interactions is minimised. Nevertheless, there is evidence that the two ground states in the low-field phase of the nearest-neighbour limit are adiabatically connected to the ground states in the low-field phase for all $(d+\sigma) > 0$ [23–26,29,31,32,111].

For the all-to-all connected case $(d+\sigma) = 0$, every state with zero z-magnetisation is a ground state of the system [30,40] at $h = 0$. The Hamiltonian (292) can then be written in terms of total spin $\sigma^{\kappa}_{\mathrm{tot}} = \sum_i \sigma^{\kappa}_i$ with $\kappa \in \{x, z\}$ as

$$\mathcal{H} = -h\sigma^{x}_{\mathrm{tot}} + \frac{J}{2}(\sigma^{z}_{\mathrm{tot}})^2 - \frac{J}{2}N \tag{306}$$

with the total number of spins $N \to \infty$ [30,40]. For all finite transverse fields $h > 0$, the ground state of Equation (306) is directly located in the x-polarised phase [40,254].

The natural questions arising for the QPT in the antiferromagnetic LRTFIM on the bipartite lattices are similar to the ferromagnetic LRTFIM: How does the algebraically decaying long-range interaction influence the critical field value of the QPT? Is there a change in the universality class of the QPT depending on the precise values of the decay exponent $(d+\sigma)$?

Regarding the critical field values in one-dimensional chains, there are MC-based studies using the SSE QMC (see Section 5) [32] and pCUT+MC (see Section 4) [25,31], as well as the DMRG [24,26,248] and matrix product states [23]. We summarise the results regarding critical field values and selected critical exponents of the studies mentioned above in Figure 24. In two-dimensional systems, there are MC-based studies for the square lattice using SSE QMC (see Section 5) [32] and pCUT+MC (see Section 4) [29] (for selected results, see Figure 25). Variational methods in two dimensions such as projected entangled pair states [255,256] or DMRG calculations [257] have not been applied for antiferromagnetic systems on bipartite lattices. Entanglement-based methods become more challenging from the increased entanglement due to the area law in two dimensions [258]. Further, there are reports about the violation of the area law for one-dimensional systems with long-range interactions [23,24]. On general grounds, one cannot efficiently represent algebraically decaying long-range interactions in these techniques [259,260]. However, there exist DMRG calculations for antiferromagnetic LRTFIMs on non-bipartite lattices [27,54,55,109].

In general, it has been numerically demonstrated that the critical field values decrease monotonically as a function of σ from the nearest-neighbour value at $\sigma = \infty$ to zero at $(d+\sigma) = 0$ [23–26,29,31,32,248]. Regarding the QPT on the antiferromagnetic chain, the established picture is that it is of the $(1+1)$D-Ising type for all $\sigma \geq 1.25$ [23–26,31,32,248]. In the regime of ultra-long-range couplings $\sigma \leq 1.25$, recent finite-size DMRG findings by G. Sun [26] and R. Puebla et al. [248] suggest that the $(1+1)$D-Ising universality even holds for any $\sigma > -1$. The only study indicating a change in criticality below the boundary of $\sigma \leq 1.25$ uses matrix product states generalising the time-dependent variational principle (TDVP) [23]. Note that there is no further theoretical motivation for this boundary besides this single numerical study, which does not provide convincing evidence for its existence. The more recent results from Refs. [26,248] challenge these findings and establish the $(1+1)$D-Ising universality for all $\sigma > -1$. The studies using MC-based techniques [25,31,32] cannot extract reliable critical exponents in this regime. In the SSE QMC study [32], the increasing competition of the long-range interactions is the major problem making the numerical simulations in this regime impractical. If the long-range interaction becomes more prominent, this leads to similar algorithmic challenges as in frustrated systems [32,40,261,262]. The autocorrelation times of the SSE QMC algorithm increase as more and more bond operators are present in the operator sequence, and the field operators diffuse only slowly [32]. Nevertheless, following the trend of the results in Ref. [32], the scenario of a single universality class across the entire σ range is plausible. Regarding the pCUT+MC studies [25,31], the main obstacle to extract information about the quantum criticality in the regime $\sigma < 1$ is the fact that the critical field value is decreasing. Refs. [25,31] use a perturbative expansion around the high-field limit; therefore, if the critical field value increases, then the extrapolation of the gap series becomes increasingly challenging.

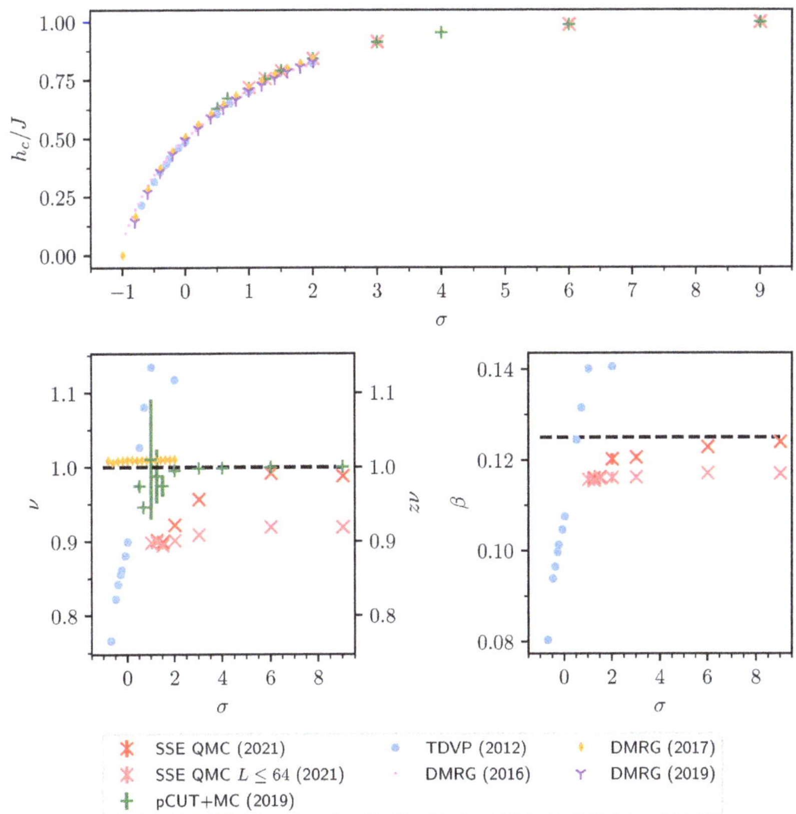

Figure 24. Critical field values and exponents from numerical studies of the antiferromagnetic LRTFIM on the chain. The upper panel displays critical field values h_c/J; the lower left panel displays values for the critical exponents ν and $z\nu$ and, the lower right panel, values for the critical exponents β. The data points "SSE QMC (2021)" and "SSE QMC $L \leq 64$ (2021)" for h_c/J, ν, and β are from Ref. [32,249]. The data points "pCUT+MC (2019)" for h_c/J and $z\nu$ are from Ref. [29]. The data points "TDVP (2012)" for h_c/J, ν, and β originate from Ref. [23]. The exponents ν and β are calculated from the scaling dimensions in Ref. [23] under the assumption of $z = 1$, which is reasonable according to [248]. The data points "DMRG (2016)" for h_c/J are from Ref. [24]. The data points "DMRG (2017)" for h_c/J and ν originate from Ref. [26]. The data points "DMRG (2019)" for h_c/J are from Ref. [248]. The values for the critical exponents of the transition in the nearest-neighbour model $\nu = 1$, $\beta = 1/8$, and $z = 1$ [246,251,252,263] are given by the black dashed lines.

To summarise, there are studies indicating a change from the $(1+1)$D-Ising criticality on the chain for small σ values [23,24]. These studies also report a possible breakdown on the area law of the entanglement entropy for the ground state in this regime [23,24], which is a vital prerequisite for the methods they implemented. Two more recent studies [26,248] demonstrate no change in the universality class, which is backed by [25,30,31]. For two-dimensional systems, the universality class is demonstrated to be $(2+1)$D-Ising for $\sigma > 0.5$ [29,32], and no trend is reported that the universality class should change for smaller values of σ.

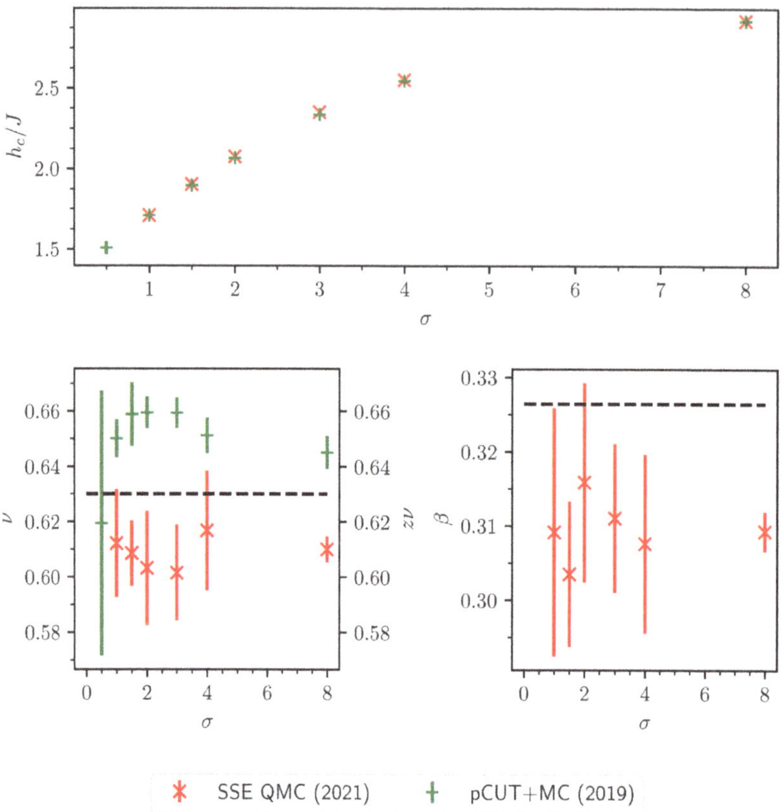

Figure 25. Critical field values and exponents from numerical studies of the antiferromagnetic LRTFIM on the square lattice. The upper panel displays critical field values h_c/J; the lower left panel displays values for the critical exponents ν and $z\nu$ and, the lower right panel, values for the critical exponents β. The data points "SSE QMC (2021)" for h_c/J, ν, and β are from Ref. [32,249]. The data points "pCUT+MC (2019)" for h_c/J and $z\nu$ originate from Ref. [29]. The values for the critical exponents of the transition in the nearest-neighbour model $\nu = 0.629971(4)$, $\beta = 0.326419(3)$, and $z = 1$ [253] are given by the dashed lines.

In general, a great algorithmic challenge that remains for the MC-based methods with regard to the antiferromagnetic LRTFIM on bipartite lattices is a reliable study of the regime of small values $(d + \sigma)$. Here, as discussed above, all of the commonly applied methods have a handicap in some way.

6.3. Antiferromagnetic Long-Range Transverse-Field Ising Models on Non-Bipartite Lattices

We start with an antiferromagnetic nearest-neighbour interacting TFIM on a non-bipartite lattice (e.g., the sawtooth chain, triangular lattice, Kagome lattice, pyrochlore lattice) at zero field. In the case of non-bipartite lattices, one can always find an odd number of Ising bonds (edges) that form a closed loop. The presence of a loop of odd length means it is impossible to satisfy all antiferromagnetic couplings simultaneously. The phenomenon of not being able to minimise all interactions due to geometrical lattice constraints is called geometrical frustration [264]. A notion for the strength of geometric frustration can be defined using the resulting ground-state degeneracy [265]. A theoretical tool to access the ground-state degeneracy is through Maxwell counting arguments [265–267]. For systems with a large degree of frustration, e.g., the antiferromagnetic nearest-neighbour TFIM on the triangular or Kagome lattice, there is a finite residual entropy per site in the thermodynamic limit [168,268–271].

In general, extensively degenerate ground-state spaces due to frustration pose a great resource for emergent exotic quantum phenomena. Quantum fluctuations introduced, e.g., by a transverse field result in a breakdown of the extensive ground-state degeneracy and the potential emergence of non-trivial ground states [270,271]. In general, it is useful to think about the low-field physics as a degenerate perturbation theory problem on the zero-field ground-state space [30,32,270,271]. We will later review that this is a reasonable framework in order to treat the breakdown of the degenerate subspace due to fluctuations and algebraically decaying long-range interactions on the same footing [30,32,112].

Perturbing extensively degenerate ground-state spaces with fluctuations may result in several distinct scenarios. First, a distinct symmetry-broken order can emerge for infinitesimal perturbations (order-by-disorder) [168,270–273]. Further, a direct realisation of a symmetry-unbroken phase may occur (disorder-by-disorder). This phase can either be trivial [168,274] or exotic, e.g., quantum spin liquids [99–101,275,276].

For the first part of this section, we review the antiferromagnetic LRTFIM on the triangular lattice [27,29,30,32,40,110,277]. In the nearest-neighbour limit of that model, the zero-field case has an extensively degenerate ground state [268,270,271,273]. When adding an infinitesimal transverse field, this ground-state degeneracy breaks down and a $\mathbb{Z}_2 \times \mathbb{Z}_3$-symmetry-broken clock order emerges from an order-by-disorder mechanism [270,271,273]. The effective Hamiltonian describing the breakdown of the degeneracy in leading order can be expressed as a quantum dimer model [270,271,273]. By investigating the degenerate ground-state space, it was observed that there are local spin configurations in which spins can be flipped without leaving the ground-state space. These configurations, called flippable plaquettes, consist of a spin surrounded by six spins with alternating orientation. It can be easily seen that flipping the spin in the middle of the flippable plaquette results in turning the plaquette by $\pi/3$ (see Figure 26).

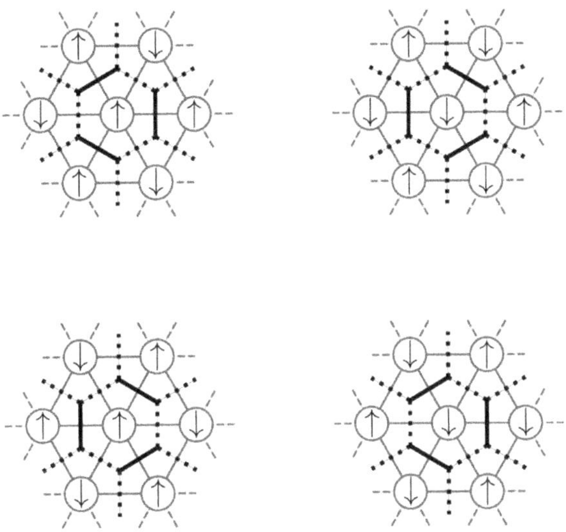

Figure 26. Illustration of the four local spin configurations called flippable plaquettes. The arrows denote the local spin orientation in the z-direction. The grey lines in the back visualise the triangular lattice. Solid (dotted) black lines are depicted on (anti)ferromagnetically aligned bonds. Note that flipping the spin in the centre of each configuration maps the left to the right configuration in each row and vice versa.

The effective low-field quantum dimer model [270,271,273], consisting of the plaquette rotation term, is, therefore, the hexagonal-lattice version of the Rokhsar–Kivelson quantum dimer model with $t = h$ and $v = 0$ [270,271,278]:

$$H_{\mathrm{QDM}} = t\Big(\big|\mathrm{I}\!>\big\rangle\big\langle <\!\mathrm{I}\big| + \mathrm{h.\,c.}\Big) + v\Big(\big|\mathrm{I}\!>\big\rangle\big\langle \mathrm{I}\!>\big| + \big|<\!\mathrm{I}\big\rangle\big\langle <\!\mathrm{I}\big|\Big)\,. \qquad (307)$$

From this mapping, the nature of the state at $h > 0$ was predicted, as well as the breakdown of the order with a 3D-XY QPT [270,271]. Further numerical studies extended the insights into this system [168,262,273]. The critical field value at which the phase transition between the symmetric x-polarised paramagnetic high-field phase and the symmetry-broken gapped clock-ordered low-field phase occurs is at $h/J = 1.65 \pm 0.05$ [168,273].

To infer the nature of the full phase diagram including the transverse field and the long-range interaction, the next step is to discuss the zero-field limit of the Ising model with long-range interactions. As the antiferromagnetic long-range interactions introduce a hierarchy of constraints due to the interaction between sites further apart, the long-range interaction breaks the ground-state degeneracy of the nearest-neighbour zero-field as well [30,110,277,279]. It is important to emphasise that the long-range interaction lifts the degeneracy in a different way and stabilises a gapped stripe phase breaking the translational symmetry in a different way than the clock order promoted by the transverse field [30,110,277,279,280]. This plain stripe state is sixfold degenerate by rotations around $\pi/3$, as well as spin flips [30,110,277,279]. The spins are aligned in straight lines with alternating z-orientation [30,110,277,279]. An important aspect is that these stripe states are gapped and, therefore, stable against finite transverse fields [30,110,277].

Regarding the limit of large transverse fields, pCUT+MC was used to investigate the QPT between the x-polarised high-field phase and the clock-ordered phase [29,30]. Similar to the antiferromagnetic models on the bipartite lattices, the critical field value is reduced by the long-range interaction towards smaller field strengths [29,30] (see Figure 27). On the triangular lattice, the universality class of the transition from the high-field to a clock-ordered phase is of the (2+1)D-XY type [270,271,273]. It is reported to remain in this category for all σ values investigated ($\sigma \geq 1$) [29]. For small σ values, the extrapolation of high-field series expansion is inconclusive [29,30]. Infinite DMRG (iDMRG) investigations on triangular lattice cylinders suggest that the clock order vanishes for small values of σ and a direct transition to a low-field stripe phase occurs [27]. A first-order phase transition between the x-polarised high-field phase and a low-field stripe phase would be consistent with the incapability of the high-field gap to track the transition [27,29,30,277]. Note that the stripe phase determined from the iDMRG study [27] does not agree with subsequent studies focusing on the low-field ground states of the model [30,32,277,279,280].

Complementary to the high-field analysis and numerical iDMRG studies, approaches in the low-field limit have also been conducted [30,110,277] (see Figure 27). As already mentioned, it was discussed in the literature for some time [30,279,280] and recently demonstrated using a unit cell-based optimisation technique [110] that gapped plain stripes are the zero-field ground state as soon as long-range interactions are present. Therefore, in a phase diagram with a σ and a transverse-field axis, the clock order phase is wedged between the high-field phase from above and the plain stripe phase from below [27,29,30,277] (see Figure 27). The extent of both the stripe phase and the high-field phase increases with stronger long-range interactions [27,29,30,277]. This behaviour was also demonstrated perturbatively from the nearest-neighbour zero-field limit by treating the long-range interaction, as well as the transverse field as perturbations on the degenerate subspace [30,277].

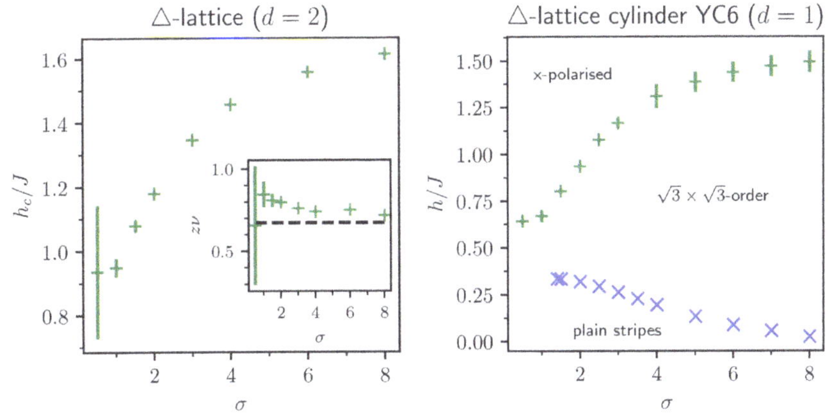

Figure 27. Phase diagrams for the long-range transverse-field Ising model on the triangular lattice (left panel) and triangular lattice cylinders with infinite extent in the x-direction and a circumference of 6 sites in the y-direction (YC6) (right panel). Right panel: The critical field values h_c/J for the triangular lattice for the QPT between the high-field polarised phase originate from pCUT+MC calculations [29]. The inset of the right panel represents the critical exponents $z\nu$ determined from the series expansion [29]. The black dashed line represents the critical value $z\nu = 0.67175(10)$ of the 3D-XY universality class [281,282]. The critical exponent $z\nu$ confirms the $(2+1)$D-XY universality class within the limitations of the series expansion. Left panel: The transitions values between the high-field x-polarised phase and the $\sqrt{3} \times \sqrt{3}$-clock-ordered phase are from the gap closing of the high-field series obtained from pCUT+MC [30]. The transition between the plain stripe low-field phase and the $\sqrt{3} \times \sqrt{3}$-clock-ordered phase is terminated via a level crossing of both ground-state energies, which were calculated perturbatively [30]. The phase diagrams for small σ values are not yet conclusively determined [27,29,30].

We sketch a qualitative picture of the quantum phase diagram in Figure 28. The precise ground state at small values of σ is still an open research question. We believe that there is some evidence that, after a certain value of σ, only a transition between the high-field phase and the low-field stripes remains, which is expected to be of first order. However, there is, by now, no numerical technique to make reliable statements in this regime. In this context, the high-field series expansions cannot detect a first-order phase transition from the elementary excitation gap. There is an SSE QMC study of the system by S. Humeniuk [40] applying the algorithm discussed in Section 5.1. The direct QMC simulation was used to determine the critical point between the high-field phase and the clock-ordered phase for $\alpha = 3$ [40]. For smaller transverse fields, the author identified a "region dominated by classical ground states [40]", which are adiabatically connected to the zero-field states. Therefore, there is a qualitatively sketch of a phase diagram in Ref. [40] similar to Figure 28. Nevertheless, the naive application of the SSE QMC approach as discussed in Section 5.1 is not as efficient as for models with interactions that are not competing [40,261,262]. However, there are ideas on how to implement efficient quantum cluster algorithms for frustrated Ising models in a transverse field [261,262], which need to be adjusted for long-range interacting Ising models.

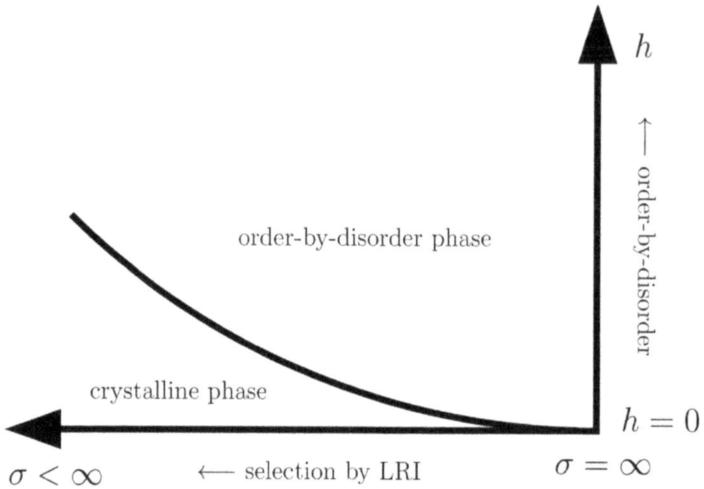

Figure 28. Sketch of a generic phase diagram of a long-range transverse-field Ising model (LRTFIM) for which a degenerate subspace at $\sigma = \infty$ and $h = 0$ breaks down in an order-by-disorder scenario for $h > 0$ and into a different crystalline state due to the long-range interactions (LRIs) $\sigma < \infty$. An example is the LRTFIM on the triangular lattice, where the order-by-disorder phase is the clock-ordered phase and the crystalline phase is the plain stripe phase [27,29,30,110,277,279,280]. The transition between the crystalline phase and the order-by-disorder phase is believed to be a first-order lever-crossing transition [29,30,110,277].

The antiferromagnetic LRTFIM on the triangular lattice is a well-studied example of the interplay between long-range interactions and order-by-disorder. Recently, several more examples arose from the field of Rydberg atom quantum simulators [54,55,109,110,112]. In Rydberg atom simulators, atoms are positioned in a desired configuration using optical tweezers [52] and are laser-driven to a Rydberg state [52]. The Rydberg blockade mechanism leads to an algebraically decaying interaction between Rydberg excitations, which decays with $(d + \sigma) = 6$ [52]. Using the Matsubara–Matsuda transformation [89], the excitations/non-excitations of a Rydberg atoms are associated with spin degrees of freedom [52]. The density–density interaction transforms into an Ising interaction [110,112]. Rydberg atom quantum simulators are capable of simulating the antiferromagnetic LRTFIM with $(d + \sigma) = 6$ and a longitudinal field [52,54,55,109]. Order-by-disorder scenarios are described for the Kagome lattice at a filling of $f = 2/9$ Rydberg excitations [54], the Ruby lattice at a filling of $f = 1/4$ Rydberg excitations [55,109], and at vanishing longitudinal field [112]. Regarding these examples, the order-by-disorder was studied for a long-range interaction truncated after the third-nearest neighbours [54,55,109,112]. Considering the remaining long-range interaction as a perturbation to the degenerate ground-state space for most of these mechanisms, a similar scenario as for the triangular lattice is expected [54,55,109,110]. Using the same unit cell-based optimisation method as for the triangular lattice, Koziol et al. [110] have calculated ground states of the zero-field model using the full untruncated long-range interaction. For the Kagome lattice at a filling of $f = 2/9$ Rydberg excitations [54] and the Ruby lattice at a filling of $f = 1/4$ Rydberg excitations [54,55,109], gapped ground states were determined [110], which do not coincide with the order arising from the order-by-disorder mechanism. Interestingly, in all these examples, the zero-field ground state determined by the long-range interaction possess none of the motifs relevant for the leading-order order-by-disorder mechanism [110]. For example, on the triangular lattice, the plain stripe states contain zero flippable plaquettes, which are the relevant motif to lower the energy at a finite transverse field in leading order. Recently, an order-by-disorder mechanism for the antiferromagnetic J_1-J_2-J_3 TFIM on the Ruby lattice was introduced by A. Duft et al. [112]. The remarkable aspect of the mechanism is that adding long-range interactions does stabilise the same order, which is also selected by the quantum fluctuations [112]. A general theory, under which conditions the long-range interactions stabilise the same order as quantum fluctuations or stabilise a completely disjoint order, is yet to be found.

Conclusively, the breakdown of extensively degenerate ground-state spaces due to an interplay of long-range interactions and quantum fluctuations is a highly vibrant and relevant field. Long-range interactions are relevant for a wide range of quantum–optical quantum–simulation platforms, including cold atoms [52,70,72] and ions [74–77,79–82, 84]. These platforms can be used to realise exotic phases of matter, e.g., quantum spin liquids [54,55,109], glassy behaviour [283], or clock-ordered states [29,30,112,277]. To understand the emergence of these exotic states of matter, simplified models are oftentimes considered [54,55,109,112,283], which truncate the long-range interaction. Therefore, it is imminent to understand the effects of the full long-range interactions on the mechanisms driving the emergence of these exotic phases. At the moment, the commonly used tools to investigate the phase diagrams of these frustrated systems consist of the following:

1. The derivation of effective models [30,54,55,109,112]. These can be used in order to predict exotic emergent phases and the nature of QPTs [30,54,55,109,112].
2. DMRG calculations [27,54,55,109], which are used to obtain a numerical insight into the ground-state phase diagrams.
3. QMC calculations, in particular the SSE, as discussed in Section 5.1, can be used for an unbiased sampling of ground-state properties [40]. In order to omit the slowdown of the algorithm due to the geometric frustration, algorithmic improvements are required [261,262]. It is still an open research question how to set up an algorithm that samples long-range interaction and frustration efficiently.
4. High-order high-field series expansions using a graph decomposition. This is also a very capable tool to track the first QPT coming from the high-field limit. With this method, the critical field value, as well as critical exponents can be determined [29,30,112]. It is also possible to infer information about the phase on the other side of the phase transition by studying the momentum at which the elementary excitation gap is located [29,30,112]. An MC embedding of white graphs can be performed to study the entire algebraically decaying long-range interaction [29,30], while an ordinary graph decomposition can be used for systems with truncated interactions [112].

7. Long-Range Transverse-Field XY Chain

In this section, we review results on the XY chain with long-range interactions in a transverse field from Ref. [31]. The Hamiltonian of the long-range transverse-field anisotropic XY model (LRTFAXYM) on a one-dimensional chain is given by

$$H = h\sum_i \sigma_i^z - \frac{J}{4}\sum_{i\neq j}\frac{1}{|i-j|^{1+\sigma}}\left[(1+\beta)\sigma_i^x\sigma_j^x + (1-\beta)\sigma_i^y\sigma_j^y\right], \tag{308}$$

with Pauli matrices σ_i^κ and $\kappa \in \{x,y,z\}$ describing spin-1/2 degrees of freedom on the i-th lattice site of the chain. The transverse-field strength is given by $h > 0$, and the strength of the coupling between sites i and j is given by $\sim J/|i-j|^{1+\sigma}$. The coupling is (anti)ferromagnetic for $J > 0$ ($J < 0$). We include a continuous interpolation parameter $\beta \in [0,1]$ to tune the system from the XY chain with isotropic interactions ($\beta = 0$) to Ising interactions ($\beta = 1$). The exponent σ determines the decay of the long-range interaction. The limit of short-range interactions is recovered for $\sigma = \infty$ and the limit of all-to-all couplings for $\sigma = -1$.

Similar to the discussion of the transverse-field Ising model, it is useful to express Equation (308) in terms of hard-core bosonic operators b_i, $b_i^\dagger$ using the Matsubara-Matsuda transformation [89]

$$\sigma_i^z = 1 - 2b_i^\dagger b_i \qquad \sigma_i^x = b_i^\dagger + b_i \qquad \sigma_i^y = i(b_i^\dagger - b_i)\,. \tag{309}$$

We obtain

$$H = \epsilon_0 N + \sum_i b_i^\dagger b_i - \sum_{i \neq j} \frac{\lambda}{4} \frac{1}{|i-j|^{1+\sigma}} [(1+\beta)(b_i^\dagger b_j + b_i^\dagger b_j^\dagger) + (1-\beta)(b_i^\dagger b_j - b_i^\dagger b_j^\dagger) + \text{h.c.}] \tag{310}$$

in units of $2h$ with $\epsilon_0 = -1/2$ being the unperturbed ground-state energy per site, N the number of sites, and $\lambda = J/(2h)$ the perturbation parameter associated with the perturbation $\mathcal{V}$ (second sum) taking the form $\mathcal{V} = T_{-2} + T_0 + T_2$ in the pCUT language, so that a high-order series expansion about the high-field limit can be performed.

In the limit $J = 0$, the ground state is a nondegenerate z-polarised state that serves as an unperturbed reference state. Elementary excitations are called local spin flips at an arbitrary site i, which can be annihilated (created) by the hard-core bosonic operator $b_i^{(\dagger)}$. Upon increasing λ, these quasiparticles (qps) become spin flips dressed by quantum fluctuations induced by the perturbation $\mathcal{V}$. In the limit of $h = 0$, the system exhibits ferromagnetic or antiferromagnetic magnetic order depending on the sign of J in the x-direction (y-direction) for $\beta > 0$ ($\beta < 0$). When tuning the interpolation parameter β from pure Ising to isotropic XY interactions, ordering in the x-direction starts to compete with ordering in the y-direction.

In order to better appreciate the results for $\sigma < \infty$, we briefly discuss the quantum phase diagram in Figure 29 of the model (308) with nearest-neighbour interactions [284–293]. For any $\beta \neq 0$, the system with nearest-neighbour interactions undergoes a (1+1)D-Ising QPT at $J = \pm h$ from the paramagnetic symmetric high-field phase to the (anti)ferromagnetically ordered low-field phase, where the ground state spontaneously breaks the $\mathbb{Z}_2$ symmetry of the Hamiltonian [284]. The transition along the $\beta = 0$ line between the two distinct $\mathbb{Z}_2$ symmetry-broken phases is of the (Ising)^2 type, which is conformally and $U(1)$ invariant [284]. These critical lines meet at $\beta = 0$ and $J = \pm h$ in multicritical points $M_{1/2}$ with critical exponents $z = 2$ and $\nu = 1/2$ [284].

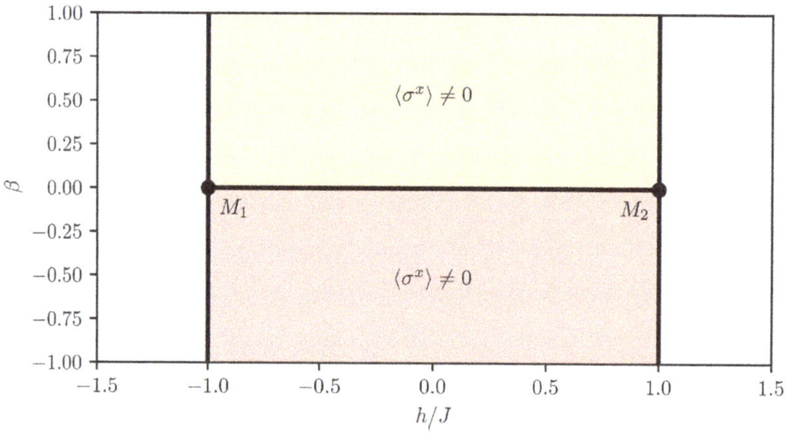

Figure 29. Quantum phase diagram of the ferromagnetic nearest-neighbour XY chain in a transverse field (see Ref. [284]). M_1 and M_2 denote multicritical points. The phase transition between the symmetric high-field polarised phases and magnetically ordered low-field phases are of (1+1)D-Ising universality for $\beta \neq 0$. For $\beta = 0$, the transition at the multicritical points M_1 and M_2 has critical exponents $z = 2$ and $\nu = 1/2$. The transition between $\langle \sigma_i^x \rangle \neq 0$ and $\langle \sigma_i^y \rangle \neq 0$ is of the (Ising)^2 type.

For $\beta = 0$, the Hamiltonian (310) becomes particle-conserving as the terms annihilating (creating) two quasiparticles $b_i^{(\dagger)} b_j^{(\dagger)}$ cancel each other, making the isotropic XY Hamiltonian $U(1)$ symmetric. Due to this particle-conserving nature of Equation (310), there are no quantum fluctuations dressing the symmetric $\lambda = 0$ ground state for $\lambda > 0$ and the perturbative treatment of the high-field dispersion becomes exact in first order of the perturbation theory [31]. This distinguished symmetry property of the $\beta = 0$ case motivates that we review the properties of the isotropic transverse-field XY chain in the ferromagnetic and antiferromagnetic case first in Section 7.1. Subsequently, we discuss the anisotropic XY chain for ferromagnetic interactions in Section 7.2 and for antiferromagnetic interactions in Section 7.3. Here, we show improved results of the ones shown in Ref. [31] since we noticed right after their publication that the Monte Carlo runs were performed with a suboptimal choice of the simulation parameters.

7.1. Isotropic Long-Range XY Chain in a Transverse Field

To consider the isotropic XY chain, $\beta = 0$ is set in the Hamiltonians in Equations (308) and (310). The quantum criticality of this model was studied in Ref. [31] in an analytical fashion, by evaluating the ground-state energy and the dispersion of the elementary excitations analytically. Setting $\beta = 0$, Equation (310) reads

$$H = \epsilon_0 N + \sum_i b_i^\dagger b_i - \sum_{i \neq j} \frac{\lambda}{2} \frac{1}{|i-j|^{1+\sigma}} [b_i^\dagger b_j + \text{h.c.}] \tag{311}$$

since the pair creation and annihilation terms $b_i^{(\dagger)} b_j^{(\dagger)}$ cancel, as stated above. The perturbation $\mathcal{V}$ acting on the unperturbed z-polarised ground state does not introduce any quantum fluctuations. Therefore, the z-polarised state becomes an exact eigenstate for arbitrary λ and stays in the ground state until a QPT occurs [31]. Analogously, the one-quasiparticle dispersion is exact in first-order perturbation theory. The one-particle dispersion in the symmetric high-field phase reads [31]

$$\omega(k) = 1 - 2\lambda \sum_{\delta=1}^{\infty} \frac{\cos(k\delta)}{\delta^{1+\sigma}} \,. \tag{312}$$

The critical value λ_c can be determined from the dispersion (312) via the closing of the quasiparticle gap $\Delta(\lambda)$. The critical exponent $z\nu$ associated with the gap closing can be extracted from the dominant power-law behaviour of the gap near λ_c:

$$\Delta(\lambda) \propto |\lambda - \lambda_c|^{z\nu} \,. \tag{313}$$

Further, it is possible to separate the dynamic z and static ν exponent by evaluating the 1qp dispersion $\omega(k)$ at the quantum-critical point such that

$$\omega(k)|_{\lambda=\lambda_c} \propto |k - k_c|^z \,, \tag{314}$$

which allows the extraction of the critical dynamic exponent z. For ferromagnetic XY interactions, the minimum of the dispersion (312) is located at the momentum $k = 0$. Therefore, the gap series is given by $\Delta(\lambda) = 1 - 2\lambda\zeta(1+\sigma)$ and $\lambda_c(\sigma) = (2\zeta(1+\sigma))^{-1}$. The Riemann-zeta function $\zeta(1+\sigma) := \sum_{n=1}^{\infty} n^{-(1+\sigma)}$ is convergent for all $\sigma > 0$ and diverges for $\sigma \to 0$. This mathematical observation coincides with the qualitative behaviour found in ferromagnetic Ising systems discussed in Section 6.1 in the sense that $\lim_{\sigma\to 0} \lambda_c(\sigma) = 0$. Since the expression for the gap $\Delta(\lambda)$ is linear in λ, the critical exponent is $z\nu = 1$, independent of the σ value. Using the expression for the dispersion (312), the knowledge about λ_c, and the definition of the dynamic critical exponent (314), Adelhardt et al. [31] evaluated z and, therefore, also ν as a function of σ. The results of their calculations are presented in Figure 30. The exponents resulting from the consideration of the dispersion can be categorised into two distinct regimes: For $\sigma > 2$, the critical exponents of the nearest-neighbour transition are found ($z = 2$ and $\nu = 1/2$). For $\sigma < 2$, the exponent z decreases linearly from 2 to 0 and ν increases like $1/z$ from $1/2$ to ∞. Note that this linear decrease of z and the divergence of ν appear similar to the behaviour of the ferromagnetic LRTFIM at small σ values (see Section 6.1) [20,21]. To explain this behaviour, a quantum field theory was proposed in Ref. [31]. The idea was to add a k^σ term (analogous to [20,21]) to the well-studied bosonic action used for the study of the isotropic short-range transition [113,231,284,294]. The suggested action reads

$$S = \frac{1}{2}\int_{k,\omega}(ak^2 + bk^\sigma + ig\omega + r)|\psi_{k,\omega}|^2 + u\int_{k,\omega}|\psi_{k,\omega}|^4 \tag{315}$$

with ψ being a complex c-number order-parameter field of the transition, $a, b > 0$, and the real constants u, g and r [31]. The notation of Equation (315) is taken from Ref. [31,231]. From power counting, one obtains the critical exponents:

$$z = \begin{cases} 2 & \text{for} \quad \sigma \geq 2 \\ \sigma & \text{for} \quad \sigma < 2 \end{cases} \tag{316}$$

$$\nu = \begin{cases} 1/2 & \text{for} \quad \sigma \geq 2 \\ 1/\sigma & \text{for} \quad \sigma < 2 \end{cases} \tag{317}$$

from the Gaussian part of the action in Equation (315) [31]. Following the arguments of Ref. [231], these exponents hold "presumably" [231] for $1 \leq d \leq d_{\mathrm{uc}}$ below the upper critical dimension $d_{\mathrm{uc}} = 2$. Refs. [231,294] provide arguments that the self-energy vanishes for every order in u and the renormalisation of u can be performed in all orders via ladder diagrams [31]. The vanishing self-energy is explained by the fact that, in each diagram, every pole in ω lies in the complex upper-half plane. The frequency integral can be deformed into the lower half-plane to give zero [294]. The fact that the free propagator of the field theory is not changed by a self-energy is a manifestation of the absence of fluctuations that do not preserve the particle number [31]. The predictions of the field theory (315) for z in Equation (316) and ν in Equation (317) are in perfect agreement with the results of the high-field excitation gap (see Figure 30) [31].

Note that, in the nearest-neighbour case, the low-field ground state does not break the $U(1)$ symmetry of the Hamiltonian due to the Hohenberg–Mermin–Wagner (HMW) theorem, ruling out continuous symmetry breaking [113,130,131,284,295,296]. Long-range interactions are a known mechanism to circumvent the HMW theorem [22,35,297–306]. With the high-field approach discussed above, it is not possible to determine the nature of the low-field ground state and if a continuous symmetry breaking occurs for sufficiently small decay exponents. We are not aware of studies addressing this topic.

For the remainder of this section, we discuss the antiferromagnetic isotropic XY chain in a transverse field along the same lines as for the ferromagnetic case. The minimum of the dispersion is at momentum $k = \pi$, and the gap can be expressed as $\Delta(\lambda) = 1 - 2\lambda\eta(1+\sigma)$ with $\eta(1+\sigma)$ being the Dirichlet eta function (also known as the alternating ζ function) [31]. The Dirichlet eta function $\eta(1+\sigma)$ is convergent for all $\sigma > -1$; therefore, the definition of an excitation gap is also possible in the regime of strong long-range interactions ($\sigma \leq 0$). Analogous to the discussion of the Ising interaction in Section 6.2, the extent of the low-field phase is diminishing for decreasing σ values. For $\sigma \to -1$, the energy gap closes at diverging $\lambda_c \to \infty$ [31]. With a similar analysis to the ferromagnetic case, the same critical exponents $z = 2$ and $\nu = 1/2$ are found along the entire σ line [31]. This is also analogous to the antiferromagnetic LRTFIM discussed in Section 6.2, where the nearest-neighbour criticality is believed to be the correct universality class for all values of σ.

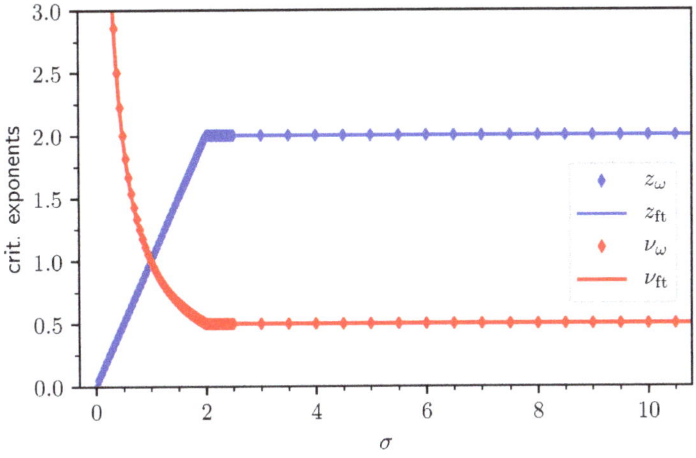

Figure 30. Critical exponents z and ν for the ferromagnetic long-range transverse-field XY model as a function of σ. The data points z_ω and ν_ω were determined in Ref. [31] by studying the dispersion of the elementary excitations at the critical point that can be determined exactly in first-order perturbation theory from the high-field limit. The lines $z_{\rm ft}$ and $\nu_{\rm ft}$ are theoretical predictions from the QFT investigated in Ref. [31]. The region with the blue-shaded background denotes the regime in which the nearest-neighbour XY universality occurs [284], while for $\sigma < 2$, the QPT is in a long-range regime with continuously varying exponents.

7.2. Ferromagnetic Anisotropic Long-Range XY Chain in a Transverse Field

The fundamental difference between the isotropic $\beta = 0$ and anisotropic $\beta > 0$ XY chain in a transverse field is that, for $\beta \neq 0$, the Hamiltonian is no longer $U(1)$-invariant (particle-conserving); see Equation (310). Therefore, a perturbative treatment of the anisotropic model with long-range interaction from the high-field limit requires the entire procedure described in Section 4 to perform pCUT+MC calculations. In Ref. [31], this procedure was applied to obtain the critical values λ_c and critical exponents $z\nu$ from the gap. We review these results in this section, followed by the results for the antiferromagnetic case in Section 7.3.

Critical values λ_c and critical exponents $z\nu$ for several β values as a function of σ are depicted in Figure 31. The overall σ dependence of λ_c for $\beta > 0$ is similar to the $\beta = 0$ case (see the inset of Figure 31). However, the low-field ferromagnetic phase becomes more stable for increasing β values [31]. Note that the perturbative parameter is $\lambda = J/2h$; therefore, a QPT at smaller λ values means a larger extent of the phase in h/J. In Ref. [307], the QPT of the anisotropic LRTFAXYM was studied with exact diagonalisation, and the critical point was determined for $\beta = 1/2$, which coincides perfectly with the value from pCUT+MC (see Figure 31). For all $\beta > 0$, Adelhardt et al. [31] identified the same three regimes for the universality of the QPT as for the ferromagnetic LRTFIM. The gap exponent behaves analogously to the LRTFIM ($\beta = 1$), as depicted in Figure 31. Solely the $\beta = 0$ exponent behaves in a special manner, being $z\nu = 1$ for all σ values, in full agreement with the existence of a multicritical point in the nearest-neighbour model. For $\beta > 0$, the first domain identified in Ref. [31] is the long-range mean-field regime for $\sigma < 2/3$ with $z\nu = 1/2$. For large values of σ, the gap exponent is $z\nu = 1$ and the transition belongs to the 2D-Ising universality class [31]. Third, in the intermediate regime, the critical exponents vary continuously between the two values in a non-trivial fashion [31].

The criticality regimes for $\beta > 0$ can be understood using the same field-theoretical argument (see Section 6.1.1) as for the ferromagnetic LRTFIM, since the underlying symmetry that is spontaneously broken is, in both cases, the same $\mathbb{Z}_2$ symmetry of the Hamiltonian. Therefore, at $\sigma = 2/3$, the upper critical dimension as a function of σ becomes smaller than one and the transition enters the regime above the upper critical dimension. The boundary between nearest-neighbour criticality and the intermediate long-range non-trivial regime is expected to be at $\sigma = 2 - \eta_{\text{SR}}$ [21]. The data from Ref. [31], presented in Figure 31, are in good agreement with these three regimes. Nevertheless, the changes between the regimes cannot be determined accurately by the pCUT+MC approach. In Figure 31, we can observe that the interfaces between the nearest-neighbour and intermediate long-range regime become more pronounced for increasing β values, while at the interface between the long-range mean-field and intermediate regime, the deviation is the same for all values of $\beta > 0$. The reason for the distinct deviations at the nearest-neighbour interface may be due to the finite nature of the perturbative series or may also be indicative of different corrections to scaling. At $\sigma = 2/3$, where the system dimension equals the upper critical dimension of the transition, there are multiplicative logarithmic corrections to the critical exponents [125,135,308–310]. From the biased DlogPadé extrapolations at $\sigma = 2/3$, Adelhardt et al. [31] determined these corrections to scaling at the upper critical dimension with a qualitatively good agreement for different β values.

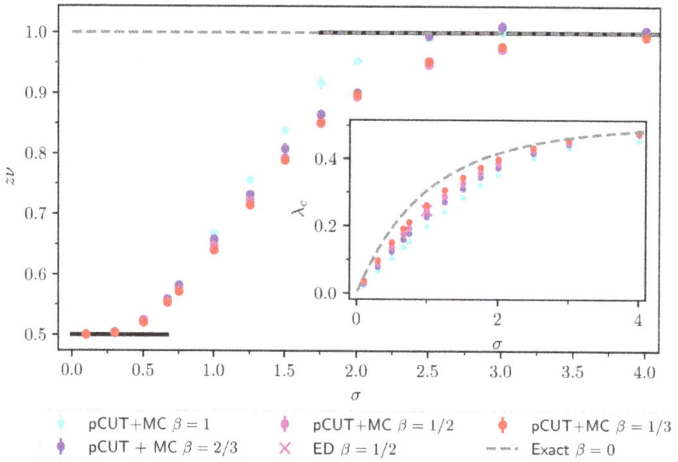

Figure 31. Critical gap exponent $z\nu$ and critical values of λ (see inset) as a function of σ for the anisotropic ferromagnetic transverse-field XY model. The data points "pCUT+MC" are improved results from Ref. [31], and the data point "ED" is from Ref. [307]. The anisotropy parameter β is tuned from $\beta = 1$ (Ising) to $\beta = 0$ (isotropic XY). As discussed in Section 7.1, the isotropic case is analytically solvable with $z\nu = 1$ and $\lambda_c = (2\zeta(\sigma+1))^{-1}$. The black lines denote the values of $z\nu$ if the ferromagnetic LRTFIM is in the nearest-neighbour and the long-range mean-field regime.

To summarise the findings, we provide the following: Similar to the ferromagnetic LRTFIM, the ferromagnetic LRTFAXYM is a paradigmatic toy model extending the LRTFIM. With the pCUT+MC method described in Section 4, the critical exponents as a function of the decay exponent σ can be determined as demonstrated in Ref. [31]. The LRTFAXYM displays three distinct regimes of quantum criticality including a mean-field regime above the upper critical dimension for $\sigma < 2/3$ for all $\beta > 0$. The sensitivity of the series expansion method suffices to even study corrections to scaling at the upper critical dimension. The limiting case $\beta = 0$ is especially interesting, as in the high-field limit, the ground state contains no quantum fluctuations, the dispersion is exactly solvable in first-order perturbation theory, the underlying QFT is different from the case $\beta > 0$, and two instead of three critical regimes were found. We would expect that the setup of an SSE QMC algorithm following the directed loop idea [215,217,311] is conceptually possible for the model.

7.3. Antiferromagnetic Anisotropic Long-Range XY Chain in a Transverse Field

For frustrated antiferromagnetic interactions, we begin our discussion with the Ising case at $\beta = 1$. Here, we have already discussed in Section 6.2 that, for the antiferromagnetic LRTFIM, there is strong numerical evidence that it remains in the 2D-Ising universality class for all $\sigma > -1$. It was also demonstrated that it is possible to determine critical gap exponents up to $\sigma \approx 0.5$ with the pCUT+MC method [29,31,35].

Following the same procedure as for the ferromagnetic model, the critical values λ_c and the critical gap exponents $z\nu$ were studied in Ref. [31], and we summarise these results in Figure 32. The critical values λ_c shift towards larger values when decreasing σ, eventually diverging for $\sigma \to -1$. It is possible to reliably determine the critical exponents for the anisotropic XY chain until $\sigma \approx 0.5$, but it becomes increasingly challenging to extract the exponents for smaller σ due to the shifting of λ_c [31]. For $\beta \neq 0$, the critical values λ_c behave qualitatively in a similar way to $\beta = 0$. However, the larger β, the more the high-field phase is stabilised (see the inset of Figure 32). As a consequence, we observe more reliable exponent estimates in the regime $\sigma < 0.5$ for β values closer to zero. In general, the observations in Ref. [31] are in line with a constant gap exponent $z\nu = 1$ within the limitations of the method.

This promotes the scenario that the nearest-neighbour 2D-Ising universality class remains the correct universality class for all considered σ values and $\beta > 0$. For $\beta = 0$, the critical exponents $z\nu = 1$ and critical values $\lambda_c = (2\eta(1+\sigma))^{-1}$ were determined analytically in Section 7.1, and it was shown that the universality class of the nearest-neighbour isotropic chain remains for all $\sigma > -1$ [31]. A recent study [312] investigated the multicritical point using exact diagonalisation for comparably small system sizes. The critical points for $\beta = 1/2$ and $\beta = 1/5$ agree well with the exact limit and the pCUT+MC values. The study further confirms 2D-Ising universality up to $\sigma > -0.4$; however, in contrast to Ref. [31], signatures of a different crossover criticality at $\sigma = -0.4$ are claimed.

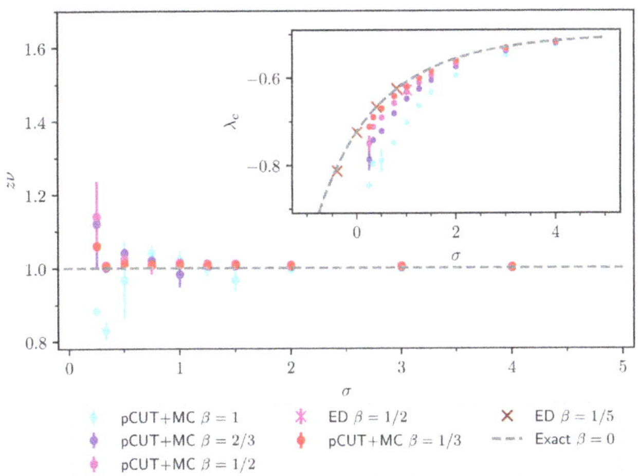

Figure 32. Critical gap exponent $z\nu$ and critical values of λ (see inset) as a function of σ for the anisotropic antiferromagnetic transverse-field XY model [31]. The anisotropy parameter β is tuned from $\beta = 1$ (Ising) to $\beta = 0$ (isotropic XY). As discussed in Section 7.1, the isotropic case is analytically solvable with $z\nu = 1$ and $\lambda_c = (2\eta(\sigma+1))^{-1}$.

8. Long-Range Heisenberg Models

We review the results obtained from SSE and pCUT+MC calculations for Heisenberg models with long-range interactions. The starting point is the nearest-neighbour Heisenberg model, which can be written as

$$\mathcal{H} = \sum_{\langle i,j\rangle} \vec{S}_i \vec{S}_j = \sum_{\langle i,j\rangle} \left(S_i^x S_j^x + S_i^y S_j^y + S_i^z S_j^z \right) \tag{318}$$

with the spin-1/2 operators $S^\kappa = \sigma^\kappa/2$ and $\kappa \in \{x, y, z\}$. Instead of a transverse field that induces quantum fluctuations in the (anti)ferromagnetic phase of the (LR)TFIM or (LR)TFAXYM, a magnetic field would immediately break the $SU(2)$-symmetry of the Heisenberg Hamiltonian. Here, we introduce a dimerisation limit. We consider two layers of lattices with Heisenberg spins on each site stacked on top of each other, where each spin couples to its nearest-neighbour. The nearest-neighbour coupling between the layers gives rise to interlayer dimers, also referred to as rungs. In one dimension, the layers are just two chains, resulting in a ladder system, while in two dimensions, we consider two square lattice layers, resulting in a square lattice bilayer model. Generalising this model for long-range interactions gives rise to intralayer, as well interlayer long-range couplings. The generic antiferromagnetic Hamiltonian is then given by

$$\mathcal{H} = J_\perp \sum_i \vec{S}_{i,1}\vec{S}_{i,2} - \frac{1}{2}\sum_{i\neq j}\Big[J_{\shortparallel}(i-j) \underbrace{\left(\vec{S}_{i,1}\vec{S}_{j,1} + \vec{S}_{i,2}\vec{S}_{j,2}\right)}_{\text{intralayer}} + J_\times(i-j) \underbrace{\left(\vec{S}_{i,1}\vec{S}_{j,2} + \vec{S}_{i,2}\vec{S}_{j,1}\right)}_{\text{interlayer}} \Big], \tag{319}$$

where the first term couples the nearest-neighbour spins between the layers with coupling strength $J_\perp > 0$, the second term describes the long-range intralayer coupling with $J_{\shortparallel}$, and the third term the long-range interlayer coupling of Heisenberg spins with $J_\times$. The long-range interaction is given by

$$J_{\mu,\nu}(i-j) = J\frac{(-1)^{||\vec{r}_{i,\mu}-\vec{r}_{j,\nu}||_1}}{|\vec{r}_{i,\mu}-\vec{r}_{j,\nu}|^{d+\sigma}}, \tag{320}$$

where J is the coupling constant, i and j are the rung dimer indices, and μ and ν are the indices of the respective legs (layers) of the ladder (bilayer) model. In Equation (319), we use the notation $J_{\shortparallel}(i-j) = J_{1,1}(i-j) = J_{2,2}(i-j)$ and $J_\times(i-j) = J_{1,2}(i-j) = J_{2,1}(i-j)$. The long-range interaction introduced is similar to the one introduced in the sections before, however with the difference that the alternating sign in the numerator gives rise to staggered, non-frustrating antiferromagnetic long-range interactions, as we chose $J > 0$. The exponent of the numerator is given by the one-norm $||\cdot||_1$, while the usual geometric distance in the denominator given by $|\cdot|$ can be identified as the two-norm $|\cdot| = ||\cdot||_2$. When increasing the overall interaction strength J of intralayer and interlayer coupling and, likewise, making the long-range decay exponent σ smaller, the Heisenberg spins will anti-align at odd distances and align at even distances, inducing a Néel-ordered antiferromagnetic phase. For $\sigma \leq 0$, the system becomes superextensive (analogous to the ferromagnetic LRTFIM in Section 6.1), and for $\sigma \to \infty$, we recover the antiferromagnetic short-range ladder or bilayer model.

In order to employ the pCUT+MC method, we need an exactly soluble limit about which we can perform the series expansion. We introduce the perturbation parameter $\lambda = J/J_\perp$ rescaling the Hamiltonian by $J_\perp$ and identify the first term of Equation (319) as the unperturbed part $\mathcal{H}_0$, which corresponds to a sum over Heisenberg dimers that can be easily diagonalised. For a single dimer, the lowest lying state with total spin $S = 0$ is the antisymmetric combination of $S = 1/2$ spins:

$$|s\rangle = \frac{1}{\sqrt{2}}(|\uparrow\downarrow\rangle - |\downarrow\uparrow\rangle), \tag{321}$$

also called a rung singlet in our case with the associated energy $\epsilon_0 = -3/4$. The total spin $S = 1$ states:

$$|t_-\rangle = |\downarrow\downarrow\rangle, \quad |t_0\rangle = \frac{1}{\sqrt{2}}(|\uparrow\downarrow\rangle + |\downarrow\uparrow\rangle), \quad |t_+\rangle = |\uparrow\uparrow\rangle \tag{322}$$

are three-fold degenerate with the energy $1/4$ and are called triplets. Alternatively, we can use an $SU(2)$-symmetric basis, where the singlet and triplet states read

$$\begin{aligned} |s\rangle &= \frac{1}{\sqrt{2}}(|\uparrow\downarrow\rangle - |\downarrow\uparrow\rangle), \\ |t_x\rangle = -\frac{1}{\sqrt{2}}(|\uparrow\uparrow\rangle - |\downarrow\downarrow\rangle), \quad |t_y\rangle &= \frac{\mathrm{i}}{\sqrt{2}}(|\uparrow\uparrow\rangle + |\downarrow\downarrow\rangle), \quad |t_z\rangle = \frac{1}{\sqrt{2}}(|\uparrow\downarrow\rangle + |\downarrow\uparrow\rangle). \end{aligned} \tag{323}$$

There is also a convenient mapping from the spin operators to bosonic, $SU(2)$-symmetric operators creating and annihilating these states introduced in Ref. [313], which can be readily adapted to hard-core bosonic operators for triplet excitations [314]. We use the mapping

$$\begin{aligned} \vec{S}^{\alpha}_{i,1} &= \frac{1}{2}\left(t^{\dagger}_{i,\alpha} + t_{i,\alpha} - \mathrm{i}\epsilon_{\alpha,\beta,\gamma} t^{\dagger}_{i,\beta} t_{i,\gamma}\right), \\ \vec{S}^{\alpha}_{i,2} &= \frac{1}{2}\left(t^{\dagger}_{i,\alpha} + t_{i,\alpha} + \mathrm{i}\epsilon_{\alpha,\beta,\gamma} t^{\dagger}_{i,\beta} t_{i,\gamma}\right), \end{aligned} \tag{324}$$

where $\epsilon_{\alpha,\beta,\gamma}$ is the Levi-Civita symbol, and insert these expressions into Equation (319). After some straightforward manipulations, we obtain

$$\begin{aligned} \mathcal{H} = \epsilon_0 N_R &+ \sum_{i,\alpha} t^{\dagger}_{i,\alpha} t_{i,\alpha} \\ &+ \frac{1}{4}\sum_{i\neq j} \lambda_{\shortparallel}(i-j)\left(t^{\dagger}_{i,\alpha}t^{\dagger}_{j,\alpha} + t^{\dagger}_{i,\alpha}t_{j,\alpha} - t^{\dagger}_{i,\alpha}t^{\dagger}_{j,\alpha}t_{i,\beta}t_{j,\beta} + t^{\dagger}_{i,\alpha}t^{\dagger}_{j,\beta}t_{i,\beta}t_{j,\alpha} + \text{h.c.}\right) \\ &+ \frac{1}{4}\sum_{i\neq j} \lambda_{\times}(i-j)\left(-t^{\dagger}_{i,\alpha}t^{\dagger}_{j,\alpha} - t^{\dagger}_{i,\alpha}t_{j,\alpha} - t^{\dagger}_{i,\alpha}t^{\dagger}_{j,\alpha}t_{i,\beta}t_{j,\beta} + t^{\dagger}_{i,\alpha}t^{\dagger}_{j,\beta}t_{i,\beta}t_{j,\alpha} + \text{h.c.}\right), \end{aligned} \tag{325}$$

where N_R is the number of rungs, $\lambda_{\shortparallel}(i-j) = J_{\shortparallel}(i-j)/J_{\perp}$, and $\lambda_{\times}(i-j) = J_{\times}(i-j)/J_{\perp}$. The second and third line constitute the perturbation $\mathcal{V} = T_{-2} + T_0 + T_2$ associated with the physical perturbation parameter λ. The T_1 and T_{-1} operators are absent, as terms contributing to these operators cancel out due to the reflection symmetry about the centre of the rung dimers.

For $\lambda = 0$, the ground state is given by a trivial product state of rung singlets, and for small, but finite λ, the ground state is still adiabatically connected to this product state. We refer to this adiabatically connected ground state as the rung-singlet ground state. Elementary excitations in this phase are called triplons [315], corresponding to dressed triplet excitations. For large $\lambda \gg 0$, we expect the non-frustrating antiferromagnetic long-range interaction to induce an antiferromagnetic phase giving rise to a quantum phase transition (QPT) between these two phases. In contrast to the transverse-field Ising model and the anisotropic XY model, where the ground state in the ordered phase spontaneously breaks the discrete $\mathbb{Z}_2$ symmetry of the Hamiltonian, the Néel-ordered antiferromagnetic ground state for strong long-range couplings must break the continuous $SU(2)$ symmetry of the Heisenberg Hamiltonian. Related to this observation, we present in the following two prominent theorems of great importance that apply to Hamiltonians with continuous symmetries:

First, there is the Hohenberg–Mermin–Wagner (HMW) theorem [130,131,295], which rules out the spontaneous breaking of continuous symmetries in one- and two dimensional systems for $T > 0$. From the quantum–classical correspondence (see Ref. [113]), we can infer that, for one-dimensional systems, a QPT at $T = 0$ breaking such symmetry should be ruled out as well. Indeed, it was shown rigorously by Pitaevskii and Stringari in 1991 [296] that it is prohibited in one dimension for $T = 0$ as well. However, there is the restriction that the interaction must be sufficiently short-ranged, i.e., the condition $\sigma > 2$ for $T > 0$ must hold for the long-range decay exponent. A stronger condition was given by Ref. [316] with $\sigma \geq d$, where d is the spatial dimension of the system. For the quantum case, it was shown in Ref. [317] for a staggered antiferromagnetic long-range Heisenberg chain ($d = 1$) that long-range order is absent for all $\sigma > 2$.

Second, Goldstone's theorem [318–320] applies. The theorem states that, when spontaneous breaking of a continuous symmetry occurs, it gives rise to massless excitations, also known as Nambu–Goldstone modes [318–320]. For instance, magnons that are quantised spin–wave excitations are a manifestation of such gapless Nambu–Goldstone modes inside the ordered phase of Heisenberg systems. The recent findings in Ref. [321] actually go beyond the conventional scenario of Goldstone modes due to long-range interactions. The authors identified three regimes depending on the decay exponent σ: for ferromagnetic (antiferromagnetic) interactions, they found standard Goldstone modes for $\sigma \geq 2$ ($\sigma \geq 0$) and an anomalous Goldstone regime with $\omega \sim |k|^s$ and $s < 2$ ($s < 1$) for $\sigma < 2$. Interestingly, a third regime for strong long-range interactions $\sigma \leq 0$ ($\sigma \leq -2$) was found where the Goldstone modes become gapped via a generalised Higgs mechanism [321]. A recent large-scale QMC study [38] investigating the dynamic properties of the non-frustrating staggered antiferromagnetic square lattice Heisenberg model confirmed these three scenarios and found evidence that the Higgs regime already occurs in the regime $\sigma \leq 0.2$ when the Hamiltonian is still extensive [38]. Let us mention at this point that previous results from linear spin–wave theory [22,322] already indicated the existence of a sublinear dispersion in the staggered antiferromagnetic long-range Heisenberg chain a decade earlier. Remarkably, the existence of gapped Goldstone modes in the Heisenberg model with long-range interactions was already pointed out in the 1960s by Refs. [323,324].

The HMW theorem, as well as Nambu–Goldstone modes are a distinguishing feature of quantum systems with continuous symmetries like Heisenberg antiferromagnets. The quantum field theory describing dimerised Heisenberg antiferromagnets [113] is given by the action

$$\mathcal{S}_\phi = \int \mathrm{d}^d x \int_0^\beta \mathrm{d}\tau \left[\{g(\partial_\tau \phi(\tau,x))^2 + (\nabla_x \phi(\tau,x))^2 + r\phi(\tau,x)^2\} + u\phi(\tau,x)^4\right], \tag{326}$$

which is the same n-component ϕ^4-theory as the one describing the transverse-field Ising model. Here, the order-parameter field $\phi(\tau,x)$ is now a 3-component field instead of a 1-component one. We can readily include long-range interactions by adding the term

$$\int \mathrm{d}^d x \int \mathrm{d}^d y \int \mathrm{d}\tau \, a \frac{\phi(\tau,x)\phi(\tau,y)}{|x-y|^{d+\sigma}} \tag{327}$$

to the action (326). For a more detailed discussion of the short-range ϕ^4 action, we refer to Appendix A and to Section 6.1.1, where the implications of long-range interactions were already discussed for $n = 1$. The classical equivalent of the $n = 3$ action can describe the finite temperature phase transition of the ferromagnetic Heisenberg model [1,36,113,325]. Interestingly, the action of zero-temperature Heisenberg ferromagnets is not given by the above action [113]. In fact, the quantum field theory describes a class of dimerised Heisenberg antiferromagnets, which can be appropriately described by Equation (326) since their low-energy physics can be mapped onto the $n = 3$ quantum rotor model because pairwise antiferromagnetically coupled Heisenberg spins are an effective low-energy representation of quantum rotors [113]. We can see this correspondence by looking at the Hamiltonian of the $O(3)$ quantum rotor model:

$$\mathcal{H} = \mathcal{H}_{\text{kin}} + \mathcal{V} = \frac{K}{2}\sum_i \boldsymbol{L}_i^2 - J\sum_{\langle i,j\rangle} \boldsymbol{n}_i \boldsymbol{n}_j, \tag{328}$$

where the first part $\mathcal{H}_{\text{kin}} = K/2\sum_i \boldsymbol{L}_i^2$ with $K > 0$ and $\boldsymbol{L}$ the angular momentum operator gives the kinetic energy of the rotors. The second part $\mathcal{V} = -J\sum_{\langle i,j\rangle} \boldsymbol{n}_i \boldsymbol{n}_i$ with $J > 0$ is the interaction between the three-component quantum rotors inducing parallel ordering of the rotors [113]. The physical picture is that the kinetic energy is minimised when the orientation of the rotors is maximally uncertain and the energy of the interaction term is minimised when the rotors are aligned [113]. The eigenvalues of the kinetic term of a single quantum rotor are given by $E_{\mathrm{kin}}(l) = K/2 l(l+1)$ with $l \in \mathbb{N}$ corresponding to a $(2l+1)$-fold degenerate state. Now, given two spins S interacting antiferromagnetically with coupling strength $J_\perp$ forming a dimer, the total spin is $0 \leq S_{\mathrm{tot}} \leq 2S$ and the eigenenergies are given by $E_{\mathrm{dimer}}(S_{\mathrm{tot}}) = J_\perp/2\,(S_{\mathrm{tot}}(S_{\mathrm{tot}}+1) - 2S(S+1))$ with a $(2S_{\mathrm{tot}}+1)$-fold degeneracy. This gives a one-to-one correspondence to the kinetic energy of a rotor, but with an upper bound for the energy. Thus, this mapping is only valid when considering the low-energy properties of the models [113] and holds certainly for large K/J. We just introduced a powerful mapping from the low-energy properties of the antiferromagnetic Heisenberg systems like ladder and bilayer models forming Heisenberg dimers on the rungs (see Equation (319)) to an $O(3)$ quantum rotor model, which is described by the action (326). It turns out that this mapping is not only valid for large K/J, but also in the ordered phase and at the quantum critical point [113]. To include algebraically decaying long-range interactions, the term (327) must be added to the action. The implications of long-range interactions for an n-component order-parameter field in the ϕ^4 field theory was already extensively discussed in the 1970s [1–3] for the classical action, but it took until 2001 for the quantum analogue to be studied by Ref. [20]. More studies followed by N. Defenu et al. employing functional RG approaches [21,244,326].

The predictions for long-range Heisenberg models are in large part the same as for the LRTFIM. We expect long-range mean-field behaviour with a Gaussian fixed point for $\sigma \leq \sigma_{\mathrm{uc}}$, a regime of nearest-neighbour criticality for $\sigma \geq \sigma_*$, and a non-trivial regime with continuously varying critical exponents in between [20,21,244,326]. For $d = 1$, however, there is the major difference that a QPT associated with the spontaneous breaking of continuous symmetries is ruled out by the HMW theorem [130,131,295,296], at least for $\sigma > 2$ [21,244,317,326]. The boundary is then called lower critical exponent σ_{lc}. For the Heisenberg model, the previously discussed action (326) with long-range interactions (327) holds, which together with the HMW theorem rules out the breaking of the order-parameter field for sufficiently short-range interactions and, therefore, a QPT in the regime $\sigma > \sigma_{\mathrm{lc}}$. For $\sigma \leq \sigma_{\mathrm{lc}}$, a QPT breaking the continuous symmetry is predicted [20,21,244,326]. Indeed, it has been long confirmed by several numerical studies [22,297–306,327] that the HMW theorem can be circumvented if the long-range interaction is sufficiently strong. Notably, in recent experiments with trapped ions [328] and Rydberg atoms [329] continuous symmetry breaking was realized in one and two dimensions which would be prohibited in the abscence of long-range interactions.

So far, most of the studies considered only Heisenberg Hamiltonians with long-range interactions in one-dimensional systems [22,35,297–306,327], however often considering various modifications to the Hamiltonian. Some studies considered a one-dimensional Heisenberg chain with non-frustrating (staggered) antiferromagnetic long-range interactions [22,300,305], while others included an anisotropy along the z-components resulting in the XXZ-model [302,303,327]. An interesting example with frustrating long-range interactions is the exactly solvable Haldane–Shastry model [330,331] with $\sim r^{-2}$, which is known to show quasi-long-range order (QLRO) just like the conventional Heisenberg chain, but with vanishing logarithmic corrections [332]. For the dimerisation transition between the QLRO and a valence bond solid (VBS), the frustrated long-range interaction seems to play only a minor role compared to the J_1-J_2 model [213]. Based on this, in Refs. [213,297–299,333], a model combining the frustrated J_1-J_2 model with additional non-frustrating staggered long-range interaction was investigated. The initial intention was to realise a one-dimensional analogue of deconfined criticality [334–337] between a Néel-order phase and a VBS. As the transition was found to be of first order, the authors in Ref. [338] later turned to the one-dimensional long-range J-Q model, indeed finding evidence for a continuous deconfined QPT. Interestingly, a very recent paper [327] also predicts a deconfined QPT in the antiferromagnetic long-range XXZ chain. There are also studies considering larger spin $S = 1$. In Ref. [304], a Heisenberg chain with single-ion anisotropy was studied, while in Ref. [301], the XXZ chain was under scrutiny. In both cases, a rich phase diagram was found with intriguing quantum critical properties. For instance, continuous symmetry breaking due to long-range interactions gives rise to a new, possibly exotic, tricritical point in the XXZ chain with no analogue in short-range one-dimensional spin systems [301].

In Sections 6 and 7, we reviewed the results of various numerical studies investigating the quantum-critical properties of the LRTFM [23–26,28–30,32–34] and the LRTFAXYM [31] with the focus on extracting the critical exponent as a function of the long-range decay σ. In comparison to the LRTFIM, for long-range Heisenberg models, just recently, efforts have been made to extract the critical exponents associated with the QPT as a function of the long-range decay exponent [22,35,37,206]. Reference [22], studying a one-dimensional long-range Heisenberg chain, is one notable exception in the sense that it was ahead of its time, performing large-scale SSE QMC simulations for long-range Heisenberg systems already in 2005, way before the rapid progress in implementing quantum simulators in quantum optics reignited the interest in long-range interacting systems.

In the following, we will first review the results from Refs. [37,206,339] for the long-range square-lattice Heisenberg bilayer model in Section 8.1. Then, we proceed with the discussion of the results from Ref. [35] for long-range Heisenberg ladders in Section 8.2. Finally, we conclude this section with the discussion of the long-range Heisenberg chain from Ref. [22] in Section 8.3.

8.1. Staggered Antiferromagnetic Long-Range Heisenberg Square Lattice Bilayer Model

In this section, we consider two square lattices stacked directly on top of each other, where the Heisenberg spins on each lattice site interact with their nearest neighbours. When the two spins on top of each other form a rung dimer interacting with coupling strength $J_\perp$, the Hamiltonian describing the bilayer system is given by Equation (319). Here, we neglect the long-range interlayer interactions, which results in the Hamiltonian

$$\mathcal{H} = J_\perp \sum_i \vec{S}_{i,1}\vec{S}_{i,2} - \frac{1}{2}\sum_{i\neq j}\left[J_{||}(i-j)\left(\vec{S}_{i,1}\vec{S}_{j,1}+\vec{S}_{i,2}\vec{S}_{j,2}\right)\right]. \tag{329}$$

We consider staggered non-frustrating long-range interactions along the layers $J_{||}(i-j)$ of the form (320). The short-range model ($\sigma \to \infty$) was the subject of several studies from the 1990s onward [162,183,184,340–346] investigating its quantum-critical properties. For a critical coupling ratio $\lambda = J_{||}/J_\perp$, the system undergoes a QPT breaking the continuous $SU(2)$ symmetry of the Hamiltonian from a ground state adiabatically connected to the product singlet state towards an antiferromagnetic Néel ground state with gapless magnon excitations (Nambu–Goldstone modes). For long-range interactions, we expect that these Goldstone modes are altered [38,321] and that the criticality of the system changes as a function of the decay exponent σ. We expect short-range criticality until $\sigma \geq \sigma_* = 2 - \eta_{\mathrm{SR}}$, a non-trivial regime of continuously varying criticality in $\sigma_{\mathrm{uc}} \leq \sigma < \sigma_*$ with $\sigma_{\mathrm{uc}} = 4/3$, and a long-range mean-field regime for $\sigma < \sigma_{\mathrm{uc}}$ [20,21,244,326].

Until recently, to the best of our knowledge, there was a complete absence of numerical confirmation of this scenario. The first data for the critical point and exponents as a function of σ were published by Song et al. [37] using the SSE QMC algorithm and is now supplemented by pCUT+MC data from Adelhardt and Schmidt [206,339].

In Figure 33, the critical values from Refs. [37,206,339] are plotted together with the results from functional renormalisation group (FRG) calculations from Ref. [21] for the $O(3)$ quantum rotor model. The critical points λ_c for both methods are in excellent agreement over the entire σ-range. However, the critical exponents cannot be directly compared since the FSS of the magnetisation curves from SSE QMC simulations give β and ν, while pCUT+MC calculations for the gap and spectral weight give $z\nu$ and $(2-z-\eta)z$. While the QMC data show relatively large error bars in the long-range mean-field regime, the pCUT approach overestimates the critical exponents in the short-range regime. For the two-dimensional long-range Heisenberg bilayer, the deviation is larger than for the LRTFIM in two dimensions. This is no surprise, as a previous series expansion for the nearest-neighbour square-lattice Heisenberg bilayer model showed similar deviations [347]. Also, the critical exponents from QMC are better at capturing the boundaries of the long-range mean-field and the short-range regime. In general, both approaches show good agreement with the FRG results from Ref. [21] for the non-trivial intermediate regime (which apparently has its own shortcomings at the boundary to the short-range regime).

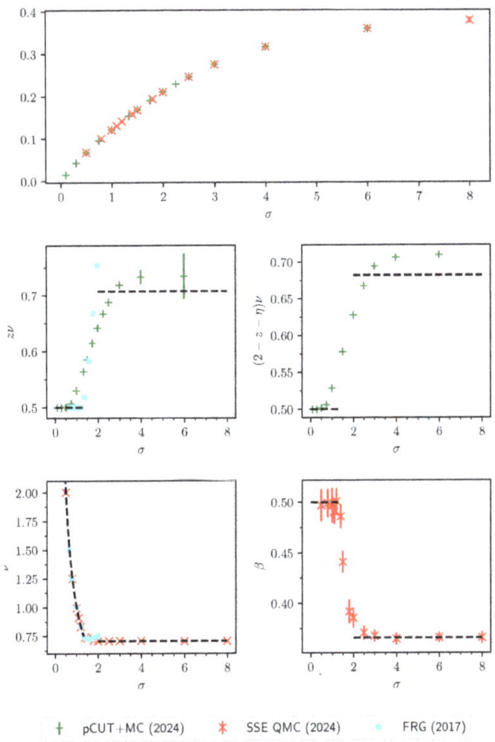

Figure 33. Critical values and exponents from numerical studies of the Néel-ordering transition in the unfrustrated antiferromagnetic long-range Heisenberg square-lattice bilayer model. The upper panel shows the critical values. The middle left panel displays critical exponent values $z\nu$, the middle right panel the exponent $(2-z-\eta)\nu$, the lower right panel the exponent ν, and the lower right panel the one-particle spectral weight exponent β. The data points "QMC (2024)" for ν and β originate from Ref. [37]. The data points "pCUT+MC (2024)" for $z\nu$ and $(2-z-\eta)\nu$ originate from Ref. [206,339]. The data points "FRG (2017)" for ν and $z\nu$ originate from Ref. [21]. The black dashed lines denote the critical exponents in the regime of short-range $\mathcal{O}(3)$ criticality ($\nu = 0.7116(10)$, $\beta = 0.36932(16)$, $z\nu = 0.7116(10)$, and $(2-z-\eta)\nu = 0.6847(10)$ [348,349]) and long-range mean-field criticality ($\nu = 1/\sigma$, $\beta = 1/2$, $z\nu = 1/2$, and $(2-z-\eta)\nu = 1/2$ [20,21]).

In conclusion, we summarised the results from SSE QMC and pCUT+MC studies for the square-lattice Heisenberg bilayer model with staggered antiferromagnetic interactions. Both approaches are in good agreement and confirm the three critical regimes [1–3,20,21,244,326] from the two-dimensional $O(3)$ quantum rotor model within their limitations.

8.2. Staggered Antiferromagnetic Long-Range Heisenberg Ladder Models

After the discussion of the square-lattice bilayer model, we can imagine the Heisenberg ladder models as effectively reducing the dimension of the square-lattice bilayer model by one now considering two linear chains coupled by rung interactions. The long-range ladder Hamiltonian then reads

$$\mathcal{H} = J_\perp \sum_i \vec{S}_{i,1}\vec{S}_{i,2} - \frac{1}{2}\sum_{i\neq j}\Big[J_{\shortparallel}(i-j)\left(\vec{S}_{i,1}\vec{S}_{j,1} + \vec{S}_{i,2}\vec{S}_{j,2}\right) + J_\times(i-j)\left(\vec{S}_{i,1}\vec{S}_{j,2} + \vec{S}_{i,2}\vec{S}_{j,1}\right)\Big], \tag{330}$$

with long-range coupling $J_{\shortparallel}(i-j)$ along the legs of the ladder and $J_\times(i-j)$ between spins of different legs. Recalling the long-range interactions of Equation (320) and the definitions of $J_{\shortparallel}(i-j)$ and $J_\times(i-j)$, thereafter, we define two distinct Hamiltonians $\mathcal{H}_{\shortparallel} = \mathcal{H}|_{J_\times=0}$ and $\mathcal{H}_{\bowtie} = \mathcal{H}$, where the first one includes long-range interactions only along the legs, while the second one is the original Hamiltonian including long-range interactions both along and in between the legs.

A sketch of these ladders is provided in Figure 34. As for the bilayer model, for small $J/J_\perp$ (cf. Equation (320)), the ground state is adiabatically connected to the product state of rung singlets (rung-singlet ground state), while for strong coupling ratios, the long-range interactions want to induce an antiferromagnetic ground state. However, in contrast to the bilayer model in Section 8.1, there is no QPT for the nearest-neighbour Heisenberg ladder [350–352], due to the HMW theorem ruling out continuous symmetry breaking for one-dimensional quantum models [296]. Note that the HMW theorem only rules out a QPT with continuous symmetry breaking and not a QPT in general. For instance, the nearest-neighbour isotropic XY model in a transverse field exhibits a QPT without breaking the $U(1)$ symmetry [284–293], while in the nearest-neighbour Heisenberg ladder, there is no QPT at all [350–352]. In fact, a QPT was ruled out until $\sigma \geq 2$ in another one-dimensional model, namely the staggered antiferromagnetic long-range Heisenberg chain in Ref. [317]. As there is a one-to-one correspondence between antiferromagnetic Heisenberg ladders and the low-energy properties of the one-dimensional $O(3)$ quantum rotor model [113], we can expect a QPT predicted by the quantum field theory given by the action (326) with Equation (327) for $\sigma \leq \sigma_{\mathrm{lc}} = 2 - \eta_{\mathrm{sr}}$ with $\eta_{\mathrm{SR}} = 0$ [20,21,244]. The existence of a lower critical decay exponent σ_{lc} has the consequence that there are only two quantum-critical regimes. One is the long-range mean-field regime for $\sigma \leq 2/3$, and the other one is a regime of a continuously varying critical exponent with a non-trivial fixed point. The third regime in these models is non-critical.

The only study investigating the full parameter space of this model is Ref. [35] using pCUT+MC and complementary linear spin–wave calculations. A previous study [306] using QMC and DMRG investigated the $\lambda = 1$ parameter line for $\mathcal{H}_{\bowtie}$. There are also known limiting cases of decoupled staggered long-range Heisenberg chains at $\lambda = \infty$, where a QPT from a QLRO phase towards the same Néel-ordered antiferromagnetic phase occurs. We can compare the critical points determined by pCUT+MC for $\mathcal{H}_{\shortparallel}$ and the linear spin–wave results to the SSE QMC data from Ref. [22] and also to the linear spin–wave calculations in Refs. [22,322] of the long-range Heisenberg chain. In fact, the linear spin–wave calculations for $\mathcal{H}_{\shortparallel}$ can be seen as a generalisation of the ones for the Heisenberg chain and, therefore, exactly including them as a limiting case.

We can find a plot in Figure 35 showing a ground-state phase diagram for both $\mathcal{H}_{\shortparallel}$ and $\mathcal{H}_{\bowtie}$ from Ref. [35,353]. The figure also includes other known values from Refs. [22,306], which fit into the overall picture of the pCUT+MC results. We also show the critical exponents $z\nu$, $(2-z-\eta)\nu$, and α determined by the pCUT+MC approach. It is easy to identify the two critical regimes of long-range mean field behaviour and continuously varying critical exponents as predicted by quantum field theory [20,21,244]. In the non-trivial regime, the exponents shown here diverge when approaching the lower critical dimension $\sigma \to \sigma_{\mathrm{lc}}$, which is in reasonable agreement with FRG results in Ref. [21,244]. The largest deviation can be seen for α, which is the hardest exponent to extract from DlogPadé extrapolations. Using the (hyper)scaling relations (3)–(9), the remaining critical exponents can be determined. See Figure A2 in Appendix D showing all critical exponents for the Heisenberg ladders. It should be noted that there are three main difficulties discussed in Ref. [35]. First, as just stated, the α exponent is difficult to determine. Second, it becomes increasingly hard to extrapolate the perturbative series in the regions $\sigma \gtrsim 1.1$ ($\sigma \gtrsim 1.2$) for $\mathcal{H}_{||}$ ($\mathcal{H}_{\bowtie}$). Third, the presence of logarithmic corrections to the dominant power-law behaviour close to the critical point spoils the exponents around the upper critical dimension at $\sigma = 2/3$. In the end, all factors play a role when determining all critical exponents due to error propagation. We can observe in Figure A2 that several exponents from the pCUT+MC approach deviate from the FRG exponents significantly in the non-trivial regime. For instance, the ν exponent seems to approach a constant value $\nu \approx 1$ for $\sigma \to \sigma_{\rm lc}$ for pCUT+MC, while the FRG predicts a diverging exponent. Another important finding is that the lower critical decay exponent $\sigma_{\rm lc}$ is apparently not universal in these two models and considerably smaller than the predicted value $\sigma_{\rm lc} = 2$ from quantum field theory [21,244]. This claim is made in Ref. [35] due to the known limiting case of the long-range Heisenberg chain [22] from SSE QMC simulations and due to linear spin–wave calculations [35].

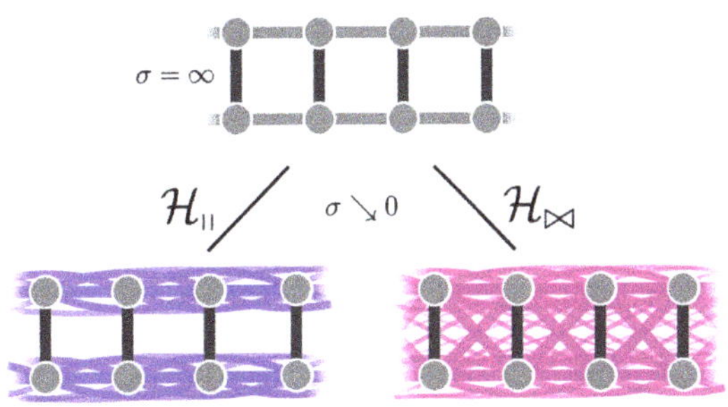

Figure 34. Illustration of the quantum spin ladders with long-range interactions. For nearest-neighbour interactions ($\sigma = \infty$), both long-range ladders $\mathcal{H}_{||}$ (left) and $\mathcal{H}_{\bowtie}$ (right) reduce to the same Heisenberg ladder. The coupling on the rungs $\sim J_\perp$ is illustrated with black lines, and the long-range coupling along the legs $\sim J_{||}(i-j)$ and in between the legs $\sim J_\times(i-j)$ is depicted in blue for $\mathcal{H}_{||}$ and in purple for $\mathcal{H}_{\bowtie}$. The figure is adapted from Ref. [35,353].

Beyond these interesting discrepancies, there was speculation about a possible deconfined quantum critical point along the $\lambda = 1$ parameter line for $\mathcal{H}_{\bowtie}$ in Ref. [306]. The reason for this was the fact that the staggered long-range Heisenberg ladder undergoes a QPT from a disordered phase with a non-local string order parameter towards a Néel-ordered phase with conventional order. Also, they found a sharp peak and a gap in the dynamic structure factor at the ordering momentum k_c in the ordered phase, which could be indicative of deconfined excitations in terms of spinons [306,354]. Usually, when there is a QPT between two competing ordered phases, the system undergoes a first-order phase transition or there must be a coexistence phase. There is also a much more exotic scenario beyond the Landau–Ginzburg–Wilson theory of phase transitions [334]. A deconfined quantum critical point [334–337] is a second-order QPT that is not described by a "confining" order parameter, but by an emergent $U(1)$-symmetric gauge field with "deconfined" degrees of freedom accompanied by a fractionalisation of the order parameters [334]. A paradigmatic example is the deconfined QPT between a Néel-ordered antiferromagnetic ground state and a VBS state on a two-dimensional lattice [334,335,337,355–360]. While deconfined criticality was originally proposed in two dimensions, similarities in terms of a conventional Luttinger Liquid theory description have been drawn in one dimension [361–368]. Interestingly, a very close analogy to a two-dimensional deconfined critical point was found in Ref. [338] using a toy model with six-spin Heisenberg interactions and long-range two-spin interactions inducing a continuous phase transition between a VBS phase and an antiferromagnetic phase. A scenario between a conventional order and non-local string order as proposed by Ref. [306] would go even beyond the one-dimensional scenarios found so far. However, it was argued in Ref. [35] that there should be no such deconfined critical point in the above long-range Heisenberg ladders due to the critical exponents found and the fact that the rung-singlet phase is adiabatically connected to the trivial product state of rung singlets not falling into the category of symmetry-protected topological phases despite the presence of a non-local string order parameter [35,369]. Another possible interpretation of the finding in Ref. [306] is probably along the lines of Refs. [38,321]. The observed gap in the dynamic structure factor is in agreement with strong finite-size artefacts arising from the altered dispersion $\omega \sim |k|^s$ of sublinear behaviour $s < 1$ in the anomalous Goldstone regime $\sigma \leq 2$.

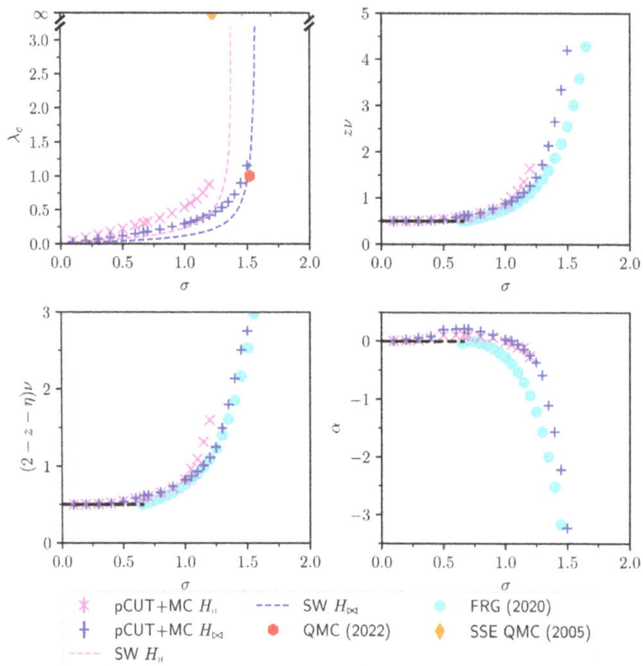

Figure 35. Critical values and exponents from numerical studies of the Néel-ordering transition in the unfrustrated antiferromagnetic long-range Heisenberg ladders. The panels show the critical values λ_c (upper left), $z\nu$ (upper right), $(2 - z - \eta)\nu$ (lower left), and α (lower right). The data points "pCUT+MC $H_{||}$", "SW $H_{||}$", "pCUT+MC $H_{\bowtie}$", and "SW $H_{\bowtie}$" originate from Ref. [35,353] and refer to parallel (||) and parallel + diagonal ($\bowtie$) interactions. The "SSE QMC (2005)" data point from Ref. [22] shows a $\lambda_c = \infty$ value on the long-range Heisenberg chain, which corresponds to the limiting case of decoupled legs. The "QMC (2022)" data [306] shows a $\lambda_c = 1$ value for $\mathcal{H}_{\bowtie}$. The data points "FRG 2020" are from Ref. [244] and show the critical exponents for the one-dimensional $O(3)$ quantum rotor model. The blue-shaded region denotes the σ regime in which long-range mean-field criticality is expected. The black-dashed lines denote long-range mean field critical exponents.

In this subsection, we have seen that the antiferromagnetic Heisenberg ladders with staggered long-range interactions show two quantum-critical regimes: one regime with long-range mean-field behaviour and a second non-trivial regime with continuously varying critical exponents. There is also a third regime that does not show any QPT because continuous symmetry breaking is ruled out by the HMW theorem. Despite the numerical confirmation of the two critical regimes predicted from quantum field theory [20,21,244,326], a discrepancy between the upper bounds of the Néel-ordered phase and the predicted lower critical exponent σ_{lc} was identified by Ref. [35]. On the other hand, the data in Ref. [35,353] are in agreement with other literature for the limiting case of long-range Heisenberg spin chains [22,322]. We also discussed briefly the possibility of a deconfined QPT and the presence of anomalous Goldstone modes in one dimension for the above ladder models. To put it briefly, the Hamiltonian (330) hosts some intriguing physics with several aspects that need further clarification.

8.3. Staggered Antiferromagnetic Long-Range Heisenberg Chain

Lastly, we consider another one-dimensional system, the Heisenberg chain with staggered non-frustrating antiferromagnetic long-range interactions:

$$\mathcal{H} = \sum_i \left[\vec{S}_i \vec{S}_{i+1} - \sum_{j=2}^{\infty} \lambda(i-j) \vec{S}_i \vec{S}_{i+j} \right]. \tag{331}$$

In the previous two models (bilayer and ladders), the unperturbed part at $\lambda = 0$ consisted of uncoupled dimers with a trivial product singlet state as its ground state and local triplet excitations above. Here, the unperturbed part is the nearest-neighbour Heisenberg chain with a ground state exhibiting quasi-long-range order (QLRO) and fractionalised elementary excitations. These excitations are referred to as spinons and can be seen as propagating domain walls carrying $S = 1/2$ degrees of freedom. The perturbation consists of long-range interactions that couple sites beyond the nearest-neighbours. This interaction is of the same algebraic form as Equation (320). Because of the non-frustrating nature of the antiferromagnetic long-range interactions, it induces a Néel-ordered antiferromagnetic phase upon increasing its coupling strength. Again, a QPT breaking the $SU(2)$-symmetry of the Hamiltonian is only allowed when the long-range decay exponent satisfies $\sigma < 2$ due to the HMW theorem [22,317], and thus, such a transition can be ruled out for larger decay exponents. In this model, a QPT from a QLRO towards an ordered phase is expected to occur, and therefore, the ϕ^4 theory of Equation (326) does not apply. The $k = 1$ Wess–Zumino–Witten non-linear σ model [370,371] is known to describe the low-energy physics of Heisenberg chains and includes topological coupling to account for the presence of QLRO in the lattice model [22,213]. As for the ϕ^4 theory, a long-range coupling analogous to Equation (327) can be added to describe the Hamiltonian (331) [22].

Reference [22] is a comprehensive study of the Heisenberg chain (331) using large-scale SSE QMC simulations to extract the critical properties of the QPT. The results for the critical point, as well as the critical exponents η and z can be found in Figure 36. The overall behaviour of the critical point as a function of the decay exponent σ is very similar to the one found in the previous subsection for the Heisenberg ladder. Here, the critical point diverges at about $\sigma \approx 1.8$ when approaching the lower critical exponent $\sigma \to \sigma_{\text{lc}}$. The hard boundary for a QPT is again given by the HMW theorem. Yet, there is a significant difference. While, for the Heisenberg ladders, the disordered rung-singlet phase exists for any $\sigma > 0$, for the Heisenberg chain, the QLRO phase only exists for $\sigma > 1$. Thus, for the Heisenberg chain, the critical line terminates in a marginal point at $\sigma = 1$ and $\lambda = 0$. For any $\sigma < 1$, the perturbation parameter λ becomes irrelevant, and the system is always in the antiferromagnetic phase [22]. The long-range Heisenberg chain (331) was also studied in Ref. [213,297,333] in the context of a Heisenberg chain with frustrated next-nearest-neighbour and non-frustrating long-range interactions, i.e., the $J_1 - J_2$ chain with staggered long-range interactions. In both models, the QLRO-Néel QPT can be identified by a level crossing between a triplet $S = 0$ and quintuplet $S = 2$ excitations (The transition was initially misidentified as a level crossing between two $S = 0$ states in Refs. [213,297] until it was later clarified to be a level crossing between the $S = 0$ and $S = 2$ states [333]). The critical point from finite-size scaling of the level crossing from the ED [297] and DMRG [333] is in very good agreement with the QMC results at $\lambda = 1$ (see Figure 36). Proceeding with the critical exponents, we can see that the critical exponent η matches well with the field-theoretical expectations $\eta = 2 - \sigma$ (linear in at least leading order, but also a simple scaling argument predicts the linear behaviour [22]) in the range $1 \leq \sigma \leq 1.3$. For $\sigma > 1.3$, the η from SSE QMC simulations starts to deviate from the linear behaviour. One interesting observation was pointed out in Ref. [35]. In both the Heisenberg ladders and the Heisenberg chain, the linear behaviour $\eta = 2 - \sigma$ is expected from the underlying quantum field theory, yet the data from pCUT+MC and SSE QMC studies indicate a deviation from this with $\eta \leq 2 - \sigma$ for the Heisenberg ladders and $\eta \geq 2 - \sigma$ for the Heisenberg chain. It should be noted, however, that the exponent from the pCUT+MC approach is determined using scaling relations and, therefore, can suffer from unfavourable error propagation, especially when the α exponent is involved. The dynamical exponent z of the Heisenberg chain was extracted from SSE QMC simulations as well. The exponent z is one at the marginal point $\sigma = 1$ and then quickly drops to $z \approx 0.75$, where it seems to be constant within the error bars up to $\sigma = 1.7$. This finding is also in contrast to the RG prediction, where $z = 1$ in leading order, and the exponent is expected to be constant even in higher orders [22]. In Ref. [213], for $\lambda = 1$, the dynamic exponent z was determined in excellent agreement with the QMC results (see Figure 36). Also, the results for the $J_1 - J_2$ model with non-frustrating long-range interactions, where the same QLRO-Néel transition is realised, the exponent is in agreement with $z \approx 0.75$ [213,297]. Further, the RG analysis in Ref. [22] gave also a prediction for ν. However, this could not be compared with QMC as it was not possible to obtain accurate estimates of ν [22]. Note, also, that neither the long-range transverse-field XY chain nor the Heisenberg chain show long-range mean-field behaviour.

The Heisenberg chain with staggered long-range interactions is another prime example of long-range models hosting intriguing critical behaviour. The remaining discrepancy between the numerical SSE QMC results and the underlying field-theoretical description shows that the critical properties are not yet fully settled and further exploration of the model is necessary.

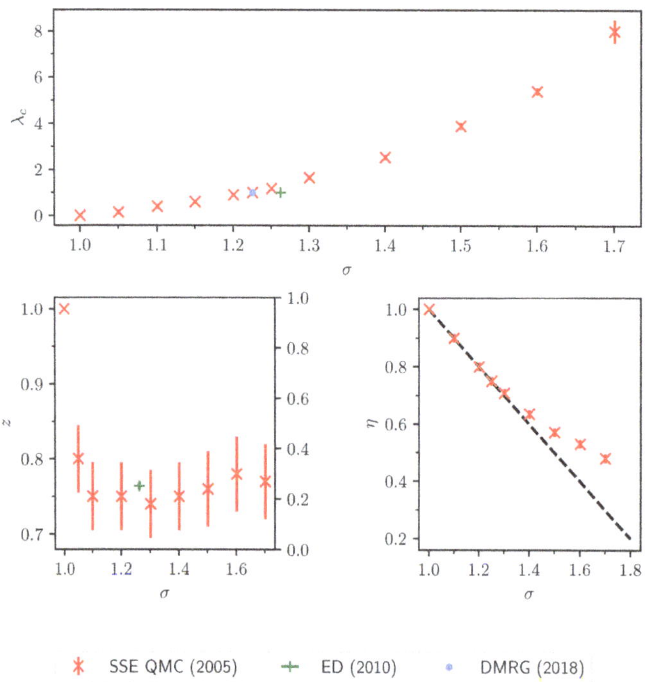

Figure 36. Critical values and exponents from numerical studies of the QLRO-Néel transition in the unfrustrated antiferromagnetic long-range Heisenberg chain. The upper panel shows critical values λ_c; the lower left shows the critical exponent z; the lower right shows η. The data points "SSE QMC 2005" are from Ref. [22]; "ED 2010" are from Refs. [213,297]; the single data point "DMRG 2018" is from Ref. [333]. The dashed line is for η and is the prediction from the first-order RG and scaling arguments provided in Ref. [22].

9. Summary and Outlook

In this review, we gave an overview of recent advances in the investigation of the quantum-critical properties of quantum magnets with long-range interactions focusing on two techniques, both based on Monte Carlo integration, but complementary in spirit. On the one hand, we described pCUT+MC, where classical Monte Carlo integration is decisive in the embedding scheme of white graphs. This allows extracting series expansions of relevant physical quantities directly in the thermodynamic limit. On the other hand, SSE QMC enables calculations on large finite systems where finite-size scaling can be used to determine the physical properties of the infinite system. Both quantitative and unbiased approaches take the full long-range interaction into account and can be used a priori in any spatial dimension for any geometry.

In recent years, both techniques, alongside other methods, have been applied successfully to one- and two-dimensional quantum magnets involving long-range Ising, XY, and Heisenberg interactions on various bipartite and non-bipartite lattices. In this work, we have summarised the obtained quantum-critical properties including quantum phase diagrams and the (full sets of) critical exponents for all these systems coherently. Further, we reviewed how long-range interactions are used to study quantum phase transitions above the upper critical dimension and how the scaling techniques are extended to extract these quantum critical properties from the numerical calculations. This is indeed generically the case for all unfrustrated systems in this review, with the exception of the one-dimensional isotropic XY and Heisenberg chain. For frustrated systems, one can apply both MC techniques successfully for Ising interactions, while in general, only the pCUT+MC method is applicable (if an appropriate perturbative limit exists) due to the sign problem of SSE QMC. Nevertheless, in all frustrated cases, the small-σ regime of long-range interactions is challenging for both approaches, and further technical developments are desirable.

In the future, several extensions and research directions are interesting. As mentioned in Section 6.3 of this review, the interplay between long-range interactions and geometric frustration is a vibrant research field at the moment. It has been demonstrated numerically and experimentally that this interplay provides a great resource to engineer exotic phases of matter [54,55,109,283,372] with the most spectacular example being the $\mathbb{Z}_2$ quantum spin liquid on the Ruby lattice [55,109,372]. We expect further rapid development in the field since many promising theoretical proposals can be realised in analogue quantum simulation platforms (e.g., programmable Rydberg atom quantum simulators [52,55]).

In terms of methods, pCUT+MC is yet to be extended to arbitrary unit cells, larger spin values, and multi-spin interactions. The access to larger unit cells will enable the investigation of the interplay between long-range interactions and frustration on even more relevant lattice structures, e.g., the Kagome or Ruby lattice. There are systems with multi-spin interactions hosting deconfined quantum criticality [338,355–359], and an introduction of long-range interactions to this type of systems seems to be an interesting research topic [338]. We hope to spark further interest in the development and application of pCUT+MC by other users.

The SSE QMC approach is a widely used numerical tool for the calculation of unbiased thermal averages of observables. A large variety of distinct QPTs is potentially accessible by these QMC simulations. We envision that, for all systems that do not suffer from a sign problem, SSE QMC, in combination with appropriate zero-temperature protocols and finite-size scaling, can be used to study how long-range interactions affect QPTs beyond the standard $\mathcal{O}(n)$-symmetry. The SSE QMC method has also been extended to tackle frustrated systems in a more efficient way [261,262,373]. However, an efficient treatment of both the long-range interaction and frustration has not been introduced yet [32,40]. Reference [374] developed an SSE QMC approach to access the toric code quantum spin liquid regime [55,109,372]. An application of the SSE QMC approach to extended long-range interacting Bose–Hubbard models (see Section 5.5) along the lines of directed loop updates [215–217] would also be a natural development. This would enable numerically calculating observables with SSE QMC for ultracold gas experiments with optical lattices. A possible application could be the study of complex crystalline phases and their breakdown in frustrated Bose–Hubbard systems with long-range interactions [64,73,111,230].

Finally, in the context of the long-range mean-field regime above the upper critical dimension, much research has been conducted regarding finite-size scaling in classical systems [12–19,41,125,128,142,145], including the study of multiplicative logarithmic corrections for the characteristic length scale at the upper critical dimension [17] and the investigation of the role of Fourier modes and boundary conditions [19]. On the contrary, its quantum counterpart has only been treated successfully in recent years [32,34]. In addition to the transfer of established concepts from classical to quantum Q-FSS, one interesting open question is a detailed understanding of the crossover regime between classical and quantum Q-FSS for small temperatures. Even though we focused on the quantum version of the Q-FSS of Ref. [34] due to the quantum nature of the models analysed in this review, the ground work has been conducted by the inventors of classical Q-FSS (Refs. [15–19,41,125]) and, in general, many other researchers who provided valuable insight into the scaling above the upper critical dimension, e.g., Refs. [10,12–14,127,128,142]. Overall, it is exciting that the abstract concept of dangerous irrelevant variables and the physics above the upper critical dimension are accessible in quantum–optical platforms realising long-range interactions.

Author Contributions: Conceptualisation: all authors; investigation: all authors; writing—original draft preparation: introduction (J.A.K. and K.P.S.), quantum phase transitions (J.A.K. and A.L.), Monte Carlo integration (J.A.K. and A.L.), series expansion Monte Carlo embedding (P.A.), stochastic series expansion quantum Monte Carlo (J.A.K. and A.L.), long-range transverse-field Ising models (J.A.K. and A.L.), long-range transverse-field XY chain (P.A. and J.A.K.), long-range Heisenberg models (P.A.), and conclusion and outlook (J.A.K. and K.P.S.); writing—review and editing: all authors; visualisation: P.A., J.A.K. and A.L.; supervision: K.P.S. All authors have read and agreed to the published version of the manuscript.

Funding: This work was funded by the Deutsche Forschungsgemeinschaft (DFG) (German Research Foundation)—Project-ID 429529648—TRR 306 QuCoLiMa (Quantum Cooperativity of Light and Matter). K.P.S. acknowledges the support by the Munich Quantum Valley, which is supported by the Bavarian state government with funds from the Hightech Agenda Bayern Plus. The authors gratefully acknowledge the scientific support and HPC resources provided by the Erlangen National High Performance Computing Center (NHR@FAU) of the Friedrich-Alexander-Universität Erlangen-Nürnberg (FAU) under the NHR project b177dc ("SELRIQS"). NHR funding is provided by federal and Bavarian state authorities. NHR@FAU hardware is partially funded by the German Research Foundation (DFG)–440719683.

Data Availability Statement: The data presented in this study are available on request from the corresponding author.

Acknowledgments: We thankfully acknowledge the valuable input on parts of the manuscript by Antonia Duft, Matthias Mühlhauser, Calvin Krämer, and Andreas Schellenberger. We also thank Jiarui Zhao and Nicolas Laflorencie for providing the data from the publications [22,37]. We are grateful for the fruitful collaboration with Sebastian Fey and Sebastian C. Kapfer regarding many of the projects reviewed in this work. It is with great dismay that we express our deepest condolences on the death of Ralph Kenna, who invited and encouraged us to write this review as part of his Special Issue on the "violation of hyperscaling above the upper critical dimension" in the journal "*Entropy*". We will always remember his encouraging and cheerful spirit in our correspondence with him.

Conflicts of Interest: The authors declare no conflicts of interest.

Abbreviations

The following abbreviations are used in this manuscript:

AF	antiferromagnetic
DIV	dangerous irrelevant variable
DMRG	density matrix renormalisation group
DOF	degrees of freedom
F	ferromagnetic
FRG	functional renormalisation group
FSS	finite-size scaling
GHF	generalised homogeneous function
HMW theorem	Hohenberg–Mermin–Wagner theorem
iDMRG	infinite density matrix renormalisation group
LCE	linked-cluster expansion
LRI	long-range interactions
LRTFAXYM	long-range transverse-field anisotropic XY model
LRTFIM	long-range transverse-field Ising model
MC	Monte Carlo
MCI	Monte Carlo integration
pCUT	perturbative continuous unitary transformation
PDF	probability density function
PI	path integral
QLRO	quasi long-range order
QMC	quantum Monte Carlo
QPT	quantum phase transition
qp	quasiparticle
RG	renormalisation group
SSE	stochastic series expansion
SPO	stochastic parameter optimisation
TDVP	time-dependent variational principle
VBS	valence bond solid

Appendix A. Short-Range $\mathcal{O}(n)$ Transition ϕ^4-Theory

For ferromagnetic short-range interacting transverse-field Ising models (see Equation (292)) the underlying $\mathbb{Z}_2$ symmetry of the Hamiltonian is spontaneously broken at the quantum phase transition [113]. The associated order parameter is the magnetisation in coupling direction which is microscopically defined as $\sum_i \sigma_i^z$ for Ising couplings in z-direction. Rewriting the partition function of the TFIM in the picture of Feynman path integrals, coarse-graining and going to the continuum limit is the general approach to derive a field theoretical description at criticality [113]. The fields of the quantum field theory $\phi(x,\tau)$ are obtained by coarse-graining the order parameters at imaginary time τ for a coarse-graining neighbourhood $\mathcal{N}(x)$ [113]

$$\phi(x,\tau) \propto \sum_{i\in\mathcal{N}(x)} \sigma_i^z \,. \tag{A1}$$

By properly [113] taking care of the microscopic degrees of freedom leading to terms of order $(\phi^2)^k$ with $k \in \mathbb{N}$, the resulting sufficiently truncated action leading to the right partition function to describe the phase transition is given by

$$\mathcal{Z} = \int \mathcal{D}\phi(x,\tau)\, e^{-\mathcal{S}_\phi} \tag{A2}$$

$$\mathcal{S}_\phi = \int \mathrm{d}^d x \int_0^\beta \mathrm{d}\tau \left[\{g(\partial_\tau\phi)^2 + (\nabla_x\phi)^2 + r\phi^2\} + u\phi^4\right] \tag{A3}$$

with periodic boundary conditions in imaginary time τ [113]. Note that the dependencies of $\phi(x,\tau)$ remain, but are dropped to lighten the notation. Gradient and time derivative terms are accounting for the spatial and temporal direction fluctuations [113,132]. The initial couplings in Equation (A3) are depending on the values of the couplings in the microscopical Hamiltonian [113]. Regarding the action in Equation (A3) the symmetry-breaking quantum phase transition between $\langle\phi\rangle = 0$ and $\langle\phi\rangle \neq 0$ becomes visualised as tuning the coupling r to a negative sign changes the minimum structure of the action into the symmetry broken Mexican hat. The field theory defined in Equation (A3) is a distilled representation of the key features of the quantum short range Ising criticality with ϕ having the interpretation of an "order parameter field".

In general, field theories as in Equation (A3) are used to study QPT with a spontaneous breaking of a $\mathcal{O}(n)$ symmetry. Here, $\mathcal{O}(n)$ refers to the orthogonal group (the group of all orthogonal $(n \times n)$-matrices). The study of a $O(1)$-symmetry breaking is equivalent to the study of a $\mathcal{Z}_{\in}$-symmetry breaking since both groups are isomorphic. Dependent on n of the orthogonal group $\mathcal{O}(n)$ the quantum fields $\phi(x,\tau)$ of the action Equation (A3) become multicomponent fields [113,132]. In order to study a $\mathcal{O}(n)$ symmetry breaking QPT real-valued fields $\phi(x,\tau)$ with n components are considered.

After introducing actions of the form Equation (A3) to describe quantum criticality, we review basic computations that can be performed within this framework. We start with the derivation of field correlators

$$G^{(n)}(x_1,\tau_1;\ldots;x_n,\tau_n) = \frac{1}{\mathcal{Z}}\int \mathcal{D}\phi(x,\tau)\,\phi(x_1,\tau_1)\ldots\phi(x_n,\tau_n)\,e^{-S_\phi} \tag{A4}$$

from a field-path-integral formulated partition function is technically accomplished by additionally coupling the order parameter field $\phi(x,\tau)$ linearly to an auxiliary field $j(x,\tau)$ [132]

$$\mathcal{S}_{\phi,j} = \int \mathrm{d}^d x \int_0^\beta \mathrm{d}\tau\,[\{g(\partial_\tau\phi)^2 + (\nabla_x\phi)^2 + r\phi^2\} + u\phi^4 - \phi(x,\tau)j(x,\tau)]\,. \tag{A5}$$

The n-field correlator can now be understood as the functional derivative of the partition function with respect to the respective auxiliary fields [132]

$$G^{(n)}(x_1,\tau_1;\ldots;x_n,\tau_n) = \frac{1}{\mathcal{Z}}\left[\frac{\delta}{\delta j(x_1,\tau_1)}\cdots\frac{\delta}{\delta j(x_n,\tau_n)}\int \mathcal{D}\phi(x,\tau)\,e^{-S_{\phi,j}}\right]_{j=0}\,. \tag{A6}$$

From the expression for the spatial correlator in Table 1 we see that at criticality the system is scale-free therefore for distances $x \gg a$ much larger than the lattice spacing the theory should be invariant under a scaling transformation of coordinates in space and time

$$x \to x' = x/b \tag{A7}$$
$$\tau \to \tau' = x/b^z \tag{A8}$$

with b being a rescaling factor [113]. We define the power by which a quantity $\mathcal{O}$ has to be rescaled in order to preserve the structure as the scaling dimension $[\mathcal{O}]$ of that quantity [113]. The Gaussian part of the action in Equation (A3) keeps its structure under scaling transformations if the field ϕ and the mass coupling r are also scaled appropriately. One can calculate the scaling dimensions of ϕ and r as follows,

$$\int \mathrm{d}^d x' \int_0^\beta \mathrm{d}\tau' (\partial_{\tau'}\phi'(x',\tau'))^2 = b^{-d-z+2z+2[\phi]} \int \mathrm{d}^d x \int_0^\beta \mathrm{d}\tau \, (\partial_\tau \phi(x,\tau))^2 \tag{A9}$$

$$\int \mathrm{d}^d x' \int_0^\beta \mathrm{d}\tau' (\nabla_{x'}\phi'(x',\tau'))^2 = b^{-d-z+2+2[\phi]} \int \mathrm{d}^d x \int_0^\beta \mathrm{d}\tau \, (\nabla_x \phi(x,\tau))^2 \tag{A10}$$

$$\int \mathrm{d}^d x' \int_0^\beta \mathrm{d}\tau' \, r'\phi'(x',\tau')^2 = b^{-d-z+[r]+2[\phi]} \int \mathrm{d}^d x \int_0^\beta \mathrm{d}\tau \, r\phi(x,\tau)^2 \tag{A11}$$

$$\Downarrow$$

$$-d - z + 2z + 2[\phi] = 0 \tag{A12}$$

$$-d - z + 2 + 2[\phi] = 0 \tag{A13}$$

$$-d - z + [r] + 2[\phi] = 0 \tag{A14}$$

$$\Downarrow$$

$$z = 1 \tag{A15}$$

$$[\phi] = (d-z)/2 = (d-1)/2 \tag{A16}$$

$$[r] = 2z = 2 \tag{A17}$$

$$\Downarrow$$

$$\phi \to \phi'(x',\tau') = b^{(d-1)/2}\phi(x,\tau) \tag{A18}$$

$$r \to r' = b^2 r \,. \tag{A19}$$

So far the following naively obtained power counting scaling dimensions were encountered,

$$[\mathrm{d}^d x] = -d \tag{A20}$$

$$[\mathrm{d}\tau] = -z = -1 \tag{A21}$$

$$[\nabla_x] = 1 \tag{A22}$$

$$[\partial\tau] = z = 1 \tag{A23}$$

$$[\phi] = (d-z)/2 = (d-1)/2 \tag{A24}$$

$$[r] = 2 \,. \tag{A25}$$

Note, one can fix $z = 1$ from the three linear equations (A12) to (A14). That means the fact that the TFIM has an exponent $z = 1$ and therefore space/imaginary time isotropy is available directly in the quantum field theoretical description by the form of the appropriate action. Requiring also scale-invariance for the ϕ^4 terms of the action in Equation (A3), one finds additionally to Equations (A12) to (A14) the relation

$$[\mathrm{d}^d x] + [\mathrm{d}\tau] + [u] + 4[\phi] = 0 \,. \tag{A26}$$

Therefore, the ϕ^4 term remains invariant for a scaling dimension of $[u] = 3z - d = 3 - d$. We can now regard the scaling in terms of differential equations representing the renormalisation group flow of the coupling constants using $b = 1 + \mathrm{d}l$ for $\mathrm{d}l \ll 1$ and building up a finite rescaling $b = e^l$ by repeated action of infinitesimal rescalings [113]

$$r + \mathrm{d}r = e^{\mathrm{d}l[r]} r = (1 + \mathrm{d}l[r])r = r + \mathrm{d}l[r]r \tag{A27}$$

$$u + \mathrm{d}u = e^{\mathrm{d}l[u]} u = (1 + \mathrm{d}l[u])u = u + \mathrm{d}l[u]u \tag{A28}$$

$$\Downarrow$$

$$\frac{\mathrm{d}r}{\mathrm{d}l} = 2r \tag{A29}$$

$$\frac{\mathrm{d}u}{\mathrm{d}l} = (3-d)u \,. \tag{A30}$$

Focusing only on the flow of r (see Equation (A29)) one sees that for $r > 0$ it flows towards the stable fixed point at $r = \infty$, and towards the stable fixed point at $-\infty$ for $r < 0$, while there is an unstable fixed point at $r = r^\star = 0$. Regarding also the coupling u one sees directly one fixed point of the coupled flow at $(r^\star = 0, u^\star = 0)$. For $d > 3$ the flow in u is stable towards the fixed point while for $d < 3$ it is unstable. That is a hint of the irrelevance of the coupling u for systems with $d > 3$ as independently of the starting condition the system is always in the attraction basin of $u^\star = 0$.

A coupling c with scaling dimension $[c] < 0$ always has one stable fixed point at $c^\star = 0$. Independent of the starting point it always flows to zero and is called irrelevant [10,127]. As long as the coupling is not dangerously irrelevant (see Section 2.3), meaning that there is no additional singular behaviour in observables approaching $c \to 0$ [127], it can be neglected. That is the reason why it is sufficient to regard an action with ϕ^4 as its highest power as all higher powers have irrelevant couplings [10].

The spatial dimension d at which the coupling of u has scaling dimension zero $[u] = 0$ is called the upper critical dimension. Below the upper critical dimension the coupling u needs to be included perturbatively and a " true " decimation of degrees of freedom is necessary [113]. In this thesis we will not deal with the technicalities, but in order to understand some arguments from other resources a basic understanding of the principle and the results of renormalisation group techniques are required. The decimation of degrees of freedom (DOF) is often done in this context by a momentum shell renormalisation [113]. Here as a first step " fast" DOF, i.e., DOF with large momentum/frequency, are partially integrated out [113,132,375]. One chooses a momentum cutoff Λ and frequency cutoff $\mathcal{J}$ and defines $\tilde{\Lambda} = \Lambda/b$ and $\tilde{\mathcal{J}} = \mathcal{J}/b^z$. After that one integrates the " fast" DOF between Λ and $\tilde{\Lambda}$ and respectively $\mathcal{J}$ and $\tilde{\mathcal{J}}$ [113,132,375]. After that one rescales $x' = x/b$ which is equivalent to $\Lambda' = b\tilde{\Lambda} = \Lambda$ and $\tau' = \tau/b^z$ which is equivalent to $\mathcal{J}' = b^z\tilde{\mathcal{J}} = \mathcal{J}$ [113,132,375]. Technicalities can be found in [113,132,375]. The most important result of this scheme are new modified couplings

$$\tilde{r} = r + f_r(u, r) \tag{A31}$$

$$\tilde{u} = u + f_u(u, r) \tag{A32}$$

which are related to the original couplings plus some contributions $f_{r/u}$ which can be calculated order by order in u from occurring Feynman diagrams [113]. The modified couplings scale as [113]

$$r' = b^2\tilde{r} \tag{A33}$$

$$u' = b^{3-d}\tilde{u}\,. \tag{A34}$$

One can now also calculate the flow equations from those quantities,

$$\frac{\mathrm{d}r}{\mathrm{d}l} = 2r + g_r(r, u) \tag{A35}$$

$$\frac{\mathrm{d}u}{\mathrm{d}l} = (3 - d)u + g_u(r, u) \tag{A36}$$

for which we see that the equations are not decoupled like in the case of the Gaussian theory with an irrelevant u (see Equations (A29) and (A30)) and there is a second fixed point $(r^{\star\star}, u^{\star\star})$ differing from the trivial one $r^\star = u^\star = 0$ [113]. The new non-trivial fixed point is called Wilson-Fisher fixed point [113]. One can express the fixed point and flow equations in terms of $\epsilon = d_{\mathrm{uc}} - d$ and perturbatively extract the values of $(r^{\star\star}, u^{\star\star})$ as well as the eigenstates and eigenvalues of the linearised flow equation around the Wilson-Fisher fixed point [113].

For calculations higher than first-order it is also necessary to rescale the field at least for the apparent ϕ^4 theory of the TFIM due to self-energy contributions changing the propagator [113]. This could be taken care of by defining a new scaling dimension of the field $[\phi]_{\text{new}} = [\phi]_{\text{naive}} + h(\eta)$ from the naive scaling dimension (see Equation (A24)) and a further functional dependence $h(\eta)$ on the anomalous dimension exponent to be consistent with the scaling of the two-point correlators η [113]. The underlying mechanism relating the anomalous dimension to the self-energy $\Sigma(q)$ is as follows: We define the self-energy as the difference between the inverse propagator of the Gaussian field theory and the propagator including the u terms $G_0^{-1}(q,\omega) - G^{-1}(q,\omega) = \Sigma(q,\omega)$ and we find,

$$G(q,\omega) = \frac{1}{G_0^{-1}(q,\omega) - \Sigma(q,\omega)} = \frac{1}{q^2 + \Sigma(q,\omega) + \tilde{g}\omega^2 + r} \propto \frac{1}{q^{2-\eta}} \tag{A37}$$

that the self-energy potentially changes the leading q dependence for small q values [113]. This difference is incorporated into the anomalous dimension [113]. The Fourier transform of the propagator gives the two-point correlators which again are proportional to two times the field scaling dimension. Concluding the scaling dimension,

$$[\phi] = [\phi]_{\text{naive}} + h(\eta) \tag{A38}$$

is always necessary if the anomalous dimension is non-zero [113].

As already seen for the anomalous dimension exponent η in Equation (A37) and the dynamical correlation length exponent z it is possible to connect the canonical exponents (see Table 1) to the results of scaling/renormalisation group theory in the field-theoretical framework. We will now discuss the relations between canonical exponents which characterise a phase transition and scaling results. We will also briefly skim the origin of the fluctuation-dissipation relation and the analogy between the auxiliary field in Equation (A5) and the conjugate field to coupling to the order parameter in the microscopic model.

- The correlation length exponent ν is given by the inverse of the relevant eigenvalue λ_1 of the linearised flow equation spectrum at the relevant fixed point of the phase transition [113]. That means $\nu = 1/\lambda_1$ given by the non-trivial eigenvalue λ_1 of the linearised flow equations at the Wilson-Fisher fixed point below the upper critical dimension and $\nu = 1/[r]$ from the Gaussian fixed point above the upper critical dimension [113].
- The dynamical correlation length exponent stands for the space–time anisotropy. So far the short-range ϕ^4-theory defined in Equation (A3) did not have such an anisotropy, but in the case of long-range interaction models [20,21,31] or the Bose–Hubbard superfluidity onset transition [31,113,231] such space–time anisotropy occurs.
- The anomalous dimension exponent η is obtained by the q proportionality of the propagator of the theory $G(q,\omega) \propto q^{\eta-2}$ and is defined to stand for the difference to two in the proportionality [113]. The anomalous dimension η is therefore affected by self-energy contributions. We will also see in the long-range theory that even in the free theory without any self-energy contributions an η can occur if the spatial fluctuations in the action go with a q^σ term with $\sigma < 2$ [20,21,31].
- The conjugate field used in the consideration of phase transitions in the microscopical model translates to the auxiliary field in Equation (A5) [113,132]. Therefore the scaling dimension of the auxiliary field is in the Gaussian theory given by the scale invariance of

$$\int d^dx \int d\tau \phi(x,\tau) j(x,\tau) \tag{A39}$$

$$[d^dx] + [d\tau] + [\phi] + [j] = 0\,. \tag{A40}$$

- According to Equation (A6) one can derive the expectation value of the field ϕ and the two-point correlator via functional derivatives with respect to the auxiliary field j and setting j to zero afterwards [132]. This is in full analogy to the derivation of the magnetisation and the susceptibility from the free energy via derivatives with respect to the longitudinal field.
- In the linear coupling formalism, Equation (A6) leads to the later used fluctuation-dissipation relation between the two-point correlator and the susceptibility [113],

$$\mathfrak{Im}[\chi(q,\omega)] = f(\omega/T)S(q,\omega) \tag{A41}$$

with $f(\omega/T)$ being a function implementing the temperature occupation, $S(q,\omega)$ the dynamic structure factor and $\mathfrak{Im}[\chi(q,\omega)]$ the imaginary part of the susceptibility in momentum and frequency space.

Appendix B. Generalised Homogenous Functions

The following definitions, theorems and comments on generalised homogeneous functions are replicated from [124]. Proofs to all presented theorems can be found in [124].

Definition A1. A function $f(x_1, x_2, \ldots, x_n)$ is a generalised homogeneous function (GHF) with n variables if there exist n numbers $a_1, a_2, \ldots, a_n$ such that for all positive $\lambda \in \mathbb{R}^+$,

$$f(\lambda^{a_1}x_1, \lambda^{a_2}x_2, \ldots, \lambda^{a_n}x_n) = \lambda^{a_f} f(x_1, x_2, \ldots, x_n)\,. \tag{A42}$$

$a_1, a_2, \ldots, a_n$ are being referred to as scaling powers of the respective variables $x_1, x_2, \ldots, x_n$ and a_f is the scaling power of $f(x_1, x_2, \ldots, x_n)$. The scaling powers $a_1, \ldots a_n$ are non-zero.

Remark A1. Regarding a GHF with n variables there are only n independent scaling powers. This can be seen by setting $\lambda = \tilde{\lambda}^p$, leading to

$$f(\tilde{\lambda}^{pa_1}x_1, \tilde{\lambda}^{pa_2}x_2, \ldots, \tilde{\lambda}^{pa_n}x_n) = \tilde{\lambda}^{pa_f} f(x_1, x_2, \ldots, x_n)\,. \tag{A43}$$

If $a_f = 0$ the statement that there are only n independent scaling powers is trivial. If $a_f \neq 0$ one can set $p = 1/a_f$ resulting in

$$f(\tilde{\lambda}^{a_1/a_f}x_1, \tilde{\lambda}^{a_2/a_f}x_2, \ldots, \tilde{\lambda}^{a_n/a_f}x_n) = \tilde{\lambda} f(x_1, x_2, \ldots, x_n)\,. \tag{A44}$$

Theorem A1. Let $f(x_1, x_2, \ldots, x_n)$ and $g(x_1, x_2, \ldots, x_n)$ be GHF with scaling powers a_1, $a_2, \ldots, a_n$ and respective a_f and a_g then,

- $f \cdot g$ is a homogeneous function with scaling power $a_f + a_g$.
- $f + g$ is only a homogeneous function if $a_f = a_g$. The scaling power of $f + g$ is then a_f.

Theorem A2. If $f(x_1, x_2, \ldots, x_n)$ is a GHF with n variables with scaling power a_f, then

$$f^{(j_1,j_2,\ldots,j_n)}(x_1, x_2, \ldots, x_n) = \frac{\partial^{j_1}}{\partial x_1^{j_1}}\frac{\partial^{j_2}}{\partial x_2^{j_2}}\cdots\frac{\partial^{j_n}}{\partial x_n^{j_n}} f(x_1, x_2, \ldots, x_n) \tag{A45}$$

is also a GHF

$$f^{(j_1,j_2,\ldots,j_n)}(\lambda^{a_1}x_1, \lambda^{a_2}x_2, \ldots \lambda^{a_n}x_n) = \lambda^{a_f - \sum_{i=1}^n j_i a_i} f^{(j_1,j_2,\ldots,j_n)}(x_1, x_2, \ldots, x_n) \tag{A46}$$

with scaling power $a_f - \sum_{i=1}^n j_i a_i$.

Theorem A3. Let $f(x_1, x_2, \ldots, x_n)$ be a GHF with n variables and scaling power a_f. The Legendre transform of $f(x_1, x_2, \ldots, x_n)$ in which x_i is replaced as a free parameter by its conjugate variable $\overline{x}_i = f^{(0,0,\ldots,0,j_i=1,0,\ldots,0)}(x_1, x_2, \ldots, x_n)$

$$\overline{f}(x_1, x_2, \ldots, \overline{x}_i, \ldots, x_n) = f(x_1, x_2, \ldots, x_n) - x_i\overline{x}_i \tag{A47}$$

is also a GHF with the same scaling power $a_{\overline{f}} = a_f$. The scaling power of $\overline{x}_i$ is given by Theorem A2 as $\overline{a}_i = a_f - a_i$.

Theorem A4. A function $f(x_1, x_2)$ is a two-variable GHF with scaling power a_f and independent scaling powers a_1, a_2 if there exists some function $g_\pm(u)$ such that

$$f(x_1, x_2) = |x_1|^{a_f/a_1} g_{\mathrm{sgn}(x_1)}\left(\frac{x_2}{|x_1|^{a_2/a_1}}\right) \tag{A48}$$

or equally, if there exists some function $h_\pm(u)$ such that

$$f(x_1, x_2) = |x_2|^{a_f/a_2} h_{\mathrm{sgn}(x_2)}\left(\frac{x_1}{|x_2|^{a_1/a_2}}\right). \tag{A49}$$

Conversely if f is a GHF then,

$$g_\pm(u) = f(\pm 1, u) \tag{A50}$$

$$h_\pm(u) = f(u, \pm 1)\,. \tag{A51}$$

The statement of this theorem is straightforward generalisable to n-variable GHF. Due to limitations in compact notation for the formulation two-variable GHF are used.

Theorem A5. All GHFs have power-law singularities at the origin when it is approached along one of the principal axes. The exponent for the path of the approach $|x_i| \to 0$ is given by a_f/a_i. For a function of n variables that results in,

$$f(0, \ldots, 0, x_i, \ldots, 0) \propto |x_i|^{a_f/a_i} f(0, \ldots, 0, 1, \ldots, 0). \tag{A52}$$

Theorem A6. Let $f(x_1, x_2, \ldots, x_n)$ be a n-variable GHF with scaling power a_f. Let

$$\tilde{f}(x_1, x_2, \ldots, \tilde{x}_j, \ldots, x_n) = \int \mathrm{d}^d x_j\, f(x_1, x_2, \ldots, x_j, \ldots, x_n) \exp(i x_j \tilde{x}_j) \tag{A53}$$

denote the Fourier transform in which x_j is replaced by $\tilde{x}_j$ as the independent variable and d being the dimensionality of the variable. $\tilde{f}(x_1, x_2, \ldots, \tilde{x}_j, \ldots x_n)$ is also a GHF with scaling power $\tilde{a}_f = a_f - d\tilde{a}_j$ and the scaling power of the transformed variable $\tilde{x}_j$ being $\tilde{a}_j = -a_j$

$$\tilde{f}(\lambda^{a_1} x_1, \lambda^{a_2} x_2, \ldots, \lambda^{\tilde{a}_j} \tilde{x}_j, \ldots, \lambda^{a_n} x_n) = \lambda^{a_f - d\tilde{a}_j} \tilde{f}(x_1, x_2, \ldots, \tilde{x}_j, \ldots x_n). \tag{A54}$$

Appendix C. Walker's Method of Alias

Walker's method of alias is a numerical technique to draw integer random numbers from a discrete PDF in constant time [212,214]. Let $P(\omega_j) = q_j$ be a discrete probability distribution with $j \in \{1, \ldots, N\}$ elements ω_j and their probabilities q_j. The key idea of Walker's method is to redistribute the unbalanced part of the probability distribution. A naive way to sample a discrete distribution would be the rejection method. This means to draw an integer $j \in \{1, \ldots, N\}$ uniformly and accept it with probability $q_j / \max(q_j)$ [214]. However, this leads to $\mathcal{O}(N)$ steps until a proposed move eventually gets accepted [214]. In the Walker's method one draws the integer j uniformly, possibly accepts it with a probability P_j and otherwise chooses its alias A_j. One therefore succeeds after one try. The probabilities P_j and the aliases A_j need to be determined in advance such that the correct distribution q_j is sampled. In Ref. [214] an algorithm to set up the tables for P_j and A_j with complexity $\mathcal{O}(N)$ is given. Note that the construction of the tables needs to be conducted only once before the actual simulation is performed. During the simulation we draw from q_j with the help of the precalculated P_j and A_j in $\mathcal{O}(1)$ time. The Walker's method therefore does not alter the complexity of the whole algorithm.

In Figure A1 we illustrate the algorithm by Fukui and Todo [214] to set up the tables P_j and A_j. The algorithm proceeds as follows: One starts by setting P_j to a preliminary distribution $P_j = N \cdot q_j$ and creates an empty alias table A_j. The table P_j is split into $P_j \geq 1$ and $P_j < 1$ (see Figure A1b). The elements of the partition $P_j < 1$ get filled up by the elements $P_j \geq 1$. Starting from the outermost right elements of both parts in Figure A1b one fills up the shortfall $1 - P_j$ (green blocks in Figure A1) of weight P_j and sets the alias A_j to the number corresponding to the weight left from the separating line of the partitions ($j = 5$ and $A_j = 6$ in Figure A1b). This alias is later chosen with probability $1 - P_j$ if j gets drawn. In order to compensate for this, one cuts the weight P_{A_j} of the alias by $1 - P_j$ (red blocks in Figure A1). If the new weight P_{A_j} falls below 1, it is transferred to the partition with $P_j < 1$ (see e.g., $j = 6$ in Figure A1d,e). This filling up of the shortfalls is performed as long as the partition $P_j < 1$ is not empty. Figure A1f therefore shows the last step.

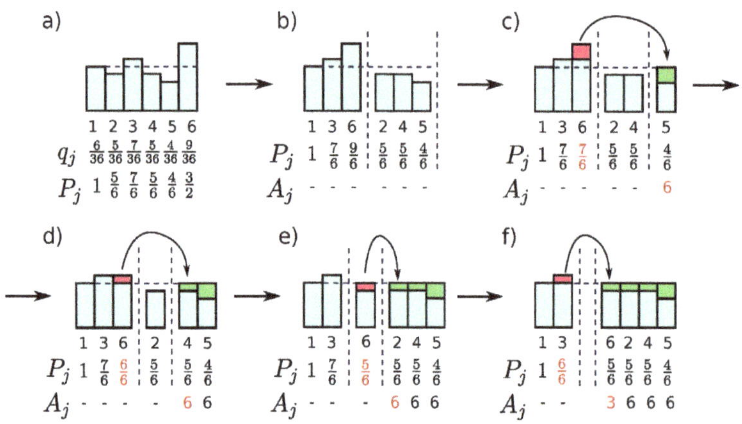

Figure A1. Illustration of Walker's method. The horizontal dashed line corresponds to the mean of the probability distribution q_j with $j \in \{1, \ldots, N\}$. (**a**) Definition of a tentative probability $P_j = n \cdot q_j$ which compares the probabilities q_j with the mean $q = 1/N$. (**b**) The probabilities q_j are discarded, a table A_j for the aliases is created and the distribution is splitted into $P_j \geq 1$ and $P_j < 1$. (**c–f**) The weights of P_j are gradually redistributed by creating aliases until the distribution is flattened. The weights that are taken away are depicted with red bars and the added weights with green bars. The values of P_j and A_j that get modified during a step are written in red. At the right part of the distributions, the weights which are already filled up are gathered. In the middle part of the distribution there are the weights which still need to be filled and the left part contains all weights that are at least full from the beginning and yet need to be redistributed. (**c**) Weight of 6 is added to 5 and $A_5 = 6$ is set. (**d**) Weight of 6 is added to 4 and $A_4 = 6$ is set. (**e**) Weight of 6 is added to 2 and $A_2 = 6$ is set. (**f**) Weight of 3 is added to 6 and $A_6 = 3$ is set. After this, the middle part is empty and we are done with the redistribution.

For drawing ω_j corresponding to its original weight q_j, one draws a candidate k for j uniformly in $1, \ldots, N$ and a uniform number $u \in [0, 1]$. If $u < P_k$ then $j = k$ and otherwise $j = A_k$.

Appendix D. Critical Exponents for Long-Range Heisenberg Ladders

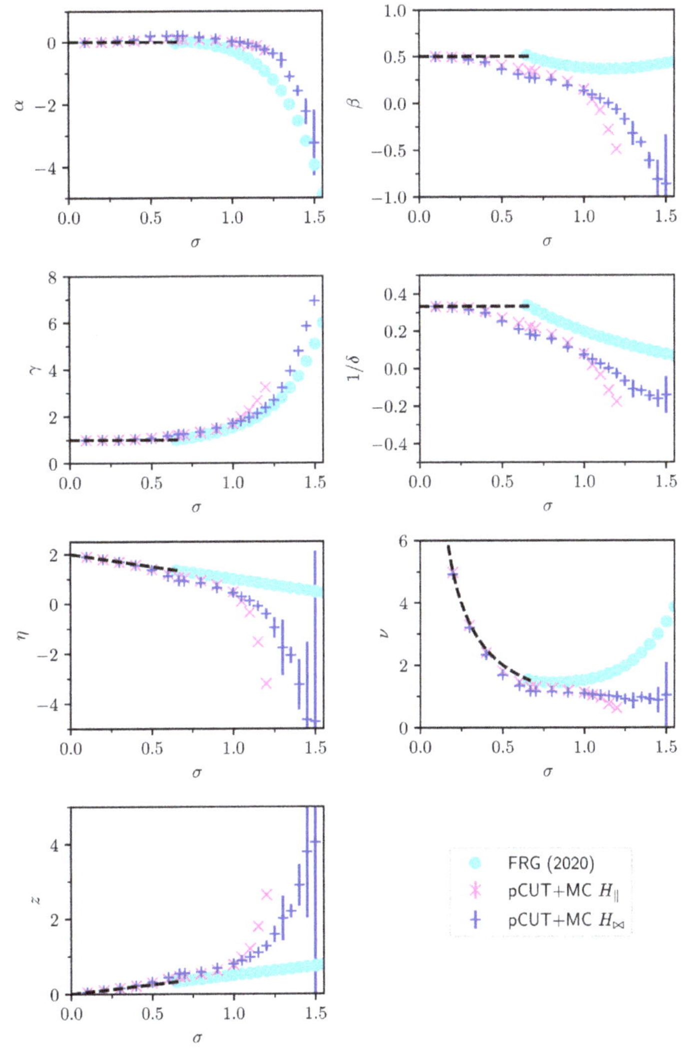

Figure A2. Critical exponents of the Néel-ordering transition in the unfrustrated antiferromagnetic long-range Heisenberg ladders. The panels display the exponents α (first row left), β (first row left), γ (second row left), δ (second row right), η (third row left), ν (third row right), and z (fourth row left). The data points 'pCUT+MC $H_{\parallel}$' and 'pCUT+MC $H_{\bowtie}$' are from Ref. [35,353] and refer to parallel ($\parallel$) and parallel + diagonal ($\bowtie$) interactions. The other data 'FRG 2020' is from Ref. [244] and show the critical exponents for the one-dimensional $O(3)$ quantum rotor model. The blue shaded region denotes the σ regime in which long-range mean-field criticality is expected. The black dashed lines denote long-range mean field critical exponents.

References

1. Sak, J. Recursion Relations and Fixed Points for Ferromagnets with Long-Range Interactions. *Phys. Rev. B* **1973**, *8*, 281–285. [CrossRef]
2. Sak, J. Low-temperature renormalization group for ferromagnets with long-range interactions. *Phys. Rev. B* **1977**, *15*, 4344–4347. [CrossRef]
3. Fisher, M.E.; Ma, S.k.; Nickel, B.G. Critical Exponents for Long-Range Interactions. *Phys. Rev. Lett.* **1972**, *29*, 917–920. [CrossRef]
4. Maity, S.; Bhattacharya, U.; Dutta, A. One-dimensional quantum many body systems with long-range interactions. *J. Phys. A Math. Theor.* **2019**, *53*, 013001. [CrossRef]
5. Defenu, N.; Donner, T.; Macrì, T.; Pagano, G.; Ruffo, S.; Trombettoni, A. Long-range interacting quantum systems. *Rev. Mod. Phys.* **2023**, *95*, 035002. [CrossRef]
6. Wilson, K.G. Renormalization Group and Critical Phenomena. I. Renormalization Group and the Kadanoff Scaling Picture. *Phys. Rev. B* **1971**, *4*, 3174–3183. [CrossRef]
7. Wilson, K.G. Renormalization Group and Critical Phenomena. II. Phase-Space Cell Analysis of Critical Behavior. *Phys. Rev. B* **1971**, *4*, 3184–3205. [CrossRef]
8. Wilson, K.G.; Kogut, J. The renormalization group and the epsilon expansion. *Phys. Rep.* **1974**, *12*, 75–199. [CrossRef]
9. Wilson, K.G. The renormalization group and critical phenomena. *Rev. Mod. Phys.* **1983**, *55*, 583–600. [CrossRef]
10. Fisher, M.E. Renormalization group theory: Its basis and formulation in statistical physics. *Rev. Mod. Phys.* **1998**, *70*, 653–681. [CrossRef]
11. Stanley, H.E. Scaling, universality, and renormalization: Three pillars of modern critical phenomena. *Rev. Mod. Phys.* **1999**, *71*, S358–S366. [CrossRef]
12. Luijten, E.; Blöte, H.W.J. Monte Carlo method for spin models with long-range interactions. *Int. J. Mod. Phys. C* **1995**, *6*, 359–370. [CrossRef]
13. Luijten, E.; Blöte, H.W.J. Classical critical behavior of spin models with long-range interactions. *Phys. Rev. B* **1997**, *56*, 8945–8958. [CrossRef]
14. Jones, J.L.; Young, A.P. Finite-size scaling of the correlation length above the upper critical dimension in the five-dimensional Ising model. *Phys. Rev. B* **2005**, *71*, 174438. [CrossRef]
15. Berche, B.; Kenna, R.; Walter, J.C. Hyperscaling above the upper critical dimension. *Nucl. Phys. B* **2012**, *865*, 115–132. [CrossRef]
16. Kenna, R.; Berche, B. Fisher's scaling relation above the upper critical dimension. *EPL (Europhys. Lett.)* **2014**, *105*, 26005. [CrossRef]
17. Kenna, R.; Berche, B. A new critical exponent koppa and its logarithmic counterpart koppa-hat. *arXiv* **2014**, arXiv:1411.2754. [CrossRef]
18. Flores-Sola, E.J.; Berche, B.; Kenna, R.; Weigel, M. Finite-size scaling above the upper critical dimension in Ising models with long-range interactions. *Eur. Phys. J. B* **2015**, *88*, 28. [CrossRef]
19. Flores-Sola, E.; Berche, B.; Kenna, R.; Weigel, M. Role of Fourier Modes in Finite-Size Scaling above the Upper Critical Dimension. *Phys. Rev. Lett.* **2016**, *116*, 115701. [CrossRef] [PubMed]
20. Dutta, A.; Bhattacharjee, J.K. Phase transitions in the quantum Ising and rotor models with a long-range interaction. *Phys. Rev. B* **2001**, *64*, 184106. [CrossRef]
21. Defenu, N.; Trombettoni, A.; Ruffo, S. Criticality and phase diagram of quantum long-range O(N) models. *Phys. Rev. B* **2017**, *96*, 104432. [CrossRef]
22. Laflorencie, N.; Affleck, I.; Berciu, M. Critical phenomena and quantum phase transition in long range Heisenberg antiferromagnetic chains. *J. Stat. Mech. Theory Exp.* **2005**, *2005*, P12001. [CrossRef]
23. Koffel, T.; Lewenstein, M.; Tagliacozzo, L. Entanglement Entropy for the Long-Range Ising Chain in a Transverse Field. *Phys. Rev. Lett.* **2012**, *109*, 267203. [CrossRef] [PubMed]
24. Vodola, D.; Lepori, L.; Ercolessi, E.; Pupillo, G. Long-range Ising and Kitaev models: Phases, correlations and edge modes. *New J. Phys.* **2015**, *18*, 015001. [CrossRef]
25. Fey, S.; Schmidt, K.P. Critical behavior of quantum magnets with long-range interactions in the thermodynamic limit. *Phys. Rev. B* **2016**, *94*, 075156. [CrossRef]
26. Sun, G. Fidelity susceptibility study of quantum long-range antiferromagnetic Ising chain. *Phys. Rev. A* **2017**, *96*, 043621. [CrossRef]
27. Saadatmand, S.N.; Bartlett, S.D.; McCulloch, I.P. Phase diagram of the quantum Ising model with long-range interactions on an infinite-cylinder triangular lattice. *Phys. Rev. B* **2018**, *97*, 155116. [CrossRef]
28. Zhu, Z.; Sun, G.; You, W.L.; Shi, D.N. Fidelity and criticality of a quantum Ising chain with long-range interactions. *Phys. Rev. A* **2018**, *98*, 023607. [CrossRef]
29. Fey, S.; Kapfer, S.C.; Schmidt, K.P. Quantum Criticality of Two-Dimensional Quantum Magnets with Long-Range Interactions. *Phys. Rev. Lett.* **2019**, *122*, 017203. [CrossRef] [PubMed]
30. Koziol, J.; Fey, S.; Kapfer, S.C.; Schmidt, K.P. Quantum criticality of the transverse-field Ising model with long-range interactions on triangular-lattice cylinders. *Phys. Rev. B* **2019**, *100*, 144411. [CrossRef]
31. Adelhardt, P.; Koziol, J.A.; Schellenberger, A.; Schmidt, K.P. Quantum criticality and excitations of a long-range anisotropic XY chain in a transverse field. *Phys. Rev. B* **2020**, *102*, 174424. [CrossRef]

32. Koziol, J.A.; Langheld, A.; Kapfer, S.C.; Schmidt, K.P. Quantum-critical properties of the long-range transverse-field Ising model from quantum Monte Carlo simulations. *Phys. Rev. B* **2021**, *103*, 245135. [CrossRef]
33. Gonzalez-Lazo, E.; Heyl, M.; Dalmonte, M.; Angelone, A. Finite-temperature critical behavior of long-range quantum Ising models. *SciPost Phys.* **2021**, *11*, 76. [CrossRef]
34. Langheld, A.; Koziol, J.A.; Adelhardt, P.; Kapfer, S.C.; Schmidt, K.P. Scaling at quantum phase transitions above the upper critical dimension. *SciPost Phys.* **2022**, *13*, 88. [CrossRef]
35. Adelhardt, P.; Schmidt, K.P. Continuously varying critical exponents in long-range quantum spin ladders. *SciPost Phys.* **2023**, *15*, 87. [CrossRef]
36. Zhao, J.; Song, M.; Qi, Y.; Rong, J.; Meng, Z.Y. Finite-temperature critical behaviors in 2D long-range quantum Heisenberg model. *npj Quantum Mater.* **2023**, *8*, 59. [CrossRef]
37. Song, M.; Zhao, J.; Qi, Y.; Rong, J.; Meng, Z.Y. Quantum criticality and entanglement for the two-dimensional long-range Heisenberg bilayer. *Phys. Rev. B* **2024**, *109*, L081114. [CrossRef]
38. Song, M.; Zhao, J.; Zhou, C.; Meng, Z.Y. Dynamical properties of quantum many-body systems with long-range interactions. *Phys. Rev. Res.* **2023**, *5*, 033046. [CrossRef]
39. Sandvik, A.W. Stochastic series expansion method for quantum Ising models with arbitrary interactions. *Phys. Rev. E* **2003**, *68*, 056701. [CrossRef] [PubMed]
40. Humeniuk, S. Quantum Monte Carlo study of long-range transverse-field Ising models on the triangular lattice. *Phys. Rev. B* **2016**, *93*, 104412. [CrossRef]
41. Berche, B.; Ellis, T.; Holovatch, Y.; Kenna, R. Phase transitions above the upper critical dimension. *SciPost Phys. Lect. Notes* **2022**, 60. [CrossRef]
42. Dauxois, T.; Ruffo, S.; Arimondo, E.; Wilkens, M. Dynamics and Thermodynamics of Systems with Long-Range Interactions: An Introduction. In *Dynamics and Thermodynamics of Systems with Long-Range Interactions*; Dauxois, T., Ruffo, S., Arimondo, E., Wilkens, M., Eds.; Springer: Berlin/Heidelberg, Germany, 2002; pp. 1–19. [CrossRef]
43. Campa, A.; Dauxois, T.; Ruffo, S. Statistical mechanics and dynamics of solvable models with long-range interactions. *Phys. Rep.* **2009**, *480*, 57–159. [CrossRef]
44. French, R.H.; Parsegian, V.A.; Podgornik, R.; Rajter, R.F.; Jagota, A.; Luo, J.; Asthagiri, D.; Chaudhury, M.K.; Chiang, Y.m.; Granick, S.; et al. Long range interactions in nanoscale science. *Rev. Mod. Phys.* **2010**, *82*, 1887–1944. [CrossRef]
45. Bouchet, F.; Gupta, S.; Mukamel, D. Thermodynamics and dynamics of systems with long-range interactions. *Phys. A Stat. Mech. Its Appl.* **2010**, *389*, 4389–4405. [CrossRef]
46. Barredo, D.; Labuhn, H.; Ravets, S.; Lahaye, T.; Browaeys, A.; Adams, C.S. Coherent Excitation Transfer in a Spin Chain of Three Rydberg Atoms. *Phys. Rev. Lett.* **2015**, *114*, 113002. [CrossRef] [PubMed]
47. Barredo, D.; de Léséleuc, S.; Lienhard, V.; Lahaye, T.; Browaeys, A. An atom-by-atom assembler of defect-free arbitrary two-dimensional atomic arrays. *Science* **2016**, *354*, 1021–1023. [CrossRef]
48. Labuhn, H.; Barredo, D.; Ravets, S.; de Léséleuc, S.; Macrì, T.; Lahaye, T.; Browaeys, A. Tunable two-dimensional arrays of single Rydberg atoms for realizing quantum Ising models. *Nature* **2016**, *534*, 667–670. [CrossRef] [PubMed]
49. Lienhard, V.; de Léséleuc, S.; Barredo, D.; Lahaye, T.; Browaeys, A.; Schuler, M.; Henry, L.P.; Läuchli, A.M. Observing the Space- and Time-Dependent Growth of Correlations in Dynamically Tuned Synthetic Ising Models with Antiferromagnetic Interactions. *Phys. Rev. X* **2018**, *8*, 021070. [CrossRef]
50. Schauss, P. Quantum simulation of transverse Ising models with Rydberg atoms. *Quantum Sci. Technol.* **2018**, *3*, 023001. [CrossRef]
51. de Léséleuc, S.; Lienhard, V.; Scholl, P.; Barredo, D.; Weber, S.; Lang, N.; Büchler, H.P.; Lahaye, T.; Browaeys, A. Observation of a symmetry-protected topological phase of interacting bosons with Rydberg atoms. *Science* **2019**, *365*, 775–780. [CrossRef] [PubMed]
52. Browaeys, A.; Lahaye, T. Many-body physics with individually controlled Rydberg atoms. *Nat. Phys.* **2020**, *16*, 132–142. [CrossRef]
53. Scholl, P.; Schuler, M.; Williams, H.J.; Eberharter, A.A.; Barredo, D.; Schymik, K.N.; Lienhard, V.; Henry, L.P.; Lang, T.C.; Lahaye, T.; et al. Quantum simulation of 2D antiferromagnets with hundreds of Rydberg atoms. *Nature* **2021**, *595*, 233–238. [CrossRef] [PubMed]
54. Samajdar, R.; Ho, W.W.; Pichler, H.; Lukin, M.D.; Sachdev, S. Quantum phases of Rydberg atoms on a kagome lattice. *Proc. Natl. Acad. Sci. USA* **2021**, *118*, e2015785118. [CrossRef] [PubMed]
55. Semeghini, G.; Levine, H.; Keesling, A.; Ebadi, S.; Wang, T.T.; Bluvstein, D.; Verresen, R.; Pichler, H.; Kalinowski, M.; Samajdar, R.; et al. Probing topological spin liquids on a programmable quantum simulator. *Science* **2021**, *374*, 1242–1247. [CrossRef] [PubMed]
56. Jaksch, D.; Bruder, C.; Cirac, J.I.; Gardiner, C.W.; Zoller, P. Cold Bosonic Atoms in Optical Lattices. *Phys. Rev. Lett.* **1998**, *81*, 3108–3111. [CrossRef]
57. Greiner, M.; Mandel, O.; Esslinger, T.; Hänsch, T.W.; Bloch, I. Quantum phase transition from a superfluid to a Mott insulator in a gas of ultracold atoms. *Nature* **2002**, *415*, 39–44. [CrossRef] [PubMed]
58. Góral, K.; Santos, L.; Lewenstein, M. Quantum Phases of Dipolar Bosons in Optical Lattices. *Phys. Rev. Lett.* **2002**, *88*, 170406. [CrossRef] [PubMed]
59. Bloch, I. Ultracold quantum gases in optical lattices. *Nat. Phys.* **2005**, *1*, 23–30. [CrossRef]

60. Bloch, I.; Greiner, M. Exploring Quantum Matter with Ultracold Atoms in Optical Lattices. In *Advances in Atomic, Molecular, and Optical Physics*; Elsevier: Amsterdam, The Netherlands, 2005; pp. 1–47. [CrossRef]
61. Kovrizhin, D.L.; Pai, G.V.; Sinha, S. Density wave and supersolid phases of correlated bosons in an optical lattice. *Europhys. Lett. (EPL)* **2005**, *72*, 162–168. [CrossRef]
62. Menotti, C.; Trefzger, C.; Lewenstein, M. Metastable States of a Gas of Dipolar Bosons in a 2D Optical Lattice. *Phys. Rev. Lett.* **2007**, *98*, 235301. [CrossRef]
63. Bloch, I.; Dalibard, J.; Zwerger, W. Many-body physics with ultracold gases. *Rev. Mod. Phys.* **2008**, *80*, 885–964. [CrossRef]
64. Capogrosso-Sansone, B.; Trefzger, C.; Lewenstein, M.; Zoller, P.; Pupillo, G. Quantum Phases of Cold Polar Molecules in 2D Optical Lattices. *Phys. Rev. Lett.* **2010**, *104*, 125301. [CrossRef]
65. Bloch, I.; Dalibard, J.; Nascimbène, S. Quantum simulations with ultracold quantum gases. *Nat. Phys.* **2012**, *8*, 267–276. [CrossRef]
66. Yamamoto, D.; Masaki, A.; Danshita, I. Quantum phases of hard-core bosons with long-range interactions on a square lattice. *Phys. Rev. B* **2012**, *86*, 054516. [CrossRef]
67. Moses, S.A.; Covey, J.P.; Miecnikowski, M.T.; Yan, B.; Gadway, B.; Ye, J.; Jin, D.S. Creation of a low-entropy quantum gas of polar molecules in an optical lattice. *Science* **2015**, *350*, 659–662. [CrossRef] [PubMed]
68. Baier, S.; Mark, M.J.; Petter, D.; Aikawa, K.; Chomaz, L.; Cai, Z.; Baranov, M.; Zoller, P.; Ferlaino, F. Extended Bose–Hubbard models with ultracold magnetic atoms. *Science* **2016**, *352*, 201–205. [CrossRef] [PubMed]
69. Moses, S.A.; Covey, J.P.; Miecnikowski, M.T.; Jin, D.S.; Ye, J. New frontiers for quantum gases of polar molecules. *Nat. Phys.* **2017**, *13*, 13–20. [CrossRef]
70. Gross, C.; Bloch, I. Quantum simulations with ultracold atoms in optical lattices. *Science* **2017**, *357*, 995–1001. [CrossRef] [PubMed]
71. Schäfer, F.; Fukuhara, T.; Sugawa, S.; Takasu, Y.; Takahashi, Y. Tools for quantum simulation with ultracold atoms in optical lattices. *Nat. Rev. Phys.* **2020**, *2*, 411–425. [CrossRef]
72. Chomaz, L.; Ferrier-Barbut, I.; Ferlaino, F.; Laburthe-Tolra, B.; Lev, B.L.; Pfau, T. Dipolar physics: A review of experiments with magnetic quantum gases. *Rep. Prog. Phys.* **2022**, *86*, 026401. [CrossRef] [PubMed]
73. Su, L.; Douglas, A.; Szurek, M.; Groth, R.; Ozturk, S.F.; Krahn, A.; Hébert, A.H.; Phelps, G.A.; Ebadi, S.; Dickerson, S.; et al. Dipolar quantum solids emerging in a Hubbard quantum simulator. *Nature* **2023**, *622*, 724–729. [CrossRef] [PubMed]
74. Friedenauer, A.; Schmitz, H.; Glueckert, J.T.; Porras, D.; Schaetz, T. Simulating a quantum magnet with trapped ions. *Nat. Phys.* **2008**, *4*, 757–761. [CrossRef]
75. Kim, K.; Chang, M.S.; Islam, R.; Korenblit, S.; Duan, L.M.; Monroe, C. Entanglement and Tunable Spin-Spin Couplings between Trapped Ions Using Multiple Transverse Modes. *Phys. Rev. Lett.* **2009**, *103*, 120502. [CrossRef] [PubMed]
76. Kim, K.; Chang, M.S.; Korenblit, S.; Islam, R.; Edwards, E.E.; Freericks, J.K.; Lin, G.D.; Duan, L.M.; Monroe, C. Quantum simulation of frustrated Ising spins with trapped ions. *Nature* **2010**, *465*, 590–593. [CrossRef]
77. Islam, R.; Edwards, E.E.; Kim, K.; Korenblit, S.; Noh, C.; Carmichael, H.; Lin, G.D.; Duan, L.M.; Joseph Wang, C.C.; Freericks, J.K.; et al. Onset of a quantum phase transition with a trapped ion quantum simulator. *Nat. Commun.* **2011**, *2*, 377. [CrossRef] [PubMed]
78. Blatt, R.; Roos, C.F. Quantum simulations with trapped ions. *Nat. Phys.* **2012**, *8*, 277–284. [CrossRef]
79. Schneider, C.; Porras, D.; Schaetz, T. Experimental quantum simulations of many-body physics with trapped ions. *Rep. Prog. Phys.* **2012**, *75*, 024401. [CrossRef] [PubMed]
80. Britton, J.W.; Sawyer, B.C.; Keith, A.C.; Wang, C.C.J.; Freericks, J.K.; Uys, H.; Biercuk, M.J.; Bollinger, J.J. Engineered two-dimensional Ising interactions in a trapped-ion quantum simulator with hundreds of spins. *Nature* **2012**, *484*, 489–492. [CrossRef]
81. Islam, R.; Senko, C.; Campbell, W.C.; Korenblit, S.; Smith, J.; Lee, A.; Edwards, E.E.; Wang, C.C.J.; Freericks, J.K.; Monroe, C. Emergence and Frustration of Magnetism with Variable-Range Interactions in a Quantum Simulator. *Science* **2013**, *340*, 583–587. [CrossRef] [PubMed]
82. Jurcevic, P.; Lanyon, B.P.; Hauke, P.; Hempel, C.; Zoller, P.; Blatt, R.; Roos, C.F. Quasiparticle engineering and entanglement propagation in a quantum many-body system. *Nature* **2014**, *511*, 202–205. [CrossRef] [PubMed]
83. Richerme, P.; Gong, Z.X.; Lee, A.; Senko, C.; Smith, J.; Foss-Feig, M.; Michalakis, S.; Gorshkov, A.V.; Monroe, C. Non-local propagation of correlations in quantum systems with long-range interactions. *Nature* **2014**, *511*, 198–201. [CrossRef]
84. Bohnet, J.G.; Sawyer, B.C.; Britton, J.W.; Wall, M.L.; Rey, A.M.; Foss-Feig, M.; Bollinger, J.J. Quantum spin dynamics and entanglement generation with hundreds of trapped ions. *Science* **2016**, *352*, 1297–1301. [CrossRef] [PubMed]
85. Monroe, C.; Campbell, W.; Duan, L.M.; Gong, Z.X.; Gorshkov, A.; Hess, P.; Islam, R.; Kim, K.; Linke, N.; Pagano, G.; et al. Programmable quantum simulations of spin systems with trapped ions. *Rev. Mod. Phys.* **2021**, *93*. [CrossRef]
86. Douglas, J.S.; Habibian, H.; Hung, C.L.; Gorshkov, A.V.; Kimble, H.J.; Chang, D.E. Quantum many-body models with cold atoms coupled to photonic crystals. *Nat. Photonics* **2015**, *9*, 326–331. [CrossRef]
87. Vaidya, V.D.; Guo, Y.; Kroeze, R.M.; Ballantine, K.E.; Kollár, A.J.; Keeling, J.; Lev, B.L. Tunable-Range, Photon-Mediated Atomic Interactions in Multimode Cavity QED. *Phys. Rev. X* **2018**, *8*, 011002. [CrossRef]
88. Dubin, D.H.E.; O'Neil, T.M. Trapped nonneutral plasmas, liquids, and crystals (the thermal equilibrium states). *Rev. Mod. Phys.* **1999**, *71*, 87–172. [CrossRef]
89. Matsubara, T.; Matsuda, H. A Lattice Model of Liquid Helium, I. *Prog. Theor. Phys.* **1956**, *16*, 569–582. [CrossRef]
90. Harris, M.J.; Bramwell, S.T.; McMorrow, D.F.; Zeiske, T.; Godfrey, K.W. Geometrical Frustration in the Ferromagnetic Pyrochlore $Ho_2Ti_2O_7$. *Phys. Rev. Lett.* **1997**, *79*, 2554–2557. [CrossRef]

91. Bramwell, S.T.; Harris, M.J. Frustration in Ising-type spin models on the pyrochlore lattice. *J. Phys. Condens. Matter* **1998**, *10*, L215–L220. [CrossRef]
92. Ramirez, A.P.; Hayashi, A.; Cava, R.J.; Siddharthan, R.; Shastry, B.S. Zero-point entropy in 'spin ice'. *Nature* **1999**, *399*, 333–335. [CrossRef]
93. Siddharthan, R.; Shastry, B.S.; Ramirez, A.P.; Hayashi, A.; Cava, R.J.; Rosenkranz, S. Ising Pyrochlore Magnets: Low-Temperature Properties, "Ice Rules," and Beyond. *Phys. Rev. Lett.* **1999**, *83*, 1854–1857. [CrossRef]
94. den Hertog, B.C.; Gingras, M.J.P. Dipolar Interactions and Origin of Spin Ice in Ising Pyrochlore Magnets. *Phys. Rev. Lett.* **2000**, *84*, 3430–3433. [CrossRef] [PubMed]
95. Melko, R.G.; den Hertog, B.C.; Gingras, M.J.P. Long-Range Order at Low Temperatures in Dipolar Spin Ice. *Phys. Rev. Lett.* **2001**, *87*, 067203. [CrossRef] [PubMed]
96. Melko, R.G.; Gingras, M.J.P. Monte Carlo studies of the dipolar spin ice model. *J. Phys. Condens. Matter* **2004**, *16*, R1277–R1319. [CrossRef]
97. Bramwell, S.T.; Gingras, M.J.P. Spin Ice State in Frustrated Magnetic Pyrochlore Materials. *Science* **2001**, *294*, 1495–1501. [CrossRef] [PubMed]
98. Ruff, J.P.C.; Melko, R.G.; Gingras, M.J.P. Finite-Temperature Transitions in Dipolar Spin Ice in a Large Magnetic Field. *Phys. Rev. Lett.* **2005**, *95*, 097202. [CrossRef] [PubMed]
99. Fennell, T.; Deen, P.P.; Wildes, A.R.; Schmalzl, K.; Prabhakaran, D.; Boothroyd, A.T.; Aldus, R.J.; McMorrow, D.F.; Bramwell, S.T. Magnetic Coulomb Phase in the Spin Ice Ho2Ti2O7. *Science* **2009**, *326*, 415–417. [CrossRef] [PubMed]
100. Castelnovo, C.; Moessner, R.; Sondhi, S.L. Magnetic monopoles in spin ice. *Nature* **2008**, *451*, 42–45. [CrossRef] [PubMed]
101. Jaubert, L.D.C.; Holdsworth, P.C.W. Signature of magnetic monopole and Dirac string dynamics in spin ice. *Nat. Phys.* **2009**, *5*, 258–261. [CrossRef]
102. Bitko, D.; Rosenbaum, T.F.; Aeppli, G. Quantum Critical Behavior for a Model Magnet. *Phys. Rev. Lett.* **1996**, *77*, 940–943. [CrossRef] [PubMed]
103. Chakraborty, P.B.; Henelius, P.; Kjønsberg, H.; Sandvik, A.W.; Girvin, S.M. Theory of the magnetic phase diagram of $LiHoF_4$. *Phys. Rev. B* **2004**, *70*, 144411. [CrossRef]
104. Ronnow, H.M.; Parthasarathy, R.; Jensen, J.; Aeppli, G.; Rosenbaum, T.F.; McMorrow, D.F. Quantum Phase Transition of a Magnet in a Spin Bath. *Science* **2005**, *308*, 389–392. [CrossRef] [PubMed]
105. Gingras, M.J.P.; Henelius, P. Collective Phenomena in the LiHoxY1-xF4Quantum Ising Magnet: Recent Progress and Open Questions. *J. Phys. Conf. Ser.* **2011**, *320*, 012001. [CrossRef]
106. Tiwari, A.; Ahn, H.; Kumar, B.; Saini, J.; Srivastava, P.K.; Singh, B.; Lee, C.; Ghosh, S. Phase transition in the two-dimensional Heisenberg ferromagnet Fe_3GeTe_2 with long-range interaction. *Phys. Rev. B* **2024**, *109*, L020407. [CrossRef]
107. Burch, K.S.; Mandrus, D.; Park, J.G. Magnetism in two-dimensional van der Waals materials. *Nature* **2018**, *563*, 47–52. [CrossRef] [PubMed]
108. Wang, Q.H.; Bedoya-Pinto, A.; Blei, M.; Dismukes, A.H.; Hamo, A.; Jenkins, S.; Koperski, M.; Liu, Y.; Sun, Q.C.; Telford, E.J.; et al. The Magnetic Genome of Two-Dimensional van der Waals Materials. *ACS Nano* **2022**, *16*, 6960–7079. [CrossRef] [PubMed]
109. Verresen, R.; Lukin, M.D.; Vishwanath, A. Prediction of Toric Code Topological Order from Rydberg Blockade. *Phys. Rev. X* **2021**, *11*, 031005. [CrossRef]
110. Koziol, J.A.; Duft, A.; Morigi, G.; Schmidt, K.P. Systematic analysis of crystalline phases in bosonic lattice models with algebraically decaying density–density interactions. *SciPost Phys.* **2023**, *14*, 136. [CrossRef]
111. Koziol, J.A.; Morigi, G.; Schmidt, K.P. Quantum phases of hard-core bosons with repulsive dipolar density–density interactions on two-dimensional lattices. *arXiv* **2023**, arXiv:2311.10632. [CrossRef]
112. Duft, A.; Koziol, J.A.; Adelhardt, P.; Mühlhauser, M.; Schmidt, K.P. Order-by-disorder in the antiferromagnetic long-range transverse-field Ising model on the ruby lattice. *arXiv* **2023**, arXiv:2312.12941. [CrossRef]
113. Sachdev, S. *Quantum Phase Transitions*, 2nd ed.; Cambridge University Press: Cambridge, UK, 2011. [CrossRef]
114. Jaeger, G. The Ehrenfest Classification of Phase Transitions: Introduction and Evolution. *Arch. Hist. Exact Sci.* **1998**, *53*, 51–81. [CrossRef]
115. Suzuki, M. Relationship between d-Dimensional Quantal Spin Systems and (d+1)-Dimensional Ising Systems: Equivalence, Critical Exponents and Systematic Approximants of the Partition Function and Spin Correlations. *Prog. Theor. Phys.* **1976**, *56*, 1454–1469. [CrossRef]
116. Sondhi, S.L.; Girvin, S.M.; Carini, J.P.; Shahar, D. Continuous quantum phase transitions. *Rev. Mod. Phys.* **1997**, *69*, 315–333. [CrossRef]
117. Vojta, M. Quantum phase transitions. *Rep. Prog. Phys.* **2003**, *66*, 2069–2110. [CrossRef]
118. Widom, B. Equation of State in the Neighborhood of the Critical Point. *J. Chem. Phys.* **1965**, *43*, 3898–3905. [CrossRef]
119. Domb, C.; Hunter, D.L. On the critical behaviour of ferromagnets. *Proc. Phys. Soc.* **1965**, *86*, 1147. [CrossRef]
120. Kadanoff, L.P. Scaling laws for ising models near T_c. *Phys. Phys. Fiz.* **1966**, *2*, 263–272. [CrossRef]
121. Patashinskii, A.; Pokrovskii, V. Behavior of ordered systems near the transition point. *Sov. Phys. JETP* **1966**, *23*, 292–297.
122. Fisher, M.E. The theory of equilibrium critical phenomena. *Rep. Prog. Phys.* **1967**, *30*, 615–730. [CrossRef]
123. Kadanoff, L.P.; Götze, W.; Hamblen, D.; Hecht, R.; Lewis, E.A.S.; Palciauskas, V.V.; Rayl, M.; Swift, J.; Aspnes, D.; Kane, J. Static Phenomena Near Critical Points: Theory and Experiment. *Rev. Mod. Phys.* **1967**, *39*, 395–431. [CrossRef]

124. Hankey, A.; Stanley, H.E. Systematic Application of Generalized Homogeneous Functions to Static Scaling, Dynamic Scaling, and Universality. *Phys. Rev. B* **1972**, *6*, 3515–3542. [CrossRef]
125. Kenna, R. Finite size scaling for O(N) Phi4-theory at the upper critical dimension. *Nucl. Phys. B* **2004**, *691*, 292–304. [CrossRef]
126. Brézin, E. An investigation of finite size scaling. *J. Phys. Fr.* **1982**, *43*, 15–22. [CrossRef]
127. Fisher, M.E. Scaling, universality and renormalization group theory. In *Critical Phenomena*; Hahne, F.J.W., Ed.; Springer: Berlin/Heidelberg, Germany, 1983; pp. 1–139.
128. Binder, K.; Nauenberg, M.; Privman, V.; Young, A.P. Finite-size tests of hyperscaling. *Phys. Rev. B* **1985**, *31*, 1498–1502. [CrossRef] [PubMed]
129. Kirkpatrick, T.R.; Belitz, D. Exponent relations at quantum phase transitions with applications to metallic quantum ferromagnets. *Phys. Rev. B* **2015**, *91*, 214407. [CrossRef]
130. Mermin, N.D.; Wagner, H. Absence of Ferromagnetism or Antiferromagnetism in One- or Two-Dimensional Isotropic Heisenberg Models. *Phys. Rev. Lett.* **1966**, *17*, 1133. Erratum in *Phys. Rev. Lett.* **1966**, *17*, 1307. [CrossRef]
131. Hohenberg, P.C. Existence of Long-Range Order in One and Two Dimensions. *Phys. Rev.* **1967**, *158*, 383–386. [CrossRef]
132. Kleinert, H.; Schulte-Frohlinde, V. *Critical Properties of Phi4-Theories*; World Scientific: Singapore, 2001. [CrossRef]
133. Joshi, D.G.; Coester, K.; Schmidt, K.P.; Vojta, M. Nonlinear bond-operator theory and $1/d$ expansion for coupled-dimer magnets. I. Paramagnetic phase. *Phys. Rev. B* **2015**, *91*, 094404. [CrossRef]
134. Joshi, D.G.; Vojta, M. Nonlinear bond-operator theory and $1/d$ expansion for coupled-dimer magnets. II. Antiferromagnetic phase and quantum phase transition. *Phys. Rev. B* **2015**, *91*, 094405. [CrossRef]
135. Coester, K.; Joshi, D.G.; Vojta, M.; Schmidt, K.P. Linked-cluster expansions for quantum magnets on the hypercubic lattice. *Phys. Rev. B* **2016**, *94*, 125109. [CrossRef]
136. Widom, B. Surface Tension and Molecular Correlations near the Critical Point. *J. Chem. Phys.* **1965**, *43*, 3892–3897. [CrossRef]
137. Essam, J.W.; Fisher, M.E. Padé Approximant Studies of the Lattice Gas and Ising Ferromagnet below the Critical Point. *J. Chem. Phys.* **1963**, *38*, 802–812. [CrossRef]
138. Fisher, M.E. Correlation Functions and the Critical Region of Simple Fluids. *J. Math. Phys.* **1964**, *5*, 944–962. [CrossRef]
139. Cardy, J. *Finite-Size Scaling*; Elsevier: Amsterdam, The Netherlands, 1988.
140. Fisher, M.E. The renormalization group in the theory of critical behavior. *Rev. Mod. Phys.* **1974**, *46*, 597–616. [CrossRef]
141. Pelissetto, A.; Vicari, E. Critical phenomena and renormalization-group theory. *Phys. Rep.* **2002**, *368*, 549–727. [CrossRef]
142. Binder, K. Critical properties and finite-size effects of the five-dimensional Ising model. *Z. Phys. B-Condens. Matter* **1985**, *61*, 13–23. [CrossRef]
143. Binder, K. Finite size effects on phase transitions. *Ferroelectrics* **1987**, *73*, 43–67. [CrossRef]
144. Binder, K.; Heermann, D. *Monte Carlo Simulation in Statistical Physics*; Graduate Texts in Physics; Springer: Cham, Switzerland, 2019; Chapter 2.3.4, pp. 47–56. [CrossRef]
145. Cardy, J. *Scaling and Renormalization in Statistical Physics*; Cambridge Lecture Notes in Physics; Cambridge University Press: Cambridge, UK, 1996; Chapter 5.4, pp. 92–94. [CrossRef]
146. Chib, S. Chapter 57—Markov Chain Monte Carlo Methods: Computation and Inference. In *Handbook of Econometrics*; Elsevier: Amsterdam, The Netherlands, 2001; Volume 5, pp. 3569–3649. [CrossRef]
147. Levin, D.A.; Peres, Y.; Wilmer, E.L. *Markov Chains and Mixing Times*; American Mathematical Society: Providence, RI, USA, 2006.
148. Krauth, W. *Statistical Mechanics: Algorithms and Computations*; Oxford University Press: Oxford, UK, 2006; Volume 13.
149. Metropolis, N.; Rosenbluth, A.W.; Rosenbluth, M.N.; Teller, A.H.; Teller, E. Equation of State Calculations by Fast Computing Machines. *J. Chem. Phys.* **1953**, *21*, 1087–1092. [CrossRef]
150. Hastings, W.K. Monte Carlo sampling methods using Markov chains and their applications. *Biometrika* **1970**, *57*, 97–109. [CrossRef]
151. Krauth, W. Event-Chain Monte Carlo: Foundations, Applications, and Prospects. *Front. Phys.* **2021**, *9*, 663457. [CrossRef]
152. Coester, K.; Schmidt, K.P. Optimizing linked-cluster expansions by white graphs. *Phys. Rev. E* **2015**, *92*, 022118. [CrossRef] [PubMed]
153. Knetter, C.; Uhrig, G. Perturbation theory by flow equations: Dimerized and frustrated S = 1/2 chain. *Eur. Phys. J. B* **2000**, *13*, 209–225. [CrossRef]
154. Knetter, C.; Schmidt, K.P.; Uhrig, G.S. The structure of operators in effective particle-conserving models. *J. Phys. A Math. Gen.* **2003**, *36*, 7889. [CrossRef]
155. Oitmaa, J.; Hamer, C.; Zheng, W. *Series Expansion Methods for Strongly Interacting Lattice Models*; Cambridge University Press: Cambridge, UK, 2006.
156. Knetter, C.; Bühler, A.; Müller-Hartmann, E.; Uhrig, G.S. Dispersion and Symmetry of Bound States in the Shastry-Sutherland Model. *Phys. Rev. Lett.* **2000**, *85*, 3958–3961. [CrossRef] [PubMed]
157. Dorier, J.; Schmidt, K.P.; Mila, F. Theory of Magnetization Plateaux in the Shastry-Sutherland Model. *Phys. Rev. Lett.* **2008**, *101*, 250402. [CrossRef] [PubMed]
158. Vidal, J.; Dusuel, S.; Schmidt, K.P. Low-energy effective theory of the toric code model in a parallel magnetic field. *Phys. Rev. B* **2009**, *79*, 033109. [CrossRef]
159. Kadanoff, L.P.; Kohmoto, M. Quantum mechanical ground states, nonlinear Schrodinger equations and linked cluster expansions. *J. Phys. Math. Gen.* **1981**, *14*, 1291. [CrossRef]

160. Marland, L.G. Series expansions for the zero-temperature transverse Ising model. *J. Phys. Math. Gen.* **1981**, *14*, 2047. [CrossRef]
161. Singh, R.R.P.; Gelfand, M.P.; Huse, D.A. Ground States of Low-Dimensional Quantum Antiferromagnets. *Phys. Rev. Lett.* **1988**, *61*, 2484–2487. [CrossRef] [PubMed]
162. Gelfand, M.P. Series expansions for excited states of quantum lattice models. *Solid State Commun.* **1996**, *98*, 11–14. [CrossRef]
163. Singh, R.R.P.; Gelfand, M.P. Spin-wave excitation spectra and spectral weights in square lattice antiferromagnets. *Phys. Rev. B* **1995**, *52*, R15695–R15698. [CrossRef] [PubMed]
164. Trebst, S.; Monien, H.; Hamer, C.J.; Weihong, Z.; Singh, R.R.P. Strong-Coupling Expansions for Multiparticle Excitations: Continuum and Bound States. *Phys. Rev. Lett.* **2000**, *85*, 4373–4376. [CrossRef]
165. Zheng, W.; Hamer, C.J.; Singh, R.R.P.; Trebst, S.; Monien, H. Linked cluster series expansions for two-particle bound states. *Phys. Rev. B* **2001**, *63*, 144410. [CrossRef]
166. Yang, H.Y.; Läuchli, A.M.; Mila, F.; Schmidt, K.P. Effective Spin Model for the Spin-Liquid Phase of the Hubbard Model on the Triangular Lattice. *Phys. Rev. Lett.* **2010**, *105*, 267204. [CrossRef] [PubMed]
167. Dusuel, S.; Kamfor, M.; Schmidt, K.P.; Thomale, R.; Vidal, J. Bound states in two-dimensional spin systems near the Ising limit: A quantum finite-lattice study. *Phys. Rev. B* **2010**, *81*, 064412. [CrossRef]
168. Powalski, M.; Coester, K.; Moessner, R.; Schmidt, K.P. Disorder by disorder and flat bands in the kagome transverse field Ising model. *Phys. Rev. B* **2013**, *87*, 054404. [CrossRef]
169. Schulz, M.D.; Dusuel, S.; Schmidt, K.P.; Vidal, J. Topological Phase Transitions in the Golden String-Net Model. *Phys. Rev. Lett.* **2013**, *110*, 147203. [CrossRef] [PubMed]
170. Schulz, M.D.; Dusuel, S.; Misguich, G.; Schmidt, K.P.; Vidal, J. Ising anyons with a string tension. *Phys. Rev. B* **2014**, *89*, 201103. [CrossRef]
171. Hörmann, M.; Schmidt, K.P. Projective cluster-additive transformation for quantum lattice models. *SciPost Phys.* **2023**, *15*, 097. [CrossRef]
172. Rutishauser, H. Ein infinitesimales Analogon zum Quotienten-Differenzen-Algorithmus. *Arch. Math.* **1954**, *5*, 132–137. [CrossRef]
173. Toda, M. Vibration of a Chain with Nonlinear Interaction. *J. Phys. Soc. Jpn.* **1967**, *22*, 431–436. [CrossRef]
174. Flaschka, H. The Toda lattice. II. Existence of integrals. *Phys. Rev. B* **1974**, *9*, 1924–1925. [CrossRef]
175. Moser, J. Finitely many mass points on the line under the influence of an exponential potential—An integrable system. In *Dynamical Systems, Theory and Applications: Battelle Seattle 1974 Rencontres*; Moser, J., Ed.; Springer: Berlin/Heidelberg, Germany, 1975; pp. 467–497. [CrossRef]
176. Brockett, R. Dynamical systems that sort lists, diagonalize matrices and solve linear programming problems. In Proceedings of the 27th IEEE Conference on Decision and Control, Austin, TX, USA, 7–9 December 1988; Volume 1, pp. 799–803. [CrossRef]
177. Głazek, S.D.; Wilson, K.G. Renormalization of Hamiltonians. *Phys. Rev. D* **1993**, *48*, 5863–5872. [CrossRef] [PubMed]
178. Wegner, F. Flow-equations for Hamiltonians. *Ann. Phys.* **1994**, *506*, 77–91. [CrossRef]
179. Stein, J. Flow equations and the strong-coupling expansion for the Hubbard model. *J. Stat. Phys.* **1997**, *88*, 487–511. [CrossRef]
180. Mielke, A. Flow equations for band-matrices. *Eur. Phys. J. B* **1998**, *5*, 605–611. [CrossRef]
181. Schulz, M.D. Topological Phase Transitions Driven by Non-Abelian Anyons. Ph.D. Thesis, TU Dortmund, Dortmund, Germany, Université Pierre-et-Marie-Curie, Paris, France, 2013.
182. Coester, K. Quasiparticle Pictures and Graphs—From Perturbative to Non-Perturbative Linked-Cluster Expansions. Ph.D. Thesis, TU Dortmund, Dortmund, Germany, 2015.
183. Weihong, Z. Various series expansions for the bilayer S=$\frac{1}{2}$ Heisenberg antiferromagnet. *Phys. Rev. B* **1997**, *55*, 12267–12275. [CrossRef]
184. Collins, A.; Hamer, C.J. Two-particle bound states and one-particle structure factor in a Heisenberg bilayer system. *Phys. Rev. B* **2008**, *78*, 054419. [CrossRef]
185. Klagges, D.; Schmidt, K.P. Constraints on Measurement-Based Quantum Computation in Effective Cluster States. *Phys. Rev. Lett.* **2012**, *108*, 230508. [CrossRef] [PubMed]
186. Boos, C.; Ganahl, C.J.; Lajkó, M.; Nataf, P.; Läuchli, A.M.; Penc, K.; Schmidt, K.P.; Mila, F. Time-reversal symmetry breaking Abelian chiral spin liquid in Mott phases of three-component fermions on the triangular lattice. *Phys. Rev. Res.* **2020**, *2*, 023098. [CrossRef]
187. Gelfand, M.P.; Singh, R.R.P. High-order convergent expansions for quantum many particle systems. *Adv. Phys.* **2000**, *49*, 93–140. [.]. [CrossRef]
188. Feiguin, A.E. The Density Matrix Renormalization Group and its time-dependent variants. *AIP Conf. Proc.* **2011**, *1419*, 5–92. [CrossRef]
189. Squires, G.L. *Introduction to the Theory of Thermal Neutron Scattering*, 3rd ed.; Cambridge University Press: Cambridge, UK, 2012. [CrossRef]
190. Zaliznyak, I.A.; Tranquada, J.M. Neutron Scattering and Its Application to Strongly Correlated Systems. In *Strongly Correlated Systems: Experimental Techniques*; Avella, A., Mancini, F., Eds.; Springer: Berlin/Heidelberg, Germany, 2015; pp. 205–235. [CrossRef]
191. Hamer, C.J.; Zheng, W.; Singh, R.R.P. Dynamical structure factor for the alternating Heisenberg chain: A linked cluster calculation. *Phys. Rev. B* **2003**, *68*, 214408. [CrossRef]

192. Hamer, C.J. Critical behaviour of structure factors at a quantum phase transition. *J. Phys. Condens. Matter* **2009**, *21*, 124205. [CrossRef] [PubMed]
193. Rahman, M.S. *Basic Graph Theory*; Springer: Berlin/Heidelberg, Germany, 2017; Volume 9.
194. Harary, F. *Graph Theory*; Addison-Wesley Series in Mathematics; Addison-Wesley Publishing Company: Boston, MA, USA, 1969.
195. Mühlhauser, M. Graph Decomposition Techniques and Their Application to Topological Systems. Ph.D. Thesis, Friedrich-Alexander-Universität Erlangen-Nürnberg, Erlangen, Germany, 2024; To be published.
196. Grohe, M.; Schweitzer, P. The graph isomorphism problem. *Commun. ACM* **2020**, *63*, 128–134. [CrossRef]
197. Bonnici, V.; Giugno, R.; Pulvirenti, A.; Shasha, D.; Ferro, A. A subgraph isomorphism algorithm and its application to biochemical data. *BMC Bioinform.* **2013**, *14*, S13. [CrossRef] [PubMed]
198. Bonnici, V.; Giugno, R. On the Variable Ordering in Subgraph Isomorphism Algorithms. *IEEE/ACM Trans. Comput. Biol. Bioinform.* **2017**, *14*, 193–203. [CrossRef] [PubMed]
199. Hartke, S.G.; Radcliffe, A. Mckay's canonical graph labeling algorithm. *Commun. Math.* **2009**, *479*, 99–111.
200. McKay, B.D.; Piperno, A. Practical graph isomorphism. *Congr. Numer.* **1981**, *30*, 45–87. [CrossRef]
201. McKay, B.D.; Piperno, A. Practical graph isomorphism, II. *J. Symb. Comput.* **2014**, *60*, 94–112. [CrossRef]
202. Hagberg, A.A.; Schult, D.A.; Swart, P.J. Exploring Network Structure, Dynamics, and Function using NetworkX. In Proceedings of the 7th Python in Science Conference, Pasadena, CA, USA, 19–24 August 2008; Varoquaux, G., Vaught, T., Millman, J., Eds.; pp. 11–15.
203. Siek, J.; Lee, L.; Lumsdaine, A. *The Boost Graph Library: User Guide and Reference Manual*; The C++ In-Depth Series; Pearson Education: London, UK, 2001.
204. Jahromi, S.S.; Hörmann, M.; Adelhardt, P.; Fey, S.; Orus, R.; Schmidt, K.P. Kitaev honeycomb antiferromagnet in a field: Quantum phase diagram for general spin. *arXiv* **2021**, arXiv:2111.06132. [CrossRef]
205. Kitaev, A. Anyons in an exactly solved model and beyond. *Ann. Phys.* **2006**, *321*, 2–111. [CrossRef]
206. Adelhardt, P.; Schmidt, K.P. *In Preparation* 2024. Friedrich-Alexander Universität Erlangen-Nürnberg, Erlangen, Bavaria, Germany.
207. Fey, S. Investigation of Zero-Temperature Transverse-Field Ising Models with Long-Range Interactions. Ph.D. Thesis, Friedrich-Alexander-Universität Erlangen-Nürnberg, Erlangen, Germany, 2020.
208. Baker, G. *Essentials of Padé Approximants*; Elsevier Science: Amsterdam, The Netherlands, 1975.
209. Guttmann, A.J. Asymptotic Analysis of Power-Series Expansions. In *Phase Transitions and Critical Phenomena*; Domb, C., Green, M.S., Lebowitz, J.L., Eds.; Academic Press: Cambridge, MA, USA, 1989; Volume 13.
210. Sandvik, A.W.; Kurkijärvi, J. Quantum Monte Carlo simulation method for spin systems. *Phys. Rev. B* **1991**, *43*, 5950–5961. [CrossRef]
211. Sandvik, A.W.; Singh, R.R.P.; Campbell, D.K. Quantum Monte Carlo in the interaction representation: Application to a spin-Peierls model. *Phys. Rev. B* **1997**, *56*, 14510–14528. [CrossRef]
212. Humeniuk, S. Quantum Monte Carlo Studies of Strongly Correlated Systems for Quantum Simulators. Ph.D. Thesis, Fakultät Mathematik und Physik, Universität Stuttgart, Stuttgart, Germany, 2018.
213. Sandvik, A.W.; Avella, A.; Mancini, F. Computational Studies of Quantum Spin Systems. *AIP Conf. Proc.* **2010**, *1297*, 135. [CrossRef]
214. Fukui, K.; Todo, S. Order-N cluster Monte Carlo method for spin systems with long-range interactions. *J. Comput. Phys.* **2009**, *228*, 2629–2642. [CrossRef]
215. Syljuåsen, O.F.; Sandvik, A.W. Quantum Monte Carlo with directed loops. *Phys. Rev. E* **2002**, *66*, 046701. [CrossRef] [PubMed]
216. Sandvik, A.W. Stochastic series expansion method with operator-loop update. *Phys. Rev. B* **1999**, *59*, R14157–R14160. [CrossRef]
217. Alet, F.; Wessel, S.; Troyer, M. Generalized directed loop method for quantum Monte Carlo simulations. *Phys. Rev. E* **2005**, *71*, 036706. [CrossRef] [PubMed]
218. Sandvik, A.W. A generalization of Handscomb's quantum Monte Carlo scheme-application to the 1D Hubbard model. *J. Phys. A Math. Gen.* **1992**, *25*, 3667–3682. [CrossRef]
219. Kubo, R. Statistical-Mechanical Theory of Irreversible Processes. I. General Theory and Simple Applications to Magnetic and Conduction Problems. *J. Phys. Soc. Jpn.* **1957**, *12*, 570–586. [CrossRef]
220. Michel, F.; Evertz, H.G. Lattice dynamics of the Heisenberg chain coupled to finite frequency bond phonons. *arXiv* **2007**, arXiv:0705.0799. [CrossRef]
221. Caracciolo, S.; Gambassi, A.; Gubinelli, M.; Pelissetto, A. Finite-size correlation length and violations of finite-size scaling. *Eur. Phys. J. B* **2001**, *20*, 594–598. [CrossRef]
222. Sadhukhan, D.; Dziarmaga, J. Is there a correlation length in a model with long-range interactions? *arXiv* **2021**, arXiv:2107.02508. [CrossRef]
223. Brankov, J.; Tonchev, N. Finite-size scaling for systems with long-range interactions. *Phys. A* **1992**, *189*, 583–610. [CrossRef]
224. Flores-Sola, E.J. Finite-Size Scaling above the Upper Critical Dimension. Ph.D. Thesis, Université de Lorraine, Nancy, France; Coventry University, Coventry, UK, 2016.
225. Sandvik, A.W. Classical percolation transition in the diluted two-dimensional $S = \frac{1}{2}$ Heisenberg antiferromagnet. *Phys. Rev. B* **2002**, *66*, 024418. [CrossRef]

226. Prokof'ev, N.; Svistunov, B.; Tupitsyn, I. "Worm" algorithm in quantum Monte Carlo simulations. *Phys. Lett. A* **1998**, *238*, 253–257. [CrossRef]
227. Pollet, L.; Houcke, K.; Rombouts, S. Engineering local optimality in quantum Monte Carlo algorithms. *J. Comput. Phys.* **2007**, *225*, 2249–2266. [CrossRef]
228. Söyler, S.G.; Capogrosso-Sansone, B.; Prokof'ev, N.V.; Svistunov, B.V. Sign-alternating interaction mediated by strongly correlated lattice bosons. *New J. Phys.* **2009**, *11*, 073036. [CrossRef]
229. Pollet, L.; Picon, J.; Büchler, H.; Troyer, M. Supersolid Phase with Cold Polar Molecules on a Triangular Lattice. *Phys. Rev. Lett.* **2010**, *104*, 125302. [CrossRef]
230. Maik, M.; Hauke, P.; Dutta, O.; Zakrzewski, J.; Lewenstein, M. Quantum spin models with long-range interactions and tunnelings: A quantum Monte Carlo study. *New J. Phys.* **2012**, *14*, 113006. [CrossRef]
231. Fisher, M.P.A.; Weichman, P.B.; Grinstein, G.; Fisher, D.S. Boson localization and the superfluid-insulator transition. *Phys. Rev. B* **1989**, *40*, 546–570. [CrossRef] [PubMed]
232. Scalettar, R.T.; Batrouni, G.G.; Zimanyi, G.T. Localization in interacting, disordered, Bose systems. *Phys. Rev. Lett.* **1991**, *66*, 3144–3147. [CrossRef]
233. Krauth, W.; Trivedi, N.; Ceperley, D. Superfluid-insulator transition in disordered boson systems. *Phys. Rev. Lett.* **1991**, *67*, 2307–2310. [CrossRef] [PubMed]
234. Batrouni, G.G.; Scalettar, R.T. Phase Separation in Supersolids. *Phys. Rev. Lett.* **2000**, *84*, 1599–1602. [CrossRef] [PubMed]
235. Hébert, F.; Batrouni, G.G.; Scalettar, R.T.; Schmid, G.; Troyer, M.; Dorneich, A. Quantum phase transitions in the two-dimensional hard-core boson model. *Phys. Rev. B* **2001**, *65*, 014513. [CrossRef]
236. Biedroń, K.; Łącki, M.; Zakrzewski, J. Extended Bose–Hubbard model with dipolar and contact interactions. *Phys. Rev. B* **2018**, *97*, 245102. [CrossRef]
237. Kraus, R.; Biedroń, K.; Zakrzewski, J.; Morigi, G. Superfluid phases induced by dipolar interactions. *Phys. Rev. B* **2020**, *101*, 174505. [CrossRef]
238. Kraus, R.; Chanda, T.; Zakrzewski, J.; Morigi, G. Quantum phases of dipolar bosons in one-dimensional optical lattices. *Phys. Rev. B* **2022**, *106*, 035144. [CrossRef]
239. Hirsch, J.E.; Sugar, R.L.; Scalapino, D.J.; Blankenbecler, R. Monte Carlo simulations of one-dimensional fermion systems. *Phys. Rev. B* **1982**, *26*, 5033–5055. [CrossRef]
240. Beard, B.B.; Wiese, U.J. Simulations of Discrete Quantum Systems in Continuous Euclidean Time. *Phys. Rev. Lett.* **1996**, *77*, 5130–5133. [CrossRef] [PubMed]
241. Sandvik, A.W. The Stochastic Series Expansion Method for Quantum Lattice Models. In Proceedings of the Computer Simulation Studies in Condensed-Matter Physics XIV, Athens, GA, USA, 19–24 February 2001; Landau, D.P., Lewis, S.P., Schüttler, H.B., Eds.; Springer: Berlin/Heidelberg, Germany, 2002; pp. 182–187.
242. Assaad, F.; Evertz, H. World-line and Determinantal Quantum Monte Carlo Methods for Spins, Phonons and Electrons. In *Computational Many-Particle Physics*; Fehske, H., Schneider, R., Weiße, A., Eds.; Springer: Berlin/Heidelberg, Germany, 2008; pp. 277–356. [CrossRef]
243. Reichsöllner, L.; Schindewolf, A.; Takekoshi, T.; Grimm, R.; Nägerl, H.C. Quantum Engineering of a Low-Entropy Gas of Heteronuclear Bosonic Molecules in an Optical Lattice. *Phys. Rev. Lett.* **2017**, *118*, 073201. [CrossRef] [PubMed]
244. Defenu, N.; Codello, A.; Ruffo, S.; Trombettoni, A. Criticality of spin systems with weak long-range interactions. *J. Phys. A Math. Theor.* **2020**, *53*, 143001. [CrossRef]
245. Sperstad, I.B.; Stiansen, E.B.; Sudbø, A. Quantum criticality in spin chains with non-Ohmic dissipation. *Phys. Rev. B* **2012**, *85*, 214302. [CrossRef]
246. Pfeuty, P.; Elliott, R.J. The Ising model with a transverse field. II. Ground state properties. *J. Phys. C Solid State Phys.* **1971**, *4*, 2370–2385. [CrossRef]
247. Langheld, A.; Koziol, J.A.; Adelhardt, P.; Kapfer, S.C.; Schmidt, K.P. Raw data to "Scaling at quantum phase transitions above the upper critical dimension". *Zenodo* **2021**. [CrossRef]
248. Puebla, R.; Marty, O.; Plenio, M.B. Quantum Kibble-Zurek physics in long-range transverse-field Ising models. *Phys. Rev. A* **2019**, *100*, 032115. [CrossRef]
249. Koziol, J.A.; Langheld, A.; Kapfer, S.C.; Schmidt, K.P. Raw data to "Quantum-critical properties of the long-range transverse-field Ising model from quantum Monte Carlo simulations". *Zenodo* **2021**. [CrossRef]
250. Shiratani, S.; Todo, S. Stochastic approximation analysis of dynamical quantum critical phenomena in long-range transverse-field Ising chain. *arXiv* **2023**, arXiv:2305.14121. [CrossRef]
251. Elliott, R.J.; Pfeuty, P.; Wood, C. Ising Model with a Transverse Field. *Phys. Rev. Lett.* **1970**, *25*, 443–446. [CrossRef]
252. Pfeuty, P. The one-dimensional Ising model with a transverse field. *Ann. Phys.* **1970**, *57*, 79–90. [CrossRef]
253. Kos, F.; Poland, D.; Simmons-Duffin, D.; Vichi, A. Precision islands in the Ising and O(N) models. *J. High Energy Phys.* **2016**, *2016*. [CrossRef]
254. Ribeiro, P.; Vidal, J.; Mosseri, R. Exact spectrum of the Lipkin-Meshkov-Glick model in the thermodynamic limit and finite-size corrections. *Phys. Rev. E* **2008**, *78*, 021106. [CrossRef] [PubMed]
255. Verstraete, F.; Wolf, M.M.; Perez-Garcia, D.; Cirac, J.I. Criticality, the Area Law, and the Computational Power of Projected Entangled Pair States. *Phys. Rev. Lett.* **2006**, *96*, 220601. [CrossRef] [PubMed]

256. Cirac, J.I.; Pérez-García, D.; Schuch, N.; Verstraete, F. Matrix product states and projected entangled pair states: Concepts, symmetries, theorems. *Rev. Mod. Phys.* **2021**, *93*, 045003. [CrossRef]
257. Stoudenmire, E.; White, S.R. Studying Two-Dimensional Systems with the Density Matrix Renormalization Group. *Annu. Rev. Condens. Matter Phys.* **2012**, *3*, 111–128. [CrossRef]
258. Eisert, J.; Cramer, M.; Plenio, M.B. Colloquium: Area laws for the entanglement entropy. *Rev. Mod. Phys.* **2010**, *82*, 277–306. [CrossRef]
259. Crosswhite, G.M.; Doherty, A.C.; Vidal, G. Applying matrix product operators to model systems with long-range interactions. *Phys. Rev. B* **2008**, *78*, 035116. [CrossRef]
260. Pirvu, B.; Murg, V.; Cirac, J.I.; Verstraete, F. Matrix product operator representations. *New J. Phys.* **2010**, *12*, 025012. [CrossRef]
261. Biswas, S.; Rakala, G.; Damle, K. Quantum cluster algorithm for frustrated Ising models in a transverse field. *Phys. Rev. B* **2016**, *93*, 235103. [CrossRef]
262. Biswas, S.; Damle, K. Singular ferromagnetic susceptibility of the transverse-field Ising antiferromagnet on the triangular lattice. *Phys. Rev. B* **2018**, *97*, 085114. [CrossRef]
263. Onsager, L. Crystal Statistics. I. A Two-Dimensional Model with an Order-Disorder Transition. *Phys. Rev.* **1944**, *65*, 117–149. [CrossRef]
264. Moessner, R.; Ramirez, A.P. Geometrical frustration. *Phys. Today* **2006**, *59*, 24–29. [CrossRef]
265. Moessner, R.; Chalker, J.T. Low-temperature properties of classical geometrically frustrated antiferromagnets. *Phys. Rev. B* **1998**, *58*, 12049–12062. [CrossRef]
266. Maxwell, J.C. L. on the calculation of the equilibrium and stiffness of frames. *Lond. Edinb. Dublin Philos. Mag. J. Sci.* **1864**, *27*, 294–299. [CrossRef]
267. Reimers, J.N.; Berlinsky, A.J.; Shi, A.C. Mean-field approach to magnetic ordering in highly frustrated pyrochlores. *Phys. Rev. B* **1991**, *43*, 865–878. [CrossRef] [PubMed]
268. Wannier, G.H. Antiferromagnetism. The Triangular Ising Net. *Phys. Rev.* **1950**, *79*, 357–364. Erratum in *Phys. Rev. B* **1973**, *7*, 5017. [CrossRef]
269. Kano, K.; Naya, S. Antiferromagnetism. The Kagome Ising Net. *Prog. Theor. Phys.* **1953**, *10*, 158–172. [CrossRef]
270. Moessner, R.; Sondhi, S.L.; Chandra, P. Two-Dimensional Periodic Frustrated Ising Models in a Transverse Field. *Phys. Rev. Lett.* **2000**, *84*, 4457–4460. [CrossRef] [PubMed]
271. Moessner, R.; Sondhi, S.L. Ising models of quantum frustration. *Phys. Rev. B* **2001**, *63*, 224401. [CrossRef]
272. Villain, J.; Bidaux, R.; Carton, J.P.; Conte, R. Order as an effect of disorder. *J. Phys.* **1980**, *41*, 1263–1272. [CrossRef]
273. Isakov, S.V.; Moessner, R. Interplay of quantum and thermal fluctuations in a frustrated magnet. *Phys. Rev. B* **2003**, *68*, 104409. [CrossRef]
274. Priour, D.J.; Gelfand, M.P.; Sondhi, S.L. Disorder from disorder in a strongly frustrated transverse-field Ising chain. *Phys. Rev. B* **2001**, *64*, 134424. [CrossRef]
275. Balents, L. Spin liquids in frustrated magnets. *Nature* **2010**, *464*, 199–208. [CrossRef] [PubMed]
276. Röchner, J.; Balents, L.; Schmidt, K.P. Spin liquid and quantum phase transition without symmetry breaking in a frustrated three-dimensional Ising model. *Phys. Rev. B* **2016**, *94*, 201111. [CrossRef]
277. Koziol, J.A.; Mühlhauser, M.; Schmidt, K.P. Order-by-disorder and long-range interactions in the antiferromagnetic transverse-field Ising model on the triangular lattice—A perturbative point of view. *arXiv* **2024**, arXiv:2402.10584. https://doi.org/10.48550/arXiv.2402.10584.
278. Rokhsar, D.S.; Kivelson, S.A. Superconductivity and the Quantum Hard-Core Dimer Gas. *Phys. Rev. Lett.* **1988**, *61*, 2376–2379. [CrossRef] [PubMed]
279. Smerald, A.; Korshunov, S.; Mila, F. Topological Aspects of Symmetry Breaking in Triangular-Lattice Ising Antiferromagnets. *Phys. Rev. Lett.* **2016**, *116*, 197201. [CrossRef] [PubMed]
280. Smerald, A.; Mila, F. Spin-liquid behaviour and the interplay between Pokrovsky-Talapov and Ising criticality in the distorted, triangular-lattice, dipolar Ising antiferromagnet. *SciPost Phys.* **2018**, *5*, 030. [CrossRef]
281. Hasenbusch, M. Monte Carlo study of an improved clock model in three dimensions. *Phys. Rev. B* **2019**, *100*, 224517. [CrossRef]
282. Chester, S.M.; Landry, W.; Liu, J.; Poland, D.; Simmons-Duffin, D.; Su, N.; Vichi, A. Carving out OPE space and precise O(2) model critical exponents. *J. High Energy Phys.* **2020**, *2020*, 142. [CrossRef]
283. Yan, Z.; Wang, Y.C.; Samajdar, R.; Sachdev, S.; Meng, Z.Y. Emergent Glassy Behavior in a Kagome Rydberg Atom Array. *Phys. Rev. Lett.* **2023**, *130*, 206501. [CrossRef]
284. Damle, K.; Sachdev, S. Multicritical Crossovers near the Dilute Bose Gas Quantum Critical Point. *Phys. Rev. Lett.* **1996**, *76*, 4412–4415. [CrossRef]
285. Lieb, E.; Schultz, T.; Mattis, D. Two soluble models of an antiferromagnetic chain. *Ann. Phys.* **1961**, *16*, 407–466. [CrossRef]
286. McCoy, B.M. Spin Correlation Functions of the $X-Y$ Model. *Phys. Rev.* **1968**, *173*, 531–541. [CrossRef]
287. Barouch, E.; McCoy, B.M.; Dresden, M. Statistical Mechanics of the XY Model. I. *Phys. Rev. A* **1970**, *2*, 1075–1092. [CrossRef]
288. Barouch, E.; McCoy, B.M. Statistical Mechanics of the *XY* Model. II. Spin-Correlation Functions. *Phys. Rev. A* **1971**, *3*, 786–804. [CrossRef]
289. Barouch, E.; McCoy, B.M. Statistical Mechanics of the XY Model. III. *Phys. Rev. A* **1971**, *3*, 2137–2140. [CrossRef]

290. McCoy, B.M.; Barouch, E.; Abraham, D.B. Statistical Mechanics of the XY Model. IV. Time-Dependent Spin-Correlation Functions. *Phys. Rev. A* **1971**, *4*, 2331–2341. [CrossRef]
291. Jimbo, M.; Miwa, T.; Môri, Y.; Sato, M. Density matrix of an impenetrable Bose gas and the fifth Painlevé transcendent. *Phys. D Nonlinear Phenom.* **1980**, *1*, 80–158. [CrossRef]
292. Santos, R.R.d.; Stinchcombe, R.B. Finite size scaling and crossover phenomena: The XY chain in a transverse field at zero temperature. *J. Phys. A Math. Gen.* **1981**, *14*, 2741–2757. [CrossRef]
293. Ray, P.; Chakrabarti, B. Exact ground-state excitations of the XY model in a transverse field in one dimension. *Phys. Lett. A* **1983**, *98*, 431–432. [CrossRef]
294. Uzunov, D. On the zero temperature critical behaviour of the nonideal Bose gas. *Phys. Lett. A* **1981**, *87*, 11–14. [CrossRef]
295. Coleman, S. There are no Goldstone bosons in two dimensions. *Commun. Math. Phys.* **1973**, *31*, 259–264. [CrossRef]
296. Pitaevskii, L.; Stringari, S. Uncertainty principle, quantum fluctuations, and broken symmetries. *J. Low Temp. Phys.* **1991**, *85*, 377–388. [CrossRef]
297. Sandvik, A.W. Ground States of a Frustrated Quantum Spin Chain with Long-Range Interactions. *Phys. Rev. Lett.* **2010**, *104*, 137204. [CrossRef]
298. Kumar, M.; Soos, Z.G. Decoupled phase of frustrated spin-$\frac{1}{2}$ antiferromagnetic chains with and without long-range order in the ground state. *Phys. Rev. B* **2013**, *88*, 134412. [CrossRef]
299. Li, Z.H.; Wang, A.M. Matrix product state approach to a frustrated spin chain with long-range interactions. *Phys. Rev. B* **2015**, *91*, 235110. [CrossRef]
300. Tang, Y.; Sandvik, A.W. Quantum Monte Carlo studies of spinons in one-dimensional spin systems. *Phys. Rev. B* **2015**, *92*, 184425. [CrossRef]
301. Gong, Z.X.; Maghrebi, M.F.; Hu, A.; Foss-Feig, M.; Richerme, P.; Monroe, C.; Gorshkov, A.V. Kaleidoscope of quantum phases in a long-range interacting spin-1 chain. *Phys. Rev. B* **2016**, *93*, 205115. [CrossRef]
302. Maghrebi, M.F.; Gong, Z.X.; Gorshkov, A.V. Continuous Symmetry Breaking in 1D Long-Range Interacting Quantum Systems. *Phys. Rev. Lett.* **2017**, *119*, 023001. [CrossRef]
303. Frérot, I.; Naldesi, P.; Roscilde, T. Entanglement and fluctuations in the XXZ model with power-law interactions. *Phys. Rev. B* **2017**, *95*, 245111. [CrossRef]
304. Ren, J.; You, W.L.; Oleś, A.M. Quantum phase transitions in a spin-1 antiferromagnetic chain with long-range interactions and modulated single-ion anisotropy. *Phys. Rev. B* **2020**, *102*, 024425. [CrossRef]
305. Yang, L.; Feiguin, A.E. From deconfined spinons to coherent magnons in an antiferromagnetic Heisenberg chain with long range interactions. *SciPost Phys.* **2021**, *10*, 110. [CrossRef]
306. Yang, L.; Weinberg, P.; Feiguin, A. Topological to magnetically ordered quantum phase transition in antiferromagnetic spin ladders with long-range interactions. *SciPost Phys.* **2022**, *13*, 60. [CrossRef]
307. Nishiyama, Y. Fidelity-mediated analysis of the transverse-field XY chain with the long-range interactions: Anisotropy-driven multi-criticality. *Eur. Phys. J. B* **2021**, *94*, 226. [CrossRef]
308. Brezin, E.; Le Guillou, J.C.; Zinn-Justin, J. Approach to Scaling in Renormalized Perturbation Theory. *Phys. Rev. D* **1973**, *8*, 2418–2430. [CrossRef]
309. Wegner, F.J.; Riedel, E.K. Logarithmic Corrections to the Molecular-Field Behavior of Critical and Tricritical Systems. *Phys. Rev. B* **1973**, *7*, 248–256. [CrossRef]
310. Weihong, Z.; Oitmaa, J.; Hamer, C.J. Series expansions for the 3D transverse Ising model at T=0. *J. Phys. A Math. Gen.* **1994**, *27*, 5425–5444. [CrossRef]
311. Wessel, S.; Troyer, M. Supersolid Hard-Core Bosons on the Triangular Lattice. *Phys. Rev. Lett.* **2005**, *95*, 127205. [CrossRef]
312. Nishiyama, Y. Transverse-field XY spin chain with the competing long-range interactions: Multi-criticality around the XX-symmetric point. *J. Stat. Mech. Theory Exp.* **2023**, *2023*, 083102. [CrossRef]
313. Sachdev, S.; Bhatt, R.N. Bond-operator representation of quantum spins: Mean-field theory of frustrated quantum Heisenberg antiferromagnets. *Phys. Rev. B* **1990**, *41*, 9323–9329. [CrossRef]
314. Hörmann, M.; Wunderlich, P.; Schmidt, K.P. Dynamic Structure Factor of Disordered Quantum Spin Ladders. *Phys. Rev. Lett.* **2018**, *121*, 167201. [CrossRef]
315. Schmidt, K.P.; Uhrig, G.S. Excitations in One-Dimensional $S = \frac{1}{2}$ Quantum Antiferromagnets. *Phys. Rev. Lett.* **2003**, *90*, 227204. [CrossRef]
316. Bruno, P. Absence of Spontaneous Magnetic Order at Nonzero Temperature in One- and Two-Dimensional Heisenberg and *XY* Systems with Long-Range Interactions. *Phys. Rev. Lett.* **2001**, *87*, 137203. [CrossRef]
317. Parreira, J.R.; Bolina, O.; Perez, J.F. Néel order in the ground state of Heisenberg antiferromagnetic chains with long-range interactions. *J.-Phys.-Lond.-Math. Gen.* **1997**, *30*, 1095–1100. [CrossRef]
318. Nambu, Y. Quasi-Particles and Gauge Invariance in the Theory of Superconductivity. *Phys. Rev.* **1960**, *117*, 648–663. [CrossRef]
319. Goldstone, J. Field theories with « Superconductor » solutions. *Il Nuovo C.* **1961**, *19*, 154–164. [CrossRef]
320. Goldstone, J.; Salam, A.; Weinberg, S. Broken Symmetries. *Phys. Rev.* **1962**, *127*, 965–970. [CrossRef]
321. Diessel, O.K.; Diehl, S.; Defenu, N.; Rosch, A.; Chiocchetta, A. Generalized Higgs mechanism in long-range-interacting quantum systems. *Phys. Rev. Res.* **2023**, *5*, 033038. [CrossRef]

322. Yusuf, E.; Joshi, A.; Yang, K. Spin waves in antiferromagnetic spin chains with long-range interactions. *Phys. Rev. B* **2004**, *69*, 144412. [CrossRef]
323. Lange, R.V. Nonrelativistic Theorem Analogous to the Goldstone Theorem. *Phys. Rev.* **1966**, *146*, 301–303. [CrossRef]
324. Lange, R.V. Interaction Range, The Goldstone Theorem, and Long-Range Order in the Heisenberg Ferromagnet. *Phys. Rev.* **1967**, *156*, 630–631. [CrossRef]
325. Wegner, F.J. Critical Exponents in Isotropic Spin Systems. *Phys. Rev. B* **1972**, *6*, 1891–1893. [CrossRef]
326. Defenu, N.; Trombettoni, A.; Codello, A. Fixed-point structure and effective fractional dimensionality for O(N) models with long-range interactions. *Phys. Rev. E* **2015**, *92*, 052113. [CrossRef]
327. Romen, A.; Birnkammer, S.; Knap, M. Deconfined quantum criticality in the long-range, anisotropic Heisenberg chain. *SciPost Phys. Core.* **2024**, *7*, 8. [CrossRef]
328. Feng, L.; Katz, O.; Haack, C.; Maghrebi, M.; Gorshkov, A.; Gong, Z.; Cetina, M. Continuous symmetry breaking in a trapped-ion spin chain. *Nature* **2023**, *623*, 713–717. [CrossRef]
329. Chen, C.; Bornet, G.; Bintz, M.; Emperauger, G.; Leclerc, L.; Liu, V.; Scholl, P.; Barredo, D.; Hauschild, J.; Chatterjee, S.; et al. Continuous symmetry breaking in a two-dimensional Rydberg array. *Nature* **2023**, *616*, 691–695. [CrossRef] [PubMed]
330. Haldane, F.D.M. Exact Jastrow-Gutzwiller resonating-valence-bond ground state of the spin-$\frac{1}{2}$ antiferromagnetic Heisenberg chain with $1/r^2$ exchange. *Phys. Rev. Lett.* **1988**, *60*, 635–638. [CrossRef]
331. Shastry, B.S. Exact solution of an S=1/2 Heisenberg antiferromagnetic chain with long-ranged interactions. *Phys. Rev. Lett.* **1988**, *60*, 639–642. [CrossRef] [PubMed]
332. Eβler, F.H.L. Dressed S matrices in models with long-range interactions. *Phys. Rev. B* **1995**, *51*, 13357–13366. [CrossRef] [PubMed]
333. Wang, L.; Sandvik, A.W. Critical Level Crossings and Gapless Spin Liquid in the Square-Lattice Spin-1/2 $J_1 - J_2$ Heisenberg Antiferromagnet. *Phys. Rev. Lett.* **2018**, *121*, 107202. [CrossRef] [PubMed]
334. Senthil, T.; Vishwanath, A.; Balents, L.; Sachdev, S.; Fisher, M.P.A. Deconfined Quantum Critical Points. *Science* **2004**, *303*, 1490–1494. [CrossRef] [PubMed]
335. Senthil, T.; Balents, L.; Sachdev, S.; Vishwanath, A.; Fisher, M.P.A. Quantum criticality beyond the Landau-Ginzburg-Wilson paradigm. *Phys. Rev. B* **2004**, *70*, 144407. [CrossRef]
336. Vishwanath, A.; Balents, L.; Senthil, T. Quantum criticality and deconfinement in phase transitions between valence bond solids. *Phys. Rev. B* **2004**, *69*, 224416. [CrossRef]
337. Senthil, T.; Balents, L.; Sachdev, S.; Vishwanath, A.; Fisher, M.P.A. Deconfined Criticality Critically Defined. *J. Phys. Soc. Jpn.* **2005**, *74*, 1–9. [CrossRef]
338. Yang, S.; Yao, D.X.; Sandvik, A.W. Deconfined quantum criticality in spin-1/2 chains with long-range interactions. *arXiv* **2020**, arXiv:2001.02821. [CrossRef]
339. Adelhardt, P.; Schmidt, K.P. Raw data from the pCUT+MC approach for the antiferromagnetic Heisenberg square lattice bilayer model with (non-frustrating) long-range interactions. *Zenodo* **2024**. [CrossRef]
340. Hida, K. Low Temperature Properties of the Double Layer Quantum Heisenberg Antiferromagnet -Modified Spin Wave Method-. *J. Phys. Soc. Jpn.* **1990**, *59*, 2230–2236. [CrossRef]
341. Hida, K. Quantum Disordered State without Frustration in the Double Layer Heisenberg Antiferromagnet–Dimer Expansion and Projector Monte Carlo Study–. *J. Phys. Soc. Jpn.* **1992**, *61*, 1013–1018. [CrossRef]
342. Sandvik, A.W.; Scalapino, D.J. Order-disorder transition in a two-layer quantum antiferromagnet. *Phys. Rev. Lett.* **1994**, *72*, 2777–2780. [CrossRef] [PubMed]
343. Sandvik, A.W.; Chubukov, A.V.; Sachdev, S. Quantum critical behavior in a two-layer antiferromagnet. *Phys. Rev. B* **1995**, *51*, 16483–16486. [CrossRef] [PubMed]
344. Wang, L.; Beach, K.S.D.; Sandvik, A.W. High-precision finite-size scaling analysis of the quantum-critical point of $S = 1/2$ Heisenberg antiferromagnetic bilayers. *Phys. Rev. B* **2006**, *73*, 014431. [CrossRef]
345. Wenzel, S.; Bogacz, L.; Janke, W. Evidence for an Unconventional Universality Class from a Two-Dimensional Dimerized Quantum Heisenberg Model. *Phys. Rev. Lett.* **2008**, *101*, 127202. [CrossRef] [PubMed]
346. Lohöfer, M.; Coletta, T.; Joshi, D.G.; Assaad, F.F.; Vojta, M.; Wessel, S.; Mila, F. Dynamical structure factors and excitation modes of the bilayer Heisenberg model. *Phys. Rev. B* **2015**, *92*, 245137. [CrossRef]
347. Coester, K. Series Expansions for Dimerized Quantum Spin Systems. Ph.D. Thesis, TU Dortmund, Dortmund, Germany, 2011.
348. Hasenbusch, M.; Vicari, E. Anisotropic perturbations in three-dimensional O(N)-symmetric vector models. *Phys. Rev. B* **2011**, *84*, 125136. [CrossRef]
349. Chester, S.M.; Landry, W.; Liu, J.; Poland, D.; Simmons-Duffin, D.; Su, N.; Vichi, A. Bootstrapping Heisenberg magnets and their cubic instability. *Phys. Rev. D* **2021**, *104*, 105013. [CrossRef]
350. Schmidt, K.P.; Monien, H.; Uhrig, G.S. Rung-singlet phase of the $S = \frac{1}{2}$ two-leg spin-ladder with four-spin cyclic exchange. *Phys. Rev. B* **2003**, *67*, 184413. [CrossRef]
351. Barnes, T.; Dagotto, E.; Riera, J.; Swanson, E.S. Excitation spectrum of Heisenberg spin ladders. *Phys. Rev. B* **1993**, *47*, 3196–3203. [CrossRef] [PubMed]
352. Shelton, D.G.; Nersesyan, A.A.; Tsvelik, A.M. Antiferromagnetic spin ladders: Crossover between spin S=1/2 and S=1 chains. *Phys. Rev. B* **1996**, *53*, 8521–8532. [CrossRef] [PubMed]

353. Adelhardt, P.; Schmidt, K.P. Raw data to "Continuously varying critical exponents in long-range quantum spin ladders". *Zenodo* **2023**. [CrossRef]
354. Ma, N.; Sun, G.Y.; You, Y.Z.; Xu, C.; Vishwanath, A.; Sandvik, A.W.; Meng, Z.Y. Dynamical signature of fractionalization at a deconfined quantum critical point. *Phys. Rev. B* **2018**, *98*, 174421. [CrossRef]
355. Sandvik, A.W. Evidence for Deconfined Quantum Criticality in a Two-Dimensional Heisenberg Model with Four-Spin Interactions. *Phys. Rev. Lett.* **2007**, *98*, 227202. [CrossRef] [PubMed]
356. Melko, R.G.; Kaul, R.K. Scaling in the Fan of an Unconventional Quantum Critical Point. *Phys. Rev. Lett.* **2008**, *100*, 017203. [CrossRef] [PubMed]
357. Jiang, F.J.; Nyfeler, M.; Chandrasekharan, S.; Wiese, U.J. From an antiferromagnet to a valence bond solid: Evidence for a first-order phase transition. *J. Stat. Mech. Theory Exp.* **2008**, *2008*, P02009. [CrossRef]
358. Lou, J.; Sandvik, A.W.; Kawashima, N. Antiferromagnetic to valence-bond-solid transitions in two-dimensional $SU(N)$ Heisenberg models with multispin interactions. *Phys. Rev. B* **2009**, *80*, 180414. [CrossRef]
359. Sandvik, A.W. Continuous Quantum Phase Transition between an Antiferromagnet and a Valence-Bond Solid in Two Dimensions: Evidence for Logarithmic Corrections to Scaling. *Phys. Rev. Lett.* **2010**, *104*, 177201. [CrossRef] [PubMed]
360. Harada, K.; Suzuki, T.; Okubo, T.; Matsuo, H.; Lou, J.; Watanabe, H.; Todo, S.; Kawashima, N. Possibility of deconfined criticality in $SU(N)$ Heisenberg models at small N. *Phys. Rev. B* **2013**, *88*, 220408. [CrossRef]
361. Nakamura, M. Mechanism of CDW-SDW Transition in One Dimension. *J. Phys. Soc. Jpn.* **1999**, *68*, 3123–3126. [CrossRef]
362. Nakamura, M. Tricritical behavior in the extended Hubbard chains. *Phys. Rev. B* **2000**, *61*, 16377–16392. [CrossRef]
363. Sengupta, P.; Sandvik, A.W.; Campbell, D.K. Bond-order-wave phase and quantum phase transitions in the one-dimensional extended Hubbard model. *Phys. Rev. B* **2002**, *65*, 155113. [CrossRef]
364. Sandvik, A.W.; Balents, L.; Campbell, D.K. Ground State Phases of the Half-Filled One-Dimensional Extended Hubbard Model. *Phys. Rev. Lett.* **2004**, *92*, 236401. [CrossRef] [PubMed]
365. Tsuchiizu, M.; Furusaki, A. Ground-state phase diagram of the one-dimensional half-filled extended Hubbard model. *Phys. Rev. B* **2004**, *69*, 035103. [CrossRef]
366. Jiang, S.; Motrunich, O. Ising ferromagnet to valence bond solid transition in a one-dimensional spin chain: Analogies to deconfined quantum critical points. *Phys. Rev. B* **2019**, *99*, 075103. [CrossRef]
367. Roberts, B.; Jiang, S.; Motrunich, O.I. Deconfined quantum critical point in one dimension. *Phys. Rev. B* **2019**, *99*, 165143. [CrossRef]
368. Huang, R.Z.; Lu, D.C.; You, Y.Z.; Meng, Z.Y.; Xiang, T. Emergent symmetry and conserved current at a one-dimensional incarnation of deconfined quantum critical point. *Phys. Rev. B* **2019**, *100*, 125137. [CrossRef]
369. Pollmann, F.; Berg, E.; Turner, A.M.; Oshikawa, M. Symmetry protection of topological phases in one-dimensional quantum spin systems. *Phys. Rev. B* **2012**, *85*, 075125. [CrossRef]
370. Affleck, I. Critical Behavior of Two-Dimensional Systems with Continuous Symmetries. *Phys. Rev. Lett.* **1985**, *55*, 1355–1358. [CrossRef] [PubMed]
371. Affleck, I.; Haldane, F.D.M. Critical theory of quantum spin chains. *Phys. Rev. B* **1987**, *36*, 5291–5300. [CrossRef] [PubMed]
372. Tarabunga, P.S.; Surace, F.M.; Andreoni, R.; Angelone, A.; Dalmonte, M. Gauge-Theoretic Origin of Rydberg Quantum Spin Liquids. *Phys. Rev. Lett.* **2022**, *129*, 195301. [CrossRef] [PubMed]
373. Emonts, P.; Wessel, S. Monte Carlo study of the discontinuous quantum phase transition in the transverse-field Ising model on the pyrochlore lattice. *Phys. Rev. B* **2018**, *98*, 174433. [CrossRef]
374. Patil, P. Quantum Monte Carlo simulations in the restricted Hilbert space of Rydberg atom arrays. *arXiv* **2023**, arXiv:2309.00482. [CrossRef]
375. Shankar, R. Renormalization-group approach to interacting fermions. *Rev. Mod. Phys.* **1994**, *66*, 129–192. [CrossRef]

entropy MDPI

Article

Ralph Kenna's Scaling Relations in Critical Phenomena

Leïla Moueddene [1,2,3], Arnaldo Donoso [4] and Bertrand Berche [1,*]

1 Laboratoire de Physique et Chimie Théoriques, CNRS—Université de Lorraine, 54000 Nancy, France; leila.moueddene@univ-lorraine.fr
2 L4 Collaboration, Leipzig-Lorraine-Lviv-Coventry, Europe
3 Centre for Fluid and Complex Systems, Coventry University, Coventry CV1 5FB, UK
4 Department of Experimental Physics, Maynooth University, R51 A021 Maynooth, Co. Kildare, Ireland; arnaldoivic2008@gmail.com
* Correspondence: bertrand.berche@univ-lorraine.fr

Abstract: In this note, we revisit the scaling relations among "hatted critical exponents", which were first derived by Ralph Kenna, Des Johnston, and Wolfhard Janke, and we propose an alternative derivation for some of them. For the scaling relation involving the behavior of the correlation function, we will propose an alternative form since we believe that the expression is erroneous in the work of Ralph and his collaborators.

Keywords: critical exponents; logarithmic corrections; scaling and renormalization; scaling laws; tricritical point; finite-size scaling

1. In Memory of Our Friend Ralph Kenna

This paper is dedicated to our friend Ralph Kenna, who passed away on 26 October 2023. Ralph was a close collaborator and a very good friend, and he was the PhD coadvisor of one of the authors of the present paper (LM). A renowned specialist in the study of phase transitions through the partition function zeros, he developed this formalism in difficult cases where critical behaviors are controlled beyond the dominant singularities by logarithmic corrections.

The present work was initiated on the occasion of the ComPhys23 workshop (http://www.physik.uni-leipzig.de/~janke/CompPhys23/ (accessed on 26 February 2024)) organized by W. Janke in Leipzig, and dedicated to the memory of Ralph. The opening talk was intended to highlight some of Ralph's most important contributions to statistical physics, notably the derivation of scaling laws between the exponents associated with logarithmic singularities in the vicinity of second-order phase transitions. Ralph was a well-known physicist in the statistical physics community, and these new scaling laws were very successful in the field of critical phenomena. It was by revisiting these scaling laws and their derivation that we noted that one of these relations was incomplete and that the work undertaken by Ralph and his coauthors around twenty years ago merited to be completed.

This paper could have been written by Des Johnston or by Wolfhard Janke, the coauthors of Ralph on this topic (and in many more works).

2. Scaling Relations and Universal Combinations of Amplitudes: A Short Primer

We assume that the reader is familiar with the notion of critical exponents that describe the singularities of various thermodynamic functions at the approach of a second-order phase transition. Otherwise, we can suggest referring, e.g., to the textbook of Kardar [1].

Standard scaling relations among the universal critical exponents are the following:

$$2\beta+\gamma = 2-\alpha, \tag{1}$$
$$\beta(\delta-1) = \gamma, \tag{2}$$
$$\nu(2-\eta) = \gamma, \tag{3}$$
$$2-\alpha = \nu d. \tag{4}$$

They are very useful, not only to obtain the values of all six fundamental critical exponents within a universality class from the knowledge of two of them, but also because they allow for the definition of other universal quantities, written as specific combinations of critical amplitudes. Let us show how this works. For that purpose, we first define the amplitudes as they enter the expressions of the leading singular behaviors of thermodynamic quantities in the vicinity of a second-order phase transition:

$$\text{Specific heat:}\quad C_\pm(t,0) \simeq A_\pm |t|^{-\alpha}, \tag{5}$$
$$\text{Low temperature magnetization:}\quad m_-(t,0) \simeq B_-(-t)^{\beta}, \tag{6}$$
$$\text{Susceptibility:}\quad \chi_\pm(t,0) \simeq \Gamma_\pm |t|^{-\gamma}, \tag{7}$$
$$\text{Critical temperature magnetization:}\quad m_c(0,h) \simeq B_c |h|^{1/\delta}, \tag{8}$$
$$\text{Correlation length:}\quad \xi_\pm(t,0) \simeq \xi_{0\pm} |t|^{-\nu}. \tag{9}$$

Here, the two arguments of the functions at the l.h.s. are, respectively, $t = (T - T_c)/T_c$ and $h = H/T_c$, and the indices $\pm$ specify the high ($t > 0$)- and low ($t < 0$)-temperature phases, meaning that the field is zero, and the index c, on the contrary implies the field behavior at the critical temperature. The symbol $\simeq$ stands for the leading singularity (i.e., the most singular part since there could be regular contributions to the thermodynamic quantities, power-law corrections to scaling, and multiplicative logarithmic corrections, all these being omitted in Equations (5)–(9)).

We can also define the singular part of the free energy density in zero field:

$$f_\pm^{\text{sing}}(t,0) \simeq F_\pm |t|^{2-\alpha}, \tag{10}$$

and since the specific heat is the second derivative of f^{sing} with regard to t, $F_\pm$ is not independent since this requires $A_\pm = (1-\alpha)(2-\alpha)F_\pm$. The Lee–Yang edge is another quantity of interest in critical phenomena, and we define

$$h_\pm^{\text{LY}}(t) \simeq h_{0\pm} |t|^{\Delta} \tag{11}$$

with the so-called gap exponent $\Delta = \beta + \gamma = \beta\delta$.

Universality is the observation that some quantities only depend on very general properties, like space dimensionality. The critical exponents are such universal quantities, but the amplitudes are not, although some combinations among them have the property of universality. To make it clear, let us write Widom's scaling assumption, i.e., the fact that the singular part of the free energy density is a homogeneous function of the scaling fields

$$f_\pm^{\text{sing}}(t,h) = b^{-d} \mathscr{F}_\pm(\kappa_t b^{y_t} t, \kappa_h b^{y_h} h), \tag{12}$$

where $\mathscr{F}_\pm(x(b), y(b))$ is a universal scaling function of its arguments, y_t and y_h are the RG dimensions of the relevant fields t and h, and κ_t and κ_h are non-universal metric factors that would differ, say, on the square lattice and the triangular lattice in $2d$. The amplitudes defined above depend on these metric factors, and this is why they are not universal, e.g., from $C_\pm(t,0) = \frac{\partial^2 f_\pm^{\text{sing}}(t,0)}{\partial t^2}$, setting $b = |t|^{-1/y_t}$ in the scaling form (12), one obtains

$$C_\pm(t,0) = \kappa_t^{2+(d-2y_t)/y_t} |t|^{(d-2y_t)/y_t} \left(\frac{\partial^2 \mathscr{F}_\pm(x,0)}{\partial x^2}\right)_{x=1}. \tag{13}$$

This identifies the exponent $\alpha = (2y_t - d)/y_t$ and the amplitude $A_\pm = \kappa_t^{2-\alpha}\left(\frac{\partial^2 \mathscr{F}_\pm(x,0)}{\partial x^2}\right)_{x=1}$.

The other exponents are similarly defined in terms of y_t and y_h by very famous relations that we do not repeat here, and the other amplitudes depend on the metric factors as $B_- \sim \kappa_h \kappa_t^{\beta}$, $\Gamma_\pm \sim \kappa_h^2 \kappa_t^{-\gamma}$, $B_c \sim \kappa_h^{1+1/\delta}$.

Simple ratios are immediately defined from the fact that the approach to criticality from above and from below is described by the same exponent for a given quantity (except for the magnetization, obviously), for example, in the case of the specific heat $C_+(|t|) \simeq A_+|t|^{-\alpha}$ and $C_-(-|t|) \simeq A_-|t|^{-\alpha'}$, where $\alpha' = \alpha$. It follows that the metric factors cancel in the ratio, and

$$R_C(|t|) = \frac{C_+(|t|)}{C_-(-|t|)} = \frac{A_+|t|^{-\alpha}}{A_-|t|^{-\alpha'}} \to R_C = \frac{A_+}{A_-} \tag{14}$$

is thus universal. The limit corresponds to the approach to criticality ($|t| \to 0$ here) since the combination $R_C(|t|)$ can be temperature-dependent due to the possible presence of different values for the amplitudes of the corrections to scaling, which has not been taken into account in Equations (5) to (9). In the same manner, one defines the universal ratios

$$R_\chi = \frac{\Gamma_+}{\Gamma_-}, \quad R_\xi = \frac{\xi_{0+}}{\xi_{0-}}. \tag{15}$$

The scaling relations are other examples of relations that allow the definition of new combinations. For example, the ratio $m_-^2/\chi_\pm$ eliminates κ_h, and κ_t is then eliminated, thanks to Equation (1), if we further divide by $C_\pm$. There is still an unwanted $|t|^2$ dependence that needs to be simplified, and for that purpose, we consider the quantity

$$R_\pm^{(1)}(t) = \frac{m_-^2(-|t|)}{C_\pm(t)\chi_\pm(t)|t|^2} = \frac{B_-^2}{C_\pm\Gamma_\pm}|t|^{2\beta+\gamma+\alpha-2} \to \frac{B_-^2}{C_\pm\Gamma_\pm}. \tag{16}$$

Thanks to Equation (1), the fact that all metric factors cancel out in this latter quantity makes the combination $B_-^2/C_\pm\Gamma_\pm$ universal. Proceeding the same way, Equations (2) and (3) suggest to contemplate the expressions

$$\begin{aligned} R_\pm^{(2)}(t) &= \chi_\pm(t)m_-^{\delta-1}(-|t|)m_c^{-\delta}(h)h\Big|_{h=h_\pm^{\rm LY}(|t|)} \\ &= \Gamma_\pm B_-^{\delta-1}B_c^{-\delta}|t|^{-\gamma+\beta(\delta-1)} \to \Gamma_\pm B_-^{\delta-1}B_c^{-\delta}, \qquad (17) \\ R_\pm^{(3)}(t) &= \frac{\chi_\pm(t)}{\xi_\pm^{2-\eta}(t)} = \frac{\Gamma_\pm}{\xi_{0\pm}^{2-\eta}}|t|^{-\gamma+\nu(2-\eta)} \to \frac{\Gamma_\pm}{\xi_{0\pm}^{2-\eta}}, \qquad (18) \end{aligned}$$

that reach their respective universal values. Eventually, Equation (4) leads to consider the following combination

$$R_\pm^{(4)}(t) = \xi_\pm^d(t)f_\pm^{\rm sing}(t) = \xi_{0\pm}^d F_\pm|t|^{-d\nu+2-\alpha} \to \xi_{0\pm}^d F_\pm \tag{19}$$

as universal also.

3. From the Universal Combinations of Amplitudes to Scaling Laws among Hatted Exponents

Having the universal combinations of amplitudes at hand, we consider now the case where the critical behavior is described, besides the leading singularities, by multiplicative logarithmic corrections. This may happen, for example, for a system at its upper critical dimension $d_{\rm uc}$, or in the case of the $2d$ four-state Potts model, or the $2d$ disordered Ising model as well. Many examples can be found in Ref. [2–4].

Let us first remind the standard definitions of some exponent combinations that will occur below: $\alpha_c = \alpha/\beta\delta$, $\gamma_c = \gamma/\beta\delta$, $\nu_c = \nu/\beta\delta$, $\epsilon_c = (1-\alpha)/\beta\delta$. The logarithmic corrections can appear either in the approach to the critical temperature when the magnetic field is fixed at zero or, on the other hand, right at T_c when the magnetic field approaches zero:

$$h = 0,\ t \to 0^{\pm}, \tag{20}$$

$$f_{\pm}^{\text{sing}}(t,0) \simeq F_{\pm}|t|^{2-\alpha}(-\ln|t|)^{\hat{\alpha}}, \tag{21}$$

$$m_{-}(t,0) \simeq B_{-}|t|^{\beta}(-\ln|t|)^{\hat{\beta}}, \tag{22}$$

$$e_{\pm}(t,0) \simeq \frac{A_{\pm}}{(1-\alpha)}|t|^{1-\alpha}(-\ln|t|)^{\hat{\alpha}}, \tag{23}$$

$$\chi_{\pm}(t,0) \simeq \Gamma_{\pm}|t|^{-\gamma}(-\ln|t|)^{\hat{\gamma}}, \tag{24}$$

$$C_{\pm}(t,0) \simeq A_{\pm}|t|^{-\alpha}(-\ln|t|)^{\hat{\alpha}}, \tag{25}$$

$$\xi_{\pm}(t,0) \simeq \xi_{0\pm}|t|^{-\nu}(-\ln|t|)^{\hat{\nu}}, \tag{26}$$

$$t = 0,\ h \to 0^{\pm}, \tag{27}$$

$$f_{c}^{\text{sing}}(0,h) \simeq F_{c}|h|^{1+1/\delta}(-\ln|h|)^{\hat{\delta}_c}, \tag{28}$$

$$m_{c}(0,h) \simeq B_{c}|h|^{1/\delta}(-\ln|h|)^{\hat{\delta}_c}, \tag{29}$$

$$e_{c}(0,h) \simeq E_{c}|h|^{\epsilon_c}(-\ln|h|)^{\hat{\epsilon}_c}, \tag{30}$$

$$\chi_{c}(0,h) \simeq \Gamma_{c}|h|^{-\gamma_c}(-\ln|h|)^{\hat{\delta}_c}, \tag{31}$$

$$C_{c}(0,h) \simeq \frac{A_c}{\alpha_c}|h|^{-\alpha_c}(-\ln|h|)^{\hat{\alpha}_c}, \tag{32}$$

$$\xi_{c}(0,h) \simeq \xi_{c}|h|^{-\nu_c}(-\ln|h|)^{\hat{\nu}_c}. \tag{33}$$

We can also define at criticality $t = h = 0$ the logarithmic correction of the correlation function, defining the exponents $\hat{\eta}$ that will play an essential role in the following of this paper:

$$G(0,0,|\mathbf{r}|) \simeq g_0|\mathbf{r}|^{-(d-2+\eta)}(\ln|\mathbf{r}|)^{\hat{\eta}}. \tag{34}$$

We mostly use the notations of Refs. [2,5], with quantities (amplitudes and exponents) at the critical temperature defined with the subscript c, except for δ in (29), which is standard according to the terminology fixed by Fisher long ago [6].

Ralph Kenna and his coworkers, Des Johnston and Wolfhard Janke, have established a series of scaling relations [2–4] among "hatted exponents", as Ralph used to call them. Their approach was based on the zeros of the partition function, either the Lee–Yang zeros (in complex magnetic field) or the Fisher zeros (in complex temperature).

Here, we offer an alternative derivation of most of these scaling laws, probably simpler in its approach. Universality assumes that the previous ratios of amplitudes are still universal when multiplicative logarithmic corrections are present, i.e.,

$$R_{\pm}^{(1)}(t) = \frac{m_{-}^{2}(-|t|)}{C_{\pm}(|t|)\chi_{\pm}(|t|)|t|^{2}} = \frac{B_{-}^{2}}{C_{\pm}\Gamma_{\pm}}(-\ln|t|)^{2\hat{\beta}-\hat{\alpha}-\hat{\gamma}}. \tag{35}$$

The fact that this quantity *must* tend to $B_{-}^{2}/C_{\pm}\Gamma_{\pm}$ now demands that

$$2\hat{\beta} = \hat{\alpha} + \hat{\gamma}. \tag{36}$$

This is the first of Ralph and his coworkers' scaling relations. Using the same method, the second ratio easily leads to a second relation:

$$\hat{\gamma} + \hat{\beta}(\delta - 1) - \delta\hat{\delta} = 0. \tag{37}$$

The amplitude of the Lee–Yang edge, $h_{0\pm}$, has a non-trivial dependence with the metric factors, $h_{0\pm} \sim \kappa_t^{\beta\delta}\kappa_h^{-1}$. This can be retrieved from the scaling relation $\Delta = \beta\delta$, and the universality of the ratio

$$R^{(5)}(t) = \frac{m_c(h_\pm^{\rm LY}(t))}{m_-(t)} = \frac{B_c h_{0\pm}^{1/\delta}}{B_-}|t|^{\Delta/\delta-\beta}(-\ln|t|)^{\hat{\Delta}/\delta+\hat{\delta}-\hat{\beta}} \to \frac{B_c h_{0\pm}^{1/\delta}}{B_-} \tag{38}$$

requires that

$$\hat{\Delta} = (\hat{\beta} - \hat{\delta})\delta. \tag{39}$$

The scaling relations (36) and (37) were first derived in Ref. [3]. Instead of (39), Ralph and his coworkers had $\hat{\Delta} = \hat{\beta} - \hat{\gamma}$, which is recovered here using (37) and (39).

In the same paper, they also derived

$$\hat{\varrho} = \hat{\nu} + \nu\hat{\alpha}/(2-\alpha) \tag{40}$$

that we will now consider. This is an analog of the hyperscaling relation for logarithmic relations since it is derived from the same ratio of amplitudes as the ordinary hyperscaling relations, even though the space dimension, fixed at its upper critical value, does not appear explicitly. A new *pseudo-critical exponent* [3,4,7] appears there, $\hat{\varrho}$, which describes the finite-size scaling (FSS) of the correlation length

$$\xi_L \simeq L(\ln L)^{\hat{\varrho}}. \tag{41}$$

This behavior is encoded in the scaling hypothesis for the correlation length, appropriately extended to account for the logarithmic correction:

$$\xi_\pm(t,h,L^{-1}) = b(\ln b)^{\hat{\varrho}}\mathscr{X}(\kappa_t b^{y_t}(\ln b)^{\hat{y}_t}t, \kappa_h b^{y_h}(\ln b)^{\hat{y}_h}h, bL^{-1}). \tag{42}$$

Like y_t (respectively, y_h) is the RG eigenvalue associated with the scaling field t (respectively, h), we denote $\hat{y}_t$ (respectively, $\hat{y}_h$) the corresponding exponent of the logarithmic correction. For the sake of clarity, we will later denote the rescaled variables as $x(b) = \kappa_t b^{y_t}(\ln b)^{\hat{y}_t}t$, $y(b) = \kappa_h b^{y_h}(\ln b)^{\hat{y}_h}h$, and $z(b) = bL^{-1}$. Equation (41) follows from the choice $b = L$ at criticality $t = h = 0$ in (42). The same scaling form is used in the thermodynamic limit $L \to \infty$, setting $x = 1$. This requires iterations

$$\begin{aligned} b &= (\kappa|t|)^{-1/y_t}(\ln b)^{-\hat{y}_t/y_t} \\ &\simeq (\kappa|t|)^{-1/y_t}(-\ln|t|)^{-\hat{y}_t/y_t}\Big(1 + \frac{\ln(-\ln|t|)}{(-\ln|t|)} + \text{higher order correction}\Big). \end{aligned} \tag{43}$$

Insertion into the expression of the correlation length leads to the leading order

$$\xi_\pm(t,0,0) \simeq |t|^{-1/y_t}(-\ln|t|)^{\hat{\varrho}-\hat{y}_t/y_t}\mathscr{X}(1,0,0) \tag{44}$$

and requires the usual relation $\nu = 1/y_t$ to conform to (26):

$$\hat{\varrho} = \hat{\nu} + \nu\hat{y}_t. \tag{45}$$

We now show that this agrees with Ralph's scaling relation (40). For that purpose, we use the compatibility with the extension of the phenomenological Widom scaling assumption for the free energy density (12) to the presence of logarithmic corrections, written as far as we know for the first time by Ralph Kenna in Ref. [8],

$$f_\pm^{\rm sing}(t,h,L^{-1}) = b^{-d}\mathscr{F}_\pm(\kappa_t b^{y_t}(\ln b)^{\hat{y}_t}t, \kappa_h b^{y_h}(\ln b)^{\hat{y}_h}h, bL^{-1}). \tag{46}$$

The second derivative with regard to t is the specific heat, and the choice $x = 1$ at $h = L^{-1} = 0$ then leads (using $\alpha = (2y_t - d)/y_t$) to

$$C_\pm(t,0,0) \simeq |t|^{-\alpha}(-\ln|t|)^{2\hat{y}_t - \alpha\hat{y}_t}\mathscr{C}_\pm(1,0,0), \tag{47}$$

(from now on, we always limit (43) to leading logarithmic order); hence, from (25),

$$\hat{\alpha} = (2-\alpha)\hat{y}_t \tag{48}$$

which completes the proof.

4. Solving a Disagreement with Our Friends

In Ref. [2], Ralph Kenna has given a complete account of these and many more scaling relations among hatted exponents. This is not our purpose here to be exhaustive, but rather to show alternative derivations, or to complete what Ralph and his coworkers did not do. With this perspective in mind, Equations (42) and (46) and an analogous homogeneity form for the correlation function (discussed later) offer an option to proceed as we show now.

In Ref. [4], two other scaling relations between hatted exponents were derived:

$$\hat{\alpha} = d(\hat{\varphi} - \hat{\nu}) \quad \text{or} \quad \hat{\alpha} = 1 + d(\hat{\varphi} - \hat{\nu}), \tag{49}$$

$$\hat{\eta} = \hat{\gamma} - \hat{\nu}(2-\eta). \tag{50}$$

Concerning the first relation (49), the second formula is valid in such circumstances where the model has $\alpha = 0$ and an impact angle $\phi \neq \pi/4$ for the Fisher zeros in the complex plane (this is the case for the pure two-dimensional Ising model). We will not consider this case, but rather the more general case of the first formula. It can be rederived by careful use of the ratio 4 (Equation (19)) in Section 2, and even requires the use of FSS of the correlation length. From $\xi_\pm(t) \simeq |t|^{-\nu}(-\ln|t|)^{\hat{\nu}}$, we first reverse to $|t| \simeq \xi_\pm^{-1/\nu}(t)(\ln\xi_\pm(t))^{\hat{\nu}/\nu}$. This expression is then incorporated into (21) to obtain

$$f_\pm^{\rm sing}(t) \simeq \xi_\pm^{-d}(t)(\ln\xi_\pm(t))^{d\hat{\nu}+\hat{\alpha}} \tag{51}$$

i.e., a modified version of (19):

$$R_\pm^{(4bis)}(t) = \frac{\xi_\pm^d(t) f_\pm^{\rm sing}(t)}{(\ln\xi_\pm(t))^{d\hat{\nu}+\hat{\alpha}}} \to \xi_{0\pm}^d F_\pm. \tag{52}$$

Now, inserting (41) into (51) leads to the FSS behavior of the free energy density at criticality

$$f_L^{\rm sing}(0) \simeq L^{-d}(\ln L)^{-d\hat{\varphi}}(\ln L)^{\hat{\alpha}+d\hat{\nu}}, \tag{53}$$

and compatibility with (46) at $t = h = 0$, $b = L$ then demands

$$\hat{\alpha} = d\hat{\varphi} - d\hat{\nu}. \tag{54}$$

which is Kenna and his coworkers' relation.

The same derivation can be carried out for the magnetic sector, considering the approach to criticality at T_c for $h \to 0$, yielding the scaling relation

$$\hat{\delta} = d\hat{\varphi} - d\hat{\nu}_c. \tag{55}$$

Concerning Equation (50), we believe that this relation is incomplete. Applied to the four-state Potts model in two dimensions [9–12], which has $\hat{\eta} = -\frac{1}{8}$, $\hat{\gamma} = \frac{3}{4}$, $\hat{\nu} = \frac{1}{2}$, and $\eta = \frac{1}{4}$, Equation (50) is fulfilled. We believe that this is because $\hat{\varphi} = 0$, and that an additional term $\hat{\varphi}\gamma/\nu$ is missing in the general case. A test is provided in the case of the Ising model in four dimensions, which has $\hat{\varphi} = \frac{1}{4}$ (models at their upper critical dimensions have $\hat{\varphi} = 1/d_{uc}$ [7,13–15]). There, Ralph and his coauthors had anticipated that $\hat{\eta} = 0$ for Equation (50) to work ($\hat{\gamma} = \frac{1}{3}, \hat{\nu} = \frac{1}{6}$, and $\eta = 0$ for the Ising model in four dimensions), but according to Luijten [16], $\hat{\eta} = \frac{1}{2}$ instead, a result that is in contradiction with Equation (50).

Let us examine the problem in more detail. In Ref. [4], the authors have questioned the relation between the correlation function and the square of the magnetization when the system decorrelates, i.e., for $|\mathbf{r}| \to \infty$:

$$G(t, h, L^{-1} \to \infty, |\mathbf{r}| \to \infty) \to m^2(t, h, L^{-1} \to \infty). \tag{56}$$

On the contrary, we assume that there is no reason why this would not be valid, so we start from the homogeneity of the (spin–spin) correlation function with logarithmic corrections as

$$G(t, h, L^{-1}, |\mathbf{r}|) = b^{-(d-2+\eta)}(\ln b)^{\hat{\eta}} G(x(b), y(b), z(b), |\mathbf{r}|/b), \tag{57}$$

Setting $x = 1$, $y = 0$ and the thermodynamic limit $z = 0$ leads to the following temperature behavior when $|\mathbf{r}| \to \infty$

$$G(t, 0, 0, \infty) \simeq |t|^{\frac{d-2+\eta}{y_t}}(-\ln|t|)^{\frac{d-2+\eta}{y_t}\hat{y}_t + \hat{\eta}} G(1, 0, 0, \infty) = m^2(t, 0, 0). \tag{58}$$

This requires the usual relation $2\beta = \frac{d-2+\eta}{y_t}$ and

$$\hat{\eta} = 2(\hat{\beta} - \beta\hat{y}_t). \tag{59}$$

The two examples given above are test grounds. For the four-state Potts model in two dimensions, we extract immediately $\hat{\eta} = -\frac{1}{8}$, which is correct. For the 4d Ising model, on the other hand, we obtain $\hat{\eta} = \frac{1}{2}$, in agreement with Luijten's result [16], later verified numerically in Ref. [17], but we are here in contradiction with the prediction of Refs. [2,4].

Since the question is of importance, we want to consider it from other perspectives also. The correlation function is linked to the susceptibility via the fluctuation–dissipation relation:

$$\chi(0, 0, L^{-1}) = \int_0^L d^d r |\mathbf{r}|^{-(d-2+\eta)}(\ln|\mathbf{r}|)^{\hat{\eta}} G(0, 0, |\mathbf{r}|L^{-1}, 1). \tag{60}$$

Setting $\rho = |\mathbf{r}|/L$, we have

$$\chi(0, 0, L^{-1}) = L^{2-\eta}(\ln L)^{\hat{\eta}} \int_0^1 d^d\rho \rho^{-(d-2+\eta)}\left(1 + \frac{\ln\rho}{\ln L}\right)^{\hat{\eta}} G(0, 0, \rho, 1) \simeq L^{2-\eta}(\ln L)^{\hat{\eta}}, \tag{61}$$

and since the susceptibility obeys, via the second derivative of (46),

$$\chi(t, h, L^{-1}) = \kappa_h^2 b^{-d+2y_h}(\ln b)^{2\hat{y}_h} \mathscr{Y}(x(b), y(b), z(b)), \tag{62}$$

its FSS compared to (61) demands that $2 - \eta = 2y_d - d = \gamma/\nu$ and

$$\hat{\eta} = 2\hat{y}_h. \tag{63}$$

Again, this confirms $\hat{\eta} = \frac{1}{2}$ for the 4d Ising model.

The fluctuation–dissipation theorem has also been used in Refs. [2,4] in the form

$$\chi_\infty(t) \sim \xi_\infty^{2-\eta}(t)(\ln\xi_\infty(t))^{\hat{\eta}}, \tag{64}$$

from where Equation (50) follows, so there is still some difficulty hidden to solve our disagreement. Let us set $b = \xi_\infty(t)$ in (62):

$$\chi_\infty(t) = \xi_\infty^{-d+2y_h}(t)(\ln\xi_\infty(t))^{\hat{\eta}} \mathscr{Y}(x(\xi_\infty(t)), 0) \tag{65}$$

where the variable x in the scaling function $\mathscr{Y}$ is evaluated at $\xi_\infty(t)$ to give

$$x(\xi_\infty(t)) = \xi_\infty(t)^{y_t}(\ln \xi_\infty(t))^{\hat{y}_t}|t| = (-\ln|t|)^{\hat{\varphi}/\nu}. \quad (66)$$

The scaling function must have the behavior $\mathscr{Y}(x,0) \sim x^{-\gamma}$ when $|t| \to 0$ to recover the temperature singularity of the susceptibility $\chi_\infty(t,0) \sim |t|^{-\gamma}(-\ln|t|)^{\hat{\gamma}}$. It follows that instead of (64), one has

$$\chi_\infty(t) \sim \xi_\infty^{2-\eta}(t)(\ln \xi_\infty(t))^{\hat{\eta}-\gamma\hat{\varphi}/\nu} \quad (67)$$

and instead of Equation (50), one has a third form for the exponent $\hat{\eta}$:

$$\hat{\eta} = \hat{\gamma} - \hat{\nu}(2-\eta) + \gamma\frac{\hat{\varphi}}{\nu}, \quad (68)$$

again compatible with $\hat{\eta} = \frac{1}{2}$ for the 4d Ising model. This suggests to use, instead of the ratio $R_\pm^{(3)}(t)$, the modified version

$$\begin{aligned} R_\pm^{(3\text{bis})}(t) &= \frac{\chi_\pm(t)}{\xi_\pm^{2-\eta}(t)(\ln \xi_\pm(t))^{\hat{\eta}-\gamma\hat{\varphi}/\nu}} \\ &= \frac{\Gamma_\pm}{\xi_{0\pm}^{2-\eta}}(-\ln|t|)^{\hat{\gamma}-\hat{\eta}-\gamma\hat{\varphi}/\nu-\hat{\nu}(2-\eta)} \to \frac{\Gamma_\pm}{\xi_{0\pm}^{2-\eta}} \end{aligned} \quad (69)$$

which, again, is universal. The four standard scaling laws and the corresponding four hatted scaling laws are listed in Table 1.

Table 1. The main scaling laws and hatted scaling laws and the associated universal combinations of amplitudes.

Ratio	Scaling Relation	Hatted Scaling Relation
$\frac{m^2(-\|t\|)}{C_\pm(\|t\|)\chi_\pm(\|t\|)\|t\|^2}$	$2\beta + \gamma = 2 - \alpha$	$2\hat{\beta} = \hat{\alpha} + \hat{\gamma}$
$\left.\frac{\chi_\pm(t)m_-^{\delta-1}(-\|t\|)h}{m_c^\delta(h)}\right\|_{h=h_\pm^{\text{LY}}(\|t\|)}$	$\beta(\delta-1) = \gamma$	$\hat{\gamma} + \hat{\beta}(\delta-1) - \delta\hat{\delta} = 0$
$\frac{\chi_\pm(t)}{\xi_\pm^{2-\eta}(t)(\ln\xi_\pm(t))^{\hat{\eta}-\gamma\hat{\varphi}/\nu}}$	$\nu(2-\eta) = \gamma$	$\hat{\gamma} - \hat{\nu}(2-\eta) + \gamma\frac{\hat{\varphi}}{\nu} = \hat{\eta}$
$\xi_\pm^d(t)f_\pm^{\text{sing}}(t)$	$2 - \alpha = \nu d$	$\hat{\alpha} = d\hat{\varphi} - d\hat{\nu}$

So far so good, but the situation is not yet clear since the case of percolation in six dimensions (its upper critical dimension) is maybe a counterexample. With $\gamma = 1$, $\nu = \frac{1}{2}$, $\eta = 0$, and the values of the logarithmic correction exponents $\hat{\gamma} = \frac{2}{7}$, $\hat{\nu} = \frac{5}{42}$, and the pseudo-critical exponent $\hat{\varphi} = \frac{1}{d_{\text{uc}}} = \frac{1}{6}$, using Equation (68), we predict $\hat{\eta} = \frac{8}{21}$, while Kenna and his coworkers predict $\hat{\eta} = \frac{1}{21}$ from Equation (50). This latter result conforms to an analytic prediction from Ref. [18], but on the other hand, our value is supported by an FSS prediction by Ruiz-Lorenzo [19], $\chi_L L^{-2} \sim (\ln L)^{8/21}$. This disagreement demands further analysis.

In Ref. [2], Kenna listed the values of the known hatted critical exponents for a series of models, and when $\hat{\eta}$ was not known, he proposed the expected value from the use of Equation (50). An interesting model is missing from the list, the tricritical Ising model, which has logarithmic corrections at its upper critical dimension $d_{\text{uc}} = 3$ and has non-zero $\hat{\varphi} = 1/3$. We will now analyze this universality class in more detail.

5. The Tricritical Ising Universality Class in the Blume–Capel Model at the Upper Critical Dimension

The spin-1 Blume–Capel model [20,21] is a lattice model defined by the Hamiltonian

$$\mathcal{H} = -J\sum_{\langle i,j\rangle}\sigma_i\sigma_j + \Delta\sum_i \sigma_i^2 - H\sum_i \sigma_i, \tag{70}$$

where the spin variables $\sigma_i = -1, 0, +1$, $J > 0$ denotes the ferromagnetic exchange interaction among nearest-neighbor sites ($\langle i,j\rangle$ indicates summation over nearest neighbors), and Δ is the crystal-field strength that controls the density of vacancies (the $\sigma_i = 0$ states can be viewed as vacancies in an ordinary $\sigma_i = \pm 1$ Ising model) [22]. For $\Delta = -\infty$, vacancies are suppressed from the partition function, and the Hamiltonian reduces to that of the Ising ferromagnet. At $\Delta = 0$ the second-order transition is in the pure Ising model universality class. When Δ increases from 0, a perturbation theory shows that the transition temperature decreases along a line that remains of Ising-like second-order phase transition. On the other hand, in the vicinity of $T = 0$, the transition is first order and persists first order at small values of T until it reaches the second-order line. Right at the limit, there is a tricritical point characterized by specific values of T_t, Δ_t (Figure 1). Tricriticality corresponds to the ϕ^6 Landau expansion [23], and the upper critical dimension is thus $d_{\rm uc} = 3$, the case that we consider now.

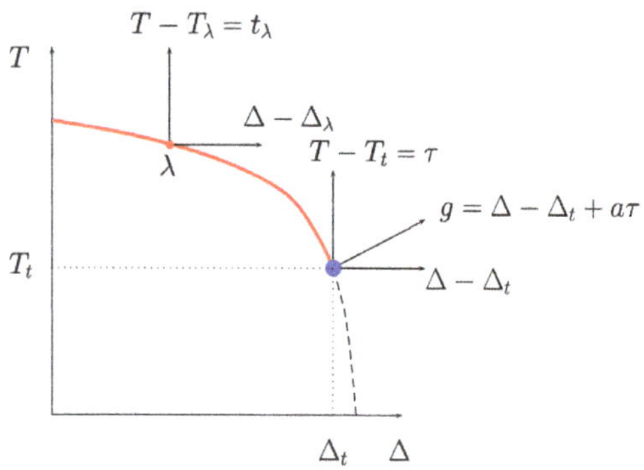

Figure 1. Typical phase diagram of the Blume–Capel model in the (T, Δ) plane. The λ line is a line of second-order phase transition in the Ising model universality class that ends at a tricritical point of coordinates (T_t, Δ_t). The dashed line is a first-order transition line. The most singular even scaling field at the tricritical point is g, a linear combination of τ and of $\Delta - \Delta_t$.

Using appropriate linear combinations of the physical parameters, determined by the geometry of the phase diagram and the fact that they have to vanish at the tricritical point, the even scaling fields are $\tau = (T - T_t)/T_t$, $g = (\Delta - \Delta_t)/T_t + a\tau$, and the odd scaling field is the usual magnetic field $h = H/T_t$. Lawrie and Sarbach [24] have shown that the free energy density at the tricritical point in $3d$ reads, in terms of these scaling fields, as

$$f_{\rm tri}^{\rm sing}(\tau, g, h) = b^{-3}\mathscr{F}_\pm(b^1(\ln b)^{\frac{4}{15}}\tau, b^2(\ln b)^{\frac{1}{3}}g(1 - a^{-1}b^{-1}(\ln b)^{-\frac{1}{15}}\tau g), b^{\frac{5}{2}}(\ln b)^{\frac{1}{6}}h), \tag{71}$$

where the non-universal metric factors κ_i have been omitted. Here, the subscript "tri" indicates that the expression is valid in the vicinity of the tricritical point where $\tau = g = h = 0$. The logarithmic corrections are explicitly given in this expression.

The notations in Ref. [24] necessitate to be adapted to be consistent with those that we have used until now. Leaving aside the logarithmic corrections for a while, let us write

$$f_{\rm tri}^{\rm sing}(\tau, g, h) = b^{-d}\mathscr{F}_\pm(b^{y_\tau}\tau, b^{y_g}g, b^{y_h}h). \tag{72}$$

The dominant even scaling field is g, the usual singularities and critical exponents are therefore defined by their behaviors with regard to g instead of τ, which brings corrections to scaling due to crossover. This means that the t and y_t of the three previous sections of this paper will now be replaced by g and y_g. At $h = 0$ and $\tau = 0$, setting $b = |g|^{-1/y_g}$ leads to $f_{\rm tri}^{\rm sing}(0, g, 0) \sim |g|^{d/y_g}$. This is compatible with $\nu = 1/y_g$. The specific heat measures the total energy fluctuations. Its most singular part is defined by

$$C(\tau, g, h) = \frac{\partial^2 f_{\rm tri}^{\rm sing}}{\partial g^2} = b^{-d+2y_g}\mathscr{C}_\pm(b^{y_\tau}\tau, b^{y_g}g, b^{y_h}h) \sim |g|^{-\alpha} \quad \text{when } \tau, h = 0 \tag{73}$$

with $\alpha = \frac{2y_g - d}{y_g}$. This shows that the exponent d/y_g of $f_{\rm tri}^{\rm sing}(0, g, 0)$ is thus equal to the usual value $2 - \alpha$, and the hyperscaling relation holds.

We can also define less singular exponents with regard to the scaling field τ (with tilde notation), e.g., $C(\tau, 0, 0) \sim |\tau|^{-\tilde{\alpha}}$. A similar analysis as above shows that $f_{\rm tri}^{\rm sing}(\tau, 0, 0) \sim |\tau|^{d/y_\tau}$ with $d/y_\tau = (d/y_g)\phi = (2 - \alpha)\phi$ with $\phi = y_g/y_\tau$ the crossover exponent. There is a caveat here since d/y_τ *is not equal to* $2 - \tilde{\alpha}$, as one can find in the literature [24]. Indeed, $\tilde{\alpha} = (2y_g - d)/y_\tau = \alpha\phi$, hence $2 - \tilde{\alpha} = 2 - \alpha\phi \neq (2 - \alpha)\phi$. This is important to collect correct expressions, and this is shown in Table 2.

Table 2. Leading and subleading singularities for the most common physical quantities and the definitions of the associated exponents. Here, the crossover exponent is $\phi = y_g/y_\tau$.

Leading even Field, g		**Subleading even Field, τ**		**Leading Odd Field, h**	
$C(0, g, 0) \sim \lvert g\rvert^{-\alpha}$	$\alpha = \frac{2y_g - d}{y_g}$	$C(\tau, 0, 0) \sim \lvert\tau\rvert^{-\tilde{\alpha}}$	$\tilde{\alpha} = \alpha\phi$		
$m(0, g, 0) \sim \lvert g\rvert^{\beta}$	$\beta = \frac{d - y_h}{y_g}$	$m(\tau, 0, 0) \sim \lvert\tau\rvert^{\tilde{\beta}}$	$\tilde{\beta} = \beta\phi$	$m(0, 0, h) \sim \lvert h\rvert^{1/\delta}$	$\delta = \frac{y_h}{d - y_h}$
$\chi(0, g, 0) \sim \lvert g\rvert^{-\gamma}$	$\gamma = \frac{2y_h - d}{y_g}$	$\chi(\tau, 0, 0) \sim \lvert\tau\rvert^{-\tilde{\gamma}}$	$\tilde{\gamma} = \gamma\phi$		

This being said, we can now incorporate the logarithmic corrections in Equation (72) to obtain

$$f_{\rm tri}^{\rm sing}(\tau, g, h) = b^{-d}\mathscr{F}_\pm(b^{y_\tau}(\ln b)^{\hat{y}_\tau}\tau, b^{y_g}(\ln b)^{\hat{y}_g}g, b^{y_h}(\ln b)^{\hat{y}_h}h), \tag{74}$$

and the comparison with Equation (71) simply leads to

$$y_g = 2, \quad y_h = \frac{5}{2}, \quad y_\tau = 1, \tag{75}$$

$$\hat{y}_g = \frac{1}{3}, \quad \hat{y}_h = \frac{1}{6}, \quad \hat{y}_\tau = \frac{4}{15}, \quad \hat{\varsigma} = \frac{1}{3}. \tag{76}$$

We can now deduce the values of the standard critical exponents and the associated logarithmic corrections exponents. They are listed in Table 3 for three universality classes, which all have non-zero values of the pseudo-critical exponent $\hat{\varsigma}$, the Ising model in four dimensions, the tricritical Ising model in three dimensions, and the problem of percolation in six dimensions.

In this table, the first six lines are not controversial. The seventh line presents the correlation function correction exponent $\hat{\eta}$, which follows from our scaling law, in any of the forms given in Equations (59), (63), or (68). These three expressions are mutually consistent, but they differ from Equation (50) used by Kenna and his coauthors. This latter formula would, respectively, predict for the three universality classes the values 0, $-\frac{1}{3}$, and $\frac{1}{21}$. We have seen that the first value, 0, is falsified in the $4d$ IM case by Refs. [16,17], but the results of Ref. [18] invalidates our third value $\frac{8}{21}$, the last entry of in Table 3, while Ref. [19], on the contrary, supports this value.

Table 3. Leading and logarithmic correction exponents for the most common physical quantities.

Quantity	Leading Exponent	IM4D	Tri. IM	Perco.	Logarithmic Correction Exponent	IM4D	Tri. IM	Perco.
$C(t,0)$	$\alpha = \frac{2y_t-d}{y_t}$	0	$\frac{1}{2}$	-1	$\hat{\alpha} = (2-\alpha)\hat{y}_t$	$\frac{1}{3}$	$\frac{1}{2}$	$\frac{2}{7}$
$m_-(t,0)$	$\beta = \frac{d-y_h}{y_t}$	$\frac{1}{2}$	$\frac{1}{4}$	1	$\hat{\beta} = \beta\hat{y}_t + \hat{y}_h$	$\frac{1}{3}$	$\frac{1}{4}$	$\frac{2}{7}$
$\chi(t,0)$	$\gamma = \frac{2y_h-d}{y_t}$	1	1	1	$\hat{\gamma} = 2\hat{y}_h - \gamma\hat{y}_t$	$\frac{1}{3}$	0	$\frac{2}{7}$
$m_c(0,h)$	$\frac{1}{\delta} = \frac{d-y_h}{y_h}$	$\frac{1}{3}$	$\frac{1}{5}$	$\frac{1}{2}$	$\hat{\delta} = \frac{1}{\delta}\hat{y}_h + \hat{y}_h$	$\frac{1}{3}$	$\frac{1}{5}$	$\frac{2}{7}$
$h_{\mathrm{LY}}(t,0)$	$\Delta = \frac{y_h}{y_t}$	$\frac{3}{2}$	$\frac{5}{4}$	2	$\hat{\Delta} = \Delta\hat{y}_t - \hat{y}_h$	0	$\frac{1}{4}$	0
$\xi(t,0)$	$\nu = \frac{1}{y_t}$	$\frac{1}{2}$	$\frac{1}{2}$	$\frac{1}{2}$	$\hat{\nu} = \hat{\varphi} - \nu\hat{y}_t$	$\frac{1}{6}$	$\frac{1}{6}$	$\frac{5}{42}$
$G(0,0,\lvert\mathbf{r}\rvert)$	$\eta = d - 2y_h + 2$	0	0	0	$\hat{\eta} = 2\hat{y}_h$	$\frac{1}{2}$	$\frac{1}{3}$	$\frac{8}{21}$

The case of the tricritical Ising model in three dimensions appears crucial, and we have to provide numerical results in support of our result. The numerical computation of the correlation function is known to be a very delicate problem, and we will approach the value of the exponent $\hat{\eta}$ differently, using FSS. Another delicate aspect is also the well-known fact that extracting logarithmic corrections in the vicinity of a critical point can be extremely difficult [25]. Recently, it was found that very accurate results can be obtained numerically in the Blume–Capel model with relatively small system sizes [26] via the analysis of the zeros of the partition function, and in particular the Lee–Yang zeros [27,28]. The Lee–Yang zeros are connected to the susceptibility [29] via

$$\chi(g,0,L^{-1}) \simeq L^{-d}\sum_{j=1}^{L^d} h_j^{-2}(g,0,L^{-1}), \tag{77}$$

where j labels the zeros in the upper-half complex plane, indexed in order of increasing distance from the critical point. The sum is dominated by the lowest zero, the Lee–Yang edge h^{LY}, and at the tricritical point, the FSS of the susceptibility is thus linked to that of h^{LY}:

$$\chi(0,0,L^{-1}) \simeq L^{-d}h^{\mathrm{LY}}(0,0,L^{-1})^{-2} \simeq L^{2-\eta}(\ln L)^{\hat{\eta}}. \tag{78}$$

In the presence of logarithmic corrections, the scaling form of the Lee–Yang edge obeys

$$h^{\mathrm{LY}}(g,h) = b^{-y_h}(\ln b)^{-\hat{y}_h}\mathscr{H}_\pm(b^{y_\tau}(\ln b)^{\hat{y}_\tau}\tau, b^{y_g}(\ln b)^{\hat{y}_g}g, b^{y_h}(\ln b)^{\hat{y}_h}h) \tag{79}$$

compatible with the behavior in terms of the thermal scaling field g, as it can be shown using the scaling laws of Table 3, $h^{\mathrm{LY}}(g,0) \sim |g|^{\Delta}(-\ln|g|)^{\hat{\Delta}}$. If one sits exactly at the tricritical point, $\tau = g = h = 0$, we can extract the FSS behavior of the zeros by setting $b = L$,

$$h^{\mathrm{LY}}(0,0,L^{-1}) \simeq L^{-y_h}(\ln L)^{-\hat{y}_h} \tag{80}$$

and it follows that we expect

$$h^{\mathrm{LY}}(0,0,L^{-1}) \simeq L^{(\eta-2-d)/2}(\ln L)^{-\hat{\eta}/2} \tag{81}$$

which agrees with Equation (78).

As we said, this form can be checked with good accuracy at the price of relatively light Monte Carlo simulations. The coordinates of the tricritical point of the Blume–Capel model in $3d$ are found in Ref. [30], $T_t \simeq 1.4182$, $\Delta_t \simeq 2.8448$, but the value of Δ_t does not seem to be as accurate as that of the temperature and for example Zierenberg et al. Ref [31] report instead $\Delta_t \simeq 2.8446$. Let us first analyze this problem ourselves. In Figure 2, we report the FSS of the magnetization at $T_t \simeq 1.4182$ for several values of Δ ranging from 2.8440 to 2.8448. The magnetization is expected to follow the FSS behavior

$$m(0,0,L^{-1}) \sim L^{-d+y_h}(\ln L)^{\hat{y}_h} \tag{82}$$

with $d - y_h = \frac{1}{2}$ and $\hat{y}_h = \frac{1}{6}$. The data points are fitted as $m(0,0,L^{-1})L^{\frac{1}{2}} = a(\ln L)^b$ with a, b free parameters. The closest to the expected result (the black dashed line) is at $\Delta = 2.8442$, where we obtain $b = 0.155 \pm 0.007$. We will thus consider this value of Δ as our estimate for the coordinate of the tricritical point Δ_t.

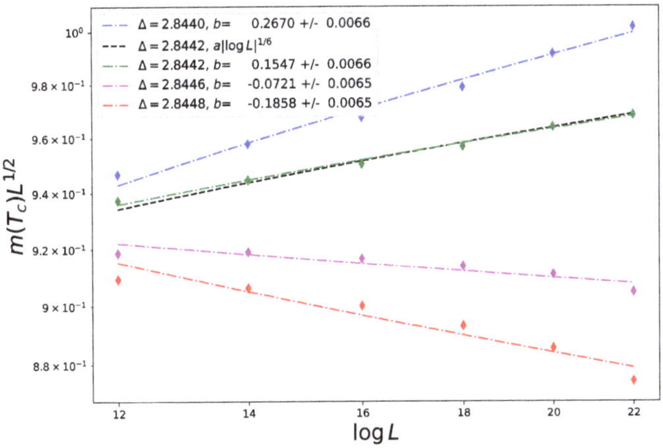

Figure 2. FSS of the magnetization for the tricritical Ising model (Blume–Capel model at its tricritical temperature) in 3d at $T_t = 1.4182$ and various values of the crystal-field parameter Δ for sizes ranging from $L = 12$ to 22. The best fit is for $\Delta = 2.8442$ ($\chi^2/\text{dof} = 30.40/4 = 7.6$ at $\Delta = 2.8440$, $\chi2/\text{dof} = 5.39/4 = 1.35$ at $\Delta = 2.8442$, $\chi2/\text{dof} = 28.36/4 = 7.09$ at $\Delta = 2.8446$ and $\chi2/\text{dof} = 76.74/4 = 19.18$ at $\Delta = 2.8448$).

The analysis of the Lee–Yang edge is presented in Figure 3 with a larger choice of values of Δ and, again, the best fit is at $\Delta_t = 2.8442$, where the estimate of $\hat{y}_h$ is now slightly larger at 0.172 ± 0.008.

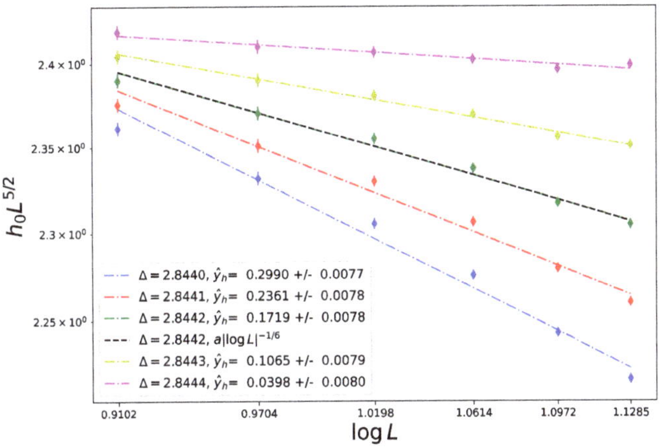

Figure 3. FSS of the Lee–Yang edge for the tricritical Ising model (Blume–Capel model at its tricritical temperature) in 3d at $T_t = 1.4182$ and various values of Δ for sizes ranging from $L = 12$ to 22.

The reader could still question the sensitivity of the value of $\hat{y}_h$ with the choice of tricritical temperature T_t. Indeed, when one looks at the FSS of the tricritical magnetization, for example, the effective exponent of the log term is either positive and close to the expected value or can differ from the expectation and even be negative, depending on the values of the crystal field Δ (see Figure 2). It makes sense to ask whether the role of T_t may also have a significant influence. We believe that the results presented in this work are reliable and to support the consistency of the numerical data, we show in Figure 4 that slight variations of T change the regime from the pure 3d Ising model at $T = 1.4197$, for which $y_h = 2.4815(15)$ [32] is expected, to first order at $T = 1.4170$, where an effective FSS $y_h = d$ is expected [33]. $T = 1.4182$ safely recovers $y_h^{\rm Tri} = 2.5$ to a very good accuracy and confirms the tricritical value of $T_t \simeq 1.4182$. Note that the transition line in the phase diagram in the vicinity of the tricritical point is almost at a fixed value of Δ, this is why the three regimes are found at the same crystal-field value of $\Delta = 2.8442$.

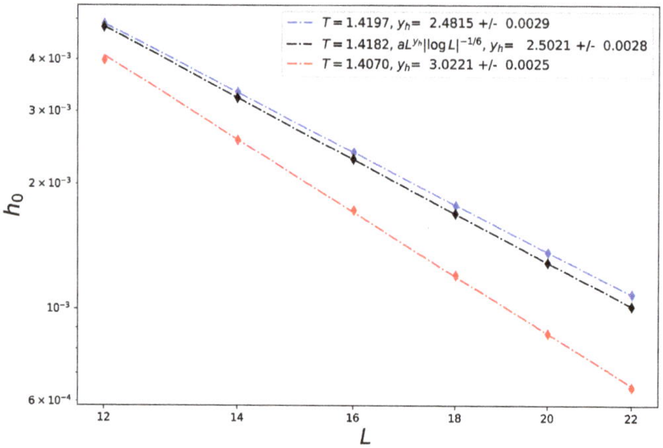

Figure 4. FSS of the Lee–Yang edge in the vicinity of the tricritical point of the Blume–Capel model in $3d$, at $T_t = 1.4170$, 1.4182 and 1.4197 and $\Delta = 2.8442$ to show the first-order, the tricritical, and the ordinary second-order regimes from the values of the corresponding RG dimensions y_h.

6. Conclusions

The numerical results obtained for the tricritical Ising model universality class in $3d$ confirm the prediction that $\hat{y}_h = \frac{1}{6}$, hence the prediction $\hat{\eta} = \frac{1}{3}$, while the scaling law of Kenna and his coworkers would have given $-\frac{1}{3}$ instead.

In Ref. [2], Ralph Kenna concluded his review with a table collecting the sets of critical exponents and hatted critical exponents for various models and predicting those that were still unknown from the use of the newly discovered scaling laws and, in particular Equation (50), which we scrutinize and propose to replace by Equation (68) or any of the equivalent forms that we have derived.

In the list, the $O(n)$ model with long-range interactions [34,35] was predicted to have $\hat{\eta} = 0$. We propose instead $\hat{\eta} = \frac{1}{2}$, following from $\hat{y}_t = (4-n)/[2(n+8)]$, $\hat{y}_h = \frac{1}{4}$.
The Lee–Yang edge in $6d$ [19] was predicted to have $\hat{\eta} = \frac{1}{9}$. We rather have $\hat{y}_t = -\frac{2}{9}$ and $\hat{y}_h = \frac{2}{9}$, hence $\hat{\eta} = \frac{4}{9}$.
For lattice animals in $8d$ [19], Kenna predicted $\hat{\eta} = \frac{1}{9}$, and we have $\hat{y}_t = \hat{y}_h = \frac{2}{9}$ and $\hat{\eta} = \frac{4}{9}$.
The case of scale-free networks [36–38] is particular in the sense that Ralph Kenna did not make any prediction for $\hat{\eta}$ because some exponents were missing. From those which are known, we can deduce that $\hat{y}_t = -\frac{1}{2}$ and $\hat{y}_h = -\frac{1}{4}$, and we deduce thus $\hat{\eta} = -\frac{1}{2}$, which is a new prediction.

Eventually, we believe that the n-color Ashkin–Teller model in $2d$ is still under question since the exponents collected by Shalaev and Jug [39] do not satisfy the "standard" scaling laws, e.g., the values reported do not obey $\hat{\alpha} + \hat{\gamma} = 2\hat{\beta}$.

To finish this paper, we would like to say that the scaling laws discovered by Ralph Kenna and his coworkers are invaluable because they make it possible to establish (or falsify) the consistency of the results obtained for various models. The case of the n-color Ashkin–Teller model in $2d$ is such an example where it seems that there are still some inconsistencies to solve. Although we happened to contradict one of these scaling laws, we admire the piece of work conducted in Refs. [2–4], where we recognize Ralph's footprint.

Author Contributions: Writing—review & editing, L.M., A.D. and B.B. All authors had equal contibutions. All authors have read and agreed to the published version of the manuscript.

Funding: This research was funded by the "Université franco-allemande" and the Collège Doctoral 02-07 Statistical Physics of Complex Systems (University of Leipzig-Université de Lorraine).

Data Availability Statement: No new data were created or analyzed in this study. Data sharing is not applicable to this article.

Acknowledgments: We gratefully acknowledge stimulating discussions with Wolfhard Janke. We also thank the L^4 collaboration Leipzig-Lviv-Lorraine-Coventry. LM thanks Coventry University and the Collège Doctoral 02-07 Statistical Physics of Complex Systems (University of Leipzig-Université de Lorraine) for financial support.

Conflicts of Interest: The authors declare no conflict of interest.

References

1. Kardar, M. *Statistical Physics of Fields;* Cambridge University Press: Cambridge, UK, 2007. [CrossRef]
2. Kenna, R. Universal scaling relations for logarithmic-correction exponents. *arXiv* **2012**, arXiv:1205.4252.
3. Kenna, R.; Johnston, D.A.; Janke, W. Scaling Relations for Logarithmic Corrections. *Phys. Rev. Lett.* **2006**, *96*, 115701. [CrossRef] [PubMed]
4. Kenna, R.; Johnston, D.A.; Janke, W. Self-Consistent Scaling Theory for Logarithmic Correction Exponents. *Phys. Rev. Lett.* **2006**, *97*, 155702. [CrossRef]
5. Privman, V.; Hohenberg, P.C.; Aharony, A. *Phase Transitions and Critical Phenomena;* Domb, C., Lebowitz, J.L., Eds.; Academic Press: New York, NY, USA, 1991; Volume 14.
6. Fisher, M.E. Notes, definitions, and formulas for critical point singularities. In *Critical Phenomena, Proceedings of a Conference Held in Washington, DC, USA, 5–8 April 1965;* Green, M.S., Sengers, J.V., Eds.; Taylor & Francis: Abingdon, UK, 1965.
7. Kenna, R.; Berche, B. A new critical exponent koppa and its logarithmic counterpart koppa-hat. *Condens. Matter Phys.* **2013**, *16*, 23601. [CrossRef]
8. Kenna, R. Finite size scaling for O(N) φ4-theory at the upper critical dimension. *Nucl. Phys. B* **2004**, *691*, 292–304. [CrossRef]
9. Nauenberg, M.; Scalapino, D.J. Singularities and Scaling Functions at the Potts-Model Multicritical Point. *Phys. Rev. Lett.* **1980**, *44*, 837. [CrossRef]
10. Cardy, J.L.; Nauenberg, M.; Scalapino, D.J. Scaling theory of the Potts-model multicritical point. *Phys. Rev. B* **1980**, *22*, 2560. [CrossRef]
11. Berche, B.; Chatelain, C. Phase transitions in two-dimensional random Potts models. *arXiv* **2004**, arXiv:cond-mat/0207421.
12. Shchur, L.N.; Berche, B.; Butera, P. A study of logarithmic corrections and universal amplitude ratios in the two-dimensional 4-state Potts model. *Europhys. Lett.* **2008**, *81*, 30008. [CrossRef]
13. Ruiz-Lorenzo J.J. Revisiting (logarithmic) scaling relations using renormalization group. *Condens. Matter Phys.* **2017**, *20*, 13601. [CrossRef]
14. Kenna, R.; Berche, B. Fisher's scaling relation above the upper critical dimension. *EPL* **2014**, *105*, 26005. [CrossRef]
15. Kenna, R.; Berche, B. On a previously unpublished work with Ralph Kenna. *arXiv* **2023**, arXiv:2312.01675.
16. Luijten, E. Interaction Range, Universality and the Upper Critical Dimension. Ph.D. Thesis, Delft University, Delft, The Netherlands, 1997.
17. Lv, J.-P.; Xu, W.; Sun, Y.; Chen, K.; Deng, Y. Finite-size scaling of O(n) systems at the upper critical dimensionality. *Natl. Sci. Rev.* **2021**, *8*, nwaa212. [CrossRef]
18. Stenull, O.; Janssen, H.-K. Logarithmic corrections to scaling in critical percolation and random resistor networks. *Phys. Rev. E* **2003**, *68*, 036129. [CrossRef] [PubMed]
19. Ruiz-Lorenzo, J.J. Logarithmic corrections for spin glasses, percolation and Lee-Yang singularities in six dimensions. *J. Phys. A Math. Gen.* **1998**, *31*, 8773. [CrossRef]

20. Capel, H.W. On the possibility of first-order phase transitions in Ising systems of triplet ions with zero-field splitting. *Physica* **1966**, *32*, 966–988. [CrossRef]
21. Blume, M. Theory of the First-Order Magnetic Phase Change in UO2. *Phys. Rev.* **1966**, *141*, 517–524. [CrossRef]
22. Cardy, J.L. *Scaling and Renormalization in Statistical Physics*; Cambridge University Press: Cambridge, UK, 1996. [CrossRef]
23. Berche, B.; Ellis, T.; Holovatch, Y.; Kenna, R. Phase transitions above the upper critical dimension. *SciPost Phys. Lect. Notes* **2022**, *60*, 1–44. [CrossRef]
24. Lawrie, I.D.; Sarbach, S. Theory of tricritical points. In *Phase Transitions and Critical Phenomena*; Domb, C., Lebowitz, J.L., Eds.; Academic Press: London, UK, 1984.
25. Shchur, L.N.; Berche, B.; Butera, P. Numerical revision of the universal amplitude ratios for the two-dimensional 4-state Potts model. *Nucl. Phys. B* **2009**, *811*, 491–518. [CrossRef]
26. Moueddene, L.; Fytas, N.G.; Holovatch, Y.; Kenna, R.; Berche, B. Critical and tricritical singularities from small-scale Monte Carlo simulations: The Blume-Capel model in two dimensions. *arXiv* **2024**, arXiv:2401.02720
27. Yang, C.N.; Lee, T.D. Statistical Theory of Equations of State and Phase Transitions. I. Theory of Condensation. *Phys. Rev.* **1952**, *87*, 404–409. [CrossRef]
28. Lee, T.D.; Yang, C.N. Statistical Theory of Equations of State and Phase Transitions. II. Lattice Gas and Ising Model. *Phys. Rev.* **1952**, *87*, 410. [CrossRef]
29. Berche, B.; Kenna, R.; Walter, J.-C. Hyperscaling above the upper critical dimension. *Nucl. Phys. B* **2012**, *865*, 115–132. [CrossRef]
30. Deserno, M. Tricriticality and the Blume-Capel model: A Monte Carlo study within the microcanonical ensemble. *Phys. Rev. E* **1997**, *56*, 5204–5210. [CrossRef]
31. Zierenberg, J.; Fytas, N.G.; Janke, W. Parallel multicanonical study of the three-dimensional Blume-Capel model. *Phys. Rev. E* **2015**, *91 032126*. [CrossRef]
32. Pelissetto, A.; Vicari, E. Critical phenomena and renormalization-group theory. *Phys. Rep.* **2002**, *368*, 549–727. [CrossRef]
33. Privman, V.; Fisher, M.E. Finite-size effects at first-order transitions. *J. Stat. Phys.* **1983**, *33*, 385–417. [CrossRef]
34. Fisher, M.E.; Ma, S.-K.; Nickel, B.G. Critical Exponents for Long-Range Interactions. *Phys. Rev. Lett.* **1972**, *29*, 917. [CrossRef]
35. Luijten, E.; Blöte, H.W.J. Classical critical behavior of spin models with long-range interactions. *Phys. Rev. B* **1997**, *56*, 8945. [CrossRef]
36. Hong, H.; Park, H.; Tang, L. Finite-size scaling of synchronized oscillation on complex networks. *Phys. Rev. E* **2007**, *76*, 066104. [CrossRef]
37. Palchykov, V.; von Ferber, C.; Folk, R.; Holovatch, Y.; Kenna, R. Critical phenomena on scale-free networks: Logarithmic corrections and scaling functions. *Phys. Rev. E* **2010**, *82*, 011145. [CrossRef] [PubMed]
38. Krasnytska, M.; Berche, B.; Holovatch, Y.; Kenna, R. Partition function zeros for the Ising model on complete graphs and on annealed scale-free networks. *J. Phys. A* **2016**, *49*, 135001. [CrossRef]
39. Jug, G.; Shalaev, B.N. Critical behavior of weakly disordered anisotropic systems in two dimensions. *Phys. Rev. B* **1996**, *54*, 3442. [CrossRef] [PubMed]

entropy MDPI

Article

Violations of Hyperscaling in Finite-Size Scaling above the Upper Critical Dimension

A. Peter Young

Physics Department, University of California, Santa Cruz, CA 95064, USA; petery@ucsc.edu

Abstract: We consider how finite-size scaling (FSS) is modified above the upper critical dimension, $d_u = 4$, due to hyperscaling violations, which in turn arise from a dangerous irrelevant variable. In addition to the commonly studied case of periodic boundary conditions, we also consider new effects that arise with free boundary conditions. Some numerical results are presented in addition to theoretical arguments.

Keywords: phase transitions; finite-size scaling; hyperscaling; upper critical dimension

1. Introduction

In the study of continuous phase transitions, finite-size scaling (FSS) is extensively used to extrapolate numerical results on a range of finite-size systems to the thermodynamic limit in order to determine properties of the transition, such as critical exponents. While the basic ideas are well established, some additional complications arise above the "upper critical dimension", $d_u = 4$, related to violations of hyperscaling relations (those relations between critical exponents which also involve the space dimension d) for bulk properties. In this article, we will discuss these additional features which arise for dimension $d > 4$. For a recent review, see Ref. [1], which is by the late Ralph Kenna and collaborators.

The plan of the paper is as follows. In Section 2, we describe the theory of FSS and briefly indicate its justification from the renormalization group in Section 3. The new aspects of FSS above $d = 4$ are given in Section 4. Section 4.1 explains how the new physics in this region is related to a "dangerous irrelevant variable". Numerical results for periodic boundary conditions in $d = 4$ are presented in Section 4.2, while a discussion of the finite-size correlation length is given in Section 4.3. The rather surprising new features that appear with periodic boundary conditions are explained in Section 4.4, and finally, we give a summary in Section 5.

2. Finite-Size Scaling

We know that a sharp phase transition can only take place in an infinite system, while for a finite system, the change in behavior at a transition is "rounded out" in some way. At a second-order transition, the correlation length ξ varies as

$$\xi \sim (T - T_c)^{-\nu}, \tag{1}$$

where T_c is the transition temperature, and ν is a critical exponent that characterizes the divergence of the correlation length. Consequently, close enough to T_c that $\xi > L$, where L is the (linear) system size, finite-size corrections must be large.

Since numerical simulations are carried out on systems with finite sizes, in order to obtain accurate results in the thermodynamic limit, it is necessary to extrapolate results from several finite sizes to infinite sizes. The technique for doing this is called finite-size scaling (FSS). We will now outline the basic ideas of FSS, which were originally due to Michael Fisher [2,3].

We consider systems that are finite in all d dimensions, as in Monte Carlo simulations. One can also study other geometries in which one or more dimensions are infinite, the most important case being *one* infinite dimension, which corresponds to transfer matrix calculations and also to quantum systems at zero temperature. However, we will not consider that situation here. For simplicity, we assume that the system has the same size L in each of d dimensions, so the total size of the system is given by

$$N = L^d. \tag{2}$$

We will consider models for magnetic materials in which a "spin" lies on each side of a hypercubic regular lattice and interacts with its neighbors. Most commonly, a spin might be a scalar that takes values $S_i = \pm 1$, which is known as the Ising model, but sometimes classical m-component vectors are taken instead. The examples considered here will all be for Ising models.

As soon as we move away from the thermodynamic limit and discuss finite-size systems, we have to specify the boundary conditions. In simulations, one would like finite-size corrections to be as small as possible, so normally, one considers period boundary conditions, where a site with coordinate $L+1$ in one of the directions is identified with site 1 in that direction. This means that there is no surface. Here, we will mainly consider periodic boundary conditions, but Section 4.4 will include the effects of surfaces by choosing free boundaries, in which a site on the surface has no coupling going outside the surface. One can convert periodic boundary conditions to free boundary conditions simply by setting to zero those interactions which "wrap around" the system.

When $T - T_c$ is sufficiently small that $\xi \gtrsim L$, the results of simulations will depend on L. By contrast, well above T_c, where $L \gg \xi$, one expects properties of a system with periodic boundary conditions to differ by only exponentially small corrections of order $\exp(-L/\xi)$ from those of an infinite system. With free boundaries, where there is a surface, corrections will be much larger in this limit, typically of $1/L$ relative to the bulk values.

Let us consider some quantity X say, which diverges at T_c as

$$X_\infty \approx A(T - T_c)^{-\lambda} \tag{3}$$

in the thermodynamic limit. A simple postulate for how X might vary on a system of size L is

$$\frac{X_L(T)}{X_\infty(T)} = f\left(\frac{L}{\xi}\right), \tag{4}$$

for some function $f(x)$. One expects that f is universal in the renormalization group sense, i.e., it does not depend on the lattice structure or irrelevant variables. One does, however, expect that it will depend on the boundary conditions.

We can deduce the form of f in two limits:

1. At a fixed T in the limit $L \to \infty$ we have $X_L \to X_\infty$ so

$$\lim_{x\to\infty} f(x) = 1. \tag{5}$$

2. At L fixed and $T \to T_c$, $X_L(T)$ is not singular because there can be no sharp transition in a finite system. This means that the behavior of $f(x)$ for $x \to 0$ should compensate for the divergence of X_∞, i.e.,

$$f(x) = C x^{\lambda/\nu} \quad \text{for} \quad x \to 0, \tag{6}$$

where we used Equation (1).

In fact, $X_L(T)$ varies smoothly around $T = T_c$, and one can incorporate this, as well as have a single formula that works both above and below T_c, by rewriting Equation (4) as

$$X_L(T) = L^{\lambda/\nu}\,\tilde{X}(L^{1/\nu}(T - T_c)), \tag{7}$$

where ignoring the amplitude A in Equation (3) for simplicity, one has

$$\tilde{X}(x) = x^{-\lambda} f(x^{\nu}). \tag{8}$$

Since $X_L(0)$ is non-zero and finite, we must have $\tilde{X}(0) = \text{const.}$, so Equation (7) gives a prediction for the size dependence at T_c, namely

$$X_L(T_c) \sim L^{\lambda/\nu}. \tag{9}$$

Equation (7) is the basic postulate of finite-size scaling. For what quantities X would we typically apply it? Naturally, we would like a quantity with a strong divergence, i.e., λ is greater than zero and not too small. An "obvious" choice is the susceptibility χ, which diverges with an exponent γ, which is exactly 7/4 in the two-dimensional Ising model and about 1.24 in the three-dimensional Ising model. However, determining χ is difficult in numerical simulations for the following reasons. It is computed from spin correlations using the relation

$$\chi = \frac{1}{T}\frac{1}{N}\sum_{i,j}[\langle S_i S_j\rangle - \langle S_i\rangle\langle S_j\rangle] \tag{10}$$

$$= \frac{1}{T}N\left[\langle m^2\rangle - \langle m\rangle^2\right], \tag{11}$$

where

$$m = \frac{1}{N}\sum_{i=1}^{N} S_i, \tag{12}$$

is the magnetization per site. However, for a finite system and in zero external magnetic fields, strictly speaking, one has $\langle m\rangle = 0$. In practice, below T_c the magnetization will very slowly fluctuate between the "up" state and the "down" state with a time that varies strongly with system size and temperature and may be much longer than the time of the simulation. Hence, at least below T_c, it is useful to calculate only quantities that are invariant under the global symmetry of the model. Here, for an Ising model in zero field, this is an inversion of all the spins.

Some proposed fixes for this problem are:

1. Replace the second term in Equation (11) by $\langle|m|\rangle^2$, i.e.,

$$\chi = \frac{1}{T}N\left[\langle m^2\rangle - \langle|m|\rangle^2\right], \tag{13}$$

(the factor of $1/T$ is often omitted.) Equation (13) gives the desirable result that the two terms in Equation (11) tend to cancel below T_c, but unfortunately, it does not give the correct susceptibility above T_c. For example, in the limit $T \to \infty$, one has $\chi = (1 - 2/\pi)/T$ instead of the correct result, which is $1/T$.

2. My preferred solution is to replace Equation (10) by

$$\chi = \frac{1}{T}\left[\frac{1}{N}\sum_{i,j}\langle S_i S_j\rangle - \sum_{i}\langle S_i S_{i+\vec{L}/2}\rangle\right], \tag{14}$$

where $S_{i+\vec{L}/2}$ indicates the spin which is $L/2$ in every direction away from i, i.e., as far as possible from i. The second term is exponentially small in L/ξ above T_c, while below T_c it tends to cancel the first term. Unfortunately, this definition of χ in a finite system has not been adopted.

3. Ignore the second term in Equations (10) and (11) completely. Typically, one also neglects the factor of $1/T$ since this varies only slightly in the vicinity of the transition. We will, therefore, define the correlation function

$$C = \frac{1}{N}\sum_{i,j}\langle S_i S_j\rangle = N\langle m^2\rangle. \tag{15}$$

Well above T_c this is the susceptibility, apart from the factor of $1/T$, but below T_c the dominant part varies as N times the magnetization squared, i.e., $N|T_c - T|^{2\beta}$ where β is the order parameter exponent defined by $\langle m\rangle \sim |T_c - T|^{\beta}$ for $T < T_c$. For χ this part is canceled by the second term in Equation (11). From Equation (7) we write

$$C = L^{2-\eta}\tilde{C}\Big(L^{1/\nu}(T - T_c)\Big), \tag{16}$$

where we used the scaling relation

$$\gamma = (2 - \eta)\nu, \tag{17}$$

in which the exponent η characterizes the power-law decay of correlations with distance at the critical point. The scaling function $\tilde{C}(x)$ must have the following limiting behaviors:

$$\tilde{C}(x) \sim \begin{cases} x^{-\gamma}, & (x \to \infty) \\ \text{const.}, & (x = 0) \\ |x|^{2\beta}, & (x \to -\infty) \end{cases}, \tag{18}$$

where, in the last line, we used the following "hyperscaling" scaling relation

$$\beta = \frac{1}{2}(d - 2 + \eta)\,\nu. \tag{19}$$

Hyperscaling relations, which are the main focus of this volume, give connections between exponents which involve the space dimension d.

Equations (15)–(19) describe the FSS of the second moment of the order parameter m. Following Binder [4], it is of interest to consider other moments and, in particular, the whole distribution $P(m)$. In order for the second moment to satisfy Equation (16) and for the distribution to be normalized, we must have

$$P(m) = L^{\beta/\nu}\tilde{P}\Big(L^{\beta/\nu}m, L^{1/\nu}(T - T_c)\Big). \tag{20}$$

The power of L in front of the scaling function $\tilde{P}$ has to be the same as the power of L multiplying m in the first argument of $\tilde{P}$ in order that the distribution is normalized, and this power has to be β/ν in order to obtain Equation (16) (to see this note Equation (19)).

Binder [4] also pointed out that it is useful to look at ratios of moments such that the total power of m in the numerator and denominator are equal because then the power of L in front of the scaling function disappears. A simple and commonly studied case is the ratio of the fourth moment to the second moment squared,

$$g = \frac{1}{2}\left(3 - \frac{\langle m^4\rangle}{\langle m^2\rangle^2}\right), \tag{21}$$

which is known as the Binder ratio or Binder cumulant. The factors 3 and $1/2$ are not essential but are often included so that g varies from 0 well above T_c (since fluctuations there are Gaussian which implies $\langle m^4\rangle = 3\langle m^2\rangle^2$) to 1 well below T_c (since $\langle m^4\rangle = \langle m^2\rangle^2$ in that region). The FSS behavior of g is

$$g = \tilde{g}\Big(L^{1/\nu}(T - T_c)\Big). \tag{22}$$

Later, we will consider the wavevector dependence of correlations so we generalize the second moment of the order parameter in Equation (15) to non-zero wavevectors,

$$C(\mathbf{k}) = \frac{1}{N}\sum_{i,j} e^{i\mathbf{k}\cdot(\mathbf{R}_i - \mathbf{R}_j)} \langle S_i S_j \rangle. \tag{23}$$

The standard FSS scaling form for $C(\mathbf{k})$ for small $\mathbf{k}$ is the same as for $\mathbf{k} = 0$, namely Equation (16).

One reason the study of critical phenomena is particularly interesting is that many quantities are "universal", i.e., they only depend on certain broad features of the problem, such as space dimensionality and the symmetry of the order parameter, but do not depend on microscopic details, such as the type of lattice and range of interactions (as long as they are not infinite range). Exponents are examples of universal quantities, while the transition temperature depends on all the microscopic details, and so does non-universal.

In addition, scaling functions, including finite-size scaling functions, have a degree of universality. Fisher and Privman [5] clarified the situation for FSS functions. Suppose we take Equation (7) for a general quantity X_L. There will be a non-universal amplitude (or metric factor) associated with the overall size of X_L and a non-universal amplitude in front of the reduced temperature in the scaling function so we can write

$$X_L(T) = AL^{\lambda/\nu}\,\tilde{X}(BL^{1/\nu}(T - T_c)), \tag{24}$$

where A and B are non-universal. According to Privman and Fisher [5], the resulting scaling function $\tilde{X}$ is universal, which means that it does not depend on microscopic details, but since we are dealing with finite-size effects, it does depend on the boundary conditions. If we include a magnetic field h, the scaling function has a second argument $CL^{y_H}h$. The amplitude C is non-universal, but the scaling function itself remains universal. Here, y_H is called the magnetic exponent and is related to exponents introduced earlier by

$$y_H = \frac{1}{2}(d + 2 - \eta)\,. \tag{25}$$

The exponent $1/\nu$, which is the power of L multiplying $T - T_c$ in the scaling functions, is called the thermal exponent and labeled y_T, so

$$y_T = \frac{1}{\nu}. \tag{26}$$

An important application of FSS functions being universal occurs for dimensionless quantities such as the Binder ratio g. Equation (22) does not have a non-universal overall scale factor like A in Equation (24) multiplying the scaling function (since g is dimensionless), and so we write

$$g = \tilde{g}(BL^{1/\nu}(T - T_c)) \tag{27}$$

with B non-universal and $\tilde{g}$ universal. Hence, the *value* of g at the intersection point (which is $T = T_c$) is a universal number; see Figure 1. Consequently, if the intersection values for two different systems known to be in the same universality class are apparently different, it follows that the asymptotic scaling regime has *not* been reached, and corrections to FSS are significant. This is a useful check.

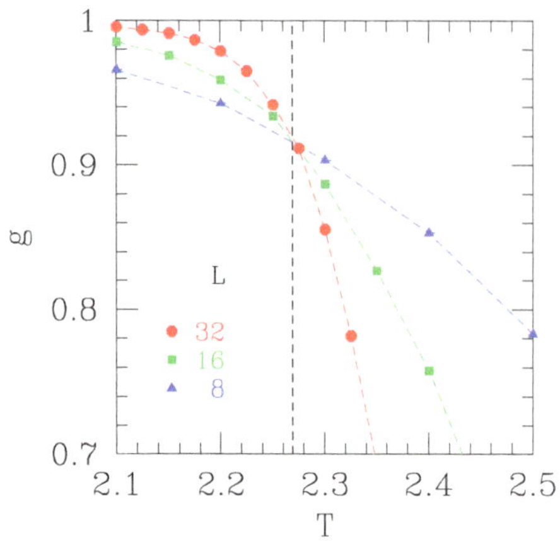

Figure 1. Calculation of the Binder ratio defined in Equation (21) for the two-dimensional Ising model on a square lattice with nearest-neighbor interactions. The dashed vertical line indicates the exact transition temperature $T_c = 2/\ln(\sqrt{2}+1) = 2.269\ldots$. The figure shows that T_c is the temperature where the data for g for different sizes intersect. Furthermore, the *value* of g at the critical (intersection) point is of interest because it is universal. From more detailed calculations, this value is found to be 0.916 in three decimal places. Universality means that the same value would occur for a triangular lattice or a square lattice with second neighbor interactions, for example, even though the transition temperatures would be different for these models. The value of $g(T_c)$ does, however, depend on the boundary conditions and is also different in different space dimensions.

3. Renormalization Group Justification for Finite-Size Scaling

To better understand FSS, we adopt a renormalization group (RG) perspective. We consider a system on a lattice with lattice spacing a with a Hamiltonian $\mathcal{H}$ which has interactions $K_1, K_2, \cdots$, into which we have absorbed the factor of $1/(k_B T)$. In the RG, we perform a coarse-graining of the system such that the new lattice spacing is $b\,a$ with $b > 1$, and the probability distribution of long wavelength fluctuations is unchanged. This distribution is governed by a new Hamiltonian $\mathcal{H}'$, of the same form as $\mathcal{H}$ but with new interactions $K_1', K_2', \cdots$.

If we are close to a second-order transition, the set of interactions will approach close to a fixed point $K_1^*, K_2^*, \cdots$. Near the fixed point is useful to consider linear combinations u_i of the deviations $K_\alpha - K_\alpha^*$, which transform simply, $u_i' = b^{y_i} u_i$, where y_i is an eigenvalue of the RG transformation linearized about the fixed point. If $y_i > 0$, then u_i is said to be a "relevant" variable since iterating the RG transformation takes u_i away from the fixed point ($u_i = 0$), whereas if $y_i < 0$, then u_i is an "irrelevant" variable since u_i decreases to zero upon iterating.

For a standard second-order transition, one has to set the temperature to the critical temperature and the magnetic field to zero in order to be at the critical point. Therefore, there are two relevant operators: the reduced temperature t defined by

$$t = \frac{T - T_c}{T_c} \tag{28}$$

and the magnetic field h. One can show that the corresponding exponents, y_T and y_H, are related to other exponents by Equations (25) and (26). In the following discussion, we will include just the most important irrelevant operator, i.e., the one with the least negative eigenvalue, and call it u. Following convention, we will write $y_u = -\omega$ where $\omega > 0$.

Under the RG transformation, quantities of interest are assumed to vary in the following way

$$X(t, h, u, L) = b^{y_X}\, X(b^{y_T} y,\, b^{y_H} h,\, b^{-\omega} u,\, L/b), \tag{29}$$

where y_X is related to the exponent, giving the divergence of X in the thermodynamic limit, as we shall see. If we fix $h = 0$, ignore the irrelevant variable, and set $b = L$ we obtain

$$X_L(t) = L^{y_X} X(L^{y_T} t, 0, 0, 1) = L^{y_X} \tilde{X}(L^{y_T}\, t), \tag{30}$$

which agrees with the basic FSS expression in Equation (7) with $y_X = \lambda/\nu = \lambda\, y_T$, where λ is the exponent for the divergence of X at the bulk critical point, see Equation (3).

Please note that we have assumed we can set the irrelevant variable u to zero to obtain Equation (7) and that including u would give corrections to scaling to that equation. However, we shall see that above the so-called upper critical dimension, $d_u = 4$, some scaling functions develop singularities in this limit. In these cases, we cannot just set $u = 0$ but need to consider the form of the singularity as u *tends* to 0. Such variables u are called "dangerous" irrelevant variables, a term first coined by Michael Fisher.

4. Finite-Size Scaling above the Upper Critical Dimension

4.1. Dangerous Irrelevant Variables

The simplest approximate theory to describe critical phenomena is mean-field theory (MFT), according to which the critical exponents we have mentioned so far have values

$$\gamma = 1, \qquad \beta = 1/2, \qquad \nu = 1/2, \qquad \eta = 0 \qquad \text{(MFT)}. \tag{31}$$

It is well established by the RG that MFT gives the exponents correctly for dimension d greater than the upper critical dimension $d = 4$. However, for $d < 4$, the exponents vary with d; for $d > 4$, they "stick" at their MF values.

Earlier, we have mentioned two scaling relations, Equations (17) and (19), relating different exponents. Equation (17) relating γ and η is satisfied by their MF values. However, the hyperscaling relation in Equation (19) does *not* work for $d > 4$. It does work, though, at the borderline dimension between MF and non-MF behavior, $d = 4$, and so, for $d > 4$, an equation analogous to Equation (19) with d replaced by d_u $(= 4)$ *is* correct [6,7]. In this section, we will discuss the physics of the violation of hyperscaling relations in the mean-field regime for $d > 4$ and show how this physics also gives rise to a form of FSS that is different from what we have discussed so far.

In RG treatments of critical phenomena, one starts with a "soft-spin" version of the problem, in which an initial coarse-graining over the discrete "fixed length" spins has been carried out. This Hamiltonian, associated with the names of Ginzburg, Landau, and Wilson, is, for scalar (Ising) spins,

$$\mathcal{H}_{GLW} = \int \left[\tfrac{1}{2} r(T)\phi^2 + \tfrac{1}{2}(\nabla\phi)^2 + \tfrac{1}{4} u\phi^4\right] d^d x, \tag{32}$$

where $\phi(x)$ is the "spin" variable, and the partition function is given by

$$Z = \int D[\phi(x)] \exp(-\mathcal{H}_{GLW})\,. \tag{33}$$

The parameter $r(T)$ varies smoothly with temperature and goes negative at low T in order to induce an ordering in ϕ. In this formulation, MFT corresponds to taking $\phi(x)$ to be independent of x and minimizing the resulting function with respect to ϕ. The transition temperature is when $r = 0$, and at lower temperatures, one finds

$$\langle\phi\rangle = \left(\frac{|r|}{u}\right)^{1/2} \qquad \text{(MFT)}\,. \tag{34}$$

Since r varies smoothly with T, i.e., $r(T) \propto T - T_c$, we have $\langle \phi \rangle \propto (T_c - T)^{1/2}$ so $\beta = 1/2$, the MF value that we quoted in Equation (31).

Renormalization group arguments show that for $d > 4$, the quartic coupling u is irrelevant, so its value iterates to zero under repeated RG transformations. However, the expression for $\langle \phi \rangle$ in Equation (34) becomes singular in this limit, so we cannot just set $u = 0$. This is important because to derive the *standard* FSS result in Equation (7), we set the irrelevant variable to zero in Equation (30). However, for $d > 4$, we cannot do this, just as we cannot set $u = 0$ in Equation (34), but must rather consider the singular nature of the scaling function $X(t, h, u, L)$ as u *tends* to 0. As mentioned in the previous section, irrelevant variables that give singularities if they are set to zero are said to be "dangerous".

Hence Equations (7), (16) and (22), for example, are not valid for $d > 4$. The way these equations have to be modified in the case of periodic boundary conditions was shown by Brézin and Zinn-Justin [8] and Binder et al. [9]. The argument of the standard FSS functions is $L^{1/\nu}(T - T_c)$. If we put the MF value $\nu = 1/2$, this would be $L^2(T - T_c)$. However Refs. [8,9] show that the correct result involves the square root of the volume, i.e., $L^2(T - T_c)$ is replaced by $L^{d/2}(T - T_c)$. The prefactor in front of the scaling function may also change in order to give the correct results for an infinite system. For $d > 4$, the expressions corresponding to Equations (7), (16) and (22) are modified to

$$X_L = L^{\lambda d/2}\tilde{X}(L^{d/2}(T - T_c)), \quad C = L^{d/2}\tilde{C}(L^{d/2}(T - T_c)), \quad g = \tilde{g}(L^{d/2}(T - T_c)), \tag{35}$$

Naively, one would expect that finite-size effects are substantial when the correlation length is of order L, which corresponds to $T - T_c \sim L^{-2}$. However, Equation (35) shows that, for periodic boundary conditions, nothing happens at this temperature and finite-size effects only occur closer (Please note that we are considering $d > 4$ so $L^{-d/2} < L^{-2}$). to T_c when $T - T_c \sim L^{-d/2}$. This surprising result is presumably due to (i) we have the periodic boundary conditions, which means there is no surface at a distance L, and (ii) the fixed point value of the coupling u in Equation (32) is zero, so long wavelength fluctuations interact very weakly.

4.2. Numerical Results for Periodic Boundary Conditions

It is possible to check the results in Equation (35) by Monte Carlo simulations on the five-dimensional Ising model. One can simulate large sizes using a clustering algorithm that was originally proposed by Swendsen and Wang [10] and subsequently modified by Wolff [11]. This algorithm considerably reduces the growth of relaxation times at the critical point (critical slowing down). Here, we will present some results found by Wittmann and the author [12].

The Hamiltonian is

$$\mathcal{H} = -\sum_{\langle i,j \rangle} J_{ij} S_i S_j, \tag{36}$$

where the $S_i = \pm 1$ are Ising spins on a five-dimensional lattice of size $N = L^5$. In this subsection, we consider periodic boundary conditions, but in Section 4.4, we will consider free boundary conditions. The J_{ij} are interactions between spins i and j, and we take $J_{ij} = 1$ for nearest neighbors and 0 otherwise. Using the Wolff algorithm, it was possible to simulate sizes up to $L = 36$.

Figure 2 shows data for the Binder ratio g, including a scaling plot according to Equation (35) with $d = 5$, which works well. An attempt to scale the data in which the power of L, namely $d/2$, is replaced by $1/\nu = 2$, as in standard FSS (see Equation (22)), is very poor.

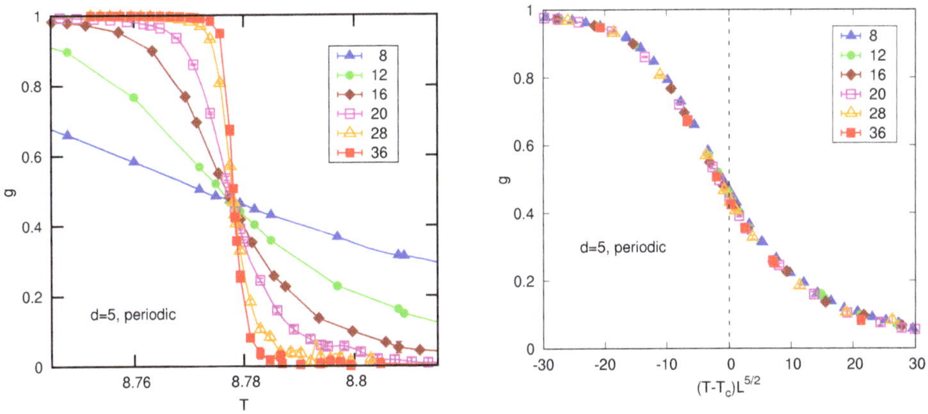

Figure 2. The left panel shows data for the Binder ratio g for the five-dimensional Ising model with periodic boundary conditions. The right panel shows the same data but scaled according to Equation (35), in which $T_c = 8.77846$ and the power of L is $d/2 (= 5/2$ here), rather than $1/\nu$, the value expected in standard FSS (see Equation (22) which equals 2. The data collapse using the exponent $5/2$ is seen to be good, whereas it is poor if the exponent value 2 is used. From Ref. [12].

We emphasize that the reason we have the power $d/2$ rather than 2 in Equation (35) ultimately comes from Equation (34), namely that the order parameter $\langle \phi \rangle$ is singular for $u \to 0$. It is, therefore, interesting to ask what form FSS would take if we look at some quantity that does not involve the order parameter (which is a $\mathbf{k} = 0$ quantity).

Ref. [12] investigated this question numerically by computing the wavevector dependent correlation function $C(\mathbf{k})$ in Equation (23) for some small but non-zero wavevectors, which, we note, are *orthogonal* to the order parameter since we still consider periodic boundary conditions. Figure 3 shows results for $C(\mathbf{k})$ where $\mathbf{k} = (2\pi/L)(1,0,0,0,0)$, the smallest non-zero wavevector. The data increases as T decreases to T_c but then decreases below T_c because these fluctuations are orthogonal to the order parameter. Apart from the smallest size presented ($L = 8$), the data scales well but with the standard FSS exponent form, Equation (16) with $\nu = 1/2, \eta = 0$, rather than the modified FSS form in Equation (35) with $d = 5$.

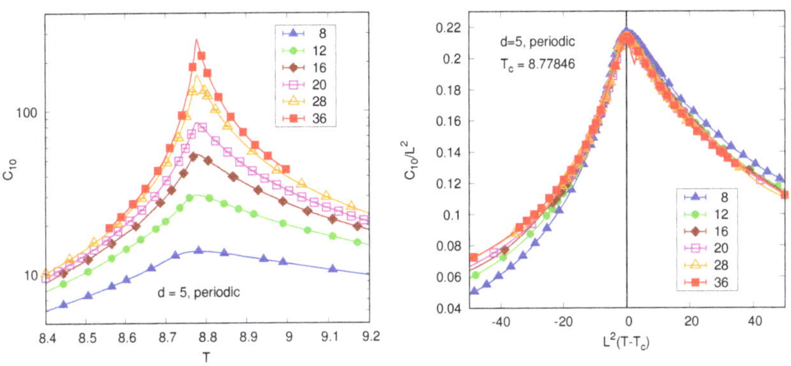

Figure 3. The left panel shows data for the wavevector dependent correlation function $C(\mathbf{k})$ for $\mathbf{k} = (2\pi/L)(1,0,0,0,0)$ for the five-dimensional Ising model with periodic boundary conditions. The right panel shows the data scaled according to the standard FSS form in Equation (16) using mean-field exponents $\nu = 1/2, \eta = 0$, rather than the modified FSS form in Equation (35) with $d = 5$. This scaling works well apart from the smallest size, $L = 8$, especially above T_c.

A natural hypothesis, then, is that *all* quantities which do not involve the order parameter satisfy standard FSS expressions even for $d > 4$.

4.3. The Finite-Size Correlation Length

The divergence of the bulk correlation length at T_c is an important feature of a second-order transition, so it is of interest to define the correlation length in a finite system as well.

The correlation length of the finite system is given by the following finite difference expression [13,14],

$$\xi_L = \frac{1}{2\sin(k_{\min}/2)}\sqrt{\frac{C(0)}{C(\mathbf{k}_{min})} - 1}, \tag{37}$$

where $\mathbf{k}_{min}$ is the smallest non-zero wavevector on the lattice. For example, in $d = 5$, $k_{min} = (2\pi/L)(1,0,0,0,0)$ and permutations. In this section, we continue to assume periodic boundary conditions. With this definition of ξ_L, the correlation function far enough above T_c obeys the expected Ornstein–Zernicke form

$$C(\mathbf{k}) = \frac{C(0)}{1 + \xi_L^2 k^2 + \cdots}, \tag{38}$$

so ξ_L is indeed a characteristic length in this region. However, it can not be interpreted as a physical length below T_c since $C(0)$ is dominated by the long-range order and so diverges as L^d, whereas $C(\mathbf{k}_{\min})$ does not have a long-range order component and so is finite below T_c. Nonetheless, ξ_L is a convenient quantity to analyze with FSS in *all* regions since, like the Binder ratio g, ξ_L/L is dimensionless, and so its FSS form does not have the power of L multiplying the scaling function.

According to [12], for $k > 0$ but $kL \ll 1$, $C(k)$ varies as L^2 at T_c, see Figure 4. This indicates that in real space, correlations fall off with distance in the expected mean-field way, namely $1/r^{d-2}$ so $\eta = 0$. However $C(0)$ goes as $L^{d/2}$ which is much bigger than $C(k_{min})$ for $d > 4$. Hence, as defined in Equation (37), we see that for large L

$$\xi_L \propto L\left(\frac{L^{d/2}}{L^2}\right)^{1/2} = L^{d/4} \tag{39}$$

at T_c which is bigger than L since $d > 4$ here. However, in contrast to some earlier results [7,14], we argue that we should not interpret ξ_L as a physical length *at* T_c. Rather the quantity defined to be ξ_L is larger than L at T_c only because $C(0)$ is anomalously large relative to $C(k > 0)$. To be more precise, correlations fall off with distance in the expected mean-field way, i.e., as $1/r^{d-2}$, so one would naively think that at a distance of order $L/2$, the correlations would be of order $1/L^{d-2}$. However, this is wrong because the $k = 0$ mode gives an additional, anomalously large, constant contribution of order $1/L^{d/2}$.

Finally, we note that the FSS form for ξ_L defined in Equation (37) is [14]

$$\xi_L = L^{d/4}\tilde{\xi}\left(L^{d/2}(T - T_c)\right), \tag{40}$$

which gives the $L^{d/4}$ behavior at T_c derived earlier in Equation (39).

4.4. Free Boundary Conditions above the Upper Critical Dimension

The above results are for periodic boundary conditions for $d > 4$, and they show that the temperature range over which there is finite-size rounding is of order $1/L^{d/2}$. This means that there is no change in behavior when $\xi \sim L$, i.e., $T - T_c \sim L^{-2}$, but only closer to T_c when $T - T_c \sim L^{-d/2}$. While surprising, this result is nonetheless possible with periodic boundary conditions. However, if we have free boundary conditions, surely *something* must happen when $\xi \sim L$ because the correlations must be affected by the surfaces.

To our knowledge, the question of what happens with free boundary conditions for $d > 4$ was first addressed by Rudnick et al. [15]. They showed that one needs to consider two exponents describing finite-size effects: a "shift" exponent and a "rounding" exponent. The possibility of two exponents to characterize finite-size effects was further studied by [12], who performed numerics on the five-dimensional Ising model with free boundary conditions and also considered possible different behavior for quantities that are orthogonal to the order parameter compared with those that involve the order parameter.

More precisely, ref. [12] computed the Fourier-transformed correlation functions

$$C(\mathbf{k}) = L^d \langle |m(\mathbf{k})|^2 \rangle \,, \tag{41}$$

in which the Fourier-transformed magnetization, $m(\mathbf{k})$, is defined differently for free boundary conditions compared with periodic boundary conditions. However, for periodic boundary conditions, the Fourier modes are plane waves, i.e.,

$$m(\mathbf{k}) = \frac{1}{N} \sum_i e^{i\mathbf{k}\cdot\mathbf{r}_i}\, S_i, \quad \text{(periodic)}, \tag{42}$$

where

$$k_\alpha = 2\pi n_\alpha / L, \quad \text{(periodic)}, \tag{43}$$

with $n_\alpha = 0, 1, \cdots, L-1$ and α denotes a Cartesian coordinate, for free boundary conditions, the Fourier modes are sine waves,

$$m(\mathbf{k}) = \frac{1}{N} \sum_i \Big[\prod_{\alpha=1}^{d} \sin(k_\alpha r_{i,\alpha}) \Big] S_i, \quad \text{(free)}, \tag{44}$$

where

$$k_\alpha = \pi n_\alpha / (L+1), \quad \text{(free)}, \tag{45}$$

with $n_\alpha = 1, 2, \cdots, L$ and the components of the lattice position, $r_{i,\alpha}$, run over values $1, 2, \cdots, L$. There is zero contribution to the sum in Equation (44) if we set $r_{i,\alpha} = 0$ or $L + 1$, so Equations (44) and (45) correctly incorporate free boundary conditions.

Please note that $\mathbf{k} = 0$ is not an allowed wavevector with free boundary conditions, so the uniform magnetization does not correspond to a single Fourier mode. Wavevectors with all n_α odd have a projection onto the uniform magnetization and so will acquire a non-zero expectation value below T_c in the thermodynamic limit. It is, therefore, natural to expect that they will obey some sort of non-standard FSS. However, if any of the n_α is even, there is no projection onto the uniform magnetization, so they will not acquire an expectation value below T_c, and hence it is expected that they will satisfy standard FSS, i.e., Equation (16) with $\eta = 0$, $\nu = 1/2$.

Ref. [12] found that corrections to FSS are larger with free boundary conditions than with periodic boundary conditions. Nonetheless, given the large sizes that can be studied with the Wolff algorithm, ref. [12] concluded that the data shows fairly convincingly that quantities that are orthogonal to the order parameter do indeed follow *standard* FSS.

What, then, is the form of non-standard FSS above the upper critical dimension with free boundary conditions for quantities that involve the order parameter? As already noted above, Rudnick et al. [15] showed that one needs both a temperature "shift" and a different temperature "rounding". For the shift, one defines for each size a characteristic temperature ($T(L)$ is called a pseudocritical temperature by Berche et al. [6], and we shall use that terminology here. In that paper, $T(L)$ is determined from the maximum of χ computed with the subtracted term shown in Equation (13), rather than from the midpoint value of g as here. Any reasonable definition of $T(L)$ should give the same exponent for the shift). $T(L)$ from which a shift exponent λ is obtained as follows:

$$\Delta T_L \equiv T_c - T(L) = \frac{A}{L^\lambda}. \tag{46}$$

The definition of $T(L)$ is not unique, but one convenient choice, the one used by [12], is the temperature where $g(T(L)) = 0.5$, see Figure 4. A convenient choice for the temperature rounding is the difference between the temperatures where $g = 0.25$ (say) and $g = 0.75$ (say), i.e.,

$$\delta t_L \equiv T(g = 0.25) - T(g = 0.75) = \frac{B}{L^{\mu}}, \tag{47}$$

see Figure 4. This defines the rounding exponent μ.

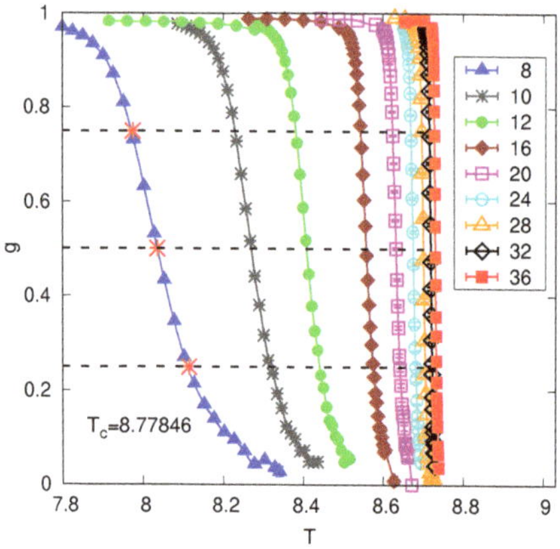

Figure 4. The Binder ratio defined in Equation (21) for the five-dimensional Ising model with free boundary conditions. The data are from Ref. [12]. The region of finite-size rounding, δt_L, is defined here, for each size, to be the range of temperature between where g has values 0.75 and 0.25. For example, for $L = 8$ we have $\delta t_8 = 8.115 - 7.975 = 0.140$ (these points are indicated by crosses). The shift ΔT_L is defined here to be the difference between $T_c(= 8.77846)$ and the value of T where $g = 0.5$. For $L = 8$, we find $g = 0.5$ at $T = 8.035$ (shown as a cross in the figure), so, taking $T_c = 8.77846$ from data on periodic boundary conditions, $\Delta T_8 = 8.77846 - 8.035 = 0.743$. Analysis of the data shows that the rounding tends to zero as $1/L^{5/2}$ while the shift tends to zero as $1/L^2$, which is bigger.

In the presence of distinct exponents for shift and rounding, FSS expressions are the same as before, except that $T - T_c$ is replaced by $T - T(L)$ so, for g for example, we have

$$g = \tilde{g}(\, L^{\mu}\,(T - T(L))\,) = \tilde{g}\Big(\, L^{\mu}\,(T - T_c) + AL^{\mu-\lambda}\,\Big), \tag{48}$$

rather than Equation (21). Please note that it is only necessary to consider the shift separately from the rounding if the size of the temperature shift is greater than the size of the temperature rounding, i.e., if $\lambda < \mu$. Suppose $\lambda > \mu$, the effects of the shift can be regarded as a correction to scaling since $AL^{\mu-\lambda} \to 0$ for $L \to \infty$ in this case. If $\lambda = \mu$, one can view Equation (48) as usual FSS but with an unimportant shift in the origin of the scaling function $\tilde{g}(x)$.

Rudnick et al. [15] showed that

$$\lambda = \frac{1}{\nu} = 2, \quad \mu = \frac{d}{2}\ (= \frac{5}{2}\ \text{here}), \tag{49}$$

so Equation (48) can be written as

$$g = \tilde{g}\Big(\, L^{5/2}\,(T - T_c) + AL^{1/2}\,\Big). \tag{50}$$

Figure 4 shows the numerical results of [12] in five dimensions with free boundary conditions. One can see by eye that the shift is bigger than the rounding, and analysis confirms the values given in Equation (49), namely a shift exponent of 2 and a rounding exponent of 5/2. These exponent values were also found numerically by Berche et al. [6].

Please note that at the critical point, the argument of the scaling function for g tends to infinity, where $g = 0$. Hence, the data for different sizes do not intersect at T_c for free boundary conditions and $d > 4$, unlike the case for periodic boundary conditions, see Figure 2, and quite generally for $d < 4$, see Figure 1.

It is also interesting to discuss the correlation function C defined in Equation (15), which serves as a proxy for the susceptibility χ. Taking the FSS form in Equation (35), which would be valid above the upper critical dimension for periodic boundary conditions, and replacing T_c by the pseudocritical temperature $T(L)$ to make it valid for free boundary conditions, we get

$$C = L^{d/2}\tilde{C}\left(L^{d/2}(T - T(L)\right), \tag{51a}$$

$$= L^{d/2}\tilde{C}\left(L^{d/2}(T - T_c) + AL^{d/2-2}\right). \tag{51b}$$

According to Equation (51a), at the pseudocritical temperature $T(L)$, C diverges as $L^{d/2}$. Berche et al. [6] computed the susceptibility with a subtracted term as in Equation (13), which is expected to have the same scaling behavior as C, and found the predicted $L^{d/2}$ behavior.

Interestingly, the behavior at the true T_c is different. Since the susceptibility diverges with an exponent $\gamma = 1$ in the mean-field region, the scaling function $\tilde{C}(x)$ must vary as $1/x$ for $x \to \infty$. Since $T = T_c$ corresponds to this large x limit, it follows from Equation (51b) that the behavior of C at T_c has the form

$$C(T_c) \propto L^{d/2}\frac{1}{L^{d/2-2}} = L^2. \tag{52}$$

A strict upper bound shows that $C(T_c)$ cannot be greater than L^2; see Refs. [16,17]. It seems, then, that this upper bound is satisfied as an equality. However, numerically, the situation is not very clear because there are large corrections to scaling due to a substantial fraction of sites being on the surface. Berche et al. [6] found an exponent of 1.71 ± 0.02 if all sites are included in the analysis. However, this changed substantially to 1.92 ± 0.02 if the outer half of the sites is removed when doing the average. Reference [6] claims that this value is incompatible with the above value of 2. However, this conclusion is not obvious to us since the error bar presumably includes only statistical errors, and very likely, some systematic errors remain for the sizes studied. Hence, in our view, the numerics of Ref. [6] does not rule out that C (or χ) varies like L^2 at T_c with free boundary conditions. We also note that Lundow et al. [18] find that $\chi \propto L^2$ at T_c for the $5d$ Ising model at T_c.

5. Conclusions

We summarize our understanding of FSS above and below the upper critical dimension, $d_u = 4$ as follows:

- *Below the upper critical dimension.*
  Here, standard FSS applies:

$$X_L = L^{\lambda/\nu}\tilde{X}\left(L^{1/\nu}(T - T_c)\right), \qquad C = L^{2-\eta}\tilde{C}\left(L^{1/\nu}(T - T_c)\right),$$
$$g = \tilde{g}\left(L^{1/\nu}(T - T_c)\right), \qquad \xi_L = L\tilde{\xi}\left(L^{1/\nu}(T - T_c)\right). \tag{53}$$

  We recall that C is the $k = 0$ correlation function in Equation (15) (a proxy for the susceptibility χ), g is the Binder ratio defined in Equation (21), ξ_L is the quantity defined in Equation (37) which above T_c is the correlation length, and X is a generic quantity which diverges like $(T - T_c)^{-\lambda}$ for $T \to T_c$.

- *Above the upper critical dimension with periodic boundary conditions.*
For quantities that involve uniform magnetization, FSS scaling is modified to

$$\begin{aligned} \chi_L &= L^{\lambda d/2}\tilde{\chi}\left(L^{d/2}(T-T_c)\right), & C &= L^{d/2}\tilde{C}\left(L^{d/2}(T-T_c)\right), \\ g &= \tilde{g}\left(L^{d/2}(T-T_c)\right), & \xi_L &= L^{d/4}\tilde{\xi}\left(L^{d/2}(T-T_c)\right). \end{aligned} \tag{54}$$

For quantities that are orthogonal to the uniform magnetization, such as $C(\mathbf{k})$ for $\mathbf{k} \neq 0$) (but $kL \ll 1$), we have standard FSS with mean-field exponents, so

$$C(\mathbf{k}) = L^2\tilde{C}\left(L^2(T-T_c)\right). \tag{55}$$

These results imply that the correlation function at the critical point varies according to Gaussian behavior, i.e., $\langle S_i S_j\rangle \sim 1/r_{ij}^{d-2}$ for $r_{ij} \equiv |\mathbf{r}_i - \mathbf{r}_j| \ll L/2$ but this is superimposed on a uniform background contribution (which only affects the $\mathbf{k} = 0$ mode) of order $1/L^{d/2}$. This was stated explicitly in Ref. [19], who demonstrated it numerically for the self-avoiding walk-in $d = 5$, which is expected to be in the same universality class as the Ising model above the upper critical dimension.
- *Above the upper critical dimension with free boundary conditions.*
For most quantities, we need to define a shift and a rounding. The shift exponent is $1/\nu\,(=2)$, and the rounding exponent is $d/2$ (the same as for periodic boundary conditions). Please note that the temperature range due to the shift is bigger than the temperature range due to rounding. Defining a pseudocritical temperature $T(L)$ for each size, most quantities behave as in Equation (54) but with T_c replaced by $T(L)$, i.e.,

$$\begin{aligned} \chi_L &= L^{\lambda d/2}\tilde{\chi}\left(L^{d/2}(T-T(L))\right), & C &= L^{d/2}\tilde{C}\left(L^{d/2}(T-T(L))\right), \\ g &= \tilde{g}\left(L^{d/2}(T-T(L))\right), & \xi_L &= L^{d/4}\tilde{\xi}\left(L^{d/2}(T-T(L))\right), \end{aligned} \tag{56}$$

where $T(L)$ is related to T_c by Equation (46) with $\lambda = 2$. However, as discussed in Section 4.4, some of the normal modes are orthogonal to the uniform magnetization, and for these, only the exponent $1/\nu\,(=2)$ appears and their correlations behave as in Equation (55).

The above conclusions are supported by theoretical arguments and numerical simulations. One of our main points is that for $d > 4$ and periodic boundary conditions, the new features of FSS, which come from hyperscaling violation and the dangerous irrelevant variable, only manifest themselves for $k = 0$ fluctuations. We note that in the simulations, corrections to the asymptotic behavior are quite large, especially for free boundary conditions, so the numerical estimates of exponents have errors. Indeed, other authors, in particular Refs. [6,7], have a somewhat different point of view from that presented here. For example, Ref. [7] argues that one should define two η-like exponents for periodic boundary conditions. However, we do not feel that this is necessary to understand our numerical results.

Finally, I will mention some more recent work. A central result in the theory of FSS for $d > d_u$ is that a complete description needs both the standard FSS exponent $1/\nu$, which is 2 here and the hyperscaling violation exponent $d/2$. An interesting and different way of understanding why both exponents are needed is discussed by Fang et al. [20]. In later work, Fang and collaborators [21] show that a complete description of FSS above the upper critical dimension is obtained by considering the Fortuin–Kasteleyn cluster representation of the states of the Ising model. These clusters show a change in behavior not only at $d_u = 4$ but also at $d = 6$.

In a recent development, Ref. [22] has shown that similar ideas to those reviewed here also apply to the random field Ising model, which has a different upper critical dimension, $d_u = 6$, as opposed to $d_u = 4$ for the more usual situation discussed here.

As a final remark, I should mention that I was invited to contribute to this Special Issue by Ralph Kenna, who sadly has since passed away. I never met Ralph, but was familiar with his work on finite-size scaling, which always contained original and stimulating ideas. I am pleased that his long-time collaborator, Bertrand Berche, has agreed to take over Ralph's role of editor and bring this Special Issue to completion.

Funding: This research received no external funding.

Institutional Review Board Statement: Not applicable.

Data Availability Statement: The data presented in this study are available on request from the corresponding author.

Acknowledgments: I have benefited from discussions on finite-size scaling with many people, including Victor Martin-Mayor, Helmut Katzgraber, and Matt Wittmann. I should particularly mention stimulating interactions on many topics, including finite-size scaling, with the late Kurt Binder.

Conflicts of Interest: The author declares no conflicts of interest.

References

1. Berche, B.; Ellis, T.; Holovatch, Y.; Kenna, R. Phase transitions above the upper critical dimension. *SciPost Phys. Lect. Notes* **2022**, *60*, 1.
2. Fisher, M.E. The theory of critical point singularities. In *Critical Phenomena, Proceedings of the 51st Enrico Fermi Summer School, Varenna*; Green, M.S., Ed.; Academic Press: New York, NY, USA, 1971; p. 1.
3. Fisher, M.E.; Barber, M.N. Scaling theory for finite-size effects in the critical region. *Phys. Rev. Lett.* **1972**, *28*, 1516. [CrossRef]
4. Binder, K. Finite size scaling analysis of Ising model block distribution functions. *Z. Phys. B* **1981**, *43*, 119. [CrossRef]
5. Privman, V.; Fisher, M.E. Universal critical amplitudes in finite-size scaling. *Phys. Rev. B* **1984**, *30*, 322. [CrossRef]
6. Berche, B.; Kenna, R.; Walter, J.C. Hyperscaling above the upper critical dimension. *Nucl. Phys. B* **2012**, *865*, 115. [CrossRef]
7. Kenna, R.; Berche, B. Fisher's scaling relation above the upper critical dimension. *Europhys. Lett.* **2014**, *105*, 26005. [CrossRef]
8. Brézin, E.; Zinn-Justin, J. Finite size effects in phase transitions. *Nucl. Phys. B* **1985**, *257*, 867. [CrossRef]
9. Binder, K.; Nauenberg, M.; Privman, V.; Young, A.P. Finite-size tests of hyperscaling. *Phys. Rev. B* **1985**, *31*, 1498. [CrossRef] [PubMed]
10. Swendsen, R.H.; Wang, J. Nonuniversal critical dynamics in Monte Carlo simulations. *Phys. Rev. Lett.* **1987**, *58*, 86. [CrossRef] [PubMed]
11. Wolff, U. Collective Monte Carlo updating for spin systems. *Phys. Rev. Lett.* **1989**, *62*, 361. [CrossRef] [PubMed]
12. Wittmann, M.; Young, A.P. Finite-size scaling above the upper critical dimension. *Phys. Rev. E* **2014**, *90*, 062137. [CrossRef] [PubMed]
13. Kim, J.K. Application of finite size scaling to Monte Carlo simulations. *Phys. Rev. Lett.* **1993**, *70*, 1735. [CrossRef] [PubMed]
14. Jones, J.L.; Young, A.P. Finite size scaling of the correlation length above the upper critical dimension. *Phys. Rev. B* **2005**, *71*, 174438. [CrossRef]
15. Rudnick, J.; Gaspari, G.; Privman, V. Effect of boundary conditions on the critical behavior of a finite high-dimensional Ising model. *Phys. Rev. B* **1985**, *32*, 7594. [CrossRef] [PubMed]
16. Gunton, J.D. Finite-size effects at the critical point. *Phys. Lett. A* **1968**, *26*, 406. [CrossRef]
17. Watson, P.G. Surface and size effects in lattice models. In *Phase Transitions and Critical Phenomena*; Domb, C., Green, M., Eds.; Academic Press: London, UK, 1973; Volume 2, p. 101.
18. Lundow, P.H.; Markström, K. Finite size scaling of the 5D Ising model with free boundary conditions. *Nucl. Phys. B* **2014**, *889*, 249. [CrossRef]
19. Grimm, J.; Elci, E.M.; Zhou, Z.; Garoni, T.M.; Deng, Y. Geometric explanation of anomalous finite-size scaling in high dimensions. *Phys. Rev. Lett* **2017**, *118*, 115701. [CrossRef] [PubMed]
20. Fang, S.; Grimm, J.; Zhou, Z.; Deng, Y. Complete graph and Gaussian fixed point asymptotics in the five-dimensional Fortuin-Kasteleyn Ising model with periodic boundaries. *Phys. Rev. E* **2020**, *102*, 022125. [CrossRef] [PubMed]

21. Fang, S.; Zhou, Z.; Deng, Y. Geometric scaling behaviors of the Fortuin-Kasteleyn Ising model in high dimensions. *Phys. Rev. E* **2023**, *107*, 044103. [CrossRef] [PubMed]
22. Fytas, N.G.; Martin-Mayor, V.; Parisi, G.; Picco, M.; Sourlas, N. Finite-size scaling of the random-field Ising model above the upper critical dimension. *Phys. Rev. E* **2023**, *108*, 044146. [CrossRef] [PubMed]

entropy

MDPI

Article

Biswas–Chatterjee–Sen Model on Solomon Networks with Two Three-Dimensional Lattices

Gessineide Sousa Oliveira [1], Tayroni Alencar Alves [1], Gladstone Alencar Alves [2], Francisco Welington Lima [1,*] and Joao Antonio Plascak [3,4,5]

1 Dietrich Stauffer Computational Physics Laboratory, Departamento de Física, Universidade Federal do Piauí, Teresina 64049-550, PI, Brazil; sousagessi21@gmail.com (G.S.O.); tay@ufpi.edu.br (T.A.A.)
2 Departamento de Física, Universidade Estadual do Piauí, Teresina 64002-150, PI, Brazil; alves.gladstone@gmail.com
3 Departamento de Física, Centro de Ciências Exatas e da Natureza (CCN), Universidade Federal da Paraíba, Cidade Universitária, João Pessoa 58051-970, PB, Brazil; pla@uga.edu
4 Departamento de Física, Universidade Federal de Minas Gerais, C. P. 702, Belo Horizonte 30123-970, MG, Brazil
5 Department of Physics and Astronomy, University of Georgia, Athens, GA 30602, USA
* Correspondence: fwslima@gmail.com

Abstract: The Biswas–Chatterjee–Sen (BChS) model of opinion dynamics has been studied on three-dimensional Solomon networks by means of extensive Monte Carlo simulations. Finite-size scaling relations for different lattice sizes have been used in order to obtain the relevant quantities of the system in the thermodynamic limit. From the simulation data it is clear that the BChS model undergoes a second-order phase transition. At the transition point, the critical exponents describing the behavior of the order parameter, the corresponding order parameter susceptibility, and the correlation length, have been evaluated. From the values obtained for these critical exponents one can confidently conclude that the BChS model in three dimensions is in a different universality class to the respective model defined on one- and two-dimensional Solomon networks, as well as in a different universality class as the usual Ising model on the same networks.

Keywords: non-equilibrium; phase transition; Monte Carlo simulations; Solomon networks

1. Introduction

In the last four decades, there has been a great increase in the study of the general behavior of people in a community. From a theoretical point of view, this rather complex dynamical system must, certainly, take into account that people can not only influence their neighbors but be influenced by them as well. According to this scenario, some early models have been introduced by Stauffer [1] and Galam [2,3] using basically local majority rule arguments, which are also now called dynamics of opinions. Other models, just to cite some examples, are the Ising model (IM) [4,5], the majority-vote model (MVM) [6], and the Biswas, Chatterjee, and Sen (BChS) [7,8] model. Regarding the Ising model, it is worth mentioning that, despite being originally proposed to treat magnetic systems, almost nine decades before the new ideas by Stauffer and Galam, the IM is so versatile that it can also be employed in the study of other interdisciplinary fields as well (see, for instance, Refs. [9–12]).

In general, the dynamical properties of the models mentioned above are better studied by considering an underlined regular lattice or network, in which every site of the lattice or node of the network are occupied by an individual (sometimes also called an agent). In this way, within each lattice or network, one has the proper type of mutual interactions that are defined by the corresponding model under study, e.g., Ising model, MVM, BChS model, and others.

In the particular case of the BChS model, which was introduced in 2012 and is the subject of the present study, the pair interactions are allowed to be either positive or negative. The interaction signs are modeled by a single noise parameter q, that represents the fraction of negative interactions. In addition, the opinion variables can be made either continuous or discrete. This non-equilibrium system has been studied from Monte Carlo simulations on different regular lattices as well as different networks. Regardless of being continuous or discrete, the model presents a second-order phase transition at q_c, with the noise parameter q playing the role of temperature in ordinary magnetic phase transitions. For $q < q_c$, the system has a non-zero order parameter, while for $q > q_c$, the order parameter is zero. However, contrary to what happens in ordinary magnetic transitions, the exponents and the universality class of the model depend on the particular chosen lattice or network. The reader is directed to Ref. [13] for a detailed relation of the universality class of the BChS model on different lattices and networks. Ref. [14] also presents a recent review of the progress already made in the study of the BChS model.

A rather more realistic situation has been proposed by Sorin Solomon [15,16], by considering two different lattices, where one lattice reflects one kind of environment (for instance, home) and the other lattice a different environment (for instance, workplace). In general, one labels the sites in the workplace lattice i and a random permutation $P(i)$ of the order already established in this lattice will provide the sites in the home lattice. In such real situations, the neighbors at home differ from the neighbors in the workplace (except, of course, when everybody works at home). Such constructions are called Solomon networks (SNs) [9,17,18], reflecting the fact that each individual is equally shared by two lattices, just as in the King Solomon biblical story. As a result, the net interaction of the relevant variables defined at site i is a sum of the corresponding interactions of site i with its neighboring sites on the workplace lattice, plus the interactions with the neighbor sites of $P(i)$ on the home lattice. Some correlation between home and workplace lattices can be introduced into the model by choosing $P(i)$ not completely at random. For example, people selecting a workplace closer to their homes. However, generally only random permutations have been considered in $P(i)$ in order not to proliferate additional theoretical parameters. The increase in the connectivity of each site i makes the SNs close to small-world networks [19,20].

The majority of the studies involving SNs have employed two linear chains and two square lattices, implying one-dimensional ($1D$) and two-dimensional ($2D$) topologies, respectively. However, three-dimensional ($3D$) lattices can be important as well, and not only from a theoretical point of view. For instance, one could make the correspondence of environments in the country as $1D$, in towns as $2D$, and in big cities, due to their building verticalization in some areas, as $3D$. This certainly implies that people living in such different environments present, as is indeed the case, different behaviors regarding the way they live and work.

The BChS model has already been studied in different lattices and networks [13] and also in $1D$ and $2D$ SNs [21]. Concerning the SNs, the critical exponents of the second-order phase transition indicate that the universality class of the model is dependent on the dimensionality of the lattices, as is expected from general renormalization group arguments. It should be stressed here that from the results one has from the literature it is not possible to definitely assert that a particular model on a particular network will undergo a first- or second-order transition, or even any transition at all. It is thus worthwhile treating the BChS model in $3D$ SNs, by defining two three-dimensional lattices, namely, simple cubic ones for simplicity. It will be seen that the model does indeed belong to a different universality class, in agreement with renormalization ideas. Moreover, the hyperscaling relation is satisfied for dimensions smaller than three, meaning that the upper critical dimensionality is certainly greater than or equal to four [22].

Thus, in the present work, the BChS dynamical system on SNs has been studied through Monte Carlo simulations, allied with finite-size scaling techniques, considering three-dimensional simple cubic lattices. The plan of the paper is the following. In the next section, the BChS model in discrete opinion dynamics is defined. In this same section, the corresponding Monte Carlo simulation details and the respective *thermodynamic* quantities used to obtain the critical behavior are presented. The results are discussed in Section 3 and some concluding remarks are summarized in the last section.

2. Model and Simulations

In this section, the BChS model is briefly defined, in its discrete form, and the thermodynamic-like variables, from which one can obtain the phase transition properties, are presented. Some details of the employed Monte Carlo simulations, together with the finite-size scaling relations of the relevant quantities to treat the model on finite-size lattices of linear size L are also discussed.

2.1. Biswas–Chatterjee–Sen Model and Relevant Thermodynamic-like Variables

The BChS model has already been described and previously studied in different regular lattices and networks (see, for instance, Refs. [7,8,21,23]). Below, we have just the main ingredients of the model directed specifically to SNs. We will keep here the same notation, for the relevant variables, already used in the previous studies of this model in other contexts.

Agents are set on each i-th node of an SN with N sites. At time step t, the agents have opinion variables defined by $o_i(t)$, that can assume three different values: -1, 0, or $+1$. The BChS rules adopted here for updating $o_i(t)$ are as follows.

(i) Begin by constructing an initial configuration, where for each site i of the SN one of the three opinion states is randomly assigned. This is the configuration of the opinion variables o_i at time $t = 0$, and can be labeled as $o_i(t = 0) = o_i$.
(ii) The workplace lattice is sequentially swept for every site i to be updated. At this point one has $o_i(t) = o_i$, where all o_i are the corresponding opinion variables that have been updated at time $t - 1$.
(iii) For a given site i, one of its nearest neighbors is randomly selected and an affinity μ_{ij} is ascribed for this bond. This affinity parameter is another discrete variable that assumes a value $+1$, but can be turned negative with a probability q. This probability q acts as an external noise, modeling local discordances.
(iv) The opinion variables of both sites i and j are then assigned new values as

$$o_i + \mu_{ij} o_j \quad \rightarrow \quad o_i \, , \tag{1}$$

$$o_j + \mu_{ij} o_i \quad \rightarrow \quad o_j \, . \tag{2}$$

where on the right-hand sides o_i and o_j are the updated opinion states.
(v) Another site ℓ, randomly distant $|\vec{R}|$ from site i, is additionally selected and an affinity $\mu_{i\ell}$ is ascribed for this bond in the same way as before.
(vi) The opinion variable of these sites are now updated as

$$o_i + \mu_{i\ell} o_\ell \quad \rightarrow \quad o_i \, , \tag{3}$$

$$o_\ell + \mu_{i\ell} o_i \quad \rightarrow \quad o_\ell \, . \tag{4}$$

(vii) If the opinion state turns out to be outside the interval $[-1, +1]$, it is automatically made $o_i = +1$ if $o_i > +1$, or $o_i = -1$ if $o_i < -1$.
(viii) One sweep through the lattice constitutes one Monte Carlo step (MCS) from time t to $t + 1$. Thus, one has $o_i(t + 1) = o_i$, with o_i the updated set of opinion variables from items (iii)–(vii).

Note that using only the workplace lattice and making a two-step selection comprising a nearest neighbor j and a distant site ℓ is a simplified, and less computer-time consuming, version of the original two-lattice SN [9].

A convenient order parameter O for this model consists of averaging the opinion variables $o_i(t)$ over all individuals as

$$O = \left| \sum_{i=1}^{N} o_i(t) \right| / N, \tag{5}$$

where t is chosen to be large enough for the system having reached the stationary state and N is the total number of sites of the SN and given by $N = L^3$. In ordinary magnetic systems one has a thermal driven phase transition. In the present model, one has a kind of random-configuration-driven phase transition instead. The phase transition means that for $q < q_c$, $O \neq 0$ and an ordered phase is established, while for $q > q_c$, $O = 0$ and the system is in a disordered phase. Exactly at $q = q_c$, a second-order phase transition takes place.

In order to fully categorize this phase transition, in addition to the above order parameter, it is also convenient to analyze its magnetic-like variables; for instance, the order parameter fluctuation $O_f(q)$ or susceptibility (which is the analogous to the magnetic susceptibility) and its reduced fourth-order Binder cumulant $O_4(q)$. These observables are formally expressed as

$$O(q) = \Big[\langle O \rangle_t \Big]_{av}, \tag{6}$$

$$O_f(q) = N \Big[\langle O^2 \rangle_t - \langle O \rangle_t^2 \Big]_{av}, \tag{7}$$

$$O_4(q) = 1 - \left[\frac{\langle O^4 \rangle_t}{3 \langle O^2 \rangle_t^2} \right]_{av}, \tag{8}$$

where $\langle \cdots \rangle_t$ stands for different time averages (which are computed after the system has reached the stationary state), and $[\cdots]_{av}$ means the averages taken over different initial configurations. The simulation details of what was used for obtaining the above quantities are given below.

2.2. Monte Carlo Simulations and Finite-Size Scaling Relations

The Monte Carlo simulations were performed following the algorithm described in items (i)–(viii) above on SNs of finite sizes L. The sizes of the lattices were chosen as $L = 6,\ 8,\ 10,\ 12,\ 14, 16, 18$, up to 20, with $N = L^3$ total sites. The relevant quantities in Equations (6)–(8) were computed as a function of the noise probability, also called the noise parameter, q. Close to the transition region, the noise parameter steps were chosen as $\Delta q = 0.001$.

In all simulations, to obtain the relevant observables we performed several MCSs on the SNs. We simulated 120 SN realizations for each network size to make quench averages. For each network replica, we considered 10^5 MCSs to let the system evolve to a stationary state, and then, another 10^5 MCSs to collect 10^5 values of the opinion variables to measure the desired observables. Error bars were calculated by using the jackknife resampling technique [24,25].

Since all the variables defined in Equations (6)–(8) depend, in addition to the noise parameter q, also on the SN sizes, the criticality exhibited by the model can be inferred by analyzing their behavior as a function of L and close to the transition point q_c. Thus, according to the finite-size scaling hypothesis one has, for large system sizes L, the following power-law expressions of the relevant variables (for further details one can see, for instance, Refs. [24,26]):

$$O(q) = L^{-\beta/\nu} f_O(x), \tag{9}$$

$$O_f(q) = L^{\gamma/\nu} f_{O_f}(x), \tag{10}$$

$$O_4(q) = f_{O_4}(x), \tag{11}$$

$$q_c(L) = q_c + cL^{-1/\nu}, \tag{12}$$

where β, ν, and γ are the critical exponents of the order parameter, correlation length, and fluctuation of the order parameter, respectively; $f_k(x)$, with $x = L^{1/\nu}(q - q_c)$ and $k = \{O, O_f, O_4\}$, are the respective scaling functions. In the above equations it is assumed that q is close to the corresponding critical noise q_c for an infinite-size network. In addition, for each finite network of L size, we can also estimate its pseudo-critical noise parameter $q_c(L)$ by looking, for example, at the peak presented by the fluctuation in the order parameter as a function of q or cumulant crossings for different network sizes. In this case, one can use Equation (12), where c is a non-universal constant.

It is implicit in Equation (11) that the order parameter cumulants, at the critical point, have no dependence on the size L. As a result, all cumulants should cross at the same value of the noise parameter q_c. However, when the finite lattices are not large enough, residual corrections to finite-size scaling make the crossing points of the cumulants suffer a systematic shift as L varies. In this case, there is a useful, although not so frequently employed, way of extrapolating the critical point from the cumulant crossings through an analysis of the scaling of $q_c(L)$ with the lattice sizes L and L_s. From phenomenological renormalization group ideas, Nightingale [27] has shown that an extrapolation procedure, where the scaling factor is given by $ln(L/L_s)$, furnishes quite good results when correlation length on strips of different widths is used. This same strategy has also proved to be suitable when considering the cumulant crossings for different lattice sizes [28–31]. The scaling form can be expressed as

$$q_c(L) = q_c + c' \Big/ \ln(L/L_s), \tag{13}$$

where $q_c(L)$ is the value of the noise parameter at the cumulant crossing of the lattice L and some reference lattice size L_s, with $L_s < L$, and c' is another non-universal constant. The above equation is convenient when one does not know the correlation length critical exponent that is necessary in Equation (12).

3. Results and Discussion

The general behavior of the relevant quantities for this model, namely, O, O_f, and O_4, given by Equations (6)–(8), are, respectively, displayed in the top panels, (a), (b), and (c), of Figure 1. The different lattice sizes used in the present work are listed in the legend. In these panels, only the lines of the data are shown for a better visualization of the behavior one has as a function of the noise parameter q. The details of extracting the critical properties of the model from these data are discussed below.

The fourth-order Binder cumulant of the order parameter O_4, as a function of the disorder parameter q close to the phase transition, is depicted in Figure 2 for several SNs having different numbers of nodes. One can clearly see from this figure that the system undergoes a second-order phase transition, since the cumulants tend to cross at the same value, corresponding to the critical disorder parameter q_c [26]. In the axes scales used in Figure 2, one can make a rough estimate of the critical noise and the universal value of the Binder cumulant, $q_c = 0.195(3)$ and $O_4^* = 0.301(5)$, respectively. In the above data, the computed errors are statistical ones and for questions of clarity are not shown in Figure 2.

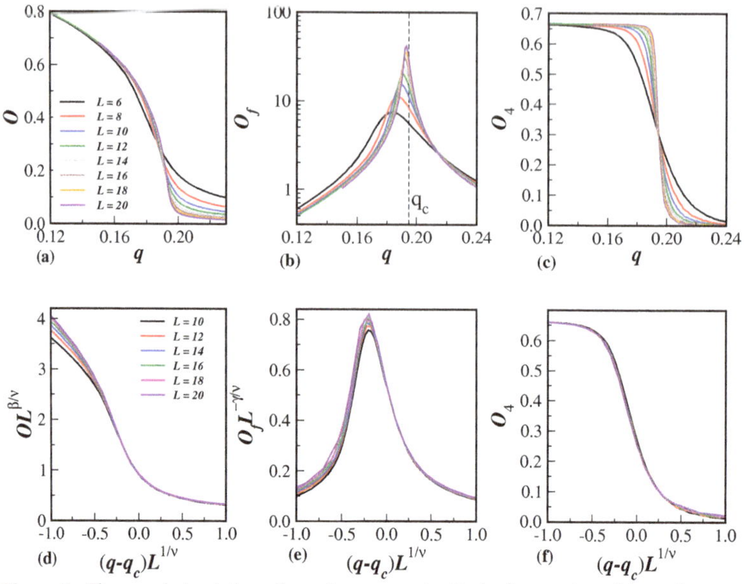

Figure 1. The panels (**a**–**c**) show the order parameter O, the fluctuation of the order parameter O_f, and the Binder fourth-order cumulant O_4 as a function of the noise parameter q. The dashed line in (**b**) locates the estimated critical noise q_c. The legends in (**a**) also apply to (**b**,**c**). The panels (**d**–**f**) show the corresponding data collapse of the quantities displayed in the top panels using the finite-size scaling relations given in Equations (9), (10), and (12) combined with numerical data in Table 1. The legends in (**d**) also apply to (**e**,**f**).

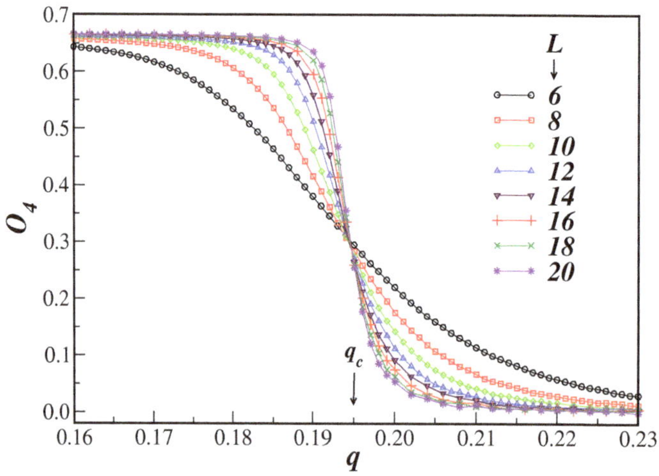

Figure 2. Fourth-order Binder cumulant of the order parameter O_4, plotted as a function of the disorder parameter q, for several values of the lattice sizes L. The vertical arrow indicates the corresponding estimate of $q_c = 0.195(3)$. For clarity, only the general trend in the cumulant behavior is shown and the error bars have been omitted.

However, looking closely at the interception region of the cumulants in Figure 2, it is noticed that the crossings do not occur at the same place due to still finite-size effects. This region has been magnified in the inset of Figure 3. Considering as a reference lattice $L_s = 6$, given by the thicker line in the inset, one can see that due to finite-size effects the value of the noise parameter at the crossings increases as the lattice size L systematically increases. Using these crossings as estimates of $q_c(L)$, and plotting them as a function of $1/\ln(L/L_s)$, one obtains the data shown in the main graph of Figure 3. The errors on $q_c(L)$ have been, in this case, estimated by using the crossings of $O_{4L} - \Delta O_{4L}$ with $O_{4L_s} + \Delta O_{4L_s}$, where ΔO_{4L} and ΔO_{4L_s} are the respective errors on O_{4L} and ΔO_{4L_s} obtained via the jackknife procedure. It is not surprising that the data are not all aligned, because quite small lattices have been considered. However, these data look similar to previous ones obtained for the Ising and Heisenberg models [28–31]. Using Equation (13) for the larger lattices $L \geq 14$, when one expects that the large lattice scaling regime has been reached, one obtains $q_c = 0.1950(7)$ and $O_4^* = 0.304(2)$ in the thermodynamic limit. These are more precise values for the transition disorder parameter and for the cumulant of the model than the rough estimate obtained from just inspecting the crossings in Figure 2. Larger values of L_s could also be considered, but the data turn out to be more disperse, making it more difficult to obtain a reasonable fit. The critical noise for the present $3D$ SNs is listed in Table 1 together with the results for $1D$ and $2D$ networks for comparison.

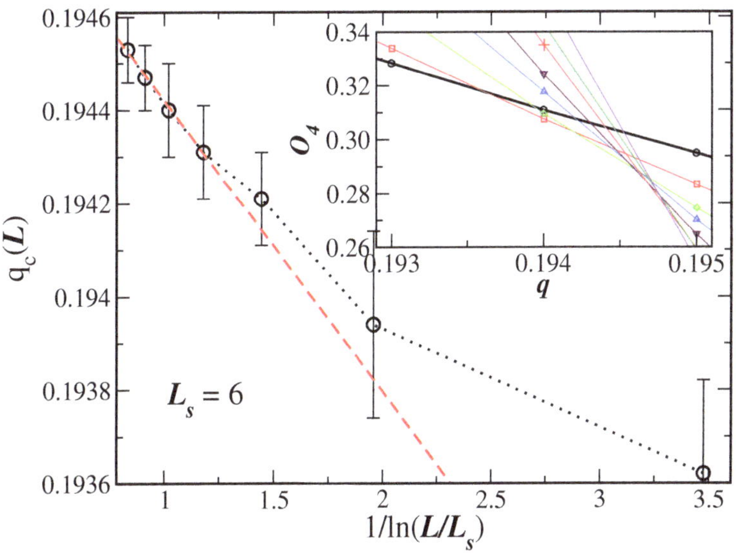

Figure 3. Cumulant disorder parameter crossings $q_c(L)$, as a function of $1/\ln(L/L_s)$, for several values of L, considering the reference lattice $L_s = 6$ (main graph). The dotted line is just a guide to the eyes and the dashed line a linear fit according to Equation (13) considering the larger lattices $L \geq 14$. The inset gives a magnified view close to the interception region of Figure 2. The legend of Figure 2 also applies to this inset and a thicker line (joining the open circles) for the reference lattice $L_s = 6$ is used.

Table 1. Critical noise parameter q_c and critical exponent ratios $1/\nu$, β/ν, γ/ν of the BChS model on three-dimensional SNs ($3D$). Also shown, in the two first rows, are the previous results obtained for the $1D$ and $2D$ on the same networks from Ref. [21] for a comparison. The exponent ratio $\gamma/\nu(q_{max})$ is the result from the maximum of the order parameter susceptibility. The last column gives the hyperscaling relation. Error bars are statistical only.

	q_c	$1/\nu$	β/ν	$\gamma/\nu(q_c)$	$\gamma/\nu(q_{max})$	$d = 2\beta/\nu - \gamma/\nu$
$1D$	0.215(2)	0.52(5)	0.238(7)	0.511(5)	0.524(6)	1.00(2)
$2D$	0.216(2)	0.92(4)	0.53(4)	1.02(4)	1.06(7)	2.1(1)
$3D$	0.1950(7)	1.63(10)	0.789(3)	1.312(8)	1.372(50)	2.95(6)

Having in hand a good estimate of the critical noise probability, Equation (9) can be used to evaluate the critical exponent ratio β/ν by computing the order parameter at q_c for different lattice sizes. Figure 4 depicts the ln–ln plot of the average opinion at the critical disorder, $O(q_c)$, as a function of the system sizes L considered herein. In this case, it is easy to see that, from Equation (9), the magnitude of the slope of the linear fit gives the critical exponent ratio β/ν. The corresponding result $\beta/\nu = 0.789(3)$ is also reproduced in Table 1 and is clearly different from the values obtained for the same model on SNs in one and two dimensions.

The very same procedure can be easily extended to the order parameter fluctuation using Equation (10). Figure 5 displays the ln–ln plot of the order parameter fluctuation (susceptibility) at the critical disorder value, $O_f(q_c)$, at the estimated q_c, as a function of the lattice size L. From Equation (10) one obtains the ratio γ/ν as the slope of the linear fit to the data. The fit provides $\gamma/\nu = 1.312(8)$ and is listed in Table 1 for a comparison to the values for other dimensions.

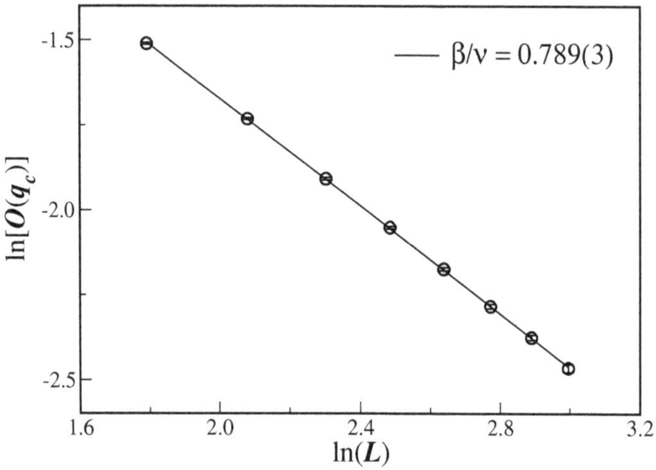

Figure 4. Ln–ln plot of the average opinion at the estimated critical disorder $O(q_c)$ as a function of the logarithm of the system sizes L. The line is the best linear fit according to Equation (9), with the slope being the critical exponent ratio β/ν given in the legend. Note that the shown error bars are smaller than the symbol sizes.

It is worthwhile to note that Equation (10) can still be used to obtain the value of the ratio γ/ν by considering the maximum value of the order parameter fluctuation, $O_f(q_{max})$ that occurs at $q = q_{max}$. It turns out the q_{max} is close, but not the same, as q_c; see Figure 1b that shows the behavior of O_f close to the transition region. Figure 5 also displays the ln–ln plot of $O_f(q_{max})$ as a function of the lattice size L. In this case, one obtains $\gamma/\nu = 1.372(50)$. The error in this exponent ratio is larger, but still comparable to that obtained from the estimate at $O(q_c)$. Moreover, both exponent ratios are different from the previous estimates on $1D$ and $2D$ networks.

As mentioned in the last paragraph, the value of q_{max}, where the fluctuation in the order parameter displays a maximum, is different from q_c. For the present model, q_{max} approaches the infinite limit q_c from below, as is apparent in Figure 1b. In fact, this noise probability is generally interpreted as a new $q_c(L)$ from which the transition point q_c is obtained by extrapolating Equation (12) to the thermodynamic limit. Of course, in this case, one needs to know the correlation length exponent ν. However, if one adopts here a different path instead, and uses $q_c(L)$ for all values of L in Equation (12), with the already estimated q_c, one is able to compute the exponent $1/\nu$ as the magnitude of the slope of the linear fit to the data. Such a procedure is shown in Figure 6, that displays the ln–ln plot of

$|q_c(L) - q_c|$ as a function of L. The respective correlation length critical exponent is given by $1/\nu = 1.63(10)$ and also displayed in Table 1.

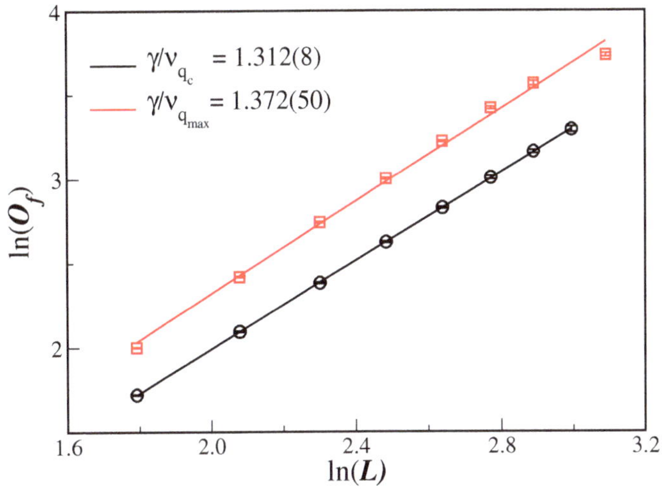

Figure 5. Ln–ln plot of the susceptibility $O_f(q_c)$ at the estimated q_c and maximum value $O_f(q_{max})$ as a function of the system sizes L. The lines are the best linear fit with the slope being the critical exponent ratio γ/ν. The displayed error bars are smaller than the symbol sizes.

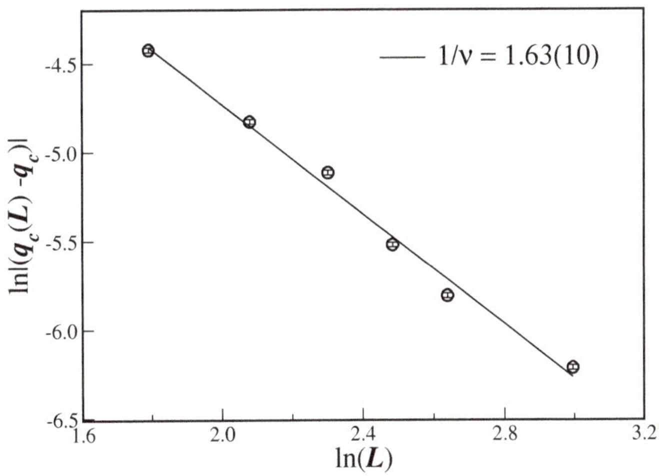

Figure 6. Ln–ln plot of the magnitude of the displacement $|q_c(L) - q_c|$, with $q_c(L)$ extracted from Figure 1b. The line is a linear fit with the corresponding exponent given in the legend.

Finally, in Figure 1, the data collapse of the rescaled order parameter $OL^{\beta/\nu}$ is shown in panel (d) , the rescaled fluctuation of the order parameter $O_f L^{-\gamma/\nu}$ in panel (e), and the rescaled reduced Binder cumulant O_4 in panel (f), all as a function of the rescaled probability displacement $(q - q_c)L^{1/\nu}$. As in the top panels, just the lines of the actual Monte Carlo data have been used for a better evaluation of the collapse and the accuracy of the critical quantities. The corresponding lattice sizes L are listed in the legend of (d), where only larger sizes $L \geq 10$ have been considered in the data collapse. It can be seen that the data collapse presents an excellent agreement with scaling relations given in Equations (9), (10) and (12), mainly for $q > q_c$ and even close to $q \lesssim q_c$, which indicates that the evaluation of the critical exponent ratios β/ν, γ/ν, and $1/\nu$ are reasonably accurate. Moreover, as can be noticed from the last column of Table 1, they obey, within the error bars, the hyperscaling relation for all lattice dimensionalities.

4. Concluding Remarks

The Biswas–Chatterjee–Sen model, in its discrete version and defined on three-dimensional Solomon networks, has been studied through extensive Monte Carlo simulations for several values of the local consensus controlling parameter. From the data collapse displayed in panels (d)–(f) of Figure 1, and from the scaling behavior depicted in Figures 4–6, it is clear that the model really undergoes a well-defined second-order phase transition.

It is clear that the present SN sizes do not ensure that the large-size scaling regime of Equations (9), (10), and (12) has been achieved. In fact, we have seen that finite-size effects are actually present, for instance, in the fourth-order Binder cumulants in Figure 3. This means that corrections to scaling could be implemented in the scaling relation. This will certainly require extra unknown correction-to-scaling exponents. However, the evaluation of the critical exponents seems to be reasonable, even for the SN sizes employed in Figures 4–6, since the data are quite aligned for a proper linear fit (the same in Figure 3, but only for the largest lattices).

In addition, looking at Figure 1d,e for the O and O_f variables, one can clearly see the finite-size behavior of the data collapse. While for $q > q_c$ all sizes collapse to the same curve, for $q < q_c$ the collapse systematically tends to the same curve as the SN size increases. For the two largest SNs, namely, $L = 18$ and $L = 20$, all data are almost superimposed in the whole range of $(q - q_c)L^{1/\nu}$. This means that we have reached the large-size regime for the universal scaling functions of the variables defined in Equations (6)–(8). It is also evident from Figure 1f that the reduced fourth-order Binder cumulant $O_4(q)$ turns out to be a more robust variable, because there has been no finite-size effects for all SN sizes used herein.

From the critical exponents of the BChS model on SNs, conveyed in Table 1, it becomes apparent that the model in three dimensions is indeed in a different universality class to the model defined in one and two dimensions. This is an expected result and in agreement with the general trend coming from renormalization group arguments. Here, we also emphasize that our results are different from the three-dimensional BChS model on regular lattices in Ref. [8]. Despite the critical exponent ν being, within the error bars, compatible to the Ising universality class at 3D, this is not true for the other exponents, which could be affected by strong finite-size effects. The present results thus indicate that the BChS model has its own universality class.

It is also interesting to note that although the critical exponents have a sensitive change with the dimension of the network, the critical noise probabilities, responsible for the phase transition in the model, seem not to vary much as the connectivity of the networks is increased by the dimensionality of the lattices.

Finally, an additional study of the BChS model on four-dimensional SN networks should be very interesting from the pure theoretical point of view. Recall that the Ising model has a lower critical dimension $1D$ and an upper critical dimension $4D$. The BChS model on SNs has a lower critical dimension $0D$ and the present results clearly indicate, if such arguments are valid here, an upper critical dimension greater than $3D$. The question of violation of the hyperscaling relation and a possible unified view of the finite-size scaling valid in all dimensions should also be addressed in this case [22].

Author Contributions: Conceptualization, T.A.A.; Methodology, G.S.O., T.A.A. and F.W.L.; Validation, G.A.A.; Formal analysis, T.A.A., G.A.A., F.W.L. and J.A.P.; Investigation, G.S.O.; Data curation, G.S.O.; Writing—original draft, F.W.L. and J.A.P.; Visualization, J.A.P. All authors have read and agreed to the published version of the manuscript.

Funding: This research received no external funding.

Institutional Review Board Statement: Not applicable.

Data Availability Statement: The data presented in this study are available on request from the corresponding author.

Acknowledgments: We thank Brazilian agencies CNPq and FAPEMIG for financial support. This work was also supported by system Silicon Graphics Internacional (SGI) Altix 1350 (CENAPAD.UNICAMP-USP, SP-BRAZIL).

Conflicts of Interest: The authors declare no conflicts of interest.

Abbreviations

The following abbreviations are used in this manuscript:

BChS	Biswas–Chatterjee–Sen
IM	Ising model
MVM	Majority-vote model
SN	Solomon network
$1D, 2D, 3D$	One dimensional, two dimensional, three dimensional

References

1. Stauffer, D. A Biased Review of Sociophysics. *J. Stat. Phys.* **2013**, *151*, 229. [CrossRef]
2. Galam, S. *Sociophysics*; Springer: Berlin/Heidelberg, Germany, 2012.
3. Galam, S.; Mod, I.J. The Trump phenomenon: An explanation from sociophysics. *Phys. B* **2017**, *31*, 1742015. [CrossRef]
4. Ising, E. Beitrag zur Theorie des Ferromagnetismus. *Z. Phys.* **1925**, *31*, 253. [CrossRef]
5. Baxter, R.J. *Exactly Solved Models in Statistical Mechanics*; Academic Press: London, UK, 1982.
6. Oliveira, M.J. Isotropic majority-vote model on a square lattice. *J. Stat. Phys.* **1992**, *66*, 273. [CrossRef]
7. Biswas, S.; Chatterjee, A.; Sen, P. Disorder induced phase transition in kinetic models of opinion dynamics. *Phys. A* **2012**, *391*, 3257. [CrossRef]
8. Mukherjee, S.; Chatterjee, A. Disorder-induced phase transition in an opinion dynamics model: Results in two and three dimensions. *Phys. Rev. E* **2016**, *94*, 062317. [PubMed]
9. Malarz, K. Social Phase Transition in Solomon Network. *Int. J. Mod. Phys. C* **2003**, *14*, 561. [CrossRef]
10. Kobe, S. Ernst Ising 1900–1998. *Braz. J. Phys.* **2000**, *30*, 649. [CrossRef]
11. Wolf, W.P. The Ising Model and Real Magnetic Materials. *Braz. J. Phys.* **2000**, *30*, 794. [CrossRef]
12. Majewski, J.; Li, H.; Ott, J. The Ising Model in Physics and Statistical Genetics. *Am. J. Hum. Genet.* **2001** *69*, 853.
13. Alencar, D.S.M.; Alves, T.F.A.; Alves, G.A.; Macedo-Filho, A.; Ferreira, R.S.; Lima, F.W.S.; Plascak, J.A. Opinion Dynamics Systems on Barabási-Albert Networks: Biswas-Chatterjee-Sen Model. *Entropy* **2023**, *25*, 183. [CrossRef] [PubMed]
14. Biswas, S.; Chatterjee, A.; Sen, P.; Mukherjee, S.; Chakrabarti, B.K. Social dynamics through kinetic exchange: The BChS model. *Front. Phys.* **2023**, *11*, 1196745. [CrossRef]
15. Erez, T.; Hohnisch, M.; Solomon, S. *Economics: Complex Windows*; Springer: Milan, Italy, 2005; p. 201.
16. Goldenberg, J.; Shavitt, Y.; Shir, E.; Solomon, S. Distributive immunization of networks against viruses using the 'honey-pot' architecture. *Nat. Phys.* **2005**, *1*, 184. [CrossRef]
17. Lima, F.W.S. Equilibrium and non-equilibrium models on Solomon networks. *Int. J. Mod. Phys. C* **2016**, *28*, 1650134. [CrossRef]
18. Lima, F.W.S. Equilibrium and nonequilibrium models on solomon networks with two square lattices. *Int. J. Mod. Phys. C* **2017**, *28*, 1750099. [CrossRef]
19. Pękalski, A. Ising model on a small world network. *Phys. Rev. E* **2001**, *64*, 057104. [CrossRef]
20. Herrero, C.P. Ising model in small-world networks. *Phys. Rev. E* **2002**, *65*, 066110. [CrossRef]
21. Alvez Filho, E.; Lima, F.W.S.; Alves, T.F.A.; Alves, G.A ; Plascak, J.A. Opinion Dynamics Systems via Biswas-Chatterjee-Sen Model on Solomon Networks. *Physics* **2023**, *5*, 873. [CrossRef]
22. Berche, B.; Ellis, T.; Holovatch, Y.; Kenna, R. Phase transitions above the upper critical dimension. *SciPost Phys. Lec. Notes* **2022**, *60*, 1–44. [CrossRef]
23. Raquel, M.T.S.A.; Lima, F.W.S.; Alves, T.F.A.; Alves, G.A.; Macedo-Filho, A.; Plascak, J.A. Non-equilibrium kinetic Biswas–Chatterjee–Sen model on complex networks. *Phys. A* **2022**, *603*, 127825. [CrossRef]
24. Landau, D.P.; Binder, K. *A Guide to Monte Carlo Simulation in Statistical Physics*, 5th ed.; Cambridge University Press: Cambridge, UK, 2021.
25. Tuckey J.W. Bias and Confidence in Not-Quite Large Sample. *Ann. Math. Stat.* **1958**, *29*, 614.
26. Binder, K.; Heermann, D.W. *Monte Carlo Simulation in Statistical Phyics*; Springer: Berlin/Heidelberg, Germany; New York, NY, USA, 1988.
27. Nightingale, M.P. Scaling theory and finite systems. *Phys. A* **1976**, *83*, 561. [CrossRef]
28. Binder, K. Finite size scaling analysis of ising model block distribution functions. *Phys. B* **1981**, *43*, 119. [CrossRef]
29. Peczak, P.; Ferrenberg, A.M.; Landau, D.P. High-accuracy Monte Carlo study of the three-dimensional classical Heisenberg ferromagnet. *Phys. Rev. B* **1991**, *43*, 6087. [CrossRef]

30. Holme C.; Janke, W. Critical exponents of the classical three-dimensional Heisenberg model: A single-cluster Monte Carlo study. *Phys. Rev. B* **1993**, *48*, 936. [CrossRef]
31. Albuquerque Silva, G.; Bergeron, A.R.; Plascak, J.A.; Landau, D.P. Three-dimensional Heisenberg model with Dzyaloshinskii-Moriya interaction: A Monte Carlo study. *Phys. Rev. E* **2024**, *109*, 064113. [CrossRef]

entropy

Article

Random Quantum Ising Model with Three-Spin Couplings

Ferenc Iglói [1,2,*] and Yu-Cheng Lin [3]

1 Wigner Research Centre for Physics, Institute for Solid State Physics and Optics, H-1525 Budapest, Hungary
2 Institute of Theoretical Physics, University of Szeged, H-6720 Szeged, Hungary
3 Graduate Institute of Applied Physics, National Chengchi University, Taipei 11605, Taiwan; yc.lin@nccu.edu.tw
* Correspondence: igloi.ferenc@wigner.hu

Abstract: We apply a real-space block renormalization group approach to study the critical properties of the random transverse-field Ising spin chain with multispin interactions. First, we recover the known properties of the traditional model with two-spin interactions by applying the renormalization approach for the arbitrary size of the block. For the model with three-spin couplings, we calculate the critical point and demonstrate that the phase transition is controlled by an infinite disorder fixed point. We have determined the typical correlation-length critical exponent, which seems to be different from that of the random transverse Ising chain with nearest-neighbor couplings. Thus, this model represents a new infinite disorder universality class.

Keywords: disordered systems; critical phenomena; renormalization group; infinite disorder fixed point

1. Introduction

Quantum phase transitions take place at zero temperature by varying a control parameter, such as the strength of a transverse field, and are signaled by a singularity in the ground state of the system [1]. Quantum phase transitions occur also in one space dimension, and the corresponding singularity is often related to the temperature-driven critical singularities in two-dimensional classical systems [2]. Disorder is an inevitable feature of real materials, and it could have a very profound effect on the critical behavior of the systems [3], and the effect of the disorder is particularly strong at a quantum critical point. The theoretical description of a random quantum phase transition is very challenging since the combined effect of disorder and quantum fluctuations, together with strong correlations, have to be considered at the same time. For some models, this type of investigation can be performed by a so called strong disorder renormalization group (SDRG) method [4,5]. In the SDRG calculation, local degrees of freedom with a large excitation energy are successively eliminated and new parameters are calculated perturbatively for the remaining degrees of freedom. In a class of models, the distribution of the renormalized parameters broaden without limit, and the random phase transition is controlled by a so-called infinite disorder fixed point (IDFP) [6]. At an IDFP, disorder fluctuations are overwhelmingly dominant over quantum fluctuations, and the obtained critical properties are expected to be asymptotically exact for large systems.

The SDRG approach was introduced by Ma, Dasgupta and Hu [7,8] and later applied by D. Fisher [9,10] to study the critical properties of the random transverse-field Ising chain. Several presumably exact results have been obtained so that the random model looks to be understood at least at the same level as its nonrandom counterpart. In one dimension, the method has been generalized for other models, resulting in exact solutions [11–15]. In higher dimensions, the method is applied numerically [16–21], and different technical simplifications are introduced in order to treat large finite systems [22–24]. For the transverse Ising model, it is demonstrated that the critical behavior is controlled by IDFPs even at higher dimensions, and it is expected that the upper critical dimension of the problem is infinite [23,24]. Regarding models with nearest neighbor interaction and having a discrete order parameter, the critical behavior is expected to be the same as that of the transverse Ising model. This result seems to hold also for random stochastic models, such as the random contact process [25–27]. For random models with long-range interactions, however, a new type of fixed point is found to control the critical behavior, which is of a conventional disorder type [28,29].

In this paper, we consider another type of transverse Ising model, which has multiple-site product interactions. The Hamiltonian is defined as

$$\mathcal{H}^{(m)} = -\sum_i J_i \prod_{l=0}^{m-1} \sigma_{i+l}^z - \sum_i h_i \sigma_i^x \,, \tag{1}$$

in terms of the Pauli-matrices $\sigma_i^{x,z}$ at site i. Here, the interactions J_i and the transverse fields h_i are both independent random variables. This model in a pure (nonrandom) case was introduced by Turban [30] and independently by Penson et al. [31]. The special case $m = 2$ is the standard transverse Ising chain with nearest neighbor interaction. According to numerical studies, the pure $m = 3$ model has a second-order quantum phase transition, which belongs to the universality class of the 4-state Potts model [32–37]. At the critical point, there are logarithmic corrections that are probably in the same form as in the 4-state Potts model [38,39]. For $m \geq 4$, the phase-transition in the pure model is of first order.

In this paper, we are going to study the phase transition in the disordered model with random $J_i > 0$ and $h_i > 0$. We do not specify the form of the disorder distributions. We assume that these are not singular so that the first and second moments of the log-variables are finite. In this respect, we have detailed information about the $m = 2$ model [4,5], but the models with $m > 2$ have not been investigated yet. Here, we study the model with $m = 3$ and consider random positive three-spin couplings and random positive transverse fields. Due to the different types of local interactions, the application of the standard SDRG method [4,5] exhibits problems: by eliminating a strong three-spin coupling, several new renormalized interactions will be generated, depending on the neighborhood of the eliminated coupling; in this case, one can not stop the proliferation of the renormalized parameters during the renormalization procedure. Therefore, we chose a different approach and used a block renormalization group method, which was introduced to the pure $m = 2$ Ising chain by Fernandez-Pacheco [40]. This type of renormalization preserves the self-duality of the model and reproduces the exact critical point and the value of the exact correlation length exponent, $\nu(m = 2) = 1$. Later, the method was used for other quantum spin chains [41–47] and was generalized for higher dimensions [48,49]. For the random transverse Ising model with nearest neighbor couplings ($m = 2$), the method was applied by Miyazaki and Nishimore [50], as well as by Cécile Monthus [51], both in one and higher dimensions.

Our paper is organized as follows. In Section 2, we introduce the basic idea of the block renormalization approach and present the duality transformation of the model. In Section 3, the method is applied to the $m = 2$ model by using the arbitrary large size of the block, while in Section 4, it is applied to the $m = 3$ model with a block size $b = 2$. In Section 5, we close our paper with a discussion.

2. Basic Idea of Block Renormalization

In the block renormalization method [53?] the spins at sites $i = nb + 1$ with $n = 0, 1, 2, \ldots$ and $b = 2, 3, \ldots$ are fixed at arbitrary positions, while the intermediate spins are integrated out. Here, b sets the scale factor, which is the size of the blocks. In this way, the Hamiltonian is divided into two parts:

$$\mathcal{H}^{(m)} = \mathcal{H}_0^{(m)} + \mathcal{V}^{(m)} \,, \tag{2}$$

where $\mathcal{H}_0^{(m)}$ represents blocks with the intra-block terms, and $\mathcal{V}^{(m)}$ is the perturbation that contains the transverse field acting on the selected spins and the couplings that couple the neighboring blocks. The block-Hamiltonians are solved either analytically or numerically and the lowest levels are retained and identified as the states of the block-spin variable. At the same time, the renormalized values of the inter-block terms are obtained in a (first-order) perturbative way.

In the use of the block renormalization approach, it is useful if the models have duality properties; in which case, the number of parameters generally does not increase during renormalization. The Hamiltonian in Equation (4) indeed has such a symmetry. Following the method by Turban [30], one can define a set of variables:

$$\mu_i^x = \prod_{l=0}^{m-1} \sigma_{i+l}^z$$
$$\mu_i^z = \prod_{n=0}^{\infty} \sigma_{i-nm}^x \sigma_{i-nm-1}^x , \tag{3}$$

that satisfy the Pauli spin algebra. In terms of these new variables, the Hamiltonian is expressed as

$$\mathcal{H}^{(m)} = -\sum_i h_i \prod_{l=0}^{m-1} \mu_{i-l}^z - \sum_i J_i \mu_i^x , \tag{4}$$

which is in the same form as that in Equation (1) with the correspondences $J_i \to h_{i+m-1}$ and $h_i \to J_i$. It follows that the pure model with $J_i = J, h_i = h, \forall i$ is self-dual and the self-dual point $J/h = 1$ corresponds to the phase-transition point.

In the following, we first present the block renormalization for the $m = 2$ model for arbitrary size of the block [47], then we consider the model with $m = 3$ and solve the renormalization approach for a block of two sites.

3. Block Renormalization Approach of the $m = 2$ Model

Here, we first turn the spin variables $\sigma_i^x \leftrightarrow \sigma_i^z$, and for convenience, the sites are relabeled as $i = (j, \alpha)$, where $j = 1, 2, \ldots$ labels the blocks and $\alpha = 0, 1, \ldots, b-1$ labels sites in a block. For the $m = 2$ model, the intra-block Hamiltonian is

$$\mathcal{H}_0^{(2)} = -\sum_j \sum_{\alpha=0}^{b-2} J_{j,\alpha+1} \sigma_{j,\alpha}^x \sigma_{j,\alpha+1}^x - \sum_j \sum_{\alpha=1}^{b-1} h_{j,\alpha} \sigma_{j,\alpha}^z , \tag{5}$$

while the perturbation is given by

$$\mathcal{V}^{(2)} = -\sum_j J_{j-1,b-1} \sigma_{j-1,b-1}^x \sigma_{j,0}^x - \sum_j h_{j,0} \sigma_{j,0}^z . \tag{6}$$

The leftmost spin of a block is fixed, thus the x-component of this spin is ± 1, which fixes the sign of the magnetization at the other sites of the block, too. We use this sign to characterize the state of the block. Solving the ground state of the block both with $+$ and $-$ leftmost spins, the interaction energy between two neighboring blocks is given in first-order perturbation theory as

$$\epsilon_j = -J_{j,b-1} \langle \sigma_{j,b-1}^x \rangle \langle \sigma_{j+1,0}^x \rangle = -J_j^R \Sigma_j^x \Sigma_{j+1}^x , \tag{7}$$

where $\langle \sigma_{j+1,0}^x \rangle = \pm 1$ is fixed, and $\langle \sigma_{j,b-1}^x \rangle = m_j^{(b)}$ is the expectation value of the end-spin magnetization in the j-th block of length b. The block-spin variables are $\Sigma_j^x = \pm 1$ and $\Sigma_{j+1}^x = \pm 1$, thus the renormalized value of the coupling is

$$J_j^R = J_{j,b-1} m_j^{(b)} . \tag{8}$$

The end-spin magnetization can be calculated exactly [47,54,55]:

$$m_j^{(b)} = \left[1 + \sum_{\alpha=1}^{b-1} \prod_{k=1}^{\alpha} \left(\frac{h_{j,b-k}}{J_{j,b-k-1}}\right)^2\right]^{-1/2}, \tag{9}$$

which, in the simplest case with $b = 2$, is given by

$$m_j^{(2)} = \left[1 + \left(\frac{h_{j,1}}{J_{j,0}}\right)^2\right]^{-1/2}. \tag{10}$$

The renormalized value of the transverse field is obtained through duality, which amounts to interchange couplings and fields, and also the two ends of the block leading to

$$h_j^R = h_{j,0}\tilde{m}_j^{(b)}, \tag{11}$$

with

$$\tilde{m}_j^{(b)} = \left[1 + \sum_{\alpha=1}^{b-1} \prod_{k=1}^{\alpha} \left(\frac{J_{j,k-1}}{h_{j,k}}\right)^2\right]^{-1/2}, \tag{12}$$

and

$$\tilde{m}_j^{(2)} = \left[1 + \left(\frac{J_{j,0}}{h_{j,1}}\right)^2\right]^{-1/2}. \tag{13}$$

Let us consider the ratio

$$K_{j,\alpha} = \frac{J_{j,\alpha}}{h_{j,\alpha}}, \tag{14}$$

and calculate its value with the renormalized parameters, which are given by

$$K_j^R = \frac{J_j^R}{h_j^R} = \frac{\prod_{\alpha=0}^{b-1} J_{j,\alpha}}{\prod_{\alpha=0}^{b-1} h_{j,\alpha}} = \prod_{\alpha=0}^{b-1} K_{j,\alpha}. \tag{15}$$

3.1. Pure Model

For the pure model with $K = K_{j,\alpha}$, $\forall j, \alpha$, Equation (15) leads to the simple relation [47]:

$$K^R = K^b \tag{16}$$

having the fixed point $K^* = (J/h)^* = 1$, which corresponds to the self-dual point of the system. Furthermore, the thermal eigenvalue of the transformation is $\lambda_t = b$ and the correlation-length critical exponent is $\nu^{\rm pure} = \ln b / \ln \lambda_t = 1$. This is the exact value, and in this way, it has been obtained for any value of b.

3.2. Random Model

Now we turn to the disordered model with random couplings and random fields. The renormalization equation in Equation (15) has a simple addition form in terms of the log-variables:

$$\ln K_j^R = \sum_{\alpha=0}^{b-1} \ln K_{j,\alpha}. \tag{17}$$

Interestingly, there are two scale factors, $\tilde{b} < b$, such that $\tilde{b}^N = b$. Repeating the iterations with the scale factor $\tilde{b}$ N-times ($N = 2, 3, \ldots$), the form of the renormalized K-parameter is the same as if the renormalization is performed with a scale factor b in one step.

According to the central-limit theorem, Equation (17) in the large-b limit has the asymptotic form

$$\ln K_b^R = b\left[\overline{\ln J_\alpha - \ln h_\alpha}\right] + b^{1/2}[Var(\ln J_\alpha) + Var(\ln h_\alpha)]^{1/2} v \tag{18}$$

where $\overline{x}$ and $Var(x)$ denote the mean value and the variance of the random variable x, respectively, and v is a Gaussian random variable with mean zero and variance unity. Introducing the quantum control parameter

$$\delta = \frac{\overline{\ln J_\alpha} - \overline{\ln h_\alpha}}{Var(\ln J_\alpha) + Var(\ln h_\alpha)}, \tag{19}$$

we have

$$\ln K_b^R = b\delta + b^{1/2}[Var(\ln J_\alpha) + Var(\ln h_\alpha)]^{-1/2} v\,. \tag{20}$$

The first term in the above equation characterizes the divergence of the typical correlation length, since by definition $b \sim \xi_{\text{typ}} \sim \delta^{-\nu_{typ}}$ near criticality. From Equation (20), we obtain the typical correlation length exponent:

$$\nu_{\text{typ}} = 1\,. \tag{21}$$

At the critical point ($\delta = 0$), the fluctuations in the log-couplings grow as $b^{1/2}$, which defines the log-excitation energy: $\ln \epsilon$. The corresponding scaling form is $\ln \epsilon \sim b^{\psi}$ with the critical exponent:

$$\psi = \frac{1}{2}\,. \tag{22}$$

This type of dynamic is relevant at an infinite disorder fixed point.

The combination of the first and the second terms in Equation (20) with $b\delta \approx -b^{1/2}$ $[Var(\ln J_\alpha) + Var(\ln h_\alpha)]^{-1/2}$ will result in the vanishing of the leading contribution of these terms, which is connected to a finite-size correlation exponent [51], ν_{FS} defined by $\delta \sim b^{-1/\nu_{\text{FS}}}$ and given by

$$\nu_{\text{FS}} = 2\,. \tag{23}$$

These results follow directly from the RG equations in Equations (8) or (11) and from the known scaling properties of the end-spin magnetization [55]. We can also recover known results in the off-critical region, in the so-called Griffith's phase [56,57]. Using Equations (11) and (12) we express the inverse square of the excitation energy as

$$\frac{1}{\epsilon^2} \equiv S = 1 + \sum_{\alpha=1}^{b-1} \prod_{k=1}^{\alpha} \left(\frac{J_{j,k-1}}{h_{j,k}}\right)^2, \tag{24}$$

which in the limit $b \to \infty$ is a Kesten variable [58], which has a tail distribution:

$$P(S) \sim \frac{1}{S^{1+\Delta}}, \tag{25}$$

for $S \gg 1$ with the exponent which is the positive root of the equation [59–61]:

$$\overline{\left(\frac{J^2}{h^2}\right)^{\Delta}} = 1\,. \tag{26}$$

Here, Δ is related to the dynamical exponent z in the Griffith's phase via

$$z = \frac{1}{2\Delta}\,, \tag{27}$$

so that length and excitation energy are related as

$$\epsilon \sim b^{-z}\,, \tag{28}$$

and the asymptotic distribution of the log-excitation energy is given by

$$P(\ln \epsilon) \sim \epsilon^{1/z} \tag{29}$$

for $\epsilon \ll 1$.

4. Block Renormalization Approach of the $m = 3$ Model

For the $m = 3$ model, the Hamiltonian is split as

$$\mathcal{H}_0^{(3)} = -\sum_j \sum_{\alpha=0}^{b-2} J_{j,\alpha+1}\sigma^x_{j,\alpha}\sigma^x_{j,\alpha+1}\sigma^x_{j,\alpha+2} - \sum_j \sum_{\alpha=1}^{b-1} h_{j,\alpha}\sigma^z_{j,\alpha}\,, \tag{30}$$

and

$$\mathcal{V}^{(3)} = -\sum_j J_{j,0}\sigma^x_{j-1,b-1}\sigma^x_{j,0}\sigma^x_{j,1} - \sum_j h_{j,0}\sigma^z_{j,0}\,. \tag{31}$$

Here, we restrict ourselves to the case where the scale factor $b = 2$, which allows the RG method to be solved analytically. For larger blocks $b > 2$, the calculations are more complicated due to the presence of three-spin interactions. The way the Hamiltonian with $m = 3$ is divided into two parts for $b = 2$ is illustrated in Figure 1.

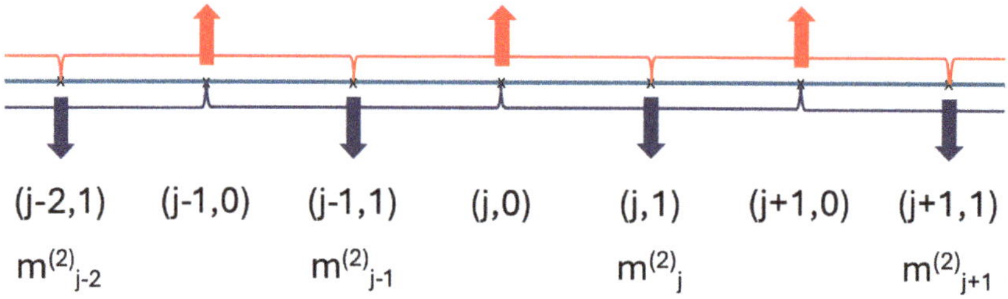

Figure 1. Illustration of the division of the Hamiltonian into the unperturbed part ($\mathcal{H}_0$: $\Downarrow$, $\smile$) and the perturbation ($\mathcal{V}$: $\Uparrow$, $\frown$), for transverse fields and three-spin couplings, respectively, for $m = 3$ and $b = 2$. The magnetization in the j-th block is denoted by $m_j^{(2)}$, see in Equation (33).

Solving the ground state of the jth-block, the interaction energy between neighboring blocks in first-order perturbation theory is given by

$$\epsilon_j = -J_{j,0}\langle\sigma^x_{j-1,1}\rangle\langle\sigma^x_{j,0}\rangle\langle\sigma^x_{j,1}\rangle = -J_j^R \Sigma_j^x \sigma^x_{j,0}\Sigma^x_{j+1}\,, \tag{32}$$

where $\langle\sigma^x_{j,0}\rangle = \pm 1$ is fixed, Σ_j^x denotes the block spin for the jth-block, and J_j^R the renormalized three-spin coupling. The expectation value $\langle\sigma^x_{j,1}\rangle = m_j^{(2)}$ is given by

$$m_j^{(2)} = \left[1 + \left(\frac{h_{j,1}}{J_{j,1}}\right)^2\right]^{-1/2}, \tag{33}$$

and similarly for $\langle \sigma^x_{j-1,1} \rangle = m^{(2)}_{j-1}$. Thus, the renormalized value of the three-spin coupling is given by

$$J_j^R = J_{j,0} m_{j-1}^{(2)} m_j^{(2)} \, . \tag{34}$$

The renormalized value of the transverse field is obtained through duality, which amounts to interchange couplings and fields, leading to

$$h_j^R = h_{j,0} \tilde{m}_{j-1}^{(2)} \tilde{m}_j^{(2)} \, , \tag{35}$$

with

$$\tilde{m}_j^{(2)} = \left[1 + \left(\frac{J_{j,1}}{h_{j,1}} \right)^2 \right]^{-1/2} \, . \tag{36}$$

Considering the ratio defined in Equation (15), we obtain

$$K_j^R = \frac{J_j^R}{h_j^R} = \frac{J_{j-1,1} J_{j,0} J_{j,1}}{h_{j-1,1} h_{j,0} h_{j,1}} = K_{j-1,1} K_{j,0} K_{j,1} \, , \tag{37}$$

as a product of three original ratios.

4.1. Pure Model

For the pure model, there is a simple relation [32]:

$$K^R = K^3 \, , \tag{38}$$

having the fixed-point $K^* = (J/h)^* = 1$, which is the self-dual point of the system. The thermal eigenvalue is $\lambda_t = 3$, and the correlation length exponent is $\nu^{pure} = \ln 2/\ln 3 = 0.631$. This value is not exact but quite close to the expected value of the 4-state Potts model [3]: $\nu^{Potts} = 2/3$.

4.2. Random Model

For the random model, repeating the renormalization in the next step, we obtain

$$\begin{aligned} K_j^{R(2)} &= K_{j-1}^R K_j^R K_{j+1}^R \\ &= K_{j-2,1} K_{j-1,0} K_{j-1,1}^2 K_{j,0} K_{j,1}^2 K_{j+1,0} K_{j+1,1} \, , \end{aligned} \tag{39}$$

which contains the product of nine terms, but two terms are represented twice. Let us introduce log-ratios and iterate the renormalization process n-times, then we obtain an additive form:

$$\ln K^{R(n)} = \sum_{i=1}^{L(n)} c_i^{(n)} \ln K_i \, , \tag{40}$$

where we have used the original notation, i, for the sites, and $c_i^{(n)}$ is the multiplicity of the term K_i. For the first few iterations, the multiplicities are given by

$$\begin{aligned} n &= 1 \quad 111 \\ n &= 2 \quad 1121211 \\ n &= 3 \quad 112132313231211 \\ n &= 4 \quad 1121323143525341435253413231211 \, , \end{aligned} \tag{41}$$

which can be generated using the following rules:

$$c_{2^n}^{(n)} = 1, \quad c_{2^n-i}^{(n)} = c_{2^n+i}^{(n)}, \quad i = 1, 2, \dots 2^n - 1 \tag{42}$$

and

$$c_i^{(n)} = c_i^{(n-1)}, \quad i = 1,2,\ldots,2^{n-1}$$
$$c_{2^{n-1}+i}^{(n)} = c_{2^{n-1}+i}^{(n-1)} + c_i^{(n-1)}, \quad i = 1,2,\ldots,2^{n-1}-1 \tag{43}$$

Due to the scale factor $b = 2$, we have the following for the number of sites involved in the renormalization:

$$L(n) = 2^{n+1} - 1\,, \tag{44}$$

and the total number of terms involving the multiplicities is

$$\sum_{i=1}^{L(n)} c_i = 3^n\,. \tag{45}$$

The average value of the multiplicities is $\bar{c} = 3^n/L(n)$, and its typical value has the same type of scaling with n:

$$c_{typ} \sim (3/2)^n\,. \tag{46}$$

Let us denote by $N_n(c)$ the number of terms having the multiplicity c after n iterations. According to Equation (45), we have

$$\sum_{c=1}^{F(n)} N_n(c)c = 3^n\,, \tag{47}$$

where the largest multiplicity after n iterations is the Fibonacci number: $F(n)$. Equations (46) and (47) follow the scaling behavior of $N_n(c)$ as

$$N_n(c) \sim (4/3)^n\,. \tag{48}$$

The numerically calculated values of $N_n(c)$ versus c in scaled variables are shown in Figure 2. The points for different values of n fit well on a master curve, confirming the scaling relations in Equations (46) and (48).

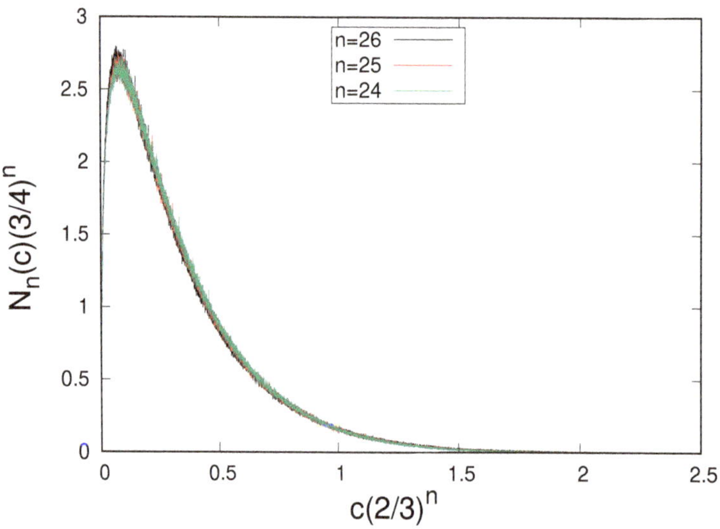

Figure 2. Numerically calculated values of terms with multiplicity c versus c in scaled variables for different values of the number of iterations n. In order to reduce noise, we averaged ten neighboring values.

Let us express the value of the renormalized ratio in Equation (40) as

$$\ln K^{R(n)} = \sum_{c=1}^{F(n)} c \sum_{k=1}^{N(c)} \ln K_k^{(c)} \approx$$
$$\sum_{c=1}^{F(n)} c \left[N(c)\overline{lnK^{(c)}} + \sqrt{N(c)(Var(\ln J) + Var(\ln h))v} \right] . \tag{49}$$

Here $\overline{\ln K^{(c)}} = \overline{\ln J} - \overline{\ln h} = \delta$ is the random quantum control parameter, which is independent of the value of c for the dominant part of the sum, and we obtain the combination

$$\ln K^{R(n)} \sim 3^n \delta \sim L^{\ln 3/\ln 2} \delta , \tag{50}$$

thus, the typical correlation-length exponent is

$$\nu_{\text{typ}} = \frac{\ln 2}{\ln 3} . \tag{51}$$

This value is different from that of the random transverse Ising chain with nearest-neighbor interactions, seen in Equation (21).

At the critical point, the fluctuations of the log-couplings grow as L^{ψ}, with an exponent:

$$\psi = \frac{3\ln 3}{2\ln 2} - 1 = 1.377 , \tag{52}$$

which corresponds to infinite disorder scaling.

We can thus conclude that the critical behavior of the random transverse Ising chain with three-spin couplings is controlled by an infinite disorder fixed point. The obtained value of the typical correlation length exponent indicates that the model belongs to a new infinite disorder universality class.

5. Discussion

In this paper, we have considered the transverse Ising chain with two-spin, nearest-neighbor interaction, as well as with three-spin product interaction in the presence of quenched disorder, and studied the quantum critical properties through a block renormalization approach. The two-spin interaction case has been well studied before, and it represents the prototype system having an infinite disorder fixed point with several presumably exact results. In this case, the new feature of our study is that the renormalization is performed analytically for any size of the block, thus the results are correct in the large block limit. Indeed, the obtained results, both in the vicinity of the critical point and in the Griffith's phase, are in agreement with the previously known exact results.

On the other hand, the model with three-spin couplings represents *terra incognita*; to our knowledge, no random quantum systems with multi-spin interactions had been studied previously. In this case, the renormalization is performed analytically for the block-size $b = 2$. By iterating the transformation, we noticed that the parameters at the sizes of the original model enter several times in the expression of the renormalized one. This is a new feature that is connected to the multispin topology of the interaction. We have calculated the position of the random critical point and demonstrated that the critical properties of the model are controlled by an infinite disorder fixed point. We have also determined the typical correlation-length critical exponent, which turned out to have a different value from that of the two-spin coupling model. The latter, however, represents a very broad universality class, involving models having a discrete order parameter and two-spin interactions.

The renormalization approach we used contains approximations; however, we may expect that some of our results are asymptotically correct for an infinite disorder fixed point. Our investigations can, in principle, be improved by using larger blocks in the transformation; in which case, however, one should involve numerical calculations. With larger blocks, the topology of the iterated renormalization process would remain the same: the original parameters of a given site appear multiple times in the similar way, as shown for $b = 2$ in this paper. Therefore, we believe that the universality class of the random model with three-spin couplings is different from that having two-site interactions.

The studies presented in this paper can be extended in several directions. One could try to apply the traditional SDRG method [4,5] by eliminating successively local degrees of freedom and explore how the topology of the renormalization equations takes a similar form, as obtained in this paper by the block renormalization approach. Another way is to perform numerical investigations of the random three-spin coupling model and check the critical properties. By performing a numerical study, other critical parameters (magnetization exponent, average correlation length exponent, etc.) and Griffith's singularities can be investigated. Studying multispin models with $m > 3$ could also be interesting, since these probably represent new universality classes. One can also think to generalize the block renormalization approach to higher dimensions, a type of study that has been quite successful for the two-spin interacting model [50,51].

This work is dedicated to the memory of Ralf Kenna. Ralf was an excellent physicist and, among others, he studied the critical behavior of systems above the upper critical dimension [62], which, in our problem, seems to be infinity. How the phase-transition takes place in our models at infinite dimension is still an open question.

Author Contributions: Conceptualization, F.I.; Investigation, Y.-C.L. All authors have read and agreed to the published version of the manuscript.

Funding: This work was supported by the Hungarian Scientific Research Fund under Grants No. K128989 and No. K146736 and by the National Research, Development and Innovation Office of Hungary (NKFIH) within the Quantum Information National Laboratory of Hungary, and by the National Science and Technology Council (NSTC) of Taiwan under Grants No. 112-2112-M-004-008.

Data Availability Statement: The data presented in this study are available on request from the corresponding author.

Acknowledgments: This work is dedicated to the memory of Ralf Kenna. F.I. is grateful to the Graduate Institute of Applied Physics, National Chengchi University for hospitality during his visit in Taipei.

Conflicts of Interest: The authors declare no conflicts of interest.

References

1. Sachdev, S. *Quantum Phase Transitions*, 2nd ed.; Cambridge University Press: Cambridge, UK, 2011.
2. Kogut, B. An introduction to lattice gauge theory and spin systems. *Rev. Mod. Phys.* **1979**, *51*, 659. [CrossRef]
3. Cardy, J. *Scaling and Renormalization in Statistical Physics*; Cambridge Lecture Notes in Physics; Cambridge University Press: Cambridge, UK, 1996.
4. Iglói, F.; Monthus, C. Strong disorder rg approach of random systems. *Phys. Rep.* **2005**, *412*, 277. [CrossRef]
5. Iglói, F.; Monthus, C. Strong disorder rg approach—A short review of recent developments. *Eur. Phys. J. B* **2018**, *91*, 290. [CrossRef]
6. Fisher, D.S. Phase transitions and singularities in random quantum systems. *Phys. A Stat. Mech. Its Appl.* **1999**, *263*, 222. [CrossRef]
7. Ma, S.-K.; Dasgupta, C.; Hu, C.-K. Random antiferromagnetic chain. *Phys. Rev. Lett.* **1979**, *43*, 1434. [CrossRef]
8. Dasgupta, C.; Ma, S.-K. Low-temperature properties of the random heisenberg antiferromagnetic chain. *Phys. Rev. B* **1980**, *22*, 1305. [CrossRef]
9. Fisher, D.S. Random transverse field ising spin chains. *Phys. Rev. Lett.* **1992**, *69*, 534. [CrossRef] [PubMed]
10. Fisher, D.S. Critical behavior of random transverse-field ising spin chains. *Phys. Rev. B* **1995**, *51*, 6411. [CrossRef]
11. Fisher, D.S. Random antiferromagnetic quantum spin chains *Phys. Rev. B* **1994**, *50*, 3799. [CrossRef]
12. Senthil, T.; Majumdar, S.N. Critical properties of random quantum potts and clock models. *Phys. Rev. Lett.* **1996**, *76*, 3001. [CrossRef]
13. Monthus, C. Golinelli, O.; Jolicœur, T. Percolation transition in the random antiferromagnetic spin-1 chain. *Phys. Rev. Lett.* **1997**, *79*, 3254. [CrossRef]
14. Hyman, R.A.; Yang, K. Impurity driven phase transition in the antiferromagnetic spin-1 chain. *Phys. Rev. Lett.* **1997**, *78*, 1783. [CrossRef]

15. Carlon, E.; Lajkó, P.; Iglói, F. Disorder induced cross-over effects at quantum critical points. *Phys. Rev. Lett.* **2001**, *87*, 277201. [CrossRef]
16. Motrunich, O.; Mau, S.-C.; Huse, D.A.; Fisher, D.S. Infinite-randomness quantum ising critical fixed points. *Phys. Rev. B* **2000**, *61*, 1160. [CrossRef]
17. Lin, Y.-C.; Kawashima, N.; Iglói, F.; Rieger, H. Numerical Renormalization Group Study of Random Transverse Ising Models in One and Two Space Dimensions. *Prog. Theor. Phys. Suppl.* **2000**, *138*, 479. [CrossRef]
18. Karevski, D.; Lin, Y.-C.; Rieger, H.; Kawashima, N.; Iglói, F. Random quantum magnets with broad disorder distribution. *Eur. Phys. J. B-Condens. Matter Complex Syst.* **2001**, *20*, 267. [CrossRef]
19. Lin, Y.-C.; Iglói, F.; Rieger, H. Entanglement entropy at infinite-randomness fixed points in higher dimensions. *Phys. Rev. Lett.* **2007**, *99*, 147202. [CrossRef]
20. Yu, R.; Saleur, H.; Haas, S. Entanglement entropy in the two-dimensional random transverse field ising model. *Phys. Rev. B* **2008**, *77*, 140402. [CrossRef]
21. Kovács, I.A.; Iglói, F. Critical behavior and entanglement of the random transverse-field ising model between one and two dimensions. *Phys. Rev. B* **2009**, *80*, 214416. [CrossRef]
22. Kovács, I.A.; Iglói, F. Renormalization group study of the two-dimensional random transverse-field ising model. *Phys. Rev. B* **2010**, *82*, 054437. [CrossRef]
23. Kovács, I.A.; Iglói, F. Infinite-disorder scaling of random quantum magnets in three and higher dimensions. *Phys. Rev. B* **2011**, *83*, 174207. [CrossRef]
24. Kovács, I.A.; Iglói, F. Renormalization group study of random quantum magnets. *J. Phys. Condens. Matter* **2011**, *23*, 404204. [CrossRef] [PubMed]
25. Hooyberghs, J.; Iglói, F.; Vanderzande, C. Strong disorder fixed point in absorbing-state phase transitions. *Phys. Rev. Lett.* **2003**, *90*, 100601. [CrossRef]
26. Hooyberghs, J.; Iglói, F.; Vanderzande, C. Absorbing state phase transitions with quenched disorder. *Phys. Rev. E* **2004**, *69*, 066140. [CrossRef] [PubMed]
27. Vojta, T.; Dickison, M. Critical behavior and griffiths effects in the disordered contact process. *Phys. Rev. E* **2005**, *72*, 036126. [CrossRef] [PubMed]
28. Juhász, R.; Kovács, I.A.; Iglói, F. Random transverse-field ising chain with long-range interactions. *Europhys. Lett.* **2014**, *107*, 47008. [CrossRef]
29. Kovács, I.A.; Juhász, R.; Iglói, F. Long-range random transverse-field ising model in three dimensions. *Phys. Rev. B* **2016**, *93*, 184203. [CrossRef]
30. Turban, L. Self-dual ising chain in a transverse field with multispin interactions. *J. Phys. C Solid State Phys.* **1982**, *15*, L65. [CrossRef]
31. Penson, K.A.; Jullien, R.; Pfeuty, P. Phase transitions in systems with multispin interactions. *Phys. Rev. B* **1982**, *26*, 6334. [CrossRef]
32. Igloi, F.; Kapor, D.V.; Skrinjar, M.; Solyom, J. The critical behaviour of a quantum spin problem with thee-spin coupling. *J. Phys. A Math. Gen.* **1983**, *16*, 4067. [CrossRef]
33. Igloi, F.; Kapor, D.V.; Skrinjar, M.; Solyom, J. Series expansion study of first- and second-order phase transitions in a model with multispin coupling. *J. Phys. A Math. Gen.* **1986**, *19*, 1189. [CrossRef]
34. Blöte, H.; Compagner, A.; Cornelissen, P.; Hoogland, A.; Mallezie, F.; Vanderzande, C. Critical behaviour of two ising models with multispin interactions. *Phys. A Stat. Mech. Its Appl.* **1986**, *139*, 395. [CrossRef]
35. Alcaraz, F.C. Order of phase transition for systems with multispin interactions: Monte carlo simulations. *Phys. Rev. B* **1986**, *34*, 4885. [CrossRef] [PubMed]
36. Kolb, M.; Penson, K.A. Conformal invariance and the phase transition of a spin chain with three-spin interaction. *J. Phys. A Math. Gen.* **1986**, *19*, L779. [CrossRef]
37. Blote, H.W.J. Equivalence of an ising model with two- and three-spin interactions with the four-state potts model. *J. Phys. A Math. Gen.* **1987**, *20*, L35. [CrossRef]
38. Vanderzande, C.; Igloi, F. Critical behaviour and logarithmic corrections of a quantum model with three-spin interaction. *J. Phys. A Math. Gen.* **1987**, *20*, 4539. [CrossRef]
39. Igloi, F. Conformal invariance and surface critical behaviour of a quantum chain with three-spin interaction. *J. Phys. A Math. Gen.* **1987**, *20*, 5319. [CrossRef]
40. Fernandez-Pacheco, A. Comment on the slac renormalization-group approach to the ising chain in a transverse magnetic field. *Phys. Rev. D* **1979**, *19*, 3173. [CrossRef]
41. Horn, D.; Karliner, M.; Yankielowicz, S. Self-dual renormalization group analysis of the potts models. *Nucl. Phys. B* **1980**, *170*, 467. [CrossRef]
42. Hu, B. Critical exponents of the four-state potts model. *J. Phys. A Math. Gen.* **1980**, *13*, L321. [CrossRef]
43. Sólyom, J.; Pfeuty, P. Renormalization-group study of the hamiltonian version of the potts model. *Phys. Rev. B* **1981**, *24*, 218. [CrossRef]
44. Sólyom, J. Duality of the block transformation and decimation for quantum spin systems. *Phys. Rev. B* **1981**, *24*, 230. [CrossRef]
45. Iglói, F.; Sólyom, J. Renormalization-group study of the hamiltonian version of the potts model. ii. self-dual renormalization-group treatment. *Phys. Rev. B* **1983**, *28*, 2785 . [CrossRef]

46. Iglói, F.; Sólyom, J. Phase diagram and critical properties of the (1+1)-dimensional ashkin-teller model. *J. Phys. A Math. Gen.* **1984**, *17*, 1531. [CrossRef]
47. Iglói, F. Renormalization for quantum systems: Generalized block transformations in the large-l limit. *Phys. Rev. B* **1993**, *48*, 58. [CrossRef] [PubMed]
48. Epele, L.N.; Fanchiotti, H.; Canal, C.A.G. Ferromagnetism and antiferromagnetism in the quantum ising model. *Phys. Rev. B* **1981**, *24*, 310. [CrossRef]
49. Miyazaki, R.; Nishimori, H.; Ortiz, G. Real-space renormalization group for the transverse-field ising model in two and three dimensions. *Phys. Rev. E* **2011**, *83*, 051103. [CrossRef] [PubMed]
50. Miyazaki, R.; Nishimori, H. Real-space renormalization-group approach to the random transverse-field ising model in finite dimensions. *Phys. Rev. E* **2013**, *87*, 032154. [CrossRef]
51. Monthus, C. Block renormalization for quantum ising models in dimension d = 2: Applications to the pure and random ferromagnet, and to the spin-glass. *J. Stat. Mech. Theory Exp.* **2015**, *2015*, P01023. [CrossRef]
52. Jullien, R. Transitions de phases a t=0 dans les systemes quantiques par le groupe de renormalisation dans l'espace reel. *Can. J. Phys.* **1981**, *59*, 605. [CrossRef]
53. Pfeuty, P.; Jullien, R.; Penson, K.A. Renormalization for quantum systems. In *Real-Space Renormalization*; Burkhardt, T.W., van Leeuwen, J.M.J., Eds.; Springer: Berlin/Heidelberg, Germany, 1982; pp. 119–147. [CrossRef]
54. Peschel, I. Surface magnetization in inhomogeneous two-dimensional ising lattices. *Phys. Rev. B* **1984**, *30*, 6783. [CrossRef]
55. Iglói, F.; Rieger, H. Random transverse ising spin chain and random walks. *Phys. Rev. B* **1998**, *57*, 11404. [CrossRef]
56. Griffiths, R.B. Nonanalytic behavior above the critical point in a random ising ferromagnet. *Phys. Rev. Lett.* **1969**, *23*, 17. [CrossRef]
57. McCoy, B.M. Incompleteness of the critical exponent description for ferromagnetic systems containing random impurities. *Phys. Rev. Lett.* **1969**, *23*, 383. [CrossRef]
58. Kesten, H. Random difference equations and Renewal theory for products of random matrices. *Acta Math.* **1973**, *131*, 207. [CrossRef]
59. Iglói, F.; Rieger, H. Anomalous diffusion in disordered media and random quantum spin chains. *Phys. Rev. E* **1998**, *58*, 4238. [CrossRef]
60. Iglói, F.; Juhász, R.; Lajkó, P. Griffiths-mccoy singularities in random quantum spin chains: Exact results through renormalization. *Phys. Rev. Lett.* **2001**, *86*, 1343. [CrossRef]
61. Iglói, F. Exact renormalization of the random transverse-field ising spin chain in the strongly ordered and strongly disordered griffiths phases. *Phys. Rev. B* **2022**, *65*, 064416. [CrossRef]
62. Berche, B.; Ellis, T.; Holovatch, Y.; Kenna, R. Phase transitions above the upper critical dimension. *SciPost Phys. Lect. Notes* **2022**, 60. [CrossRef]

entropy

MDPI

Article

Partition Function Zeros of the Frustrated J_1–J_2 Ising Model on the Honeycomb Lattice

Denis Gessert [1,2], Martin Weigel [3,*] and Wolfhard Janke [1]

1 Institut für Theoretische Physik, Leipzig University, IPF 231101, 04081 Leipzig, Germany; denis.gessert@itp.uni-leipzig.de (D.G.); wolfhard.janke@itp.uni-leipzig.de (W.J.)

2 Centre for Fluid and Complex Systems, Coventry University, Coventry CV1 5FB, UK; gessertd@uni.coventry.ac.uk

3 Institut für Physik, Technische Universität Chemnitz, 09107 Chemnitz, Germany

* Correspondence: martin.weigel@physik.tu-chemnitz.de

Abstract: We study the zeros of the partition function in the complex temperature plane (Fisher zeros) and in the complex external field plane (Lee–Yang zeros) of a frustrated Ising model with competing nearest-neighbor ($J_1 > 0$) and next-nearest-neighbor ($J_2 < 0$) interactions on the honeycomb lattice. We consider the finite-size scaling (FSS) of the leading Fisher and Lee–Yang zeros as determined from a cumulant method and compare it to a traditional scaling analysis based on the logarithmic derivative of the magnetization $\partial \ln\langle|M|\rangle/\partial\beta$ and the magnetic susceptibility χ. While for this model both FSS approaches are subject to strong corrections to scaling induced by the frustration, their behavior is rather different, in particular as the ratio $\mathcal{R} = J_2/J_1$ is varied. As a consequence, an analysis of the scaling of partition function zeros turns out to be a useful complement to a more traditional FSS analysis. For the cumulant method, we also study the convergence as a function of cumulant order, providing suggestions for practical implementations. The scaling of the zeros convincingly shows that the system remains in the Ising universality class for $\mathcal{R}$ as low as -0.22, where results from traditional FSS using the same simulation data are less conclusive. Hence, the approach provides a valuable additional tool for mapping out the phase diagram of models afflicted by strong corrections to scaling.

Keywords: Ising model; frustration; partition function zeros; scaling laws; critical exponents

1. Introduction

Frustrated systems [1] are characterized by interactions that cannot all be satisfied simultaneously. The resulting internal competition leads to quite interesting critical properties such as re-entrant phase behavior [2] and non-zero ground state entropy [3,4]. One of the most well-studied systems in this class is the Ising model with competing first- and second-neighbor interactions on the square lattice [5–9]. Noting that in the presence of frustration the lattice geometry is of fundamental importance for the occurrence and symmetry of ordered phases, it is somewhat surprising that much less is known about the analogous system on the honeycomb lattice, which only recently started to attract some attention [10–13]. Depending on the ratio $\mathcal{R}$ of the next-nearest-neighbor and nearest-neighbor interaction strengths, the system has a ferromagnetic ground state for $\mathcal{R} > -1/4$ or a largely degenerate ground state of known energy for $\mathcal{R} < -1/4$ [10]. The understanding of the critical properties for $\mathcal{R} \lesssim -0.2$ remains limited. Here, we focus on $\mathcal{R} > -1/4$, for which the system remains ferromagnetic at zero temperature.

Metastable states are a common feature in frustrated systems and their presence is a challenge for standard simulation techniques since runs become trapped in local minima. Particularly difficult are systems with rugged free-energy landscapes [14]. One contender among generalized-ensemble simulation techniques suitable for such problems is population annealing (PA) [15,16], which has recently shown its versatility in a range of applications [17–19]. PA is particularly well suited for studying such systems with many competing minima, as the large number of replicas allows sampling many local minima simultaneously. This is in contrast to its one-replica counterpart, equilibrium simulated annealing [20], which is much more likely to become trapped in a single local minimum and, hence, to fail to sample the equilibrium distribution.

Systems with competing interactions often are subject to strong corrections to scaling (commonly of unknown shape). For the above-mentioned frustrated Ising model on the square lattice, for example, which has been discussed since the 1970s [5], the study of the tricritical point required simulations of rather large system sizes, but notwithstanding this effort certain aspects remain unclear [6–9].

An alternate approach for studying phase transitions revolves around considering zeros of the partition function in the complex external field and temperature planes, based on the pioneering work by Lee and Yang [21,22] and by Fisher [23]. Using these zeros allows one to distinguish between first- and second-order transitions as well as to extract estimates of their strengths [24–27], and to examine the peculiar properties of a model with special boundary conditions [28,29]. The cumulative density of zeros and their impact angle onto the real temperature axis encode the strength of higher-order phase transitions [30–34]. This can also be used as a medium for deriving scaling relations among logarithmic correction exponents [35,36]. For a recent discussion along an alternative route, see Ref. [37].

For non-frustrated systems, the scaling of partition function zeros has been shown to yield quite accurate results for critical exponents, even when using rather small system sizes [38,39], suggesting that this might also be the case for more complex systems. Previous work studying partition function zeros for frustrated spin systems includes Refs. [40–43].

Despite some experimental measurements of zeros in complex external magnetic fields [44–46], the most established approach of studying complex partition function zeros requires an accurate estimate of the full density of states, which is difficult to obtain both in experiments as well as in simulations (but see Ref. [47]). For simulations, this commonly limits the maximum system size that can be studied. More recently, an alternative method for obtaining the leading partition function zero based on cumulants of the energy and magnetization distributions has gained some traction [48], thus enabling determinations of partition function zeros from more easily accessible observables in simulations. We use this method to analyze the finite-size scaling (FSS) of the zeros for system sizes exceeding those for which we could determine the full density of states.

For the frustrated Ising model on the honeycomb lattice, initial work using effective field theory (EFT) [10] suggested the existence of a tricritical point near $\mathcal{R} \approx -0.1$. This was later challenged through a Monte Carlo study [12], showing that the system remains within the Ising universality class at least down to $\mathcal{R} = -0.2$. Cluster mean-field theory [13] suggests that the transition may remain of second order down to $\mathcal{R} = -1/4$, but this has not been verified in the actual model. With this paper, we aim at a better understanding of the critical properties of this system, particularly close to the special point $\mathcal{R} = -1/4$, at which the critical temperature vanishes. To this end, we consider the scaling of the partition function zeros.

2. Materials and Methods

2.1. Model

We study the well-known two-dimensional Ising model, but placed on the honeycomb lattice and equipped with competing nearest- and next-nearest-neighbor interactions, resulting in the Hamiltonian

$$\mathcal{H} = -J_1 \sum_{\langle ij \rangle} \sigma_i \sigma_j - J_2 \sum_{[ik]} \sigma_i \sigma_k - h \sum_i \sigma_i \equiv -J_1 \Sigma_1 - J_2 \Sigma_2 - hM, \tag{1}$$

where $\langle ij \rangle$ and $[ik]$ denote sums over nearest neighbors and next-nearest neighbors, respectively, $J_1 > 0$ is the ferromagnetic nearest-neighbor interaction strength, $J_2 < 0$ is the competing antiferromagnetic next-nearest-neighbor interaction strength, and h the magnetic field. Σ_1 and Σ_2 refer to the sums over nearest- and next-nearest-neighbor interactions, respectively, and M is the (total) magnetization. The quantity relevant for the nature of the ordered phase and the transition is the ratio $\mathcal{R} = J_2/J_1$ of couplings. Here, we only consider the case of $\mathcal{R} \in (-1/4, 0]$, where the ground state is ferromagnetic [10]. We study systems of linear lattice size L, which due to the two-atom basis of the honeycomb lattice contain $N = 2L^2$ spins. Note that all simulations were carried out in the absence of an external magnetic field. We include the magnetic term here as it is necessary for the discussion of the Lee–Yang zeros.

2.2. Partition Function Zeros

In terms of the density of states $\Omega(\Sigma_1, \Sigma_2, M)$, the partition function at inverse temperature β and external magnetic field h is given by

$$\mathcal{Z}(\beta, h) = \sum_{\Sigma_1}\sum_{\Sigma_2}\sum_{M} \Omega(\Sigma_1, \Sigma_2, M) e^{\beta J_1 \Sigma_1 + \beta J_2 \Sigma_2 + \beta h M}. \tag{2}$$

We choose both J_1 and J_2 to be integers, such that for $h = 0$ the partition function is a polynomial in $x = e^{-\beta}$. For $h \neq 0$, on the other hand, the partition function may be written as a polynomial in $e^{2\beta h}$ for arbitrary choices of J_1 and J_2. The complex inverse temperatures $\{\beta_k\}$ solving the equation $\mathcal{Z}(\beta = \beta_k, h = 0) = 0$, i.e., in the absence of an external magnetic field h, are called Fisher zeros. Once calculated, the Fisher zeros will be studied as a function of $z = x^{J_1}$ to allow for better comparability of the results for different values of $\mathcal{R} = J_2/J_1$. Note that, as is well known, different variables yield very different visual impressions for the locations of zeros (see, for example, Appendix A). The pair of zeros closest to the positive real axis approaches $(\beta_c, 0) = \beta_c + 0i$ as $L \to \infty$. The real and imaginary parts of these leading zeros β_0 usually scale as [49]

$$\Re(\beta_0) - \beta_c \propto L^{-y_t}, \tag{3}$$

and

$$\Im(\beta_0) \propto L^{-y_t}, \tag{4}$$

respectively, with y_t being the renormalization group (RG) eigenvalue related to the temperature variable, which is connected to the critical exponent of the correlation length by $y_t = 1/\nu$.

The complex magnetic fields $\{h_k\}$ that solve the equation $\mathcal{Z}(\beta, h = h_k) = 0$ for some fixed β are the so-called Lee–Yang zeros that lie on the unit circle $e^{2\beta h} = e^{i\varphi}$ for the Ising ferromagnet, implying that all solutions for h_k are purely imaginary [21,22]. This circle theorem has been extended to many more models [50], but it is not universally valid, see, e.g., Refs. [51,52]. For the Ising model with competing interactions placed on a square lattice, the circle law was found to apply in the regime with ferromagnetic ground state [53]. As $L \to \infty$, the Lee–Yang zeros closest to the positive real axis approach zero. At the inverse critical temperature β_c, the imaginary part of the leading zeros h_0 scales as [49]

$$\Im(h_0) \propto L^{-y_h}, \tag{5}$$

with y_h being the RG eigenvalue related to the external magnetic field, which is connected to the standard critical exponents by $y_h = (\beta + \gamma)/\nu$.

To numerically estimate partition function zeros, writing $\beta = \Re(\beta) + i\Im(\beta)$ one notes that for zero field, $h = 0$, the Fisher zeros of $\mathcal{Z}(\beta, h = 0)$ are identical to the zeros of

$$\begin{aligned} Z(\Re(\beta), \Im(\beta)) &\equiv \frac{\mathcal{Z}(\Re(\beta) + i\Im(\beta), h = 0)}{\mathcal{Z}(\Re(\beta), h = 0)} \\ &= \langle \cos(\Im(\beta)\mathcal{H}) \rangle_{\Re(\beta), h=0} - i \langle \sin(\Im(\beta)\mathcal{H}) \rangle_{\Re(\beta), h=0}, \end{aligned} \tag{6}$$

and likewise the Lee–Yang zeros for fixed (real) β might be extracted from

$$\begin{aligned} Z_\beta(\Re(h),\Im(h)) &\equiv \frac{\mathcal{Z}(\beta,\Re(h)+i\Im(h))}{\mathcal{Z}(\beta,\Re(h))} \\ &= \langle\cos(\Im(h)M)\rangle_{\beta,\Re(h)} - i\langle\sin(\Im(h)M)\rangle_{\beta,\Re(h)}. \end{aligned} \tag{7}$$

While the evaluation of (6) and (7) and a systematic search for zeros requires the availability of the density of states Ω for reweighting [39,54–56], more recently a computationally lighter method based on cumulants of thermodynamic observables $\Phi(q)$, i.e.,

$$\langle\langle\Phi^n(q)\rangle\rangle = \frac{\partial^n}{\partial q^n}\ln\mathcal{Z}(q) \tag{8}$$

has been suggested [38,48,57–59]. Here, the notation $\Phi(q)$ refers to the thermodynamic observables $\mathcal{H}$ and M as a function of their control parameters $q=-\beta$ and $q=\beta h$, respectively, unifying the discussion of Fisher and Lee–Yang zeros. The method relies on the fact that the partition function can be factorized in a regular (non-zero) part $\tilde{\mathcal{Z}}(q)=\mathcal{Z}(0)e^{cq}$ for some constant c and the product of its complex roots in q, i.e.,

$$\mathcal{Z}(q) = \tilde{\mathcal{Z}}(q)\prod_k(1-q/q_k). \tag{9}$$

Note that by Equation (2) $\mathcal{Z}(q)$ is real for real q, and that thus the roots q_k in (9) appear in complex conjugate pairs. Plugging (9) into (8) yields the expression

$$\langle\langle\Phi^n(q)\rangle\rangle = -\sum_k\frac{(n-1)!}{(q_k-q)^n}, \quad n>1 \tag{10}$$

for the cumulants. The key point of the method is that the contribution of non-leading zeros in the expression above is suppressed by powers of n for the n-th order cumulant. Thus, one neglects the non-leading zeros, which allows the calculation of the leading partition function zeros using only the first few cumulants of the energy and magnetization, $\langle\langle\mathcal{H}^n\rangle\rangle$ and $\langle\langle M^n\rangle\rangle$, and hence does not require knowledge of Ω. Cumulants can be calculated from the central moments, $\langle\Phi^n\rangle_c = \langle(\Phi-\langle\Phi\rangle)^n\rangle$. The first four cumulants are given by

$$\langle\langle\Phi\rangle\rangle = \langle\Phi\rangle, \quad \langle\langle\Phi^2\rangle\rangle = \langle\Phi^2\rangle_c, \quad \langle\langle\Phi^3\rangle\rangle = \langle\Phi^3\rangle_c, \quad \langle\langle\Phi^4\rangle\rangle = \langle\Phi^4\rangle_c - 3\left(\langle\Phi^2\rangle_c\right)^2. \tag{11}$$

Relations for higher-order cumulants can be found using computer algebra systems. (We use the MATHEMATICA function `MomentConvert` to obtain the relations up to the 20-th cumulant. The first ten cumulants are listed in Ref. [60].) Within this framework, the leading zeros q_0 can be extracted in a vector-matrix notation from [48]

$$\begin{pmatrix} 2\,\Re(q_0-q) \\ |q_0-q|^2 \end{pmatrix} \approx \begin{pmatrix} 1 & -\frac{\mu_n^{(+)}}{n} \\ 1 & -\frac{\mu_{n+1}^{(+)}}{n+1} \end{pmatrix}^{-1} \begin{pmatrix} (n-1)\mu_n^{(-)} \\ n\,\mu_{n+1}^{(-)} \end{pmatrix}, \tag{12}$$

where n is the approximation order, and $\mu_n^{(\pm)}$ denotes the ratio of two cumulants of consecutive orders, i.e., $\mu_n^{(\pm)} \equiv \langle\langle\Phi^{n\pm1}\rangle\rangle/\langle\langle\Phi^n\rangle\rangle$. $(\Phi,q)=(E,-\beta)$ corresponds to Fisher zeros, and $(\Phi,q)=(M,\beta h)$ to Lee–Yang zeros. Since odd cumulants of M vanish for $h=0$, when setting $n=2k$ the expression for the Lee–Yang zeros simplifies to [39,59]

$$\Im(h_0) \approx \pm\frac{1}{\beta}\sqrt{2k(2k+1)\left|\frac{\langle\langle M^{2k}(0)\rangle\rangle}{\langle\langle M^{2(k+1)}(0)\rangle\rangle}\right|}. \tag{13}$$

2.3. Population Annealing

Population annealing (PA) [15,16] is a simulation scheme designed for parallel calculations for systems with complex free-energy landscapes in which the state space is sampled by a population of replicas that are cooled down collectively. It consists of alternating resampling and spin-update steps. As the temperature is lowered, the population is resampled according to the Boltzmann distribution at the new temperature [61]. Spin updates then help to equilibrate the replicas at this temperature, and to increase the diversity of the population [62]. The implementation of PA as used here can be summarized as follows:

1. Initialize the population by drawing $R_0 = R$ random spin configurations corresponding to the initial inverse temperature $\beta_0^{\mathrm{PA}} = 0$.
2. Set the iteration counter $i \leftarrow 0$.
3. Determine the next inverse temperature $\beta_{i+1}^{\mathrm{PA}}$ such that the energy histogram overlap between β_i^{PA} and $\beta_{i+1}^{\mathrm{PA}}$, given by [63]

$$\alpha(\beta_i^{\mathrm{PA}}, \beta_{i+1}^{\mathrm{PA}}) = \frac{1}{R_i} \sum_{j=1}^{R_i} \min\left(1, \frac{R e^{-(\beta_{i+1}^{\mathrm{PA}} - \beta_i^{\mathrm{PA}}) E_j}}{\sum_{k=1}^{R_i} e^{-(\beta_{i+1}^{\mathrm{PA}} - \beta_i^{\mathrm{PA}}) E_k}}\right), \tag{14}$$

   is approximately equal to the target value of α^*, where E_j refers to the current energy of replica j.
4. Increment i by 1.
5. Resample the replicas according to their relative Boltzmann weights, that is, make on average

$$\tau(E_j) = \frac{R e^{-(\beta_{i+1}^{\mathrm{PA}} - \beta_i^{\mathrm{PA}}) E_j}}{\sum_{k=1}^{R_i} e^{-(\beta_{i+1}^{\mathrm{PA}} - \beta_i^{\mathrm{PA}}) E_k}} \tag{15}$$

   copies of replica j with energy E_j.
6. Carry out Metropolis updates on the replicas until the effective population size R_{eff} (see [62] for definition and discussion) exceeds the threshold value of $R^* = \rho^* R$.
7. Calculate estimates for observables $\mathcal{O}$ as the population averages $\sum_j \mathcal{O}_j / R_i$, where $\mathcal{O}_j$ is the value of the observable for the j-th replica. Note that we calculate central moments of the energy directly during the simulation after calculating the average energy, because using raw moments to calculate higher-order central moments and cumulants leads to a complete loss of numeric precision in the latter.
8. Unless the lowest temperature of interest is reached, go to step 3.

Some comments are in order at this point. The above implementation contains numerous parameters of relevance to the performance of the algorithm, namely, the population size R, the target energy distribution overlap α^*, and the sweep schedule given by the threshold value ρ^* for the relative effective population size. These have been (resp. will be) discussed elsewhere [61,62,64], and here we choose R = 20,000 throughout, α^* is set to at least 80%, in some cases 90%, and $\rho^* \geq 90\%$. There are in fact many possibilities of how the resampling step can be realized. Here, we use the so-called nearest-integer resampling which has been shown to lead to optimal results in many scenarios [61]. For the spin updates we employ the Metropolis method implemented on a GPU, drawing on a domain decomposition [63] into four sublattices, adapting and extending the publicly available code of Ref. [63]. For small system sizes, we obtain an estimate for the density of states by using multi-histogram reweighting, also implemented in the source code of Ref. [63].

3. Results

3.1. Solving for all Fisher Zeros in the Complex Temperature Plane

For finite systems, the partition function for $h = 0$ is a polynomial of finite order in $x = e^{-\beta}$ (for suitable choices of J_1 and J_2). In principle, once the density of states $\Omega(E)$ is known one can solve for all partition function zeros numerically. Due to the presence of both $J_1\Sigma_1$ and $J_2\Sigma_2$ in the Hamiltonian, there are more than the usual $\mathcal{O}(N)$ distinct

function of n, which appear to decay exponentially. This observation is in line with the fact that the contributions of sub-leading partition function zeros are suppressed with power n, see Equation (10).

Next, we repeat the same analysis with data for the density of states obtained by MHR of PA data with system size $L = 16$. For each value of $\mathcal{R}$ we carried out an independent simulation. Although reweighting in J_2 is in principle possible, the reweighting range is rather limited. The results of this exercise are shown in Figure 3, and they are found to be qualitatively quite similar to the previous case. As is to be expected, the leading Fisher zero is closer to the real axis as compared to $L = 4$. As before, the cumulants are evaluated at $\beta = \Re(\beta_0^{(d)})$ from the direct method, which is possible thanks to the estimate of the density of states Ω from MHR. In particular, the exponential decay of the differences between the cumulant and direct method is more clearly visible in this case. Note that the bottom row only shows the systematic deviation of the two methods when using the same data for $\Omega(\Sigma_1, \Sigma_2)$ (subject to statistical errors), and not the actual error for β_0. For an estimate of the statistical errors encountered in the simulation, see the error bars in Figure A2. This demonstrates that the cumulant method is a viable replacement for the direct method whenever the statistical error exceeds the systematic deviation shown above (which it typically does). However, it does not say anything about the actual accuracy of the obtained results. Also, note that even on the logarithmic scale, the difference decays monotonously and no noise is visible even when using higher-order cumulants. This is because despite the fact that higher-order cumulants are noisy, the fluctuations in their ratios do not increase noticeably with n, which is due to the cross-correlation between the terms.

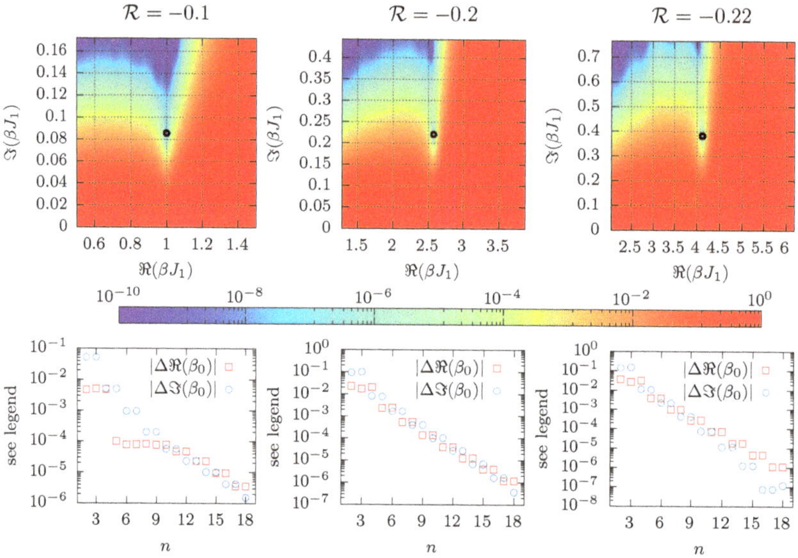

Figure 3. Same as Figure 2 for $L = 16$, but using PA simulation data instead of exact enumeration

3.3. *Determining the Leading Lee–Yang Zero Directly and by the Cumulant Method*

We now turn to an analysis of the Lee–Yang zeros. We consider the partition function in the complex-field plane at our best estimate for the infinite-volume critical temperature for different values of $\mathcal{R}$ (see Section 3.4 for details). Analogous to the Fisher zeros discussed in the previous section, Lee–Yang zeros are obtained using the direct method (utilizing the LM algorithm). As their calculation requires the density of states $\Omega(\Sigma_1, \Sigma_2, M)$, we only compute them for $L = 4$.

Similar to Figure 2, the top panel of Figure 4 shows the absolute value of the partition function in the complex field plane at the infinite-volume inverse critical temperature β_c, i.e., $|Z_{\beta_c}(\Re(h), \Im(h))|$. The found zeros are consistent with being purely imaginary, suggesting that the Lee–Yang circle theorem may also hold for the model considered here. (Note that we are unaware of any rigorous proof of the circle theorem for the model at hand.) We again also considered the alternative approach provided by the cumulant method, and the bottom panel of Figure 4 depicts the deviation of the estimate of the cumulant method from the zero determined via the direct method as a function of n. As was the case for the Fisher zeros, the deviation vanishes exponentially in n. Note that the range of the real part of the external magnetic field is the same in all three panels. For smaller values of $\mathcal{R}$, the minimum of $|Z_{\beta_c}(h)|$ becomes broader, making it more difficult to find the root numerically. The imaginary part of the leading Lee–Yang zero vanishes as $\mathcal{R}$ approaches $-1/4$.

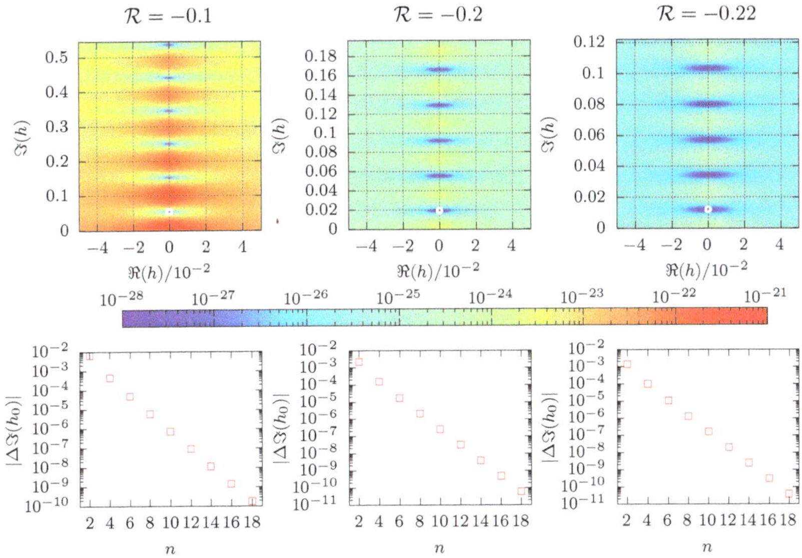

Figure 4. The top row shows $|Z_{\beta_c}(\Re(h), \Im(h))|$ for complex h, as obtained from Equation (7) for $\mathcal{R} \in \{-0.1, -0.2, -0.22\}$, calculated by exact enumeration for $L = 4$ at β_c. For β_c we used our best estimate, see Section 3.4. The leading Lee–Yang zero is denoted by an open white circle. The bottom row shows the difference in $\Im(h_0)$ for the approximation given by Equation (13) of order n and the directly obtained value.

3.4. Comparison of Standard FSS and Scaling of Partition Function Zeros

One recently proposed advantage of using the partition function zeros to obtain critical exponents is that already rather small system sizes may yield quite accurate estimates for the exponents [39]. In the following, we present tables for different values of $\mathcal{R}$ and different fit intervals from $L_{\min}$ to $L_{\max}$, thus clearly demonstrating for which ranges of system sizes reliable FSS fits can be performed. We consider the system sizes $L \in \{8, 16, 24, 32, 48, 64, 88\}$ and values of $\mathcal{R}$ equal to -0.1, -0.2, -0.21, and -0.22. As the results for $\mathcal{R} = -0.21$ are analogous to the other values of $\mathcal{R}$, they are only included in the Supplementary Materials, and not presented in the main text. For every system size L and for every value of $\mathcal{R}$, ten independent PA runs were carried out in order to improve statistics and to obtain error bars. The first run determined the temperature set for the remaining runs. In principle, the energy histogram overlap defines the temperature set uniquely. However, the actually determined temperatures by PA are subject to statistical fluctuations. Thus, to avoid every realization having its own temperatures, the first run is used to fix the temperature set.

The same simulation data were used for the comparison of standard FSS and the scaling of partition function zeros obtained via the cumulant method using up to the 20-th cumulant (corresponding to $n = 18$ in Equation (12) for Fisher zeros, and $k = 9$ in Equation (13) for Lee–Yang zeros). As shown above, we did not observe any loss in numeric precision using higher-order cumulants. Therefore, we use the largest cumulant that we measured; see also Appendix B.

3.4.1. Scaling of Fisher Zeros

Table 1 summarizes the FSS results for $\mathcal{R} = -0.1$, -0.2, and -0.22 using the leading Fisher zero as well as the logarithmic derivative of the magnetization,

$$\mathrm{dln}|m| \equiv \frac{\partial}{\partial\beta}\ln\langle|M|\rangle = \frac{\langle|M|\mathcal{H}\rangle}{\langle|M|\rangle} - \langle\mathcal{H}\rangle, \tag{16}$$

whose maximum follows the FSS relation $\mathrm{dln}|m|_{\max}(L) \propto L^{y_t}$ [67]. As the precise inverse temperature at which $\Re(\beta - \beta_0)$ is zero is not in general contained in the annealing schedule of PA, we first determine the zero crossing in the estimate for $\Re(\beta - \beta_0)$ with positive slope closest to the location of the local maximum of the second energy cumulant (ignoring the $\beta = 0$ maximum). Next, we determine the β for which the linear interpolation of $\Re(\beta - \beta_0)$ of Equation (12) between the two closest inverse temperatures is zero. $\Im(\beta_0)$ is also obtained through linear interpolation to the same β which we found to be more stable than to evaluate $\Re(\beta_0)$ and $\Im(\beta_0)$ directly using Equation (12). This is subsequently used for FSS. Details from all fits yielding estimates for y_t can be found in Section 2.1 of the Supplementary Materials. The expected exponent is the Onsager value, $y_t = 1$. The estimates y_t^{F} and $y_t^{\mathrm{dln}|m|}$ for y_t resulting from both methods are close to the expected value for the considered ranges of system sizes for all $\mathcal{R}$, albeit not always within error bars.

For $\mathcal{R} = -0.1$ it comes as a surprise that the values from standard FSS using $\mathrm{dln}|m|$ are closer to 1 and appear to have weaker corrections to scaling. Specifically, the value obtained for $y_t^{\mathrm{dln}|m|}$ is within 0.5% of the expected value for all fit ranges, whereas the value for y_t^{F} differs by as much as 3% when using the fitting range 8–24 for L. While when $L_{\min} = 8$ the value for y_t using the leading Fisher zero is far below 1, it is consistent with 1 for larger $L_{\min}$, suggestive of the differing value for smaller system sizes being due to stronger corrections to scaling.

In contrast, for $\mathcal{R} = -0.2$ for almost any (fixed) fitting range the two estimates are compatible with each other within error bars, and the error bars are comparable in magnitude. Most of the y_t estimates fall well below the expected value $y_t = 1$, and increase with both $L_{\min}$ and $L_{\max}$, being compatible with 1 only for fitting ranges limited to the largest system sizes studied here. This effective variation in the exponent with L is also reflected in the poor quality of fit: In most ranges $[L_{\min}, L_{\max}]$ the Q value falls below 0.1 [68], where Q refers to the probability of drawing a χ^2 from the χ^2 distribution that is even larger than the value calculated from the fit. Unusually small values for Q correspond to poor quality of fit.

For $\mathcal{R} = -0.22$, the estimates for y_t are even further below the Onsager value of $y_t = 1$, and again increase with $L_{\min}$ and $L_{\max}$, suggestive of the presence of strong corrections to scaling. Only in the range $[48, 88]$ is the result within error bars of the Onsager value. Here too, the effectively changing exponent is reflected in poor fit qualities. Differently from the previous cases, however, the value for y_t from the Fisher zeros is consistently closer to the expected value than the value from regular FSS, suggesting that corrections to scaling are weaker for the location of the Fisher zeros in this case. The overall worse fit quality for $\mathrm{dln}|m|$ is also in agreement with stronger correction terms.

Table 1. Estimates for y_t from FSS fits using the imaginary part of the leading Fisher zero as well as $\mathrm{d}\ln|m|$ for different $L_{\min}$ and $L_{\max}$, and using $\mathcal{R} = -0.1$, -0.2, and -0.22.

$\mathbf{L}_{\max}$ / $\mathbf{L}_{\min}$	24	32	48	64	88
			$\mathcal{R} = -0.1$		
8	$y_t^{\mathrm{F}} = 0.9668(61)$ $y_t^{\mathrm{d}\ln\lvert m\rvert} = 0.9957(24)$	$y_t^{\mathrm{F}} = 0.9667(48)$ $y_t^{\mathrm{d}\ln\lvert m\rvert} = 0.9943(17)$	$y_t^{\mathrm{F}} = 0.9796(31)^{\dagger}$ $y_t^{\mathrm{d}\ln\lvert m\rvert} = 0.9941(15)$	$y_t^{\mathrm{F}} = 0.9817(29)^{\dagger}$ $y_t^{\mathrm{d}\ln\lvert m\rvert} = 0.9939(13)$	$y_t^{\mathrm{F}} = 0.9829(29)^{\dagger}$ $y_t^{\mathrm{d}\ln\lvert m\rvert} = 0.9941(11)$
16	–	$y_t^{\mathrm{F}} = 0.972(15)$ $y_t^{\mathrm{d}\ln\lvert m\rvert} = 0.9867(96)$	$y_t^{\mathrm{F}} = 1.0001(79)^{\dagger}$ $y_t^{\mathrm{d}\ln\lvert m\rvert} = 0.9908(51)$	$y_t^{\mathrm{F}} = 1.0026(70)$ $y_t^{\mathrm{d}\ln\lvert m\rvert} = 0.9916(35)$	$y_t^{\mathrm{F}} = 1.0052(68)^{\dagger}$ $y_t^{\mathrm{d}\ln\lvert m\rvert} = 0.9932(23)$
24	–	–	$y_t^{\mathrm{F}} = 1.017(13)$ $y_t^{\mathrm{d}\ln\lvert m\rvert} = 0.9908(51)$	$y_t^{\mathrm{F}} = 1.018(11)$ $y_t^{\mathrm{d}\ln\lvert m\rvert} = 0.9916(35)$	$y_t^{\mathrm{F}} = 1.022(11)$ $y_t^{\mathrm{d}\ln\lvert m\rvert} = 0.9933(23)$
32	–	–	–	$y_t^{\mathrm{F}} = 1.042(21)$ $y_t^{\mathrm{d}\ln\lvert m\rvert} = 0.9935(50)$	$y_t^{\mathrm{F}} = 1.049(19)$ $y_t^{\mathrm{d}\ln\lvert m\rvert} = 0.9948(31)$
48	–	–	–	–	$y_t^{\mathrm{F}} = 1.048(38)$ $y_t^{\mathrm{d}\ln\lvert m\rvert} = 0.9959(65)$
			$\mathcal{R} = -0.2$		
8	$y_t^{\mathrm{F}} = 0.9215(35)$ $y_t^{\mathrm{d}\ln\lvert m\rvert} = 0.9211(27)^{\dagger}$	$y_t^{\mathrm{F}} = 0.9263(31)^{\dagger}$ $y_t^{\mathrm{d}\ln\lvert m\rvert} = 0.9284(18)^{\dagger}$	$y_t^{\mathrm{F}} = 0.9369(26)^{\dagger}$ $y_t^{\mathrm{d}\ln\lvert m\rvert} = 0.9344(15)^{\dagger}$	$y_t^{\mathrm{F}} = 0.9400(23)^{\dagger}$ $y_t^{\mathrm{d}\ln\lvert m\rvert} = 0.9380(14)^{\dagger}$	$y_t^{\mathrm{F}} = 0.9453(21)^{\dagger}$ $y_t^{\mathrm{d}\ln\lvert m\rvert} = 0.9484(10)^{\dagger}$
16	–	$y_t^{\mathrm{F}} = 0.9500(86)$ $y_t^{\mathrm{d}\ln\lvert m\rvert} = 0.9493(47)$	$y_t^{\mathrm{F}} = 0.9663(54)^{\dagger}$ $y_t^{\mathrm{d}\ln\lvert m\rvert} = 0.9581(34)^{\dagger}$	$y_t^{\mathrm{F}} = 0.9639(42)^{\dagger}$ $y_t^{\mathrm{d}\ln\lvert m\rvert} = 0.9625(29)^{\dagger}$	$y_t^{\mathrm{F}} = 0.9702(37)^{\dagger}$ $y_t^{\mathrm{d}\ln\lvert m\rvert} = 0.9707(19)^{\dagger}$
24	–	–	$y_t^{\mathrm{F}} = 0.994(12)$ $y_t^{\mathrm{d}\ln\lvert m\rvert} = 0.9694(64)^{\dagger}$	$y_t^{\mathrm{F}} = 0.9780(83)$ $y_t^{\mathrm{d}\ln\lvert m\rvert} = 0.9737(48)$	$y_t^{\mathrm{F}} = 0.9872(68)^{\dagger}$ $y_t^{\mathrm{d}\ln\lvert m\rvert} = 0.9788(26)$
32	–	–	–	$y_t^{\mathrm{F}} = 0.968(15)^{\dagger}$ $y_t^{\mathrm{d}\ln\lvert m\rvert} = 0.9829(65)$	$y_t^{\mathrm{F}} = 0.988(11)^{\dagger}$ $y_t^{\mathrm{d}\ln\lvert m\rvert} = 0.9833(32)$
48	–	–	–	–	$y_t^{\mathrm{F}} = 0.990(20)^{\dagger}$ $y_t^{\mathrm{d}\ln\lvert m\rvert} = 0.9838(67)$
			$\mathcal{R} = -0.22$		
8	$y_t^{\mathrm{F}} = 0.9328(30)^{\dagger}$ $y_t^{\mathrm{d}\ln\lvert m\rvert} = 0.8781(25)^{\dagger}$	$y_t^{\mathrm{F}} = 0.9356(23)^{\dagger}$ $y_t^{\mathrm{d}\ln\lvert m\rvert} = 0.8907(19)^{\dagger}$	$y_t^{\mathrm{F}} = 0.9364(21)^{\dagger}$ $y_t^{\mathrm{d}\ln\lvert m\rvert} = 0.9019(15)^{\dagger}$	$y_t^{\mathrm{F}} = 0.9410(18)^{\dagger}$ $y_t^{\mathrm{d}\ln\lvert m\rvert} = 0.9112(13)^{\dagger}$	$y_t^{\mathrm{F}} = 0.9461(15)^{\dagger}$ $y_t^{\mathrm{d}\ln\lvert m\rvert} = 0.9167(12)^{\dagger}$
16	–	$y_t^{\mathrm{F}} = 0.9525(76)$ $y_t^{\mathrm{d}\ln\lvert m\rvert} = 0.9302(42)$	$y_t^{\mathrm{F}} = 0.9513(63)$ $y_t^{\mathrm{d}\ln\lvert m\rvert} = 0.9365(28)$	$y_t^{\mathrm{F}} = 0.9588(45)$ $y_t^{\mathrm{d}\ln\lvert m\rvert} = 0.9431(22)^{\dagger}$	$y_t^{\mathrm{F}} = 0.9644(32)$ $y_t^{\mathrm{d}\ln\lvert m\rvert} = 0.9484(19)^{\dagger}$
24	–	–	$y_t^{\mathrm{F}} = 0.948(13)$ $y_t^{\mathrm{d}\ln\lvert m\rvert} = 0.9437(54)$	$y_t^{\mathrm{F}} = 0.9634(71)$ $y_t^{\mathrm{d}\ln\lvert m\rvert} = 0.9525(36)^{\dagger}$	$y_t^{\mathrm{F}} = 0.9697(46)$ $y_t^{\mathrm{d}\ln\lvert m\rvert} = 0.9600(30)^{\dagger}$
32	–	–	–	$y_t^{\mathrm{F}} = 0.971(11)$ $y_t^{\mathrm{d}\ln\lvert m\rvert} = 0.9612(53)$	$y_t^{\mathrm{F}} = 0.9766(68)$ $y_t^{\mathrm{d}\ln\lvert m\rvert} = 0.9702(42)^{\dagger}$
48	–	–	–	–	$y_t^{\mathrm{F}} = 0.995(16)$ $y_t^{\mathrm{d}\ln\lvert m\rvert} = 0.9905(84)$

† Q value below 0.1.

3.4.2. Scaling of Lee–Yang Zeros

In the following, we obtain y_h from the scaling behavior (see Equation (5)) of the leading Lee–Yang zero at the (fixed) inverse temperature β_c, as well as from the value of the magnetic susceptibility χ at β_c, which follows the scaling relation

$$\chi_L(\beta_c) = L^{-D}\beta_c\langle M^2\rangle_{\beta_c} \propto L^{-D+2y_h} = L^{\gamma/\nu}, \tag{17}$$

with D being the spatial dimension. In principle, one may also use the pseudo-critical points of the magnetic susceptibility with subtraction and its peak heights, which follow the same scaling relation. However, as we use the critical temperature for the Lee–Yang zeros, this would result in an unfair comparison and the values from the magnetic susceptibility by design would be subject to stronger corrections to scaling.

We consider the Lee–Yang zeros at our estimate for the inverse critical temperature β_c obtained from the scaling of the real part of the Fisher zeros for the different values of $\mathcal{R}$. This estimate is found by applying the fit ansatz $\Re(\beta_0(L)) = \beta_c - aL^{-y_t} - bL^{-2y_t}$, allowing for a first-order correction term and assuming $y_t = 1$. Using the full range of system sizes, i.e., $L = 8, \ldots, 88$, we obtain $\beta_c J_1 = 1.02350(32), 2.65740(54), 3.25721(42), 4.27017(59)$ for $\mathcal{R} = -0.1$, -0.2, -0.21, and -0.22, respectively (cf. Supplementary Tables S9–S12). The Lee–Yang zeros $\Im(h_0)$ and the susceptibility χ_L together with their statistical errors are then evaluated at this value for β_c. Note that as the critical temperature was calculated a posteriori, we have no data for the cumulants at β_c. In order to estimate the value of the cumulants at β_c, we use Lagrange interpolation between the four inverse temperatures closest to β_c, potentially resulting in a small systematic error. As all realizations use the same temperature set, the effect of this is not accounted for in the quoted error bars. Next, one carries out the FSS analysis on the calculated values for the Lee–Yang zeros and the susceptibility at the obtained β_c. In this process the statistical errors in $\Im(h_0)$ and χ_L are correctly reflected in the statistical error of the critical exponent y_h, whereas the uncertainty in β_c is not accounted for. To estimate the influence of the error of β_c on the estimate for y_h, one may repeat this analysis at $\beta_c - \epsilon(\beta_c)$ and $\beta_c + \epsilon(\beta_c)$, with $\epsilon(\beta_c)$ corresponding to the quoted error bars. This yields estimates $y_h^{(-)}$ and $y_h^{(+)}$, thus giving rise to a second contribution $|y_h^{(+)} - y_h^{(-)}|/2$ to the error of y_h. Alternatively, to combine both error contributions in the quoted error bars, we carry out jackknifing [69] over the whole FSS procedure. Specifically, we use ten jackknife blocks (each containing nine out of ten PA simulation runs), for which we then perform the entire analysis, resulting in slightly varying values for the Fisher zeros and the inverse critical temperatures. The Lee–Yang zeros and the magnetic susceptibilities are analyzed at the respective β_c of each jackknife block, and finally, result in different estimates for the exponent y_h. The final estimates of y_h quoted in Table 2 and Supplementary Tables S13–S20 are then the plain averages of the jackknife blocks.Their standard errors are calculated via the usual rescaled variance of the jackknife blocks, which accounts for their trivial correlation [69,70]. For further details including results for all fitting parameters, see Sections S2.2 and S2.3 of the Supplementary Materials. We have checked that the jackknife estimates for $\beta_c J_1$ ($\beta_c J_1 = 1.02351(22), 2.65740(68), 3.25722(84)$, and $4.27017(48)$) are in very good agreement with our aforementioned final values.

The expected value for y_h is the $D = 2$ Ising value $y_h = 1.875$. Table 2 summarizes the FSS results using the leading Lee–Yang zero and the magnetic susceptibility for $\mathcal{R} = -0.1$, -0.2, and -0.22. As before, the results are affected by corrections to scaling reflected in effectively varying exponents that approach the expected value as $L_{\min}$ and $L_{\max}$ increase. Both methods yield values for y_h well compatible with the expected exponent of 1.875.

For $\mathcal{R} = -0.1$, the estimate y_h^{LY} for y_h from the scaling of the Lee–Yang zeros even on the smallest range of system sizes, i.e., $L \in \{8, 16, 24\}$, is within two error bars of 1.875, as opposed to the estimate y_h^{χ} from the magnetic susceptibility, which is far outside the error margin. This indicates stronger corrections to scaling for the magnetic susceptibility, which is also reflected in the overall poorer quality-of-fit value Q. Despite the big difference when including the value for $L = 8$, for $L_{\min} > 8$ the value for y_h from the Lee–Yang zeros is only marginally closer to the Ising value as compared to the results from the scaling of the magnetic susceptibility.

Also, for $\mathcal{R} = -0.2$ the difference between the methods shows most clearly when including the value for $L = 8$. Here, the estimate from the partition function zeros is much closer to the expected one, albeit still not within error bars. When choosing $L_{\min} = 16$, the value from the Lee–Yang zeros is always within at most two error bars of the expected value, which is not the case for the value from the magnetic susceptibility. However, this observation may not be significant as both values are within the error bars. For $L \geq 24$, both methods yield values compatible with each other and with 1.875. Similarly, for $\mathcal{R} = -0.22$ the Lee–Yang value for y_h is much closer to 1.875 (but again not within error bars) than the magnetic susceptibility one when including $L = 8$. When excluding the smallest system size, both methods yield results compatible with the Ising exponent, and both methods appear to perform equally well.

Table 2. Jackknife estimates at β_c (see text) for y_h from FSS fits using the leading Lee–Yang zero as well as from the value of the magnetic susceptibility χ_L at β_c for different $L_{\min}$ and $L_{\max}$, and using $\mathcal{R} = -0.1$, -0.2, and -0.22.

$\mathbf{L}_{\max}$ $\mathbf{L}_{\min}$	24	32	48	64	88
			$\mathcal{R} = -0.1$		
8	$y_h^{LY} = 1.87629(97)$ $y_h^{\chi} = 1.87813(86)$	$y_h^{LY} = 1.87564(69)$ $y_h^{\chi} = 1.87730(57)^{\dagger}$	$y_h^{LY} = 1.8756(13)$ $y_h^{\chi} = 1.8769(11)^{\dagger}$	$y_h^{LY} = 1.8754(20)$ $y_h^{\chi} = 1.8764(17)^{\dagger}$	$y_h^{LY} = 1.8754(25)$ $y_h^{\chi} = 1.8763(21)^{\dagger}$
16	–	$y_h^{LY} = 1.8747(13)$ $y_h^{\chi} = 1.8755(11)^{\dagger}$	$y_h^{LY} = 1.8751(23)$ $y_h^{\chi} = 1.8757(20)$	$y_h^{LY} = 1.8750(30)$ $y_h^{\chi} = 1.8754(26)$	$y_h^{LY} = 1.8751(35)$ $y_h^{\chi} = 1.8755(31)$
24	–	–	$y_h^{LY} = 1.8745(38)$ $y_h^{\chi} = 1.8751(34)$	$y_h^{LY} = 1.8747(42)$ $y_h^{\chi} = 1.8750(37)$	$y_h^{LY} = 1.8750(46)$ $y_h^{\chi} = 1.8753(40)$
32	–	–	–	$y_h^{LY} = 1.8758(49)$ $y_h^{\chi} = 1.8759(42)$	$y_h^{LY} = 1.8757(50)$ $y_h^{\chi} = 1.8759(43)$
48	–	–	–	–	$y_h^{LY} = 1.8752(49)$ $y_h^{\chi} = 1.8754(43)$
			$\mathcal{R} = -0.2$		
8	$y_h^{LY} = 1.8780(13)$ $y_h^{\chi} = 1.8828(13)^{\dagger}$	$y_h^{LY} = 1.8772(16)$ $y_h^{\chi} = 1.8813(14)^{\dagger}$	$y_h^{LY} = 1.8771(18)$ $y_h^{\chi} = 1.8805(16)^{\dagger}$	$y_h^{LY} = 1.8769(20)$ $y_h^{\chi} = 1.8801(17)^{\dagger}$	$y_h^{LY} = 1.8767(30)^{\dagger}$ $y_h^{\chi} = 1.8789(28)^{\dagger}$
16	–	$y_h^{LY} = 1.8750(26)$ $y_h^{\chi} = 1.8767(23)$	$y_h^{LY} = 1.8757(28)$ $y_h^{\chi} = 1.8767(24)$	$y_h^{LY} = 1.8757(32)$ $y_h^{\chi} = 1.8765(28)$	$y_h^{LY} = 1.8759(44)$ $y_h^{\chi} = 1.8763(38)$
24	–	–	$y_h^{LY} = 1.8759(35)$ $y_h^{\chi} = 1.8761(30)$	$y_h^{LY} = 1.8757(38)$ $y_h^{\chi} = 1.8759(33)$	$y_h^{LY} = 1.8760(48)$ $y_h^{\chi} = 1.8761(42)$
32	–	–	–	$y_h^{LY} = 1.8765(41)$ $y_h^{\chi} = 1.8765(36)$	$y_h^{LY} = 1.8763(52)$ $y_h^{\chi} = 1.8763(46)$
48	–	–	–	–	$y_h^{LY} = 1.8758(66)$ $y_h^{\chi} = 1.8759(57)$
			$\mathcal{R} = -0.22$		
8	$y_h^{LY} = 1.88294(89)$ $y_h^{\chi} = 1.88778(65)^{\dagger}$	$y_h^{LY} = 1.8818(11)^{\dagger}$ $y_h^{\chi} = 1.8860(11)^{\dagger}$	$y_h^{LY} = 1.8808(12)^{\dagger}$ $y_h^{\chi} = 1.8845(12)^{\dagger}$	$y_h^{LY} = 1.8804(10)^{\dagger}$ $y_h^{\chi} = 1.88300(95)^{\dagger}$	$y_h^{LY} = 1.8803(10)^{\dagger}$ $y_h^{\chi} = 1.88288(95)^{\dagger}$
16	–	$y_h^{LY} = 1.8777(22)$ $y_h^{\chi} = 1.8791(19)$	$y_h^{LY} = 1.8766(16)$ $y_h^{\chi} = 1.8777(16)$	$y_h^{LY} = 1.8781(14)$ $y_h^{\chi} = 1.8786(13)$	$y_h^{LY} = 1.8781(13)$ $y_h^{\chi} = 1.8786(12)$
24	–	–	$y_h^{LY} = 1.8752(17)$ $y_h^{\chi} = 1.8758(16)$	$y_h^{LY} = 1.8779(18)$ $y_h^{\chi} = 1.8781(15)$	$y_h^{LY} = 1.8780(17)$ $y_h^{\chi} = 1.8782(15)$
32	–	–	–	$y_h^{LY} = 1.8792(23)$ $y_h^{\chi} = 1.8792(19)$	$y_h^{LY} = 1.8793(21)$ $y_h^{\chi} = 1.8792(18)$
48	–	–	–	–	$y_h^{LY} = 1.8827(36)$ $y_h^{\chi} = 1.8817(30)$

† Q value below 0.1.

4. Conclusions

We have studied the Fisher and Lee–Yang zeros for the frustrated J_1–J_2 Ising model on the honeycomb lattice. The partition function zeros are obtained using a recently suggested cumulant method [38,48,57,58] that does not require knowledge of the density of states Ω. For small systems where Ω was available, we compared the values for the leading Fisher and Lee–Yang zeros from the cumulant method against the directly obtained estimates and observed only small deviations that vanish exponentially in the cumulant order n, regardless of the value of $\mathcal{R}$. For larger systems, we also saw an exponential convergence of the cumulant estimates to their asymptotic values when evaluating the cumulants at β close to $\Re(\beta_0)$.

We compared FSS using the location of the leading partition function zeros with a traditional FSS protocol. Both approaches indicate that the model remains in the Ising universality class for all studied values of $\mathcal{R} > -1/4$. For the temperature exponent y_t, our numerical results do not favor one method over the other. Instead, both approaches seem to be subject to non-trivial corrections to scaling, such that dependent on $\mathcal{R}$ one or the other approach appears preferable. For $\mathcal{R} = -0.1$ conventional FSS shows practically no signs of corrections to scaling even for very small systems, whereas the values obtained for y_t from the partition function zeros have a clear system-size dependence. On the other hand, for values of $\mathcal{R}$ closer to $-1/4$, conventional FSS is subject to very strong corrections to scaling, whereas the values obtained using the partition function zeros show only slightly stronger corrections as compared to $\mathcal{R} = -0.1$. Therefore, our data for the partition function zeros for $\mathcal{R} = -0.22$ convincingly indicate that the system remains in the Ising universality class, whereas the results from traditional FSS alone are much less conclusive. The field exponent y_h, as obtained from the Lee–Yang zeros, in most cases was marginally closer to the expected value than the estimate derived from the magnetic susceptibility, although only when including the smallest system size of $L = 8$ the former approach performed significantly better than the latter.

Thus, based on our results, studying the critical behavior using the partition function zeros does not in general promise to yield results less afflicted by scaling corrections but, as expected, in different regimes one or the other approach might have an edge in this respect. Both techniques (as well as other scaling paradigms) can hence be used with good success in a complementary fashion.

Supplementary Materials: The following supporting information can be downloaded at: https://www.mdpi.com/article/10.3390/e26110919/s1, it contains a video illustrating the motion of Fisher zeros on varying $\mathcal{R}$ and tables for the finite-size scaling studies.

Author Contributions: Conceptualization, D.G. and W.J.; methodology, D.G.; software, D.G.; validation, W.J. and M.W.; formal analysis, D.G.; writing—original draft preparation, D.G.; writing—review and editing, M.W. and W.J.; visualization, D.G.; supervision, M.W. and W.J.; project administration, W.J.; funding acquisition, W.J. All authors have read and agreed to the published version of the manuscript.

Funding: This project was supported by the Deutsch-Französische Hochschule (DFH-UFA) through the Doctoral College "$\mathbb{L}^4$" under Grant No. CDFA-02-07. We further acknowledge the resources provided by the Leipzig Graduate School of Natural Sciences "BuildMoNa".

Data Availability Statement: The data that support the findings of this study are available from the corresponding author upon reasonable request.

Acknowledgments: We thank Leïla Moueddene for interesting discussions. This work is dedicated to the memory of Ralph Kenna who passed away at the end of October 2023. Ralph was fascinated by partition function zeros since his PhD Thesis work [54,55] and revisited this versatile tool for obtaining properties of phase transitions during his scientific career many times [39]. We will remember Ralph as good friend and inspiring collaborator. His deep insights, witty comments, and captivating presentations will be dearly missed.

Conflicts of Interest: The authors declare no conflicts of interest. The funders had no role in the design of the study; in the collection, analyses, or interpretation of data; in the writing of the manuscript; or in the decision to publish the results.

Abbreviations

The following abbreviations are used in this manuscript:

MC	Monte Carlo
PA	Population annealing
LY	Lee–Yang
F	Fisher
FSS	Finite-size scaling
RG	Renormalization group

Appendix A. Fisher Zeros for $\mathcal{R} = 0$

As a consistency check, we carried out the partition function zero analysis for $\mathcal{R} = 0$ in the complex temperature plane, where exact Fisher zeros are known in the thermodynamic limit [71,72]. The Fisher zeros of the field-free Ising model on the honeycomb lattice without next-nearest-neighbor interactions lie on the partial circle $z^2 = e^{i\varphi}$ for $\varphi \in [\pi/3, 5\pi/3]$, and the heart-shape-like curve given by [72]

$$\left[\Im(z^2)\right]^2 = 1 + 2\Re(z^2) - \left[\Re(z^2)\right]^2 \pm \sqrt{8\Re(z^2)}. \tag{A1}$$

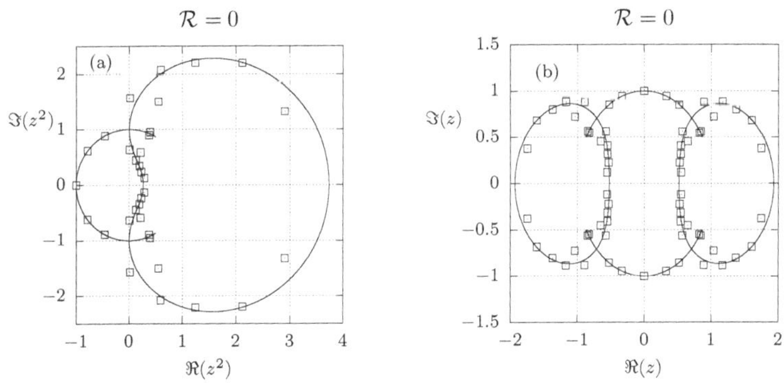

Figure A1. Fisher zeros in the complex temperature planes (**a**) $z^2 = e^{-2\beta J_1}$ and (**b**) $z = e^{-\beta J_1}$ for $\mathcal{R} = 0$. Points denote our data for $L = 4$ obtained by exact enumeration. The solid lines are the exact zeros for $L \to \infty$ [71,72].

Figure A1 shows the Fisher zeros for $L = 4$ (points) and for $L \to \infty$ (solid lines). Despite the rather small system size, both appear to be already somewhat compatible, with the closest zero for the finite system size, of course, not being located on the positive real axis. As in the absence of next-nearest-neighbor interactions, the partition function is an even polynomial in z, the natural variable to consider the Fisher zeros is z^2 [see panel (a)]. When using z as variable, the $z \to -z$ symmetry is visible [see panel (b)], which was absent for $\mathcal{R} \neq 0$; cf. Figure 1.

Appendix B. Convergence of the Cumulant Method for Larger System Sizes

We have tested the convergence of the cumulant method for small system sizes extensively to assure its convergence. For larger system sizes we did not have access to the density of states. To test the convergence for these larger system sizes, we therefore consider the predicted value of the partition function zero as a function of cumulant order n.

Figure A2 shows the estimated location of the Fisher zero using the cumulant method. The β_0 is found by determining the zero crossing of $\Re(\beta - \beta_0)$, and the imaginary part is calculated using the equation for $|\beta - \beta_0|^2$ at that value for β. For all system sizes and for all considered choices of $\mathcal{R}$, the estimated value from the cumulant method quickly approaches a constant with increasing n.

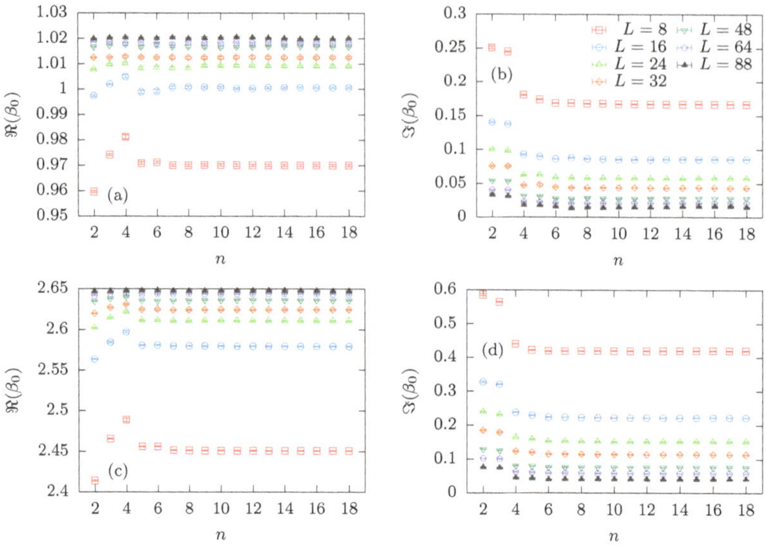

Figure A2. Real and imaginary parts of the Fisher zeros β_0 obtained through the cumulant method as a function of n [see Equation (12)] for different L. In (**a**,**b**) $\mathcal{R} = -0.1$, and in (**c**,**d**) $\mathcal{R} = -0.2$.

Similarly, in Figure A3 we show the values for the complex field h_0 of the Lee–Yang zeros at β_c estimated by the cumulant method for the different values of $\mathcal{R}$ as a function of n. These also converge quickly. Note that even on the logarithmic scale of Figure A3, error bars do not appear to grow significantly as the order of cumulants is increased. As mentioned above, we attribute this to the fact that despite increasing statistical errors of the individual cumulants, their ratios show only little statistical fluctuation due to the cross-correlation between the terms.

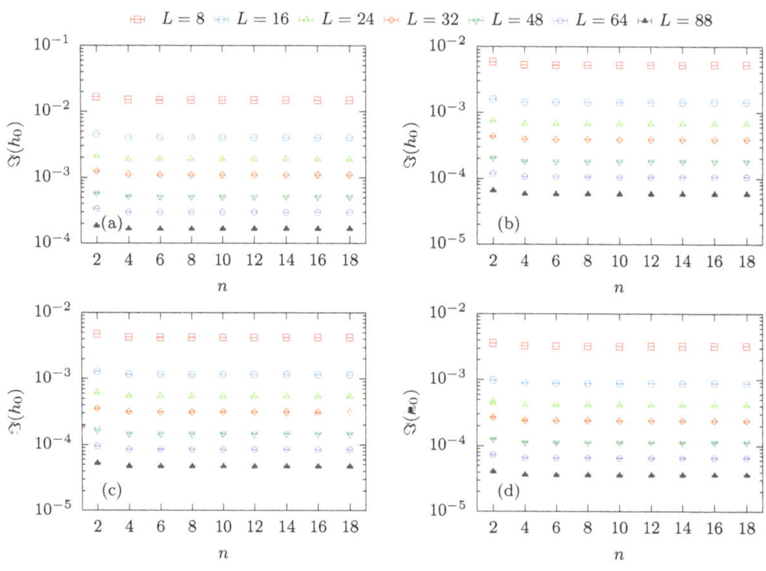

Figure A3. Imaginary part of the Lee–Yang zeros h_0 for different system sizes L and coupling strengths $\mathcal{R}$. (**a**) $\mathcal{R} = -0.1$, (**b**) $\mathcal{R} = -0.2$, (**c**) $\mathcal{R} = -0.21$, and (**d**) $\mathcal{R} = -0.22$.

References

1. Diep, H.T. (Ed.) *Frustrated Spin Systems*, 2nd ed.; World Scientific: Singapore, 2013. [CrossRef]
2. Azaria, P.; Diep, H.T.; Giacomini, H. Coexistence of order and disorder and reentrance in an exactly solvable model. *Phys. Rev. Lett.* **1987**, *59*, 1629–1632. [CrossRef] [PubMed]
3. Espriu, D.; Prats, A. Dynamics of the two-dimensional gonihedric spin model. *Phys. Rev. E* **2004**, *70*, 046117. [CrossRef] [PubMed]
4. Mueller, M.; Johnston, D.A.; Janke, W. Multicanonical analysis of the plaquette-only gonihedric Ising model and its dual. *Nucl. Phys. B* **2014**, *888*, 214–235. [CrossRef]
5. Nightingale, M. Non-universality for Ising-like spin systems. *Phys. Lett. A* **1977**, *59*, 486–488. [CrossRef]
6. Jin, S.; Sen, A.; Sandvik, A.W. Ashkin-Teller criticality and pseudo-first-order behavior in a frustrated Ising model on the square lattice. *Phys. Rev. Lett.* **2012**, *108*, 045702. [CrossRef]
7. Kalz, A.; Honecker, A.; Moliner, M. Analysis of the phase transition for the Ising model on the frustrated square lattice. *Phys. Rev. B* **2011**, *84*, 174407. [CrossRef]
8. Kalz, A.; Honecker, A. Location of the Potts-critical end point in the frustrated Ising model on the square lattice. *Phys. Rev. B* **2012**, *86*, 134410. [CrossRef]
9. Yoshiyama, K.; Hukushima, K. Higher-order tensor renormalization group study of the J_1-J_2 Ising model on a square lattice. *Phys. Rev. E* **2023**, *108*, 054124. [CrossRef]
10. Bobák, A.; Lučivjanský, T.; Žukovič, M.; Borovský, M.; Balcerzak, T. Tricritical behaviour of the frustrated Ising antiferromagnet on the honeycomb lattice. *Phys. Lett. A* **2016**, *380*, 2693–2697. [CrossRef]
11. Acevedo, S.; Arlego, M.; Lamas, C.A. Phase diagram study of a two-dimensional frustrated antiferromagnet via unsupervised machine learning. *Phys. Rev. B* **2021**, *103*, 134422. [CrossRef]
12. Žukovič, M. Critical properties of the frustrated Ising model on a honeycomb lattice: A Monte Carlo study. *Phys. Lett. A* **2021**, *404*, 127405. [CrossRef]
13. Schmidt, M.; Godoy, P. Phase transitions in the Ising antiferromagnet on the frustrated honeycomb lattice. *J. Magn. Magn. Mater.* **2021**, *537*, 168151. [CrossRef]
14. Janke, W. (Ed.) *Rugged Free Energy Landscapes—Common Computational Approaches to Spin Glasses, Structural Glasses and Biological Macromolecules*; Lecture Notes in Physics; Springer: Berlin/Heidelberg, Germany, 2007; Volume 736. [CrossRef]
15. Iba, Y. Population Monte Carlo algorithms. *Trans. Jpn. Soc. Artif. Intell.* **2001**, *16*, 279–286. [CrossRef]
16. Hukushima, K.; Iba, Y. Population annealing and its application to a spin glass. *AIP Conf. Proc.* **2003**, *690*, 200–206. [CrossRef]
17. Machta, J. Population annealing with weighted averages: A Monte Carlo method for rough free-energy landscapes. *Phys. Rev. E* **2010**, *82*, 026704. [CrossRef]
18. Wang, W.; Machta, J.; Katzgraber, H.G. Population annealing: Theory and application in spin glasses. *Phys. Rev. E* **2015**, *92*, 063307. [CrossRef]
19. Christiansen, H.; Weigel, M.; Janke, W. Accelerating molecular dynamics simulations with population annealing. *Phys. Rev. Lett.* **2019**, *122*, 060602. [CrossRef]
20. Rose, N.; Machta, J. Equilibrium microcanonical annealing for first-order phase transitions. *Phys. Rev. E* **2019**, *100*, 063304. [CrossRef]
21. Yang, C.N.; Lee, T.D. Statistical theory of equations of state and phase transitions. I. Theory of condensation. *Phys. Rev.* **1952**, *87*, 404–409. [CrossRef]
22. Lee, T.D.; Yang, C.N. Statistical theory of equations of state and phase transitions. II. Lattice gas and Ising model. *Phys. Rev.* **1952**, *87*, 410–419. [CrossRef]
23. Fisher, M.E. The nature of critical points. In *Lecture Notes in Theoretical Physics*; University of Colorado Press: Boulder, CO, USA, 1965; Volume VII C.
24. Janke, W.; Kenna, R. The strength of first and second order phase transitions from partition function zeroes. *J. Stat. Phys.* **2001**, *102*, 1211–1227. [CrossRef]
25. Janke, W.; Kenna, R. Analysis of the density of partition function zeroes: A measure for phase transition strength. In *Computer Simulation Studies in Condensed-Matter Physics XIV*; Landau, D.P., Lewis, S.P., Schüttler, H.B., Eds.; Springer: Berlin/Heidelberg, Germany, 2002; pp. 97–101. [CrossRef]
26. Janke, W.; Kenna, R. Density of partition function zeroes and phase transition strength. *Comput. Phys. Commun.* **2002**, *147*, 443–446. [CrossRef]
27. Janke, W.; Kenna, R. Phase transition strengths from the density of partition function zeroes. *Nucl. Phys. B Proc. Suppl.* **2002**, *106–107*, 905–907. [CrossRef]
28. Janke, W.; Kenna, R. Finite-size scaling and corrections in the Ising model with Brascamp-Kunz boundary conditions. *Phys. Rev. B* **2002**, *65*, 064110. [CrossRef]
29. Janke, W.; Kenna, R. Exact finite-size scaling with corrections in the two-dimensional Ising model with special boundary conditions. *Nucl. Phys. B Proc. Suppl.* **2002**, *106–107*, 929–931. [CrossRef]
30. Janke, W.; Johnston, D.A.; Kenna, R. New methods to measure phase transition strength. *Nucl. Phys. B Proc. Suppl.* **2003**, *119*, 882–884. [CrossRef]
31. Janke, W.; Johnston, D.A.; Kenna, R. Phase transition strength through densities of general distributions of zeroes. *Nucl. Phys. B* **2004**, *682*, 618–634. [CrossRef]

32. Janke, W.; Johnston, D.A.; Kenna, R. Critical exponents from general distributions of zeroes. *Comput. Phys. Commun.* **2005**, *169*, 457–461. [CrossRef]
33. Janke, W.; Johnston, D.A.; Kenna, R. Properties of phase transitions of higher-order. *PoS (LAT2005)* **2005**, *244*. [CrossRef]
34. Janke, W.; Johnston, D.A.; Kenna, R. Properties of higher-order phase transitions. *Nucl. Phys. B* **2006**, *736*, 319–328. [CrossRef]
35. Kenna, R.; Johnston, D.A.; Janke, W. Scaling relations for logarithmic corrections. *Phys. Rev. Lett.* **2006**, *96*, 115701. [CrossRef] [PubMed]
36. Kenna, R.; Johnston, D.A.; Janke, W. Self-consistent scaling theory for logarithmic-correction exponents. *Phys. Rev. Lett.* **2006**, *97*, 155702. [CrossRef] [PubMed]
37. Moueddene, L.; Donoso, A.; Berche, B. Ralph Kenna's scaling relations in critical phenomena. *Entropy* **2024**, *26*, 221. [CrossRef] [PubMed]
38. Deger, A.; Flindt, C. Determination of universal critical exponents using Lee-Yang theory. *Phys. Rev. Res.* **2019**, *1*, 023004. [CrossRef]
39. Moueddene, L.; G Fytas, N.; Holovatch, Y.; Kenna, R.; Berche, B. Critical and tricritical singularities from small-scale Monte Carlo simulations: The Blume-Capel model in two dimensions. *J. Stat. Mech. Theory Exp.* **2024**, *2024*, 023206. [CrossRef]
40. Monroe, J.L.; Kim, S.Y. Phase diagram and critical exponent ν for the nearest-neighbor and next-nearest-neighbor interaction Ising model. *Phys. Rev. E* **2007**, *76*, 021123. [CrossRef]
41. Kim, S.Y. Ising antiferromagnet on a finite triangular lattice with free boundary conditions. *J. Korean Phys. Soc.* **2015**, *67*, 1517–1523. [CrossRef]
42. Sarkanych, P.; Holovatch, Y.; Kenna, R. On the phase diagram of the 2d Ising model with frustrating dipole interaction. *Ukr. J. Phys.* **2015**, *60*, 334–338. [CrossRef]
43. Kim, S.Y. Study of the frustrated Ising model on a square lattice based on the exact density of states. *J. Korean Phys. Soc.* **2021**, *79*, 894–902. [CrossRef]
44. Peng, X.; Zhou, H.; Wei, B.B.; Cui, J.; Du, J.; Liu, R.B. Experimental observation of Lee-Yang zeros. *Phys. Rev. Lett.* **2015**, *114*, 010601. [CrossRef]
45. Ananikian, N.; Kenna, R. Imaginary magnetic fields in the real world. *Physics* **2015**, *8*, 2. [CrossRef]
46. Fläschner, N.; Vogel, D.; Tarnowski, M.; Rem, B.; Lühmann, D.S.; Heyl, M.; Budich, J.; Mathey, L.; Sengstock, K.; Weitenberg, C. Observation of dynamical vortices after quenches in a system with topology. *Nat. Phys.* **2018**, *14*, 265–268. [CrossRef]
47. Macêdo, A.R.S.; Vasilopoulos, A.; Akritidis, M.; Plascak, J.A.; Fytas, N.G.; Weigel, M. Two-dimensional dilute Baxter-Wu model: Transition order and universality. *Phys. Rev. E* **2023**, *108*, 024140. [CrossRef]
48. Flindt, C.; Garrahan, J.P. Trajectory phase transitions, Lee-Yang zeros, and high-order cumulants in full counting statistics. *Phys. Rev. Lett.* **2013**, *110*, 050601. [CrossRef]
49. Itzykson, C.; Pearson, R.B.; Zuber, J.B. Distribution of zeros in Ising and gauge models. *Nucl. Phys. B* **1983**, *220*, 415–433. [CrossRef]
50. Bena, I.; Droz, M.; Lipowski, A. Statistical mechanics of equilibrium and nonequilibrium phase transitions: The Yang–Lee formalism. *Int. J. Mod. Phys. B* **2005**, *19*, 4269–4329. [CrossRef]
51. Fröhlich, J.; Rodriguez, P.F. Some applications of the Lee-Yang theorem. *J. Math. Phys.* **2012**, *53*, 095218. [CrossRef]
52. Krasnytska, M.; Berche, B.; Holovatch, Y.; Kenna, R. Violation of Lee-Yang circle theorem for Ising phase transitions on complex networks. *Europhys. Lett.* **2015**, *111*, 60009. [CrossRef]
53. Katsura, S.; Abe, Y.; Yamamoto, M. Distribution of zeros of the partition function of the Ising model. *J. Phys. Soc. Jpn.* **1971**, *30*, 347–357. [CrossRef]
54. Kenna, R.; Lang, C. Finite size scaling and the zeroes of the partition function in the Φ^4_4 model. *Phys. Lett. B* **1991**, *264*, 396–400. [CrossRef]
55. Kenna, R.; Lang, C.B. Scaling and density of Lee-Yang zeros in the four-dimensional Ising model. *Phys. Rev. E* **1994**, *49*, 5012–5017. [CrossRef] [PubMed]
56. Hong, S.; Kim, D.H. Logarithmic finite-size scaling correction to the leading Fisher zeros in the p-state clock model: A higher-order tensor renormalization group study. *Phys. Rev. E* **2020**, *101*, 012124. [CrossRef] [PubMed]
57. Deger, A.; Brandner, K.; Flindt, C. Lee-Yang zeros and large-deviation statistics of a molecular zipper. *Phys. Rev. E* **2018**, *97*, 012115. [CrossRef]
58. Deger, A.; Flindt, C. Lee-Yang theory of the Curie-Weiss model and its rare fluctuations. *Phys. Rev. Res.* **2020**, *2*, 033009. [CrossRef]
59. Deger, A.; Brange, F.; Flindt, C. Lee-Yang theory, high cumulants, and large deviation statistics of the magnetization in the Ising model. *Phys. Rev. B* **2020**, *102*, 174418. [CrossRef]
60. Janke, W.; Kappler, S. Monte-Carlo study of pure-phase cumulants of 2D q-state Potts models. *J. Phys. I* **1997**, *7*, 663–674. :1997183 [CrossRef]
61. Gessert, D.; Janke, W.; Weigel, M. Resampling schemes in population annealing—Numerical and theoretical results. *Phys. Rev. E* **2023**, *108*, 065309. [CrossRef]
62. Weigel, M.; Barash, L.; Shchur, L.; Janke, W. Understanding population annealing Monte Carlo simulations. *Phys. Rev. E* **2021**, *103*, 053301. [CrossRef]

63. Barash, L.Y.; Weigel, M.; Borovský, M.; Janke, W.; Shchur, L.N. GPU accelerated population annealing algorithm. *Comput. Phys. Commun.* **2017**, *220*, 341–350. [CrossRef]
64. Gessert, D.; Janke, W.; Weigel, M. (in preparation).
65. Levenberg, K. A method for the solution of certain non-linear problems in least squares. *Q. Appl. Math.* **1944**, *2*, 164–168. [CrossRef]
66. Marquardt, D.W. An algorithm for least-squares estimation of nonlinear parameters. *J. Soc. Ind. Appl. Math.* **1963**, *11*, 431–441. [CrossRef]
67. Ferrenberg, A.M.; Landau, D.P. Critical behavior of the three-dimensional Ising model: A high-resolution Monte Carlo study. *Phys. Rev. B* **1991**, *44*, 5081–5091. [CrossRef] [PubMed]
68. Press, W.H.; Teukolsky, S.A.; Vetterling, W.T.; Flannery, B.P. *Numerical Recipes*; Cambridge University Press: New York, NY, USA, 2007.
69. Efron, B. *The Jackknife, the Bootstrap, and Other Resampling Plans*; Society for Industrial and Applied Mathematics: Philadelphia, PA, USA, 1982. [CrossRef]
70. Janke, W. Monte Carlo Simulations in Statistical Physics—From Basic Principles to Advanced Applications. In *Order, Disorder and Criticality*; Holovatch, Y., Ed.; World Scientific: Singapore, 2012; pp. 93–166. [CrossRef]
71. Matveev, V.; Shrock, R. Complex-temperature singularities in the $d = 2$ Ising model: Triangular and honeycomb lattices. *J. Phys. A* **1996**, *29*, 803–823. [CrossRef]
72. Kim, S.Y.; Hwang, C.O.; Kim, J.M. Partition function zeros of the antiferromagnetic Ising model on triangular lattice in the complex temperature plane for nonzero magnetic field. *Nucl. Phys. B* **2008**, *805*, 441–450. [CrossRef]

entropy

MDPI

Article

Finite-Size Scaling in the Ageing Dynamics of the 1*D* Glauber–Ising Model †

Malte Henkel [1,2]

[1] Laboratoire de Physique et Chimie Théoriques (CNRS UMR 7019), Université de Lorraine Nancy, B.P. 70239, F-54506 Vandœuvre lès Nancy Cedex, France; malte.henkel@univ-lorraine.fr
[2] Centro de Física Teórica e Computacional, Universidade de Lisboa, Campo Grande, P-1749-016 Lisboa, Portugal
† In memoriam Ralph Kenna.

Abstract: Single-time and two-time correlators are computed exactly in the 1*D* Glauber-Ising model after a quench to zero temperature and on a periodic chain of finite length N, using a simple analytical continuation technique. Besides the general confirmation of finite-size scaling in non-equilibrium dynamics, this allows for testing the scaling behaviour of the plateau height $C_{\infty}^{(2)}$, to which the two-time auto-correlator converges when deep in the finite-size regime.

Keywords: ageing dynamics; Glauber-Ising model; finite-size scaling

1. Introduction

An important class of physical phenomena arises in the context of *ageing phenomena* [1,2] which occur after a many-body system has been quenched from some prescribed initial state, either onto a critical point where at least two physical phases become indistinguishable or into a phase co-existence region where two macroscopic physical phases coexist. In either case, the post-quench dynamics are slow, which may come from the effects of critical-point fluctuations or from the competition between relaxation towards at least two distinct physical states. Microscopically, the system separates into many (correlated or ordered) clusters whose mean size, $\ell(t)$, grows with time. Phenomenologically, one observes the three defining properties of *physical ageing* on a macroscopic scale [3]:

1. Slow dynamics (relaxations are slower than might be described by simple exponentials);
2. Absence of time-translation invariance;
3. Dynamical scaling.

These properties manifest themselves in the typical behaviour of correlation functions, which might be thought of in terms of a coarse-grained order-parameter $\phi = \phi(t, \boldsymbol{r})$ which depends on the time t and the space coordinates $\boldsymbol{r}$. For example, notably in situations where the average order parameter $\langle \phi(t, \boldsymbol{r}) \rangle = 0$, one often considers single-time or two-time correlators (which depend on both the waiting time s and the observation time $t > s$)

$$C(t; \boldsymbol{r}) = \langle \phi(t, \boldsymbol{r}) \phi(t, \boldsymbol{0}) \rangle, \;\; C(t, s; \boldsymbol{r}) = \langle \phi(t, \boldsymbol{r}) \phi(s, \boldsymbol{0}) \rangle \tag{1}$$

We restrict this work to *phase ordering*, which occurs for a non-conserved order parameter quenched to $T < T_c$. Then, one generically finds, for large enough times (spatial translation and rotation invariance are implicitly admitted)

$$C(t;r) := C(t;|r|) = F_C\left(1, \frac{|r|}{t^{1/z}}\right), \quad C(t,s;r) := C(t,s;|r|) = F_C\left(\frac{t}{s}, \frac{|r|}{s^{1/z}}\right) \tag{2}$$

which is specified here for systems where the typical domain size $\ell(t) \sim t^{1/z}$ increases algebraically for large times. This defines the *dynamical exponent* z. For a non-conserved order parameter and short-ranged interactions, to which we shall restrict ourselves throughout this work, one has $z = 2$ [4–6] (For a conserved order parameter, one speaks of *phase separation* and z takes different values [4–6]. Long-range interactions lead to further modifications [5,7–12]). Up to metric scale factors, the form of the scaling function F_C is generically expected to be universal, meaning it is independent of most of the 'details' of the underlying microscopic physics; see [4,13–16] for reviews. Knowing the form of F_C is an important theoretical task and is also of practical importance, as a priori knowledge of F_C allows us to make long-time predictions on the basis of short-time data. These expectations (2) are also confirmed experimentally [17,18].

In practical situations, it may be difficult to achieve spatially totally homogeneous samples without any kind of interfaces and/or granular effects. It is therefore of interest to study situations of physical ageing in geometries of finite extent, such as a hyper-cubic form with N^d sites. A simple example illustrates a typical kind of finite-size effect (see Figure 1). If one considers the auto-correlation function (please distinguish carefully between the two-time auto-correlator $C(t,s)$, Equation (3a) below, and the time-space correlator $C(t;r)$, Equation (2)). $C(t,s) := C(t,s;\mathbf{0})$ of a phase-ordering (or phase-separating) system, one finds for a spatially infinite system and for sufficiently large times, where $t, s \gg \tau_{\text{micro}}$ and $t - s \gg \tau_{\text{micro}}$, that (i) a data collapse occurs and (ii) the characteristic power-law behaviour

$$C(t,s) = F_C\left(\frac{t}{s}, |\mathbf{0}|\right) = f_C\left(\frac{t}{s}\right), \quad f_C(y) \overset{y \gg 1}{\sim} y^{-\lambda/z} \tag{3a}$$

for large time ratios $y = t/s > 1$. Here, λ is the *auto-correlation exponent* and is independent of the equilibrium critical exponents. A recent list of estimates of z and λ for phase-ordering systems can be found in [19]. On the other hand, in a fully finite system, even if the auto-correlator is still close to that of the spatially infinite system for not-too-large y, deviations from (3a) will arise (see again Figure 1). Generically, in a finite system, the auto-correlator should first decrease more rapidly as a function of y than it would in the infinite system. For even larger values of y, the auto-correlator saturates at a plateau of height

$$\lim_{y \to \infty} C(ys, s; 0; N) = C_\infty^{(2)}(s, N) \tag{3b}$$

which, in principle, should depend on the waiting time s and the system size N. Qualitative discussions on this go back a long time (see [20,21]). The systematic study of finite-size effects—by which we mean the consequences of the system being in a restricted spatial volume of linear size N, and the associated finite-size scaling—has a long history indeed, both for equilibrium phase transitions [22–25] and equilibrium dynamics [26]. Early examples of finite-size studies in non-equilibrium systems include [27–29]. Finite-size effects in glassy dynamics are studied theoretically in [30–32] and experimentally in [20,21, 33,34]. They are also one possible way to artificially create spurious sub-ageing effects [35]). Another early observation of this saturation effect occurs in the Kuramoto model of self-synchronisation [36].

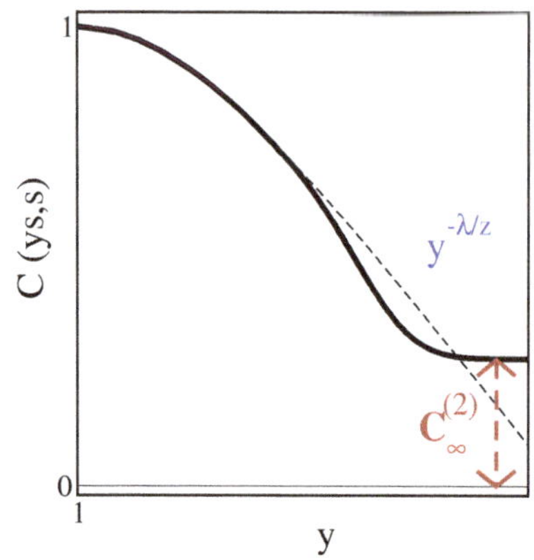

Figure 1. Qualitative dependence of the scaled two-time auto-correlator $C(t,s)$ on the time ratio $y = t/s$ for (i) a spatially infinite system (dashed line) with the power-law behaviour $\sim y^{-\lambda/z}$ and (ii) in a fully finite system (full line) which converges to a characteristic plateau $C_\infty^{(2)}$.

The expected limit behaviour of (3b) can be understood heuristically [19,37]. We recall the argument for quenches to $T < T_c$. For large times, one expects that the auto-correlator in terms of domain sizes $\ell(t)$ and $\ell(s)$ reads $C(t,s) \sim \left(\frac{\ell(t)}{\ell(s)}\right)^{-\lambda}$ (see Equation (3a)). If the observation time t becomes so large that the domain size $\ell(t) \sim N$ crosses over into the saturation regime, while the waiting time s is still small enough that the infinite-system rule $\ell(s) \sim s^{1/z}$ applies (hence $\ell(s) \ll N$), one would find

$$C_\infty^{(2)} \sim N^{-\lambda} \quad \text{if } s \text{ is fixed}, \quad C_\infty^{(2)} \sim s^{\lambda/z} \quad \text{if } N \text{ is fixed} \tag{4}$$

More formally, (4) is one of the several consequences of the hypothesis of generalised time-translation invariance [38]. Furthermore, one can write generalisations for quenches to all temperatures $T \leq T_c$, for both conserved and non-conserved order parameters, and so on. One interesting aspect of (4) is that it offers a new way to numerically estimate the exponents λ and λ/z, respectively.

The present work strives to obtain a test of (4) in the context of an exactly solvable model. Since there already exists an exact confirmation of (4) in the spherical model for dimensions $2 < d < 4$ and $T < T_c$ [19], we consider here the case of the $1D$ Glauber–Ising model (for a short summary of the rôle of the Ising model in equilibrium phase transitions, discovered by Cagniard de la Tour about 200 years ago, see, e.g., [39] and refs. therein), quenched to temperature $T = 0$ from a fully disordered initial state (initial correlations are irrelevant at large times in the $1D$ Glauber–Ising model [40]) and whose single- and two-time correlators obey the scaling form (2). Tests of (4) in the $2D$ Glauber–Ising model quenched to $T < T_c$ will be presented elsewhere [37]. We shall be interested in deriving the full size-dependent single-time and two-time spin–spin correlators which will ultimately allow us to perform an explicit test of (4).

This work is organised as follows: In Section 2, we shall introduce the analytic continuation technique and confirm that it reproduces the known exact results. In Section 3, we shall use this technique to compute the finite-size effects in the ageing dynamics and finally confirm (4). Section 4 gives our conclusions. Technical details of the exact solution are given in Appendices A–E.

2. Critical Relaxations in Infinite-Size Systems

2.1. The 1D Glauber–Ising Model

The nearest-neighbour Ising model on a chain $\Lambda \subset \mathbb{Z}$ is defined through the hamiltonian

$$\mathcal{H} = -\sum_{n\in\Lambda} \sigma_n \sigma_{n+1} \tag{5}$$

for the Ising spins $\sigma_n = \pm 1$ and the exchange coupling is normalised to unity. In a heat-bath formulation, at temperature T, at each time step Δt, a randomly chosen site $n \in \Lambda$ is updated according to Glauber dynamics [41] with the rates [42]

$$\sigma_n(t) \mapsto \pm 1 \quad \text{with the probability } \tfrac{1}{2}\left(1 \pm \tanh \tfrac{\sigma_{n-1}(t)+\sigma_{n+1}(t)}{T}\right) \tag{6}$$

On a discrete chain, the single-time correlator is $C_n(t) := \langle \sigma_n(t)\sigma_0(t)\rangle$, where the average is taken over the thermal histories defined by Equation (6). The correlator obeys the equation of motion [41–44]

$$\partial_t C_n(t) = -2C_n(t) + \gamma\big(C_{n-1}(t) + C_{n+1}(t)\big) \quad \text{when } n \neq 0 \ , \ \ C_0(t) = 1 \tag{7}$$

with the abbreviation $\gamma = \tanh(2/T)$ (such that $0 \leq \gamma \leq 1$), and where the microscopic rate constant is normalised to unity. An initial condition must still be given; for an initially fully disordered system, one has $C_n(0) = \delta_{n,0}$ [41,42]. Throughout this work, spatial translation invariance is implicitly admitted. We shall give the periodicity conditions for systems on a lattice of finite size N later. Similarly, with the time difference $\tau = t - s$, the two-time correlator is defined as $C_n(\tau, s) := \langle \sigma_n(t)\sigma_0(s)\rangle = \langle \sigma_n(\tau + s)\sigma_0(s)\rangle$ and obeys the equation of motion [41–44]

$$\partial_\tau C_n(\tau,s) = -C_n(\tau,s) + \frac{\gamma}{2}\big(C_{n-1}(\tau,s) + C_{n+1}(\tau,s)\big), \ \ C_n(0,s) = C_n(s) \tag{8}$$

Herein, the single-time correlator $C_n(s)$ serves as initial value. Solving Equations (7) and (8) constitutes the mathematical problem for the determination of the single-time and two-time correlators.

2.2. The Discrete Case

There are many well-known ways [40–53] to solve Equations (7) and (8), using, for example, generating functions [41], free fermions [48], a scaling ansatz [45,50], Laplace transforms [42], grassmannian variables [46], or systems of equations of motion [40,43,44,51–53]. Here, we shall adopt a method which easily generalises to finite systems as well.

We begin with the single-time correlator (For a disordered initial condition $\langle \sigma_n(0)\rangle = 0$, the site-dependent magnetisation $\langle \sigma_n(t)\rangle = 0$ for all times). In principle, one wishes to decouple the equations of motion (7) by a Fourier transform, but because of the boundary condition $C_0(t) = 1$, this is not straightforward. Rather, we observe first that the single-time correlator $C_n(t) = \langle \sigma_n(t)\sigma_0(t)\rangle = \langle \sigma_0(t)\sigma_n(t)\rangle = \left\langle \sigma_{|n|}(t)\sigma_0(t)\right\rangle$ is even in n and does not depend on the sign of n. It is enough to restrict the physical interpretation to positive values of $n > 0$. In particular, the equation of motion (7) is needed for $n > 0$ only. Therefore, we consider that $C_{-n}(t)$ can be thought of as an entity devoid of physical significance. It can therefore be used for purely calculational, mathematical purposes. We shall *define* $C_{-n}(t)$ in such a way that the equation of motion (7) holds for all values of $n \in \mathbb{Z}$. To do so, we write down the ansatz

$$C_{-n}(t) = \alpha_n - C_n(t); \ \ n \geq 0 \tag{9}$$

and try to choose α_n such that the boundary condition $C_0(t) = 1$ and the equation of motion (7) become valid for all $n \in \mathbb{Z}$. This kind of analytical continuation will be used several times below (it had already been used in the exact treatment of $1D$ coagulation-diffusion processes [54–56] to find the single-time and two-time correlation and response functions). It is easy to check that this is indeed possible if the α_n obey the (time-independent) recursion relation

$$\alpha_{n+1} = \frac{2}{\gamma}\alpha_n - \alpha_{n-1}, \;\; \alpha_0 = 2, \;\; \alpha_1 = \frac{2}{\gamma} \tag{10}$$

Lemma 1. *The solution of the recursion relation (10) is for all $n \geq 0$*

$$\alpha_n = \left(\frac{\gamma}{1-\sqrt{1-\gamma^2}}\right)^n + \left(\frac{\gamma}{1+\sqrt{1-\gamma^2}}\right)^n \tag{11}$$

Proof. This is easily checked by induction. For $n = 0$ and $n = 1$, the initial values in (10) are immediately reproduced. Then, it is straightforward to check that (11) obeys indeed the recurrence relation (10). □

It then follows that one has for all $n \in \mathbb{Z}$ the linear equation of motion

$$\partial_t C_n(t) = -2C_n(t) + \gamma\big(C_{n-1}(t) + C_{n+1}(t)\big) \tag{12}$$

but now in the absence of any boundary condition. This is the goal we wanted to achieve. The solution of Equation (12) is described in Appendix A. It follows that the physical correlator is

$$C_n(t) = \left\langle \sigma_{|n|}(t)\sigma_0(t)\right\rangle = \eta_-^{|n|} + \sum_{m=1}^{\infty}(C_m(0) - \eta_-^m)e^{-2t}\Big[I_{|n|-m}(2\gamma t) - I_{|n|+m}(2\gamma t)\Big] \tag{13}$$

where $I_n(t)$ is a modified Bessel function [57] and

$$C_{n,\text{eq}} = \eta_-^{|n|} = \left(\frac{1-\sqrt{1-\gamma^2}}{\gamma}\right)^{|n|} \tag{14}$$

is the equilibrium correlator $C_{n,\text{eq}} = \left\langle \sigma_{|n|}\sigma_0\right\rangle_{\text{eq}}$. This is indeed nothing else than Glauber's time-honoured result [41] (Equation (63)). For any $0 \leq \gamma < 1$, it shows the rapid relaxation towards the equilibrium correlator (14) on a finite time-scale $\tau_{\text{rel}}^{-1} = 2(1-\gamma)$. Formally, τ_{rel} diverges if $\gamma \to 1$ (or the temperature $T \to 0$), and ageing is only possible in this latter case (see Section 1).

This whole calculation only has the purpose of showing that the analytical continuation technique used here does reproduce the well-known results in the literature [40–42,44,51]. Having verified this, we can now proceed towards the derivation of new results, notably on critical dynamics, with $\gamma = 1$

For the two-time correlator, it is straightforward to solve (8) by a Fourier transform and then insert the single-time correlator (13). Since this leads to lengthy expressions [51], which are not needed below, we do not carry this out explicitly.

2.3. The Continuum Limit

1. Having seen that ageing can only occur for $\gamma = 1$, we restrict ourselves to this case from now on. The calculations become shorter in the continuum limit (This is valid in the scaling limit of (7) where $t \to \infty$, $n \to \infty$ but n^2/t is kept fixed. This will always be taken in what follows, where t and x are re-scaled time and space coordinates), where, for the single-time correlator $C(t;x) = \langle \sigma(t,x)\sigma(t,0)\rangle$, we have from (7) the equation of motion

$$\partial_t C(t;x) = \partial_x^2 C(t;x), \quad C(t;0) = 1 \tag{15}$$

where the diffusion constant was scaled to unity. Since the physical correlator $C(t;x)$ is even in x, we can restrict the problem (15) to the half-line $x \geq 0$ and use the function $C(t;-x)$ for computational purposes. The analytic continuation (9) then simplifies to

$$C(t;-x) = 2 - C(t;x) \tag{16}$$

for $x \geq 0$. In Appendix B, we show that (with the complementary error function erfc (x) [57])

$$C(t;x) = \mathrm{erfc}\left(\frac{|x|}{2t^{1/2}}\right) + \frac{e^{-x^2/4t}}{\sqrt{\pi t}} \int_0^\infty \mathrm{d}x'\, C(0;x')\, e^{-x'^2/4t} \sinh\left(\frac{|x|\,x'}{2t}\right) \tag{17}$$

remains for any initial correlations $C(0;x)$. For an initially fully disordered lattice, $C(0;x) = 0$ for $x > 0$ and the second term in (17) vanishes, as stated countless times in the literature (see e.g., [4,41–45,50] and references therein). In order to further clarify its importance, consider as an example $C(0;x) = \left(1+x^2\right)^{-\aleph/2}$ with a positive exponent $\aleph > 0$. For large times, $C(t;x) = \mathfrak{T}_1 + \mathfrak{T}_2$, where the contribution of the second term $\mathfrak{T}_2$ will be of the order

$$\mathfrak{T}_2 \sim e^{-x^2/4t} \int_0^\infty \mathrm{d}x'\, t^{-\aleph/2} e^{-x'^2/4} \sinh\left(\frac{|x|x'}{2t^{1/2}}\right) \sim t^{-\aleph/2} \to 0$$

when $t \to \infty$ but where x^2/t is kept fixed. This is much smaller than the scale-invariant and finite first term $\mathfrak{T}_1 = \mathrm{erfc}\left(|x|/2t^{1/2}\right)$. We conclude that, in the $1D$ Glauber–Ising model at $T = 0$, initially decaying correlations are always irrelevant for $t \to \infty$. This merely confirms long-standing results for the two-time correlator in this model [40]. Consequently, only the universal first term in (17) is important in the long-time scaling limit and the initial-condition-dependent second term can be discarded. The leading term in (17) might have been found more readily by inserting the scaling ansatz (2) into (15) [45].

Using the notation of (2), and for an initially fully disordered lattice, the scaled correlator $C(t;x) = F_C(1, |x|/2\sqrt{t})$ is shown in the right panel of Figure 2 as the full black curve, labelled $y = 1.0$, as a function of the scaling variable $\xi^2 = x^2/2t$. For small arguments, one observes a sharp peak at $\xi = 0$. This indicates that the interfaces between domains are sharp, as expected for kinetic Ising models and in agreement with Porod's law [4,14].

2. In Appendix B, we also compute the characteristic length scale $\ell(t) \sim t^{1/z}$ from the second moment of $C(t;x)$ and find

$$\ell^2(t) = \frac{\int_0^\infty \mathrm{d}x\, x^2\, C(t;x)}{\int_0^\infty \mathrm{d}x\, C(t;x)} = \frac{4}{3}t \tag{18}$$

which is exact for a fully disordered initial state. Equation (18) confirms the expected [5] dynamical exponent $z = 2$. One might use it in (17) to achieve a data collapse $C(t;x) = \mathrm{erfc}\left(|x|/(\sqrt{3}\,\ell(t))\right)$, up to irrelevant terms.

3. We now turn to the two-time correlator

$$C(\tau,s;x) := \langle \sigma(\tau+s,x)\sigma(s,0)\rangle \tag{19}$$

For the zero-temperature case $\gamma = 1$, we have from (8) the equation of motion

$$\partial_\tau C(\tau,s;x) = \frac{1}{2}\partial_x^2 C(\tau,s;x) \;,\;\; C(0,s;x) = C(s;x) = \mathrm{erfc}\left(\frac{|x|}{2s^{1/2}}\right) \tag{20}$$

where, in the initial condition, we merely retain from (17) the most relevant term for large times or use a totally disordered initial state. In what follows, we shall use the scaling variables

$$y = \frac{t}{s} > 1 \;,\;\; \xi^2 = \frac{x^2}{2s} \tag{21}$$

for the ratio y of the two times and the re-scaled length ξ. Then, the time difference $\tau = t - s = s(y-1)$. From now on, we shall always work in the long-time limit $s \to \infty$, with y and ξ being kept fixed. In Appendix B, we show that (20) is solved by

$$C(\tau,s;x) = 1 - \frac{1}{\pi}e^{-x^2/(2\tau)}\left(\frac{2\tau}{s}\right)^{1/2}\Psi_1\left(1,\frac{1}{2};\frac{3}{2},\frac{1}{2};-\frac{\tau}{2s},\frac{x^2}{2\tau}\right) \tag{22a}$$

$$= 1 - \frac{2}{\pi}e^{-\xi^2/(y-1)}\left(\frac{y-1}{2}\right)^{1/2}\Psi_1\left(1,\frac{1}{2};\frac{3}{2},\frac{1}{2};-\frac{y-1}{2},\frac{\xi^2}{y-1}\right) \tag{22b}$$

where Ψ_1 is a Humbert confluent hypergeometric function of two variables [58,59]. In the scaled expression (22b), we observe the independence from the waiting time s. In what follows, with a slight change of notation with respect to (2), we shall write more shortly $F_C(y,\xi) := C(\tau,s;x)$ in terms of the scaling variables y and ξ, defined above in (21). To the best of our knowledge, previous efforts have concentrated on the two-time auto-correlator $C(\tau,s;0)$ (When considering the scaling influence of the temperature through γ, the two-time *auto*-correlator can be expressed via another Humbert function Φ_1 [42]. Taking into account all three scaling variables will likely involve a three-argument Lauricella/Horn hypergeometric series).

To recover the known two-time auto-correlator $C(ys,s) = F_C(y,0)$, we now recall that $\Psi_1(a,b;c,c';x,0) = {}_2F_1(a,b;c;x)$ reduces to Gauß' hypergeometric function, which, in turn, is related to the elementary functions in our example [58] (7.3.1.123), [58] (7.3.2.83). This gives

$$F_C(y,0) = 1 - \frac{\sqrt{2}}{\pi}\left(\frac{\tau}{s}\right)^{\frac{1}{2}}\left(-\frac{\tau}{2s}\right)^{-\frac{1}{2}}\mathrm{artanh}\sqrt{-\frac{\tau}{2s}} = \frac{2}{\pi}\arctan\sqrt{\frac{2s}{\tau}} = \frac{2}{\pi}\arctan\sqrt{\frac{2}{y-1}} \tag{23}$$

where [57] (4.4.42) was also used. The last two expressions reproduce the well-known [42] (Equation (4.24)), [43] (Equation (17)), as expected. Specifically, for $y \gg 1$, one has $F_C(y,0) \simeq \frac{2\sqrt{2}}{\pi}y^{-1/2}$, and by comparison with the standard expected asymptotics (3a), one reads off $\frac{\lambda}{z} = \frac{1}{2}$, or, since $z = 2$ [5], the auto-correlation exponent $\lambda = 1$ [40,42,43], as it should be. In the left panel of Figure 2, the auto-correlator is the full black line labelled $\xi = 0.0$.

Several mathematical identities, listed in Appendix E, can be used to check the expressions (22) and to extract a few consequences. First, the limit relation given in Lemma A1 allows us to confirm that in the limit $\tau \to 0$ or $y \to 1$, the expressions (22) indeed reduce to the single-time correlator (17), without, however, the irrelevant and sub-dominant terms coming from the initial conditions. This is illustrated in the left panel of Figure 2, where the curves become more close to the auto-correlator when $\xi \to 0$. Second, the asymptotic relation given in Lemma A2, especially the independence of the two leading terms on the second argument, describes what happens when $y = t/s$ becomes large. In Figure 2, the left panel shows that the asymptotic behaviour is

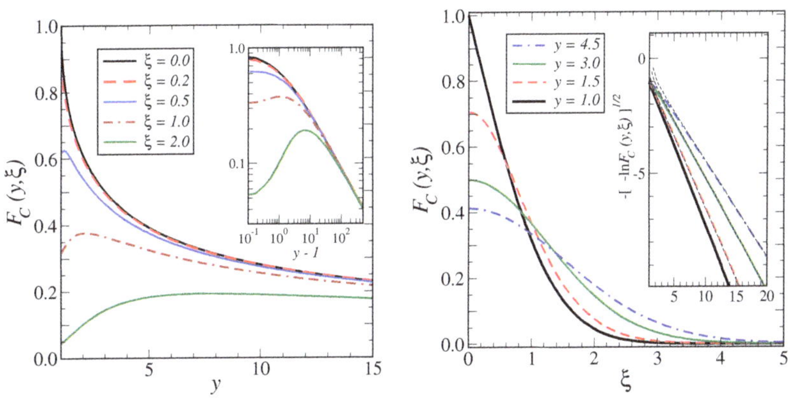

Figure 2. Properties of the scaled two-time correlator (2) in the $1D$ Glauber–Ising model. Left panel: Universal decay of the correlator $F_C(y,\xi)$ for large y, with $\xi = [0.0, 0.2, 0.5, 1.0, 2.0]$ from top to bottom. The inset shows the expected universal power-law decay (24) for large values of y. Right panel: Decay of the correlator $F_C(y,\xi)$ as a function of ξ for for $y = [1.0, 1.5, 3.0, 4.5]$ from bottom to top on the right of the figure. The inset highlights the expected Gaussian decay for large ξ and the dashed lines indicate the leading decay behaviour (25).

$$F_C(y,\xi) \overset{y\gg 1}{\simeq} \frac{\sqrt{8}}{\pi}\frac{1}{y^{1/2}}\left(1 + \mathrm{O}(y^{-1/2})\right) \tag{24}$$

and that this leading behaviour is universal in the sense that it is independent of the scaled distance $\xi^2 = \frac{x^2}{2s}$. This is reproduced from (A35), since the leading term cancels, but the next order does give (24). Third, one may also investigate the leading behaviour of the correlator when ξ becomes large but y is kept fixed. This is shown in the right panel of Figure 2. While the main figure merely displays a rather rapid decay with increasing ξ, the inset shows that the plot of the square root of the logarithm of F_C, namely $-\sqrt{-\ln F_C(y,\xi)}$ against ξ, produces an almost straight line as a function of ξ, but with a y-dependent slope. Lemma A3 gives the mathematical justification of this observation and, again, the second term in (A36) establishes that for $\xi \gg 1$

$$F_C(y,\xi) \overset{\xi\gg 1}{\simeq} \frac{1}{\sqrt{2\pi}}\frac{y+1}{\xi}\exp\left(-\frac{\xi^2}{y+1}\right) \tag{25}$$

The dashed lines in the inset represent these asymptotic predictions for several values of $y \geq 1$. Again, if one takes the limit $y \to 1$ in (25), one recovers the leading large-distance asymptotics of the single-time correlator erfc $(\xi/\sqrt{2})$, as it should be.

3. Critical Relaxations in Finite-Size Systems

Having seen for the infinite system that our technique of analytical continuation, via (9) or (16), reproduces the known results, we shall now apply it to extract the finite-size scaling properties of the ageing dynamics.

1. Consider the Glauber–Ising model on a periodic ring with N sites (see Figure 3). In the continuum limit, the equation of motion for the single-time correlator is now

$$\partial_t C(t;x) = \partial_x^2 C(t;x), \quad C(t;0) = C(t;N) = 1 \tag{26}$$

The first of these boundary conditions will be treated as before by analytic continuation. For the second one, in the case of periodic boundary conditions, we expect to have

$$C(t;x) = C(t;N-x,\ \ C(t;-x) = 2 - C(t;x) \tag{27}$$

The first of these is an inversion relation, as suggested by Figure 3. Both together analytically continue $C(t;x)$ from the physical region $x \in [0,N]$ towards $x \in \mathbb{R}$, via (27). Together, they imply the periodicity relation

$$\begin{aligned} C(t;2N+x) &= C(t;N-(-N-x)) = C(t;-N-x) \\ &= 2 - C(t;N+x) \qquad = 2 - C(t;N-(-x)) = 2 - C(t;-x) \\ &= C(t;x) \end{aligned} \tag{28}$$

where, in the first line, we used the inversion condition (27). In the second line, we applied the second continuation condition (27) and then the inversion once more. Finally, a last application of the second condition (27) brought us to the end result in the third line.

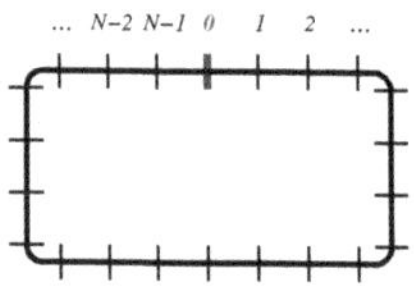

Figure 3. Periodic ring with N sites. Starting from an arbitrary site labelled 0, the property $C(t;x) = C(t;N-x)$ becomes intuitive for $x \geq 1$.

Consequently, while the physical correlator is obtained for values $0 \leq x \leq N$, the analytically continued function $C(t;x)$ is a function of period $2N$ in x, but only half of it has a physical meaning. We therefore have the Fourier representation

$$C(t;x) = \sum_{k=-\infty}^{\infty} \widetilde{C}(t;k)\, e^{i\pi k \frac{x}{N}},\ \ \widetilde{C}(t;k) = \frac{1}{2N}\int_{-N}^{N} dx\, C(t;x) e^{-i\pi k \frac{x}{N}} \tag{29}$$

Because of the periodicity condition (28), we specifically have

$$C(t;N) = C(t;-N),\ \ \partial_x C(t;N) = \partial_x C(t;-N) \tag{30}$$

which are, of course, compatible with the required boundary conditions in (26).

In this way, the boundary conditions are taken into account, such that the analytically continued function $C(t;x)$ is a periodic solution of period $2N$, of the simple diffusion equation $\partial_t C(t;x) = \partial_x^2 C(t;x)$. In Appendix C, we show that for $|x| \leq N$, the physical correlator can be expressed in terms of a Jacobi theta function $\vartheta_3(z,q)$

$$\begin{aligned} C(t;x) &= \frac{1}{2N}\int_0^N dx' \left\{ ?\vartheta_3\left(\frac{\pi}{2}\frac{|x|+x'}{N}, e^{-\pi^2 t/N^2}\right) \right. \\ &\left. + C(0;x')\left[\vartheta_3\left(\frac{\pi}{2}\frac{|x|-x'}{N}, e^{-\pi^2 t/N^2}\right) - \vartheta_3\left(\frac{\pi}{2}\frac{|x|+x'}{N}, e^{-\pi^2 t/N^2}\right)\right]\right\} \end{aligned} \tag{31}$$

and that the required periodicity properties (27) are indeed satisfied, if they only hold for the initial correlator $C(0;x)$. The term depending on the initial conditions $C(0;x)$ should be irrelevant. Then, or for a fully disordered initial state, one may alternatively re-write the physical single-time correlator (2) as

$$C(t;x) = F_C\left(1, \frac{|x|}{2\sqrt{t}}\right) = \int_0^1 du\, \vartheta_3\left(\frac{\pi}{2}u + \frac{\pi}{2}\frac{|x|}{N}, e^{-\pi^2 t/N^2}\right) = 1 - \frac{2}{\pi}\int_0^{|x|/N} dv\, \vartheta_2\left(\pi v, e^{-4\pi^2 t/N^2}\right) \quad (32)$$

where $\vartheta_{2,3}$ are distinct Jacobi theta functions [57]. The expressions (32) give finite-size scaling expressions for the single-time correlator in terms of the finite-size scaling variables x/N and t/N^2.

In Figure 4a, we show the analytically continued function $C(t;x)$ as computed in Appendix C. By construction, for $t > 0$, it is an analytic function of x and periodic with period $2N$, but it only has a physical meaning in the interval $0 \leq x \leq N$. That function is quite distinct from the physical scaling function $F_C(1,\xi)$ of the single-time correlator (2), with the scaling variables (21), which is shown in Figure 4b on the same scale. It becomes periodic in x with period N, as expected intuitively for a periodic lattice of N sites, but its derivative has jumps at $x = nN$ with $n \in \mathbb{Z}$. For $N \to \infty$, it converges towards the infinite-size correlator; however, at $|x| = \frac{N}{2}$, it has a minimum value which converges exponentially to zero as $N \to \infty$. For a fully disordered initial state, the correlator is non-vanishing only at $x = nN$.

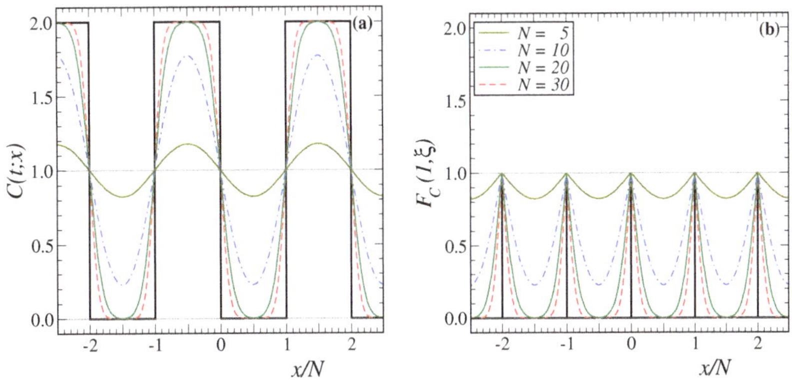

Figure 4. (**a**) Analytically continued function $C(t;x)$, as computed in Appendix C, for $t = 5$ and $N = [5, 10, 20, 30]$ from top to bottom, in the interval $0 \leq x \leq N$. It satisfies the periodicity conditions (27). The full black line gives the initial function $C(0;x)$ for a completely disordered initial lattice. The thin horizontal lines indicate the values $C(t;x) = 0$ and $C(t;x) = 1$, respectively. (**b**) Physical scaling function $F_C(1,\xi)$ of Equation (2), for the same values of t and N. The full black line corresponds to a completely disordered initial state.

2. A finite-size generalisation of the second moment can be given as follows

$$\begin{aligned} \ell^2(t) &= \frac{\int_0^N dx\, 8\left(\frac{N}{\pi}\sin\frac{\pi x}{N}\right)^2 C(t;x)}{\int_0^N dx\, C(t;x)} = \frac{2N^2}{\pi^2}\left(1 - \frac{1}{2\pi}\frac{\int_0^1 dw\, \sin(2\pi w)\vartheta_3\left(\frac{\pi}{2}w, q\right)}{\int_0^1 dw\, w\, \vartheta_3\left(\frac{\pi}{2}w, q\right)}\right) \\ &= \begin{cases} \frac{2}{\pi^2}N^2 & \text{if } t \gg N^2 \\ \frac{4}{3}t & \text{if } t \ll N^2 \end{cases} \end{aligned} \quad (33)$$

with the only variable $q = e^{-\pi\left(\pi t/N^2\right)}$ (see Appendix C for details). This reproduces the infinite lattice result (18) and has the expected finite-size scaling form $\ell^2(t) = N^2 F_\ell\left(tN^{-2}\right)$. Deep into the finite-size saturation regime, one has $\ell(t) \approx 0.45N$.

3. The critical two-time correlator $C(\tau, s; x)$ satisfies the equation of motion

$$\partial_\tau C(\tau, s; x) = \frac{1}{2}\partial_x^2 C(\tau, s; x) \ , \ \ C(0, s; x) = C(s; x) \tag{34}$$

with the single-time correlator, restricted to the relevant term, from (32). In Appendix D, we show that

$$C(\tau, s; x) = \frac{1}{2}\iint_0^1 \mathrm{d}v\mathrm{d}u\, \vartheta_3\left(\frac{\pi}{2}(v+u), e^{-\frac{\pi^2}{N^2}s}\right)\left[\vartheta_3\left(\frac{\pi}{2}\frac{|x|}{N} - \frac{\pi}{2}v, e^{-\frac{\pi^2}{N^2}\frac{\tau}{2}}\right) + \vartheta_3\left(\frac{\pi}{2}\frac{|x|}{N} + \frac{\pi}{2}v, e^{-\frac{\pi^2}{N^2}\frac{\tau}{2}}\right)\right] \tag{35}$$

and hence, the auto-correlator becomes

$$C(\tau, s; 0) = \int_0^1 \mathrm{d}v \int_0^1 \mathrm{d}u\, \vartheta_3\left(\frac{\pi}{2}(v+u), e^{-\frac{\pi^2}{N^2}s}\right)\vartheta_3\left(\frac{\pi}{2}v, e^{-\frac{y-1}{2}\frac{\pi^2}{N^2}s}\right) \tag{36}$$

which only depends on the finite-size scaling variables s/N^2 and $\tau/N^2 = \frac{s}{N^2}(y-1)$.

In Figure 5a, the scaled two-time auto-correlator $C(ys, s) = F_C(y, 0)$ given in (36) is shown for several values of N and $s = 5$ fixed. Analogous with the generic expectation formulated in Section 1, we find a cross-over to a plateau when $y \gg 1$, with its height $C_\infty^{(2)}$ decreasing as N increases for fixed s. However, in the $1D$ Glauber–Ising model, the cross-over towards to the plateau occurs directly, and we do not initially see the more rapid decay observed in the spherical model in $2 < d < 4$ dimensions and $T < T_c$ [19] or in the $2D$ Glauber–Ising model at $T < T_c$ [37] (see Figure 1). From the heuristics of Section 1, this plateau should be found when the observation time $t \gg N^2$ is deeply into the finite-size regime, but the waiting time s is not, thus $s \ll N^2$. When we apply this to the auto-correlator (36) and let $y \gg 1$, the second theta function should converge rapidly towards unity, whereas the evaluation of the first one requires us to take many terms of the defining series [57] into account. This is treated by using the modular identity (A23), such that the auto-correlator becomes, in the plateau region

$$\begin{aligned}
C(\tau, s; 0) &\overset{y\gg 1}{\simeq} \int_0^1 \mathrm{d}v \int_0^1 \mathrm{d}u\, \frac{1}{\sqrt{\pi}}\frac{N}{\sqrt{s}} e^{-(u+v)^2 N^2/4s}\, \vartheta_3\left(\mathrm{i}\frac{u+v}{2}\frac{N^2}{s}, e^{-N^2/s}\right)(1+\ldots) \\
&= \frac{1}{\sqrt{\pi}}\frac{N}{\sqrt{s}}\left(\frac{N}{\sqrt{s}}\right)^{-2}\int_0^{N/\sqrt{s}} \mathrm{d}\alpha \int_0^{N/\sqrt{s}} \mathrm{d}\beta\, e^{-(\alpha+\beta)^2/4}\, \vartheta_3\left(\mathrm{i}\frac{\alpha+\beta}{2}\frac{N}{\sqrt{s}}, e^{-N^2/s}\right) \\
&\overset{N\gg\sqrt{s}}{\simeq} \frac{1}{\sqrt{\pi}}\frac{s^{1/2}}{N}\int_0^\infty \mathrm{d}\alpha \int_0^\infty \mathrm{d}\beta\, e^{-(\alpha+\beta)^2/4}\,(2+\ldots) \\
&= \frac{4}{\sqrt{\pi}}\frac{s^{1/2}}{N}
\end{aligned} \tag{37}$$

where we changed the integration variables and then see that, in the theta function ϑ_3, only the leading terms are not strongly suppressed for $N^2 \gg s$. Because of the exponential suppression in the integrand, the integration limits can be extended to infinity without significantly changing the value of the integral, which rapidly converges to 2. Then, we have the finite-size plateau height, for $s \ll N^2$

$$C_\infty^{(2)} = \lim_{y\to\infty} C(s(y-1), s; 0; N) = \frac{4}{\sqrt{\pi}}\frac{s^{1/2}}{N} \tag{38}$$

Given that we have previously seen that $z = 2$ and $\lambda = 1$, we confirm the generic expectations (4). That was the main aim of this work. Both scaling laws quoted in (4) are tested in Figure 5, respectively. Figure 5a shows that $C_\infty^{(2)}$ decreases with N while Figure 5b shows that $C_\infty^{(2)}$ increases with s, as long as $s \ll N^2$. Figure 5 also serves as an illustration that the asymptotic form (38) already holds to good approximation for relatively small values of s and N. After the kinetic spherical model [19], this is only the second exactly solvable example where this kind of test was carried out exactly.

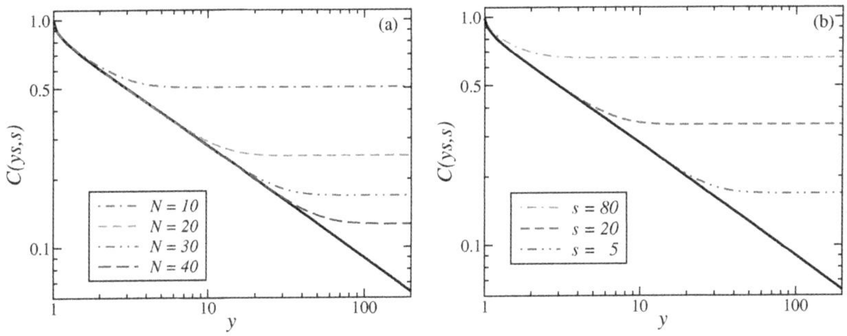

Figure 5. Two-time scaled auto-correlator $C(ys,s) = F_C(y,0)$ in the $1D$ Glauber–Ising model quenched to $T = 0$, as a function of $y = t/s$ for (**a**) finite systems of sizes $N = [10,20,30,40]$ from top to bottom and for a waiting time $s = 5$, and (**b**) the waiting times $s = [5,20,80]$ from bottom to top and of finite size $N = 30$. The full black line is the scaled infinite-size auto-correlator (23).

4. Conclusions

We evaluated the scaled single-time and two-time correlators in the $1D$ Glauber–Ising model, quenched to $T = 0$, on a finite-size and periodic chain. This allows us to check for the validity of the generically expected finite-size scaling forms [26] and, in particular, to confirm the novel expectation (4) of the plateau height $C_\infty^{(2)}$. Both scaling laws quoted in (4) are exemplified in Figure 5a and 5b, respectively. We notice that the precise form of the finite-size auto-correlator of the $1D$ Glauber–Ising model, as shown in Figure 5, does not reproduce all of the generic expectations in coarsening systems, as schematised in Figure 1.

In view of the general derivation [38], we expect the result (4) to hold in general, but further tests, including non-exactly-solvable models, are welcome. One forthcoming case is the detailed test of (4) in the $2D$ Glauber–Ising model at $T < T_c$ [37]. Since the exact solution of the $1D$ Glauber–Ising model is restricted to nearest-neighbour interactions, any tests of (4) in the currently intensively studied long-range coarsening [7–11,60–70] or phase-separating Ising models [12] will require numerical studies. Here, it might also be of interest to study non-integrable generalisations of Glauber dynamics (6), for example, by combining non-conserved Glauber-type dynamics with conserved Kawasaki-type dynamics at a different temperature, which no longer satisfies detailed balance [47]. At the least, the use of (4) gives a different computational tool for the determination of λ and λ/z, although it still has to be seen how precise a numerical technique it will turn out to be.

In addition, there is not yet any test of the generalisation of (4) [38] for quenches onto $T = T_c$. Furthermore, eventual extensions to quantum systems have not yet been tried.

Funding: This work was supported by the French ANR-PRME UNIOPEN (ANR-22-CE30-0004-01).

Data Availability Statement: The data that support the findings of this study are available from the corresponding author(s) upon reasonable request.

Acknowledgments: It is a pleasure to thank D. Warkotsch, W. Janke, and P.-C. Hang for their interesting discussions and correspondence.

Conflicts of Interest: The author declare no conflicts of interest.

Appendix A. Analytical Derivations: Discrete Case

For the infinite system, on the discrete chain, the analytically continued single-time correlator $C_n(t)$ obeys, for all $n \in \mathbb{Z}$, the equation of motion

$$\partial_t C_n(t) = -2C_n(t) + \gamma\big(C_{n-1}(t) + C_{n+1}(t)\big) \tag{A1}$$

and without any boundary condition. Equation (A1) is solved by a standard Fourier transform, namely

$$\widetilde{C}(t,k) = \sum_{n\in\mathbb{Z}} e^{-ikn}\, C_n(t) \quad , \quad C_n(t) = \frac{1}{2\pi}\int_{-\pi}^{\pi} \mathrm{d}k\, e^{+ikn}\, \widetilde{C}(t,k) \tag{A2}$$

In Fourier space, we have

$$\partial_t \widetilde{C}(t,k) = -2(1-\gamma\cos k)\widetilde{C}(t,k) \implies \widetilde{C}(t,k) = \widetilde{C}(0,k)e^{-2(1-\gamma\cos k)t} \tag{A3}$$

This can be re-expressed in direct space, but now we must remember from (9) that the initial conditions $C_{-n}(0) = \alpha_n - C_n(0)$ for the 'negative' position must be specified in terms of the physical initial correlators $C_n(0)$ with $n > 0$. This gives

$$\begin{aligned}
C_n(t) &= \frac{1}{2\pi}\int_{-\pi}^{\pi} \mathrm{d}k\, e^{ikn}\, \widetilde{C}(0,k)\, e^{-2(1-\gamma\cos k)t} \\
&= \sum_{m\in\mathbb{Z}} C_m(0)e^{-2t} I_{n-m}(2\gamma t) \qquad \text{(A4)} \\
&= \sum_{m\geq 0} C_m(0)e^{-2t} I_{n-m}(2\gamma t) + \sum_{m>0} C_{-m}(0)e^{-2t} I_{n+m}(2\gamma t) \\
&= e^{-2t} I_n(2\gamma t) + \sum_{m>0} C_m(0)e^{-2t} I_{n-m}(2\gamma t) \\
&\quad + \sum_{m>0}\left\{\left[\left(\frac{\gamma}{1+\sqrt{1-\gamma^2}}\right)^m + \left(\frac{\gamma}{1+\sqrt{1-\gamma^2}}\right)^m\right] e^{-2t} I_{n+m}(2\gamma t) - C_m(0)e^{-2t} I_{n+m}(2\gamma t)\right\} \\
&= e^{-2t} I_n(2\gamma t) + \sum_{m>0} C_m(0)e^{-2t}\Big[I_{n-m}(2\gamma t) - I_{n+m}(2\gamma t)\Big] \\
&\quad + \sum_{m>0}\left[\left(\frac{\gamma}{1+\sqrt{1-\gamma^2}}\right)^m + \left(\frac{\gamma}{1+\sqrt{1-\gamma^2}}\right)^m\right] e^{-2t} I_{n+m}(2\gamma t) \qquad \text{(A5)}
\end{aligned}$$

where, in the second line, we used the integral representation of the modified Bessel function $I_n(t)$ [57]. In the forth line, we re-used the ansatz (9), along with (11), for the initial condition at $t = 0$. In particular, the critical zero-temperature case $\gamma = 1$ reproduces [40] (Equation (2.9)).

Sometimes, it is advantageous to reformulate this result with respect to stationary (or equilibrium) properties. Making the ansatz $C_{n,\text{st}} = \eta^n$, the equilibrium constant η is found from the stationarity condition $\partial_t C_{n,\text{st}} = 0$ in (A1). This gives the condition [41] $\eta^2 - \frac{2}{\gamma}\eta + 1 = 0$. It has the two solutions $\eta_\pm = \frac{1}{\gamma}\left(1 \pm \sqrt{1-\gamma^2}\right)$ which are related by $\eta_+\eta_- = 1$. Clearly, the stationary (equilibrium only for $n > 0$) correlator is then

$$C_{n,\text{st}} = \eta_-^n = \left(\frac{1-\sqrt{1-\gamma^2}}{\gamma}\right)^n \tag{A6}$$

Because of the identity, derived from [57] (9.6.33)

$$S := e^{-2t} \sum_{m\in\mathbb{Z}} \eta^m I_{n-m}(2\gamma t) = e^{-2t} \sum_{m\in\mathbb{Z}} \eta^{m-n+n} I_{n-m}(2\gamma t) = e^{-2t}\eta^n e^{\frac{1}{2}2\gamma t(\eta^{-1}+\eta)} = \eta^n \tag{A7}$$

(since $\eta^{-1} + \eta = \eta_- + \eta_+ = \frac{2}{\gamma}$) one may re-write the single-time correlator as follows, starting again at (A4)

$$\begin{aligned} C_n(t) &= \sum_{m\in\mathbb{Z}} (C_m(0) - \eta_-^m + \eta_-^m) e^{-2t} I_{n-m}(2\gamma t) \\ &= \eta_-^n + \sum_{m\in\mathbb{Z}} (C_m(0) - \eta_-^m) e^{-2t} I_{n-m}(2\gamma t) \\ &= \eta_-^n + \underbrace{\left(C_0(t) - \eta_-^0\right)}_{=0} e^{-2t} I_n(2\gamma t) + \sum_{m=1}^{\infty} (C_m(0) - \eta_-^m) e^{-2t} [I_{n-m}(2\gamma t) - I_{n+m}(2\gamma t)] \end{aligned} \tag{A8}$$

where we also used $C_{-m}(0) - \eta_-^{-m} = C_{-m}(0) - \alpha_{-m} + \alpha_{-m} - \eta_-^{-m} = (-1)(C_m(0) - \eta_-^m)$ because $\alpha_{-m} = \alpha_m$ from the Lemma in Section 2. The physical validity is restricted to non-negative values of n because of (9). Then, we obtain the result quoted in (13) in the text.

Appendix B. Analytical Derivations: Continuum Limit

The calculations for the infinite system in the continuum limit are described below.

1. With the analytic continuation (15), the equation of motion of the single-time correlator has become the simple diffusion equation $\partial_t C(t;x) = \partial_x^2 C(t;x)$. It is solved by using the Fourier representation $C(t;x) = \frac{1}{\sqrt{2\pi}} \int_{\mathbb{R}} \mathrm{d}k\, e^{ikx} \widetilde{C}(t;k)$, which gives

$$\partial_t \widetilde{C}(t;k) = -k^2 \widetilde{C}(t;k) \implies \widetilde{C}(t;k) = \widetilde{C}(0;k) e^{-k^2 t} \tag{A9}$$

This is re-expressed in direct space via

$$\begin{aligned} C(t;x) &= \frac{1}{2\pi} \int_{\mathbb{R}} \mathrm{d}y\, C(0;y) \int_{\mathbb{R}} \mathrm{d}k\, e^{ik(x-y)-k^2 t} \\ &= \frac{1}{\sqrt{4\pi t}} \int_{-\infty}^{\infty} \mathrm{d}y\, C(0;y)\, e^{-(x-y)^2/(4t)} \\ &= \frac{1}{\sqrt{4\pi t}} \left[\int_0^{\infty} \mathrm{d}y\, C(0;y)\, e^{-(x-y)^2/(4t)} + \int_0^{\infty} \mathrm{d}y\, C(0;-y)\, e^{-(x+y)^2/(4t)} \right] \\ &= \mathrm{erfc}\left(\frac{x}{2t^{1/2}}\right) + \frac{\exp(-\frac{x^2}{4t})}{\sqrt{\pi t}} \int_0^{\infty} \mathrm{d}y\, C(0;y)\, e^{-y^2/(4t)} \sinh\left(\frac{xy}{2t}\right) \end{aligned} \tag{A10}$$

where, in the first line, we use the solution (A9) in Fourier space. In the second line, we carry out the k-integration. In third line, we insert the analytical continuation (15) and then compute the remaining integrals using the definition of the complementary error function $\mathrm{erfc}\,(x)$ [57]. From this, the physical correlator is retrieved by restricting to positive values of x. This gives (17) in the text (this might as well have been derived from (A8) or (A5) by the asymptotic expansion [57] of the modified Bessel function $I_n(t)$).

2. A characteristic length scale $\ell(t)$ is found as a second moment of $C(t;x)$ as follows

$$\ell^2(t) = \frac{\int_0^{\infty} \mathrm{d}x\, x^2 C(t;x)}{\int_0^{\infty} \mathrm{d}x\, C(t;x)} = 4t \frac{\int_0^{\infty} \mathrm{d}x\, x^2 \mathrm{erfc}\,(x)}{\int_0^{\infty} \mathrm{d}x\, \mathrm{erfc}\,(x)} = \frac{4}{3} t \tag{A11}$$

for a completely disordered initial state. Herein, the integrals are computed with the help of the identities [57] (7.2.7) and [57] (7.2.14). This is (18) in Section 2.

3. The equation of motion of the two-point correlator $C(\tau, s; x) = \langle\sigma(\tau+s, x)\sigma(s, 0)\rangle$ is again the simple diffusion Equation (20). In Fourier space, this means $\widetilde{C}(\tau, s; k) = \widetilde{C}(0, s; k)\exp\left(-\frac{1}{2}k^2\tau\right)$. From the single-time correlator (17), we only retain the most relevant term (or restrict to a disordered initial state), which gives

$$
\begin{aligned}
C(\tau, s; x) &= \frac{1}{2\pi}\int_{\mathbb{R}} dy\, \mathrm{erfc}\left(\frac{|y|}{2s^{1/2}}\right)\int_{\mathbb{R}} dk\, e^{\mathrm{i}k(x-y)-\frac{1}{2}\tau k^2} \\
&= \frac{1}{\sqrt{2\pi\tau}}\int_{\mathbb{R}} dy\, \mathrm{erfc}\left(\frac{|y|}{2s^{1/2}}\right) e^{-(x-y)^2/(2\tau)} \\
&= \frac{\exp\left(-\frac{x^2}{2\tau}\right)}{\sqrt{2\pi\tau}}\left[\int_0^\infty dy\, \mathrm{erfc}\left(\frac{y}{2s^{1/2}}\right) e^{-y^2/(2\tau)+xy/\tau} + \int_0^\infty dy\, \mathrm{erfc}\left(\frac{y}{2s^{1/2}}\right) e^{-y^2/(2\tau)-xy/\tau}\right] \\
&= \frac{2\exp\left(-\frac{x^2}{2\tau}\right)}{\sqrt{2\pi\tau}}\int_0^\infty dy\, \left(1-\mathrm{erf}\left(\frac{y}{2s^{1/2}}\right)\right)e^{-y^2/(2\tau)}\cosh\left(\frac{xy}{\tau}\right)
\end{aligned}
\tag{A12}
$$

where, in the second line, the k-integration was carried out. In the third line, the contributions of positive and negative y are separated. These are then re-united in a cosh-function in the forth line. The second of the integrals in the forth line is carried out with the help of the identity [71] (2.8.6.5)

$$
\int_0^\infty dx\, e^{-bx^2}\cosh(px)\mathrm{erf}(cx) = \frac{c}{\sqrt{\pi}\, b}\Psi_1\left(1, \frac{1}{2}; \frac{3}{2}, \frac{1}{2}; -\frac{c^2}{b}, \frac{p^2}{4b}\right) \tag{A13}
$$

involving the Humbert function Ψ_1 (see Appendix E), which can be proven by series expansion of the cosh- and erf-functions [57] and by subsequent term-wise integration. The first integral in (A12) is elementary (see e.g., [72]) (2.4.15.2). Combining all this, we arrive at (22) in the text.

Appendix C. Finite-Size Single-Time Correlator

In order to find the finite-size single-time correlator $C(t; x)$, we need the Fourier-transformed equation of motion. From the Fourier representation (29), we have

$$
\partial_t\widetilde{C}(t; k) = \frac{1}{2N}\int_{-N}^{N} dx\, (\partial_t C(t; x))e^{-\mathrm{i}\pi kx/N} = \frac{1}{2N}\int_{-N}^{N} dx\, (\partial_x^2 C(t; x))e^{-\mathrm{i}\pi kx/N} \tag{A14}
$$

This is evaluated by repeated partial integrations and taking the periodicity conditions (30) into account

$$
\begin{aligned}
\partial_t\widetilde{C}(t; k) &= \frac{1}{2N}\underbrace{\left(e^{-\mathrm{i}\pi k}\partial_x C(t; N) - e^{\mathrm{i}\pi k}\partial_x C(t; -N)\right)}_{=-2\mathrm{i}\,\sin(\pi k)\,\partial_x C(t;N)=0} + \frac{\mathrm{i}\pi k}{N}\int_{-N}^{N} dx\, e^{-\mathrm{i}\pi k\frac{x}{N}}\partial_x C(t; x) \\
&= \frac{\mathrm{i}\pi k}{2N^2}\underbrace{\left(e^{-\mathrm{i}\pi k}C(t; N) - e^{\mathrm{i}\pi k}C(t; -N)\right)}_{=-2\mathrm{i}\,\sin(\pi k)\,C(t;N)=0} - \frac{\pi^2 k^2}{2N^3}\int_{-N}^{N} dx\, e^{-\mathrm{i}\pi k\frac{x}{N}}C(t; x) \\
&= -\left(\frac{\pi k}{N}\right)^2\widetilde{C}(t; k)
\end{aligned}
\tag{A15}
$$

where, in the first and second lines, the periodicity conditions (30) were used and the boundary terms vanish since $k \in \mathbb{Z}$. The obvious solution

$$
\widetilde{C}(t; k) = \widetilde{C}(0; k)\exp\left[-\left(\frac{\pi k}{N}\right)^2 t\right] \tag{A16}
$$

is readily transformed back into real space, with the help of (16)

$$
\begin{aligned}
C(t;x) &= \sum_{k=-\infty}^{\infty} \widetilde{C}(0;k)\exp\left(-\frac{\pi^2k^2t}{N^2}+\mathrm{i}\pi k\frac{x}{N}\right) \\
&= \frac{1}{2N}\int_{-N}^{N}\mathrm{d}y\, C(0;y) \underbrace{\sum_{k=-\infty}^{\infty}\exp\left(-\frac{\pi^2k^2t}{N^2}+\mathrm{i}\pi k\frac{x-y}{N}\right)}_{=\vartheta_3\left(\frac{\pi}{2}\frac{x-y}{N},e^{-\pi^2t/N^2}\right)} \\
&= \frac{1}{2N}\int_0^N \mathrm{d}y\left[C(0;y)\vartheta_3\left(\frac{\pi}{2}\frac{x-y}{N},e^{-\pi^2t/N^2}\right)+C(0;-y)\vartheta_3\left(\frac{\pi}{2}\frac{x+y}{N},e^{-\pi^2t/N^2}\right)\right] \\
&= \frac{1}{2N}\int_0^N \mathrm{d}y\left\{2\vartheta_3\left(\frac{\pi}{2}\frac{x+y}{N},e^{-\pi^2t/N^2}\right)\right. \\
&\quad \left.+C(0;y)\left[\vartheta_3\left(\frac{\pi}{2}\frac{x-y}{N},e^{-\pi^2t/N^2}\right)-\vartheta_3\left(\frac{\pi}{2}\frac{x+y}{N},e^{-\pi^2t/N^2}\right)\right]\right\} \qquad \text{(A17)}
\end{aligned}
$$

where $\vartheta_3(z,q)$ is a Jacobi theta function [57] (16.27.4). In view of what was said in Section 2 about the infinite-size system, we expect that the second line in (A17) is negligible for times $t \gg 1$ if only $C(0;x) \to 0$ decreases with x (see Figure 4a for an illustration).

A convenient reformulation of this result applies the following identities, namely

$$
\begin{aligned}
\frac{1}{N}\int_0^N \mathrm{d}y\,\vartheta_3\left(\pi\frac{x+y}{2N},e^{-\pi^2t/N^2}\right) &= \frac{1}{N}\int_0^N \mathrm{d}y \sum_{k=-\infty}^{\infty}\exp\left(-\frac{\pi^2k^2t}{N^2}+\mathrm{i}\pi k\frac{x+y}{N}\right) \\
&= 1+\frac{2}{\pi}\sum_{k=1}^{\infty}e^{-\pi^2k^2t/N^2}\frac{1}{k}\left(\sin\left(\pi k\frac{x-N}{N}\right)-\sin\left(\pi k\frac{x}{N}\right)\right) \\
&= 1+\frac{2}{\pi}\sum_{k=1}^{\infty}e^{-\pi^2k^2t/N^2}\frac{1}{k}\sin\left(\pi k\frac{x}{N}\right)\left((-1)^k-1\right) \\
&= 1-\frac{4}{\pi}\sum_{k=0}^{\infty}e^{-\pi^2(2k+1)^2t/N^2}\frac{1}{2k+1}\sin\left(\pi(2k+1)\frac{x}{N}\right) \qquad \text{(A18)} \\
&= 1-\frac{2}{\pi N}\sum_{k=0}^{\infty}e^{-\pi^2(2k+1)^2t/N^2}\int_0^x \mathrm{d}x'\cos\left(\pi(2k+1)\frac{x'}{N}\right) \\
&= 1-\frac{2}{\pi}\int_0^{x/N}\mathrm{d}v\,\vartheta_2\left(\pi v,e^{-\left(\frac{2\pi}{N}\right)^2t}\right) \qquad \text{(A19)}
\end{aligned}
$$

where the auxiliary integration over y is carried out first. Then, we see that only the odd values of k contribute before comparing with the other Jacobi theta function $\vartheta_2(z,q)$ [57] (16.27.2); and then

$$
\begin{aligned}
&\frac{1}{2N}\int_0^N \mathrm{d}y\, C(0;y)\left[\vartheta_3\left(\pi\frac{x-y}{2N},e^{-\pi^2t/N^2}\right)-\vartheta_3\left(\pi\frac{x+y}{2N},e^{-\pi^2t/N^2}\right)\right] \\
&= \frac{1}{2N}\int_0^N \mathrm{d}y\, C(0;y)\sum_{k\in\mathbb{Z}}e^{-\pi^2k^2t/N^2}\left(e^{\mathrm{i}\pi k(x-y)/N}-e^{-\mathrm{i}\pi k(x-y)/N}\right) \\
&= \frac{1}{N}\sum_{k=1}^{\infty}e^{-\pi^2k^2t/N^2}\int_0^N \mathrm{d}y\, C(0;y)\left(\cos\left(\pi k\frac{x-y}{N}\right)-\cos\left(\pi k\frac{x+y}{N}\right)\right) \\
&= \frac{2}{N}\sum_{k=1}^{\infty}e^{-\pi^2k^2t/N^2}\sin\left(\pi k\frac{x}{N}\right)\int_0^N \mathrm{d}y\, C(0;y)\sin\left(\pi k\frac{y}{N}\right) \qquad \text{(A20)}
\end{aligned}
$$

which follows from trigonometric addition theorems. Using these identities in (A17), we find

$$\begin{aligned} C(t;x) &= 1-\frac{4}{\pi}\sum_{k=0}^{\infty}\frac{e^{-\pi^2(2k+1)^2t/N^2}}{2k+1}\sin\frac{\pi(2k+1)x}{N} \\ &+\frac{2}{N}\sum_{k=1}^{\infty}e^{-\pi^2k^2t/N^2}\sin\frac{\pi kx}{N}\int_0^N \mathrm{d}y\, C(0;y)\sin\frac{\pi ky}{N} \end{aligned} \tag{A21}$$

for any initial correlator $C(0;x)$. Together, this gives (32) in the text. Clearly, these representations are periodic in x, with period $2N$. In addition, we also have from (A21) that

$$\begin{aligned} C(t;N-x) &= 1-\frac{4}{\pi}\sum_{k=0}^{\infty}\frac{e^{-\pi^2(2k+1)^2t/N^2}}{2k+1}\sin\frac{\pi(2k+1)(N-x)}{N} \\ &+\frac{2}{N}\sum_{k=1}^{\infty}e^{-\pi^2k^2t/N^2}\sin\frac{\pi k(N-x)}{N}\int_0^N \mathrm{d}y\, C(0;y)\sin\frac{\pi ky}{N} \\ &= 1-\frac{4}{\pi}\sum_{k=0}^{\infty}\frac{e^{-\pi^2(2k+1)^2t/N^2}}{2k+1}\cos\big(\pi(2k+1)\big)\sin\frac{\pi(2k+1)(-x)}{N} \\ &+\frac{2}{N}\sum_{k=1}^{\infty}e^{-\pi^2k^2t/N^2}\cos(\pi k)\sin\frac{\pi k(-x)}{N}\int_0^N \mathrm{d}y\, C(0;N-y)\sin\frac{\pi k(N-y)}{N} \\ &= 1-\frac{4}{\pi}\sum_{k=0}^{\infty}\frac{e^{-\pi^2(2k+1)^2t/N^2}}{2k+1}\big(-1\big)^2\sin\frac{\pi(2k+1)x}{N} \\ &+\frac{2}{N}\sum_{k=1}^{\infty}e^{-\pi^2k^2t/N^2}\big((-1)^k\big)^2\big(-1\big)^2\sin\frac{\pi kx}{N}\int_0^N \mathrm{d}y\, C(0;y)\sin\frac{\pi ky}{N} \\ &= C(t;x) \end{aligned} \tag{A22}$$

which proves the required inversion relation. In this, we changed the integration variable in the second line, applied the trigonometric addition theorem, and used the inversion relation $C(0;N-x)=C(0;x)$ for the initial configuration in the third line.

Finally, the Jacobi theta function $\vartheta_3(\frac{\pi}{2}u,q) = \vartheta_3(\frac{\pi}{2}(2-u),q)$ obeys the following modular transformation identity, which is shown via Poisson's resummation formula [73]

$$\vartheta_3\big(\pi u, e^{-\pi t}\big) = t^{-1/2}\exp\left(-\pi\frac{u^2}{t}\right)\vartheta_3\left(\mathrm{i}\pi\frac{u}{t}, e^{-\pi/t}\right) \tag{A23}$$

For the proof of (33), we need a few preparations. First, with the change of variables $w=u+v$ and $w'=u-v$ and $f(u,v)=g(w,w')$, the square domain of integration $[0,1]^2$ is decomposed into two triangles, giving us

$$\begin{aligned} \int_0^1 \mathrm{d}u\int_0^1 \mathrm{d}v\, f(u,v) &= \frac{1}{2}\int_0^1 \mathrm{d}w\int_{-w}^{w}\mathrm{d}w'\, g(w,w') + \frac{1}{2}\int_1^2 \mathrm{d}w\int_{-(2-w)}^{2-w}\mathrm{d}w'\, g(w,w') \\ &= \frac{1}{2}\int_0^1 \mathrm{d}w\int_{-w}^{w}\mathrm{d}w'\, \big(g(w,w')+g(2-w,w')\big) \end{aligned} \tag{A24}$$

Then, with (A17) for a disordered initial state, we have $q = e^{-\pi\left(\pi t/N^2\right)}$ and the properties of ϑ_3

$$\begin{aligned}\frac{1}{N}\int_0^N \mathrm{d}x\, C(t;x) &= \int_0^1 \mathrm{d}v \int_0^1 \mathrm{d}u\, \vartheta_3\left(\frac{\pi}{2}(v+u),q\right) \\ &= \frac{1}{2}\int_0^1 \mathrm{d}w \int_{-w}^{w} \mathrm{d}w' \left[\vartheta_3\left(\frac{\pi}{2}w,q\right) + \vartheta_3\left(\frac{\pi}{2}(2-w),q\right)\right] = 2\int_0^1 \mathrm{d}w\, w\, \vartheta_3\left(\frac{\pi}{2}w,q\right) \end{aligned} \quad \text{(A25)}$$

and, again using (A24)

$$\begin{aligned}\frac{1}{N}\int_0^N \mathrm{d}x \left(\frac{N}{\pi}\sin(\frac{\pi x}{N})\right)^2 C(t;x) &= \frac{N^2}{\pi^2}\int_0^1 \mathrm{d}v \int_0^1 \mathrm{d}u\, \sin^2(\pi v)\, \vartheta_3\left(\frac{\pi}{2}(v+u),q\right) \\ &= \frac{N^2}{4\pi^2}\int_0^1 \mathrm{d}w\, \vartheta_3\left(\frac{\pi}{2}w,q\right) \int_{-w}^{w} \mathrm{d}w' \left(1-\cos(\pi(w-w'))\right) \\ &= \frac{N^2}{4\pi^2}\int_0^1 \mathrm{d}w\, \vartheta_3\left(\frac{\pi}{2}w,q\right)\left(2w - \frac{1}{\pi}\sin 2\pi w\right) \end{aligned} \quad \text{(A26)}$$

The quotient of these gives the integral form for $\ell(t)$ in (18). Deep into the finite-size regime, one has $t \gg N^2$, so that $\vartheta_3 \to 1$. This establishes the first asymptotic form. For the second one, we use the modular identity (A23) and find

$$\ell^2(t) = \frac{2N^2}{\pi^2}\left[1 - \frac{1}{2\pi}\frac{\int_0^1 \mathrm{d}w\, \sin(2\pi w)\, e^{-N^2w^2/t}\, \vartheta_3\left(\mathrm{i}\pi N^2 w/(2t), e^{-N^2/t}\right)}{\int_0^1 \mathrm{d}w\, w\, e^{-N^2w^2/t}\, \vartheta_3\left(\mathrm{i}\pi N^2 w/(2t), e^{-N^2/t}\right)}\right] \quad \text{(A27)}$$

For early times and large systems, $t \ll N^2$ such that now $e^{-N^2/t} \to 0$ and again $\vartheta_3 \to 1$. Then

$$\begin{aligned}\ell^2(t) &\simeq \frac{2N^2}{\pi^2}\left[1 - \frac{\int_0^1 \mathrm{d}w\, \sin(2\pi w) e^{-N^2w^2/t}(1+\ldots)}{\int_0^1 \mathrm{d}w\, 2\pi w\, e^{-N^2w^2/t}(1+\ldots)}\right] \\ &\simeq \frac{2N^2}{\pi^2}\left[1 - 1 + \frac{\int_0^1 \mathrm{d}z\, \frac{4\pi^2}{6} z e^{-N^2 z/t}}{\int_0^1 \mathrm{d}z\, e^{-N^2 z/t}}\right]\left(1 + \mathrm{o}(tN^{-2})\right) \\ &\simeq \frac{4}{3}t\left(1 + \mathrm{o}(tN^{-2})\right) \end{aligned} \quad \text{(A28)}$$

reproduces (18), as claimed.

Appendix D. Finite-Size Two-Time Correlator

As a first step, we must derive the Fourier-transformed equation of motion (34). Because of the periodicity (30) of the single-time correlator $C(0,s;x) = C(s;x)$, one can integrate forward in time to establish that

$$C(\tau,s;N) = C(\tau,s;-N) \quad , \quad \partial_x C(\tau,s;N) = \partial_x C(\tau,s;-N) \quad \text{(A29)}$$

and the analytically continued two-time correlator is a periodic function in x of period $2N$. With the Fourier representation (29), we immediately repeat the steps of Appendix C to arrive at the equation

$$\partial_\tau \widetilde{C}(\tau,s;k) = -\frac{1}{2}\left(\frac{\pi k}{N}\right)^2 \widetilde{C}(\tau,s;k) \quad \text{(A30)}$$

The obvious solution is then re-transformed into direct space, in analogy with (A12) in Appendix B

$$
\begin{aligned}
C(\tau,s;x) &= \sum_{k=-\infty}^{\infty} \widetilde{C}(0,s;k)\, e^{-\frac{1}{2}\left(\frac{\pi k}{N}\right)^2 \tau} \\
&= \frac{1}{2N}\sum_{k=-\infty}^{\infty} \int_0^N \mathrm{d}y\, C(0,s;y)\left(e^{\mathrm{i}\pi k(x-y)/N} + e^{\mathrm{i}\pi k(x+y)/N}\right) e^{-\frac{1}{2}\left(\frac{\pi k}{N}\right)^2 \tau} \\
&= \frac{1}{2N}\int_0^N \mathrm{d}y \int_0^1 \mathrm{d}u\, \vartheta_3\left(\frac{\pi}{2}\frac{y}{N} + \frac{\pi}{2}u, e^{-\pi^2 s/N^2}\right) \times \\
&\quad \times \left[\vartheta_3\left(\frac{\pi}{2}\frac{x-y}{N}, e^{-\frac{1}{2}\frac{\pi^2}{N}\tau}\right) + \vartheta_3\left(\frac{\pi}{2}\frac{x+y}{N}, e^{-\frac{1}{2}\frac{\pi^2}{N}\tau}\right)\right] \\
&= \frac{1}{2}\int_0^1 \mathrm{d}v \int_0^1 \mathrm{d}u\, \vartheta_3\left(\frac{\pi}{2}(v+u), e^{-\frac{\pi^2 s}{N^2}}\right)\left[\vartheta_3\left(\frac{\pi}{2}\frac{x}{N} - \frac{\pi}{2}v, e^{-\frac{\pi^2}{N^2}\frac{\tau}{2}}\right) + \vartheta_3\left(\frac{\pi}{2}\frac{x}{N} + \frac{\pi}{2}v, e^{-\frac{\pi^2}{N^2}\frac{\tau}{2}}\right)\right]
\end{aligned} \tag{A31}
$$

which is (35) in Section 3. Herein, we used the solution of the equation of motion (A30) in the second line, inserted the relevant part of the single-time correlator (A17) in the third line, and then re-scaled the integration variable. Repeating the argument from Appendix C, we also have $C(\tau,s;N-x) = C(\tau,s;x)$.

Appendix E. On Humbert Function Identities

The Humbert function Ψ_1 is defined [59,74–76] in terms of a power series (converging for $|x| < 1$ and $|y| < \infty$), to which we add a useful integral representation [77] for the analytical continuation to all negative arguments

$$
\Psi_1(a,b;c,c';x,y) = \sum_{n=0}^{\infty}\sum_{m=0}^{\infty} \frac{(a)_{n+m}(b)_n}{(c)_n(c')_m}\frac{x^n}{n!}\frac{y^m}{m!} \tag{A32a}
$$

$$
= \frac{1}{\Gamma(a)}\int_0^\infty \mathrm{d}u\, e^{-u}u^{a-1}\, {}_1F_1(b;c;xu)\, {}_0F_1(c';yu) \tag{A32b}
$$

with the Pochhammer symbol $(a)_n$ and the generalised hypergeometric functions ${}_1F_1$ and ${}_0F_1$ [58]. In view of (22), several asymptotic identities on the Humbert function $\Psi_1\left(1,\frac{1}{2};\frac{3}{2},\frac{1}{2};x,y\right)$ are stated, which are needed in the text. These asymptotic relations are special cases of more general theorems proven in [78]. In the following results, it is important that the second argument is always positive.

Lemma A1. *One has the identity, where $\xi > 0$ is kept fixed*

$$
\lim_{z\to 0^+} e^{-\xi^2/2z} z^{1/2}\Psi_1\left(1,\frac{1}{2};\frac{3}{2},\frac{1}{2};-z,\frac{\xi^2}{2z}\right) = \frac{\pi}{2}\operatorname{erf}(\xi/\sqrt{2}) \tag{A33}
$$

This Lemma mathematically expresses the reduction from the two-time correlator to the single-time one, in the equal-time limit $\tau = t - s \to 0$.

Proof. This straightforward calculation starts from the integral representation (A32b). Then

$$
\begin{aligned}
\Psi_1\left(1,\frac{1}{2};\frac{3}{2},\frac{1}{2};-z,\frac{y^2}{z}\right) &= \int_0^\infty \mathrm{d}u\, e^{-u}\, {}_1F_1\left(\frac{1}{2};\frac{3}{2};-zu\right){}_0F_1\left(\frac{1}{2};\frac{y^2}{z}u\right) \\
&= \frac{1}{z}\int_0^\infty \mathrm{d}v\, e^{-v/z}\, {}_1F_1\left(\frac{1}{2};\frac{3}{2};-v\right)\cosh\left(2\sqrt{\frac{y^2}{z^2}v}\right) \\
&= \frac{1}{z}\int_0^\infty \mathrm{d}w\, w\, {}_1F_1\left(\frac{1}{2};\frac{3}{2};-w^2\right)\left(e^{-\frac{1}{z}[(w-y)^2-y^2]} + e^{-\frac{1}{z}[(w+y)^2-y^2]}\right)
\end{aligned}
$$

where, in the second line, we used the identity [58] (7.13.1.5) and tacitly admit $y > 0$. When $z \to 0^+$, the first term will have its main contribution from $w \approx y > 0$ and the second one from $w \approx -y < 0$, but this will be out of the integration domain. We can then write

$$e^{-y^2/z} z^{1/2} \Psi_1\left(1, \frac{1}{2}; \frac{3}{2}, \frac{1}{2}; -z, \frac{y^2}{2z}\right) = \int_0^\infty dw\, w\, {}_1F_1\left(\frac{1}{2}; \frac{3}{2}; -w^2\right) \frac{\sqrt{\pi}}{\sqrt{\pi z}} \left(e^{-\frac{1}{z}(w-y)^2} + e^{-\frac{1}{z}(w+y)^2}\right)$$

For $z \to 0^+$, we recognise that the first term in the bracket will give a representation of the Dirac delta function

$$\delta(w) = \lim_{z\to 0^+} \frac{1}{\sqrt{\pi z}} e^{-w^2/z} \tag{A34}$$

so that we finally have

$$\begin{aligned}
\lim_{z\to 0^+} e^{-y^2/z} z^{\frac{1}{2}} \Psi_1\left(1, \frac{1}{2}; \frac{3}{2}, \frac{1}{2}; -z, \frac{y^2}{z}\right) &= \int_0^\infty dw\, w\, {}_1F_1\left(\frac{1}{2}; \frac{3}{2}; -w^2\right) \sqrt{\pi}\, (\delta(w-y) + \delta(w+y)) \\
= \sqrt{\pi}\, y\, {}_1F_1\left(\frac{1}{2}; \frac{3}{2}; -y^2\right) &= \frac{\pi}{2} \operatorname{erf}(y)
\end{aligned}$$

where, in the last step, the identities [58] (7.11.3.1) and [57] (6.5.17) were used. □

Lemma A2. *[78] (Theorem 3.6) In the limit where $x \to +\infty$ and $y \in \mathbb{R}$ is kept fixed, one has*

$$\Psi_1\left(1, \frac{1}{2}; \frac{3}{2}, \frac{1}{2}; -x, \frac{y^2}{x}\right) \simeq \frac{\pi}{2} \frac{1}{x^{1/2}} - \frac{1}{x} + \frac{y^2}{x^{3/2}} + \frac{1}{3} \frac{1}{x^2} + \dots \tag{A35}$$

The two leading terms do not depend on y. This is needed for deriving (24).

Lemma A3. *In the limit where $y \to \infty$ and $x > 0$ is kept fixed, to leading order in $1/y$, one has*

$$\Psi_1\left(1, \frac{1}{2}; \frac{3}{2}, \frac{1}{2}; -x, \frac{y^2}{x}\right) \simeq \frac{\pi}{2} \frac{1}{x^{1/2}} e^{y^2/x} - \frac{\sqrt{\pi}}{2} \frac{1}{y} \frac{1+x}{x^{1/2}} e^{y^2/[(1+x)x]} \tag{A36}$$

We notice the presence of two distinct exponential terms, both of which are needed in the text, for the derivation of (25).

Heuristic argument. We start from the integral representation (A32b) [77]. It turns out that the contributions of interest to us come from the upper limit of the integration, whereas the lower limit merely contributes terms of algebraic size, which can be discarded. Therefore, we may split the relevant integral $\int_0^\infty = \int_0^{\eta x/y^2} + \int_{\eta x/y^2}^\infty$, where η is some constant to be fixed later. We concentrate on the second term, which reads

$$\begin{aligned}
\mathfrak{T}_2^{(\eta)} &= \frac{1}{\Gamma(a)} \int_{\eta x/y^2}^\infty du\, e^{-u} u^{a-1} {}_1F_1(b; c; -xu) {}_0F_1(c'; y^2 u/x) \\
&= \frac{2\Gamma(c')}{\Gamma(a)} y^{1-c'} x^{c'/2-1/2} \int_{(\eta x)^{1/2}/y}^\infty dw\, e^{-w^2} w^{2a-c'} {}_1F_1(b; c; -xw^2) I_{c'-1}\left(2\frac{y}{x^{1/2}} w\right)
\end{aligned}$$

where [58] (7.13.1.1) was used. Since y will be large, we need the large-argument expansions of ${}_1F_1$ and $I_{c'-1}$, Equations (13.5.1) and (9.7.1) of [57], of which we shall only write the first term, respectively. Formally,

$$
\begin{aligned}
\mathfrak{T}_2^{(\eta)} &\simeq \frac{2\Gamma(c')}{\Gamma(a)} y^{1-c'} x^{c'/2-1/2} \int_{(\eta x)^{1/2}/y}^{\infty} \mathrm{d}w\, e^{-w^2} w^{2a-c'} \times \\
&\times \left\{ \frac{\Gamma(c)}{\Gamma(c-b)} \left(xw^2\right)^{-b} (1+\ldots) + \frac{\Gamma(c)}{\Gamma(b)} e^{-xw^2} \left(-xw^2\right)^{b-c} (1+\ldots) \right\} \cdot \frac{\exp\left[2y\,x^{-1/2}w\right]}{\sqrt{2\pi \cdot 2y\,x^{-1/2}w}} (1+\ldots) \\
&\simeq \frac{\Gamma(c)\Gamma(c')}{\Gamma(a)\Gamma(c-b)} y^{-1/2-c'} x^{c'/2-b-1/4} \int_0^{\infty} \mathrm{d}w\, w^{2a-2b-c'-1/2} e^{-[(w-yx^{-1/2})^2 - y^2/x]} \frac{y}{\sqrt{\pi}} \\
&+ \frac{\Gamma(c)\Gamma(c')}{\Gamma(a)\Gamma(b)} y^{1/2-c'} x^{b-c+c'/2-1/4} \int_0^{\infty} \mathrm{d}w\, w^{2a+2b-2c-c'-1/2} \times \\
&\times \exp\left(-\left[\left((1+x)^{1/2} w - \frac{y}{[(1+x)x]^{1/2}} \right)^2 - \frac{y^2}{(1+x)x} \right] \right) \frac{y}{\sqrt{\pi}} e^{\mathrm{i}\pi(b-c)} \\
&= \frac{\Gamma(c)\Gamma(c')}{\Gamma(a)\Gamma(c-b)} y^{-1/2-c'} x^{c'/2-b-1/4} \int_0^{\infty} \mathrm{d}v\, y^{2a-2b-c'+1/2} v^{2a-2b-c'-1/2} \underbrace{e^{-y^2\left(v-x^{-1/2}\right)^2}}_{\to\delta(v-x^{-1/2})\sqrt{\pi}/y} e^{y^2/x} \frac{y}{\sqrt{\pi}} \\
&+ \frac{\Gamma(c)\Gamma(c')}{\Gamma(a)\Gamma(b)} y^{1/2-c'} x^{b-c+c'/2-1/4} \int_0^{\infty} \mathrm{d}v\, y^{2a+2b-2c-c'+1/2} v^{2a+2b-2c-c'-1/2} \times \\
&\times (1+x)^{-1/4-a-b+c+c'/2} \underbrace{e^{-y^2\left(v-[(1+x)x]^{-1/2}\right)^2}}_{\to\delta(v-[(1+x)x]^{-1/2})\sqrt{\pi}/y} e^{y^2/[(1+x)x]} \frac{y}{\sqrt{\pi}} e^{\mathrm{i}\pi(b-c)} \\
&\simeq \frac{\Gamma(c)\Gamma(c')}{\Gamma(a)\Gamma(c-b)} e^{y^2/x} \left(\frac{y^2}{x}\right)^{a-b-c'} x^{-b} \\
&+ \frac{\Gamma(c)\Gamma(c')}{\Gamma(a)\Gamma(b)} e^{y^2/[(1+x)x]} \left(\frac{y^2}{x}\right)^{a+b-c-c'} (-x)^{b-c} (1+x)^{-2a-2b+2c+c'+1/2}
\end{aligned}
$$

and specialising to $a = 1$, $b = \frac{1}{2}$, $c = \frac{3}{2}$ and $c' = \frac{1}{2}$ gives the assertion. In the second line, the integrals were extended down to zero, admissible for $y \gg 1$. In the third line, after a change of variables, we recognise the Dirac delta function (A34) for $y \gg 1$.
This is merely a heuristic argument. In principle, both terms in (A36) should be the first ones of an asymptotic series. It remains an open mathematical problem why there should be no such series associated with the first term. Mathematically, (A36) remains a conjecture.

For completeness, we add the following. It is well-known [75] (Theorem 3.6), [76] (Theorem 3.4 and 3.6) that if $x > 0$ and $y > 0$ are fixed constants, and with the necessary conditions on a, b, c, c', one has

$$
\lim_{t\to\infty} \left(e^{-ty} t^{2b+c'-a}\, \Psi_1(a,b;c,c';-tx,ty) \right) = \frac{\Gamma(c)\Gamma(c')}{\Gamma(a)\Gamma(c-b)} \left(\frac{y}{x}\right)^b y^{a-2b-c'} \tag{A37}
$$

This result also leads to the kind of singularity of Ψ_1 when $x \to 1^-$ from the left, namely

Lemma A4. *If $y > 0$ and with the necessary conditions on a, b, c, c', we have*

$$
\lim_{x\to 1^-} \left(e^{-y/(1-x)} (1-x)^{2a+2b-2c-c'}\, \Psi_1(a,b;c,c';x,y) \right) = \frac{\Gamma(c)\Gamma(c')}{\Gamma(a)\Gamma(b)}\, y^{a+b-c-c'} \tag{A38}
$$

Proof. For $x < 1$, this is based on the Kummer-type relation [76] (2.8)

$$
\Psi_1(a,b;c,c';x,y) = (1-x)^{-a} \Psi_1\left(a, c-b; c, c'; -\frac{x}{1-x}, \frac{y}{1-x}\right) \tag{A39}
$$

In view of (A39), we define a new variable t and observe that when $x \to 1^-$ from the left

$$t := \frac{1}{1-x} \to \infty \ , \ \ \frac{x}{1-x} = t\left(1 - \frac{1}{t}\right) = t - 1 \to \infty \tag{A40}$$

so that we can write, with $y > 0$ being kept fixed,

$$\lim_{x\to 1^-} e^{-y/(1-x)}(1-x)^{2a-2(c-b)-c'}\,\Psi_1(a,b;c,c';x,+y) \ = \ \lim_{t\to\infty} e^{-ty}t^{2(c-b)+c'-a}\Psi_1(a,c-b;c,c';-t,+ty)$$

Using (*), we recast (A37) into the form presently required. □

As an example, as $x \to 1^-$ and $y > 0$, read off:
$\Psi_1\left(1,\frac{1}{2};\frac{3}{2},\frac{1}{2};x,y\right) \approx \sqrt{\frac{\pi}{4y}} \cdot (1-x)^{\frac{1}{2}}\, e^{y/(1-x)}$.

References

1. Arceri, F.; Landes, F.P.; Berthier, L.; Biroli, G. A statistical mechanics perspective on glasses and ageing. In *Encyclopedia of Complexity and Systems Science*; Meyers, R.A., Ed.; Springer: Berlin/Heidelberg, Germany, 2021. [CrossRef]
2. Vincent, E. *Encyclopedia of Condensed Matter Physics*; Chakraborty, T., Ed.; Oxford University Press: Oxford, UK, 2024; Volume 2, pp. 371–387. [CrossRef]
3. Struik, L.C.E. *Physical Ageing in Amorphous Polymers and Other Materials*; Elsevier: Amsterdam, The Netherlands, 1978.
4. Bray, A.J. Theory of Phase Ordering Kinetics. *Adv. Phys.* **1994**, *43*, 357. [CrossRef]
5. Bray, A.J.; Rutenberg, A.D. Growth Laws for Phase Ordering. *Phys. Rev.* **1994**, *E49*, R27. [CrossRef] [PubMed]
6. Bray, A.J.; Rutenberg, A.D. The Energy-Scaling Approach to Phase-Ordering Growth Laws. *Phys. Rev.* **1994**, *E51*, 5499. [CrossRef]
7. Christiansen, H.; Majumder, S.; Janke, W. Phase-ordering kinetics of the long-range Ising model. *Phys. Rev.* **2019**, *E99*, 011301. [CrossRef] [PubMed]
8. Christiansen, H.; Majumder, S.; Henkel, M.; Janke, W. Ageing in the long-range Ising model. *Phys. Rev. Lett.* **2020**, *125*, 180601. [CrossRef]
9. Corberi, F.; Lippiello, E.; Politi, P. One dimensional phase-ordering in the Ising model with space decaying interactions. *J. Stat. Phys.* **2019**, *176*, 510. [CrossRef]
10. Corberi, F.; Lippiello, E.; Politi, P. Universality in the time correlations of the long-range 1d Ising model. *J. Stat. Mech.* **2019**, 2019, 074002.
11. Corberi, F.; dello Russo, S.; Smaldone, L. Ordering kinetics with long-range interactions: Interpolating between voter and Ising models. *J. Stat. Mech.* **2024**, 2024, 093206. [CrossRef]
12. Müller, F.; Christiansen, H.; Janke, W. Non-universality in ageing during phase-separation of the two-dimensional long-range Ising model. *Phys. Rev. Lett.* **2024**, *133*, 237102. [CrossRef]
13. Cugliandolo, L.F. *Slow Relaxations and Non-Equilibrium Dynamics in Condensed Matter*; Barrat, J.-L., Feiglman, M., Kurchan, J., Dalibard, J., Eds.; Les Houches LXXVII; Springer: Berlin/Heidelberg, Germany, 2003; pp. 367–521. [CrossRef]
14. Henkel, M.; Pleimling, M. *Non-Equilibrium Phase Transitions Vol. 2: Ageing and Dynamical Scaling Far from Equilibrium*; Springer: Berlin/Heidelberg, Germany, 2010.
15. Puri, S.; Wadhawan, V. (Eds.) *Kinetics of Phase Transitions*; Taylor and Francis: London, UK, 2009.
16. Täuber, U.C. *Critical Dynamics: A Field-Theory Approach to Equilibrium and Non-Equilibrium Scaling Behaviour*; Cambridge University Press: Cambridge, UK, 2014. . [CrossRef]
17. Mason, N.; Pargellis, A.N.; Yurke, B. Scaling behavior of two-time correlations in a twisted nematic liquid crystal. *Phys. Rev. Lett.* **1993**, *70*, 190. [CrossRef]
18. Almeida, R.; Takeuchi, K. Phase-ordering kinetics in the Allen-Cahn (Model A) class: Universal aspects elucidated by electrically-induced transition in liquid crystals. *Phys. Rev.* **2021**, *E104*, 054103. [CrossRef]
19. Henkel, M. Non-equilibrium relaxations: Ageing and finite-size effects. *Condens. Matt. Phys.* **2023**, *26*, 13501. [CrossRef]
20. Joh, Y.G.; Orbach, R.; Wood, G.G.; Hammann, J.; Vincent, E. Finite-size effects on spin-glass dynamics. *J. Phys. Soc. Jpn.* **2004**, *69* (Suppl. A), 215. [CrossRef]
21. Kenning, G.G.; Tennant, D.M.; Rost, C.M.; Garrote da Silva, F.; Walters, B.J.; Zhai, Q.; Harrison, D.C.; Dahlberg, E.D.; Orbach, R.L. End of ageing as a probe of finite-size effects near the spin-glass transition temperature. *Phys. Rev.* **2018**, *B98*. . [CrossRef]

22. Barber, M.N. Finite-size scaling. In *Phase Transitions and Critical Phenomena*; Domb, C., Lebowitz, J.L., Eds.; Academic Press: London, UK, 1983; Volume 8, p. 146.
23. Brankov, J.G.; Danchev, D.M.; Tonchev, N.S. *Theory of Critical Phenomena in Finite-Size Systems*; World Scientific: Singapore, 2000.
24. Brézin, É. An investigation of finite size scaling. *J. Phys.* **1982**, *43*, 15–22. [CrossRef]
25. Fisher, M.E. *Critical Phenomena, Proceedings of the 51st International School of Physics "Enrico Fermi"*; Green, M.S., Ed.; Academic Press: London, UK, 1971; p. 1.
26. Suzuki, M. Static and Dynamic Finite-Size Scaling Theory Based on the Renormalization Group Approach. *Prog. Theor. Phys.* **1977**, *58*, 1142. [CrossRef]
27. Alcaraz, F.C.; Droz, M.; Henkel, M.; Rittenberg, V. Reaction-diffusion processes, critical dynamics and quantum chains. *Ann. Phys.* **1994**, *230*, 250. [CrossRef]
28. Krebs, K.; Pfannmüller, M.P.; Wehefritz, B.; Hinrichsen, H. Finite-Size Scaling Studies of One-Dimensional Reaction-Diffusion Systems. Part I. Analytical Results. *J. Stat. Phys.* **1994**, *78*, 1429. [CrossRef]
29. Krebs, K.; Pfannmüller, M.P.; Simon, H.; Wehefritz, B.Finite-Size Scaling Studies of One-Dimensional Reaction-Diffusion Systems. Part II. Numerical Methods. *J. Stat. Phys.* **1994**, *78*, 1471. [CrossRef]
30. Barbier, D.; de Freitas Pimenta, P.H.; Cugliandolo, L.F.; Stariolo, D.A. Finite size effects and loss of self-averageness in the relaxational dynamics of the spherical Sherrington-Kirkpatrick model. *J. Stat. Mech.* **2021**, arXiv:2103.12654
31. Fernandez, L.A.; Marinari, E.; Martin-Mayor, V.; Paga, I.; Ruiz-Lorenzo, J.J. Dimensional cross-over in the ageing dynamics of spin glasses in a film geometry. *Phys. Rev.* **2019**, *B100*, 184412. [CrossRef]
32. Zamponi, N.; Zamponi, E.; Cannas, S.A.; Chialvo, D.R. Universal dynamics of mitochondrial networks: A finite-size scaling analysis. *Sci. Rep.* **2022**, *12*, 17074. [CrossRef] [PubMed]
33. Zhai, Q.; Harrison, D.C.; Tennant, D.; Dahlberg, E.D.; Kenning, G.G.; Orbach, R.L. Glassy dynamics in CuMn thin-film multilayers. *Phys. Rev.* **2017**, *B95*, 054304. [CrossRef]
34. Zhai, Q.; Martin-Mayor, V.; Schlagel, D.L.; Kenning, G.G.; Orbach, R.L. On the Slowing Down of Spin Glass Correlation Length Growth: Simulations meet experiments. *Phys. Rev.* **2019**, *B100*, 094202. [CrossRef]
35. Christiansen, H.; Majumder, S.; Janke, W.; Henkel, M. Finite-size effects in ageing can be interpreted as sub-ageing. *Phys. Rev. Lett.* **2025**. [CrossRef]
36. Ionita, F.; Meyer-Ortmanns, H. Physical ageing of classical oscillators. *Phys. Rev. Lett.* **2014**, *112*, 094101. [CrossRef]
37. Warkotsch, D. (Institute for Theoretical Physics, University of Leipzig, Leipzig, Germany); Henkel, M. (Groupe de Physique Statistique, Institut Jean Lamour, Universite de Lorraine Nancy, B.P. 70239, F-54506 Vandoeuvre-les-Nancy Cedex, France); Janke, W. (Institute for Theoretical Physics, University of Leipzig, Leipzig, Germany). Unpublished work. 2025.
38. Henkel, M. Generalised time-translation-invariance in simple ageing. In *Springer Proceedings in Mathematics and Statistics*; Dobrev, V., Ed.; Springer: Berlin/Heidelberg, Germany, 2025; Volume 473.
39. Berche, B.; Henkel, M.; Kenna, R. Critical phenomena: 150 years since Cagniard de la Tour. *J. Phys. Stud.* **2009**, *13*, 3201. . Reprint in *Rev. Bras. Ensino Física* **2009**, *31*, 2602. [CrossRef]
40. Henkel, M.; Schütz, G.M. On the universality of the fluctuation-dissipation ratio in non-equilibrium critical dynamics. *J. Phys.* **2004**, *A37*, 591. [CrossRef]
41. Glauber, R. Time-dependent statistics of the Ising model. *J. Math. Phys.* **1963**, *4*, 294. [CrossRef]
42. Godrèche, C.; Luck, J.-M. Response of non-equilibrium systems at criticality: Exact results for the Glauber-Ising chain. *J. Phys.* **2000**, *A33*, 1151. [CrossRef]
43. Lippiello, E.; Zannetti, M. Fluctuation dissipation ratio in the one-dimensional kinetic Ising model. *Phys. Rev.* **2000**, *E61*, 3369. [CrossRef]
44. Mayer, P.; Sollich, P. General solution for multispin two-time correlation and response functions in the Glauber-Ising chain. *J. Phys.* **2004**, *A37*, 9. [CrossRef]
45. Bray, A.J. *Nonequilibrium Statistical Mechanics in One Dimension*; Privman, V., Ed.; Cambridge University Press: Cambridge, UK, 1997; pp. 143–165.
46. Aliev, M.A. Generating function of spin correlations functions for kinetic Glauber-Ising model with time-dependent transition rates. *J. Math. Phys.* **2009**, *50*, 083302. [CrossRef]
47. Droz, M.; Rácz, Z.; Schmidt, J. One-dimensional kinetic Ising model with competing dynamics: Steady-state correlations and relaxation times. *Phys. Rev.* **1989**, *A39*, 2141. [CrossRef] [PubMed]
48. Felderhof, B.U. Spin relaxation in the Ising chain. *Rep. Math. Phys.* **1971**, *1*, 215; Erratum in *Rep. Math. Phys.* **1971**, *2*, 151. [CrossRef]
49. Godrèche, C.; Luck, J.-M. The Glauber-Ising chain under low-temperature protocols. *J. Phys.* **2022**, *A55*, 495001. [CrossRef]
50. Krapivsky, P.L.; Redner, S.; Ben-Naim, E. *A Kinetic View of Statistical Physics*; Cambridge University Press: Cambridge, UK, 2010.

51. Mayer, P.; Berthier, L.; Garrahan, J.P.; Sollich, P. Fluctuation-dissipation relations in the non-equilibrium critical dynamics of Ising models. *Phys. Rev.* **2003**, *E68*, 016116. [CrossRef]
52. Mayer, P.; Sollich, P.; Berthier, L.; Garrahan, J.P. Dynamic heterogeneity in the Glauber-Ising chain. *J. Stat. Mech.* **2005**, 2005, P05002. [CrossRef]
53. Verley, G.; Chétrite, R.; Lacoste, D. Modified fluctuation-dissipation theorem near non-equilibrium states and applications to the Glauber-Ising chain. *J. Stat. Mech.* **2011**. [CrossRef]
54. Durang, X.; Fortin, J.-Y.; del Biondo, D.; Henkel, M.; Richert, J. Exact correlations in the one-dimensional coagulation-diffusion process by the empty-interval method. *J. Stat. Mech.* **2010**, 2010, P04002. . [CrossRef]
55. Durang, X.; Fortin, J.-Y.; Henkel, M. Exact two-time correlation and response functions in the one-dimensional coagulation-diffusion process by the empty-interval-particle method. *J. Stat. Mech.* **2011**, 2011, P02030. [CrossRef]
56. Fortin, J.-Y. Crossover properties of a one-dimensional reaction-diffusion process with a transport current. *J. Stat. Mech.* **2010**. [CrossRef]
57. Abramowitz, M.; Stegun, I.A. *Handbook of Mathematical Functions*; Dover: New York, NY, USA, 1965.
58. Prudnikov, A.P.; Brychkov, Y.A.; Marichev, O.I. *Integrals and Series, Vol. 3: More Special Functions*; Gordon and Breach: New York, NY, USA, 1990.
59. Srivastava, H.M.; Karlsson, P.W. *Multiple Gaussian Hypergeometric Series*; Ellis Horwood: Chichester, UK, 1985.
60. Christiansen, H.; Majumder, S.; Janke, W. Coarsening and Aging of Lattice Polymers: Influence of Bond Fluctuations. *J. Chem. Phys.* **2017**, *147*, 094902. [CrossRef]
61. Agrawal, R.; Corberi, F.; Lippiello, E.; Politi, P.; Puri, S. Kinetics of the Two-dimensional Long-range Ising Model at Low Temperatures. *Phys. Rev.* **2021**, *E103*, 012108. [CrossRef]
62. Agrawal, R.; Corberi, F.; Insalata, F.; Puri, S. Asymptotic States of Ising Ferromagnets with Long-range Interactions. *Phys. Rev.* **2022**, *E105*, 034131. [CrossRef] [PubMed]
63. Agrawal, R.; Corberi, F.; Lippiello, E.; Puri, S. Ordering dynamics of the random-field long-range Ising model in one dimension. *Phys. Rev.* **2023**, *E108*, 044131. [CrossRef]
64. Christiansen, H.; Majumder, S.; Janke, W. Zero-temperature coarsening in the two-dimensional long-range Ising model. *Phys. Rev.* **2021**, *E103*, 052122. [CrossRef]
65. Corberi, F.; Lippiello, E.; Zannetti, M. Scaling and universality in the aging kinetics of the two-dimensional clock model. *Phys. Rev.* **2006**, *E74*, 04106. [CrossRef] [PubMed]
66. Corberi, F.; Iannone, A.; Kumar, M.; Lippiello, E.; Politi, P.Coexistence of coarsening and mean-field relaxation in the long-range Ising chain. *Sci. Post Phys.* **2021**, *10*, 109. [CrossRef]
67. Gessert, D.; Christiansen, H.; Janke, W. Ageing following a zero-temperature quench in the $d = 3$ Ising model. *Phys. Rev.* **2024**, *E109*, 044148. [CrossRef]
68. Janke, W.; Christiansen, H.; Majumder, S. The role of magnetisation in phase-ordering kinetics of the short-range and long-range Ising model. *Eur. Phys. J. ST* **2023**, *232*, 1693. [CrossRef]
69. Majumder, S.; Janke, W. Evidence of aging and dynamic scaling in the collapse of a polymer. *Phys. Rev.* **2016**, *E93*, 032506. [CrossRef]
70. Majumder, S.; Christiansen, H.; Janke, W. Understanding nonequilibrium scaling laws governing collapse of a polymer. *Eur. Phys. J.* **2020**, *93*, 142. [CrossRef]
71. Prudnikov, A.P.; Brychkov, Y.A.; Marichev, O.I. *Integrals and Series, Vol. 2: Special Functions*; Gordon and Breach: New York, NY, USA, 1986.
72. Prudnikov, A.P.; Brychkov, Y.A.; Marichev, O.I. *Integrals and Series, Vol. 1: Elementary Functions*; Gordon and Breach: New York, NY, USA, 1986.
73. Itzykson, C.; Drouffe, J.-M. *Théorie Statistique des Champs*; CNRS: Paris, France, 1989; Volume 2.
74. Hang, P.-C.; Luo, M.-J. Note on the Humbert function Ψ_1. *arXiv* **2024**, arXiv:2410.21985. [CrossRef]
75. Hang, P.-C.; Luo, M.-J. Asymptotics of the Humbert Function Ψ_1 for Two Large Arguments. *SIGMA* **2024**, *20*, 074. [CrossRef]
76. Hang, P.-C.; Luo, M.-J. Asymptotics of Saran's hypergeometric function F_K. *J. Math. Anal. Appl.* **2025**, *541*, 128707. [CrossRef]
77. Wald, S.; Henkel, M. Lindblad dynamics of the quantum spherical model. *Int. Transf. Spec. Funct.* **2018**, *29*, 95. [CrossRef]
78. Hang, P.-C.; Henkel, M.; Luo, M.-J. Asymptotics of the Humbert functions Ψ_1 and Ψ_2. *J. Approx. Theory*, **2025**, arXiv:2501.07281. [CrossRef]

entropy

MDPI

Article

Partition Function Zeros of Paths and Normalization Zeros of ASEPS

Zdzislaw Burda [1,†] and Desmond A. Johnston [2,*,†]

1 Faculty of Physics and Applied Computer Science, AGH University of Krakow, al. Mickiewicza 30, 30-059 Kraków, Poland; zdzislaw.burda@agh.edu.pl
2 School of Mathematical and Computer Sciences, Heriot-Watt University, Riccarton, Edinburgh EH14 4AS, UK
* Correspondence: d.a.johnston@hw.ac.uk
† These authors contributed equally to this work.

Abstract: We exploit the equivalence between the partition function of an adsorbing Dyck walk model and the Asymmetric Simple Exclusion Process (ASEP) normalization to obtain the thermodynamic limit of the locus of the ASEP normalization zeros from a conformal map. We discuss the equivalence between this approach and using an electrostatic analogy to determine the locus, both in the case of the ASEP and the random allocation model.

Keywords: Lee–Yang and Fisher zeroes; critical exponents; first-order phase transitions; second-order phase transitions

1. Introduction

The question of how the non-analyticity in the thermodynamic potential that signals a phase transition could arise from the corresponding finite-size partition functions in the thermodynamic limit was first elegantly answered by Lee and Yang [1,2] who considered expanding these in a complex fugacity. The appearance of non-analyticity was seen to be due to the accumulation of the zeros of the partition functions on the positive real axis of the complex fugacity plane at the transition point.

An appealing way to understand these results is to employ an electrostatic analogy [3,4], where the zeros are regarded as point charges in two dimensions. A finite size partition function $Z_S(u)$ for a system of size S can be expanded as a polynomial in the complexified fugacity of interest, u,

$$Z_S(u) = \sum_{n=0}^{S} a_n u^n \,, \tag{1}$$

where the coefficients a_n will all be positive, which can be written in terms of the polynomial zeros u_i as

$$Z_S(u) = A \prod_{n=1}^{S} (1 - u/u_i) \,, \tag{2}$$

where A is some constant. The corresponding free energy/thermodynamic potential in the thermodynamic limit

$$\psi(u) = \lim_{S\to\infty} \frac{\ln Z_S(u)}{S} \tag{3}$$

can be written in terms of $\rho(z)$, the local density of zeros

$$\psi(u) = \int dz \rho(z) \ln(1 - u/z) \tag{4}$$

and will in general be complex for complex z. If we consider $\Phi(u) = \Re\psi(u)$

$$\Phi(u) = \int dz\rho(z) \ln|(1 - u/z)| \tag{5}$$

we find

$$\nabla^2\Phi(u) = \frac{1}{2\pi}\rho(u) \tag{6}$$

so $\Phi(u)$ is the equivalent to the two-dimensional electrostatic potential associated to a distribution of charges $\rho(u)$.

In the thermodynamic limit, Lee–Yang zeros typically cluster in wedge-shaped regions that pinch to a critical point located on the real axis or coalesce to form a critical line γ that separates different phases in the complex fugacity plane. The latter case occurs in the models discussed here. The functional form of $\psi(u)$ may be different on different sides of the critical line, say $\psi_-(u)$ and $\psi_+(u)$, which correspond to different phases, but their real parts $\Re\psi_-(u) = \Re\psi_+(u)$ must be equal to each other for any u on the phase boundary γ. If the critical curve is parameterized by a real parameter s, the line density $\mu(s)$ of the zeros (charge density) on γ is obtained from the difference in the imaginary part of $\psi'(u)$ on both sides of the boundary $u = \gamma(s)$

$$\mu(s) = \frac{1}{2\pi i}\Im\left[\psi'_+(\gamma(s)) - \psi'_-(\gamma(s))\right] . \tag{7}$$

More precisely, the derivatives are calculated as the limiting values at the points $\gamma(s) \pm n(s)\delta$ for $\delta \to 0^+$ approaching the critical curve from both sides. $n(s)$ is a unit vector perpendicular to the curve γ and $\delta > 0$ is the distance from this curve.

At a physical phase transition, which appears at a real fugacity value, the complex zeros "pinch" the transition point from the upper and lower complex half-plane as the system size increases. This mechanism applies widely to statistical mechanical models [1,2,5–10] for both field- and temperature-driven transitions (though at least one example is known where the free energy remains analytic in the presence of such accumulating complex zeros [11]). The nature of the transition and critical exponents can be extracted from the finite-size scaling of the zeros. Typically, these ideas have been applied numerically in finite-sized systems, with partition function zeros extracted from series expansions, transfer matrix calculations, or numerical simulations [12]. Analytically determining the thermodynamic limit of the locus of zeros and their density has only been possible in a limited number of models (unsurprisingly, since it is equivalent to exactly solving the models in question) [1,2,7,13–22].

In [23] the locus of zeros and the line density of zeros on the phase boundary were calculated in the thermodynamic limit for a class of random allocation models exhibiting a condensation transition by what appears, at first sight, to be a rather different approach drawing from analytic combinatorics. The locus of zeros was found to be simply given by the image of the unit circle under a conformal transformation determined by the generating function of the statistical weights in the model.

In the current paper, we first summarize the general approach espoused in [24] and elsewhere [25–28] to calculating the thermodynamic limit of the free energy/partition function in various models using such analytic combinatorial methods. We briefly review the application of these methods to the canonical and grand canonical random allocation models. The form of the generating function for the grand canonical model allows the locus of partition function zeros for these to be read off directly.

We then apply the methods of [23] to calculate the locus of zeros in a class of lattice path models and, noting the equivalence of their partition functions to the normalization of the ASEP (Asymmetric Simple Exclusion Process), obtain the locus of the ASEP normalization zeros, which had been obtained numerically in [29] and using the "physical"/electrostatic approach in [30]. We note that the results on Lee–Yang zeros of the ASEP normalization obtained within different frameworks are in perfect agreement.

2. General Observations

The weighty Flajolet and Sedgewick tome "Analytic Combinatorics" [24] is a veritable treasure trove of analytical methods applied to combinatorial enumeration problems. To quote the book's manifesto, as stated in its preface, "Analytic Combinatorics aims at predicting the properties of large structured combinatorial configurations precisely through an approach based extensively on analytic methods. Generating functions are the central objects of study of the theory". A similar perspective was taken in the comprehensive studies by Godrèche [25,26] of condensation and extremes for fixed and fluctuating numbers of random variables and in works such as [27,28].

To set the scene, we first recall some of the relevant results from [24] with regard to the generating functions and analytical methods. A central role is played by the (ordinary) generating function for some set $\{f_n\}$ of combinatorial objects of size n

$$f(z) = \sum_n f_n z^n \,. \tag{8}$$

The coefficient f_n in the formal power series, denoted by $[z^n]f(z)$, can be extracted by using a contour integral if $f(z)$ is viewed as a complex function which is analytic near the origin

$$[z^n]f(z) = f_n = \oint_{|z|=\epsilon} \frac{dz}{2\pi i} \frac{f(z)}{z^{n+1}} \tag{9}$$

The evaluation of the integral for a large n can be facilitated in many cases by exponentiating the integrand and then using saddle point methods

$$[z^n]f(z) = f_n = \oint_{|z|=\epsilon} \frac{dz}{2\pi i} \exp[\ln f(z) - (n+1)z] \,. \tag{10}$$

Observation of various combinatorial structures suggests that in many cases the asymptotic behavior of the coefficients f_n is given by

$$[z^n]f(z) \sim A^n \theta(n) \tag{11}$$

i.e., exponential growth A^n and a subexponential factor $\theta(n)$. It is further observed that

- The location of $f(z)$'s singularities determine A. Specifically, if the closest singularity to the origin along the real axis is σ, then $A = 1/\sigma$
- The nature of the singularities determine $\theta(n)$.

Additionally, it is often useful to rescale so that the generating functions are singular at 1 using

$$[z^n]f(z) = \sigma^{-n}[z^n]f(\sigma z) \,. \tag{12}$$

Under suitable conditions [24], it is possible to replace an $f(z)$ singular at 1 with a local approximation near the singularity $s(z) \sim (1-z)^{a_1} \log^{a_2}(1-z)$ such that

$$[z^n]f(z) \sim [z^n]s(z) \tag{13}$$

which allows the asymptotic behavior of the f_n to be extracted from estimating $f(z)$ around the singular point(s) and performing the contour integral in Equation (9).

In a similar manner to $f(z)$, where z^n marks objects of size n in the power series $f(z) = \sum_n f_n z^n$ it is possible to define bivariate generating functions

$$f(u,z) = \sum_{n,k} f_{n,k}\, z^n u^k \tag{14}$$

where u^k now additionally marks some scalar parameter with value k. A particular case of this is given by the functional composition $F(u,z) = g(uf(z))$. Assume that f, g are analytic at the origin and let z_F, z_f and z_g be the radii of convergence of F, f and g. If we define the values of f and g at these points to be $\tau_f = f(z_f)$, $\tau_g = g(z_g)$ then (again with suitable conditions on f and g) there are three possible regimes to consider:

- *Supercritical* $u\tau_f > z_g$. As z is increased from 0 there will be some value $\tilde{z}$ strictly less than z_f such that $uf(\tilde{z}) = z_g$. In this case, the singularity type is that of the external function g.
- *Subcritical* $u\tau_f < z_g$. In this case, the singularity of the composition is driven by that of the internal function f.
- *Critical* $u\tau_f = z_g$. Here, there is a confluence of the two singularities.

In the sequel, we apply these methods first to the random allocation model and then to lattice path models/ASEPs.

3. Random Allocation Model: Canonical Ensemble

The random allocation model in the canonical ensemble is a particular example of a partition function given by a product form with a constraint where analytic combinatorial methods may be gainfully employed. The model consists of weighted distributions of S indistinguishable particles over N boxes, whose partition function is simply given by [31–35]

$$Z_{S,N} = \sum\nolimits_{(s_1,\ldots,s_N)} w(s_1)\ldots w(s_N)\delta_{S-(s_1+\ldots+s_N)}, \tag{15}$$

where the Kronecker delta constraint enforces the correct total number of particles. The single box weights $w(s)$ are non-negative and for convenience when considering power law weights, $w(s) = 1/s^\beta$, we shall take $s = 1, 2, \ldots$ We will consider the thermodynamic limit $S, N \to \infty$ with the ratio $S/N = \rho$ fixed, where ρ is the mean density of particles.

Consider the generating function for the weights in Equation (15)

$$f(z) = \sum\nolimits_{s=1}^{\infty} w(s) z^s\ . \tag{16}$$

Following Equation (9), we can extract $w(s)$ from $f(z)$ by carrying out the appropriate contour integral

$$w(s) = \oint_{|z|=\epsilon} \frac{dz}{2\pi i} \frac{f(z)}{z^{s+1}} \tag{17}$$

where a suitably small radius $\epsilon > 0$ for the circular contour about the origin will ensure that the function $f(z)$ is holomorphic on $|z| < \epsilon$ and only the $w(s)z^s$ term is picked out by the contour integration.

This approach may be immediately extended to evaluate the canonical partition function itself using the contour integral

$$Z_{S,N} = \oint_{|z|=\epsilon} \frac{dz}{2\pi i} \frac{f(z)^N}{z^{S+1}}\ . \tag{18}$$

A term in the expansion of $f(z)^N$ of the form $w(s_1)z^{s_1} \cdot w(s_2)z^{s_2} \cdots w(s_N)z^{s_N}$ will only contribute to the integral if $s_1 + s_2 + \cdots s_N = S$, thus enforcing the constraint. The expansion of $f(z)^N = (\sum_{s=1}^{\infty} w(s)z^s)^N$ ensures that all the products $\prod_{i=1}^{N} w(s_i)$ with permissible s_i assignments are included in the expression for the partition function. For instance, for the power-law weights $w(s) = s^{-\beta}$ with $\beta \in (1, \infty)$ the generating function (16) is the polylogarithm $f(z) = \mathrm{Li}_\beta(z)$.

We can now use saddle point methods to extract the asymptotic behavior in the thermodynamic limit $N \to \infty$ with $S/N = \rho$ in Equation (18). If we exponentiate the terms in the integrand in Equation (18) and neglect those terms of $O(1/N)$, taking $(S+1)/N \sim S/N$, we find

$$Z_{S,N} = \oint \frac{dz}{2\pi i} \exp[N(\ln f(z) - \rho \ln z)] \tag{19}$$

which may be evaluated by saddle point methods to give

$$Z \sim \exp[N\phi(\rho)] \tag{20}$$

where

$$\phi(\rho) = -\ln z^*(\rho)\rho + \ln f(z^*(\rho)) \tag{21}$$

and

$$z^* \frac{f'(z^*)}{f(z^*)} = \rho\,. \tag{22}$$

The phase transition manifests itself as a breakdown in the saddle point solution. As ρ is increased the saddle point value of z, i.e., $z^*(\rho)$, also increases and for some families of weights reaches the radius of convergence, typically scaled to be 1, of the series for $f(z)$ at a finite critical density ρ_c. This is, indeed, the case for the power law distributions $w(s) = 1/s^\beta$ where the critical density is $\rho_c = \zeta(\beta - 1)/\zeta(\beta)$ for $\beta > 2$, where ζ is the Riemann zeta function.

It is possible to use an exact recursion relation for $Z_{S,N}$

$$Z_{S,N} = \sum_{s=1}^{S-N+1} w(s) Z_{S-s,N-1} \tag{23}$$

for $S \geq N \geq 1$, and

$$Z_{S,0} = \delta_S \tag{24}$$

for $N = 0$ to show that the breakdown of the saddle point equation signals a condensation transition. When $\rho > \rho_c$ the excess balls condense into a single box.

4. Random Allocation Model: Other Ensembles

So far we have discussed the (S, N) ensemble, where the number of particles S and the number of boxes N are fixed, but we can consider three other ensembles: (z, N), where N is fixed but S is variable; (S, u) where S is fixed but N is variable; (z, u) where both S and N are variable. They have the corresponding partition functions

$$\tilde{Z}_N(z) = \sum_{S=N}^{\infty} Z_{S,N} z^S, \tag{25}$$

$$Z_S(u) = \sum_{N=1}^{S} Z_{S,N} u^N. \tag{26}$$

and

$$F(z,u) = 1 + \sum_{N=1}^{\infty} \sum_{S=1}^{\infty} Z_{S,N} z^S u^N \tag{27}$$

respectively. For convenience, we added the term 1 before the sum. As a constant, it does not play any physical role and can be interpreted as a vacuum contribution $Z_{S,0} = \delta_S$ (24). The fugacity z can be interpreted as an exponent $z = e^{-\mu}$ of the chemical potential μ, and if N is interpreted as the volume of the system, u can be interpreted as an exponent $u = e^{-p}$ of pressure p. Using the explicit form of the coefficients $Z_{S,N}$ (15) in Equation (25)) we see that

$$\tilde{Z}_N(z) = f(z)^N, \tag{28}$$

and hence

$$F(z,u) = 1 + \sum_{N=1}^{\infty} f(z)^N u^N = \frac{1}{1 - uf(z)}. \tag{29}$$

On the other hand, it follows from (27) that

$$F(z,u) = 1 + \sum_{S=1}^{\infty} Z_S(u) z^S \tag{30}$$

so

$$Z_S(u) = \oint_{|z|=\epsilon} \frac{dz}{2\pi i z^{S+1}} F(z,u) = \oint_{|z|=\epsilon} \frac{dz}{2\pi i z^{S+1}} \frac{1}{1 - uf(z)} \tag{31}$$

where ϵ must be small enough not to capture any additional singularity outside the origin of the complex plane.

We now focus on the partition function $Z_S(u)$ of the (S,u)-ensemble [26,36–39], viewed as a polynomial in u (26). In particular, we are interested in the locus of zeros, $Z_S(u_j) = 0$, $j = 1, \ldots, S$ in the complex plane in the limit $S \to \infty$. Paradoxically, it is easier to determine the locus of the zeros of the polynomial $Z_S(u)$ not by solving the problem directly in the u variable, but by examining the asymptotic properties of the integrand (31) in the integral variable z. As follows from the general discussion on electrostatic analogy in the Introduction, in the limit $S \to \infty$, when the zeros coalesce, they form a critical line at which the derivative of the thermodynamic potential $\psi'(u)$ has a discontinuity. The magnitude of the discontinuity is related to the density of zeros on this line (7). The critical line can be derived from the asymptotic properties of the integral (31) as a borderline (branch cut of $\psi(u)$) between two different solutions $\psi_1(u)$ and $\psi_2(u)$ [23]. If the function $z \to f(z)$ is injective within the radius of convergence σ, that is for $|z| < \sigma$, the critical line and thus also the limiting locus as $S \to \infty$ of the Lee–Yang zeros in the complex u plane is given by a remarkably simple formula [23]

$$u = \gamma(s) = \frac{1}{f(\sigma e^{is})}, \quad 0 < s \leq 2\pi\,. \tag{32}$$

In other words, the critical line is the image of the boundary of the convergence disc (circle or radius σ) under the conformal mapping $z \to u = 1/f(z)$. The critical line γ arises at the interface of two different asymptotic regimes of the integral (31). One regime corresponds to the situation where the position of the pole $1/(1 - uf(z))$ (31) viewed from the perspective of the variable z

$$z_0(u) = f^{-1}(1/u) \tag{33}$$

is in the convergence disc, that is for u outside the critical curve γ. In this case, the main contribution to the integral comes from the saddle point $z_{SP}(u)$, which lies very close to

$z_0(u)$ inside the convergence disc. The distance $|z_{SP}(u) - z_0(u)|$ tends to zero as $S \to \infty$, so in the limit one can replace $z_{SP}(u)$ by $z_0(u)$. The leading contribution to the integral $Z_S(u)$ is $\exp(S(z_0(u) + o(1)))$. The second regime is for u inside the critical line. In this case the position $z_0(u)$ of the pole $1/(1 - uf(z))$ is outside the convergence disc. The two cases give the result

$$\psi(u) = \begin{cases} \psi_1(u) = -\ln(f^{-1}(1/u)), & \text{for } u \text{ outside } \gamma, \\ \psi_2(u) = -\ln\sigma, & \text{for } u \text{ inside } \gamma. \end{cases} \tag{34}$$

The saddle point solution $\psi_1(u)$ corresponds to the fluid phase, while $\psi_2(u)$ corresponds to the condensed phase. This result is consistent with the result that was first derived for real fugacity $u > 0$ [37]. The real parts of the solution are continuous on the critical curve, while the imaginary parts are not. The discontinuity $\Im\psi_2'(u) - \Im\psi_1'(u)$ gives the density of zeros on the critical line (7). The density of zeros on the critical line $u = \gamma(s)$ can be calculated from the equation [23]

$$\mu(u)du = \frac{1}{2\pi}ds \tag{35}$$

for $s \in (0, 2\pi]$, which just reflects the fact that the density on γ is an image of uniform density on the circle under the conformal map (32).

These results may also be derived in the general framework of [24] discussed in Section 2 for the functional composition of the generating functions

$$g(z) = \frac{1}{1-z} \tag{36}$$

and $f(z)$ to obtain $F(z, u) = g(uf(z))$. In this language, the condensed phase corresponds to the supercritical regime of the composition (the pole is the dominant singularity) and the fluid phase corresponds to the subcritical regime (the singularity of $f(z)$ is dominant).

5. Binomial Weights and Lattice Paths

In this section, we recall the problem of enumeration of Dyck walks, which are lattice paths in the plane that start and end on the horizontal line (x-axis), do not go below it and consist of diagonal steps $(1, 1)$ or $(1, -1)$. This problem can be mapped onto the random allocation model. This map can, in particular, be used to apply the machinery developed for this model to determine Yang–Lee zeros for Dyck walks, see Figure 1.

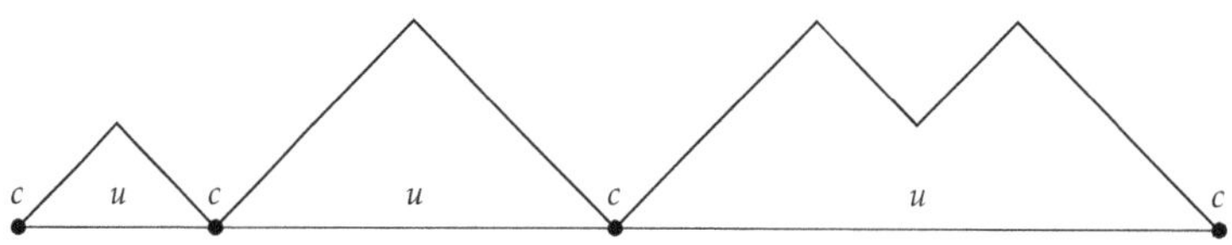

Figure 1. A Dyck walk with contact fugacity c, and additionally with excursion fugacity u.

The generating function for Dyck walks with a fugacity c for each contact with the horizontal axis is given by [40]

$$G_D(z, c) = \frac{c}{1 - cf_D(z)} = \frac{c}{1 - c(1 - \sqrt{1-4z})/2}, \tag{37}$$

in which $z^{1/2}$ is the fugacity for each step. This can be obtained by noting that $f_D(z) = (1 - \sqrt{1-4z})/2$ is the generating function for a single excursion above the horizontal axis so $G_D(z, c)$ concatenates multiple excursions with a contact weight of c every time they hit the horizontal axis. If we use the fugacity u which is associated with the number of excursions between the contacts rather than the fugacity c associated with the number of contacts, then the generating function (37) will take the form (29) that we discussed for the random allocation model

$$F_D(z,u) = \frac{1}{1-uf_D(z)}, \tag{38}$$

because the number of excursions is one less than the number of contacts. The corresponding weights (16) for $f_D(z) = (1-\sqrt{1-4z})/2$ are

$$w_D(s) = \frac{1}{2}4^s(-1)^{s+1}\binom{1/2}{s} = \frac{1}{2(2s-1)}\binom{2s}{s}. \tag{39}$$

They count the number of possible shapes of a single excursion consisting of $2s$ steps.

The weights for the Dyck walks (39) are a special case of binomial weights

$$w(s) = (-1)^{s+1}\frac{a}{\sigma^s}\binom{\theta}{s} \tag{40}$$

for $\theta \in (0,1)$, $a, \sigma \in (0,\infty)$ and $s = 1, 2, \ldots$ considered in [23]. The corresponding generating function is

$$f(z) = a\left(1-\left(1-\frac{z}{\sigma}\right)^{\theta}\right) \tag{41}$$

This function can be explicitly inverted for $|z| < \sigma$. The inverse function is

$$f^{-1}(z) = \sigma\left(1-\left(1-\frac{z}{a}\right)^{1/\theta}\right). \tag{42}$$

For Dyck walks (39) the parameters are $\theta = 1/2$, $a = 1/2$ and $\sigma = 1/4$:

$$f_D(z) = \frac{1}{2}\left(1-\sqrt{1-4z}\right) \tag{43}$$

and

$$f_D^{-1}(z) = \frac{1}{4}\left(1-(1-2z)^2\right). \tag{44}$$

In the (S,u)-ensemble, the Dyck walks have two phases, depending on the parameter u. In the thermodynamic limit $S \to \infty$, the phase transition is at $u_{cr} = 1/f_D(1/4) = 2$. For $u > u_{cr}$ the sum (26) is dominated by large values of N, that is, many short excursions, while for $u < u_{cr}$ by small N, that is, fewer but longer excursions. Actually, one of them is much longer than the others. The equivalence with the random allocation model allows one to easily understand the results in the language of random allocations [37] but also to extend the analysis to the complex plane to calculate the locus of the zeros of the partition function $Z_S(u)$ for the Dyck walks for $S \to \infty$ using the method described in Section 4. We will study the locus of the Yang–Lee zeros in the variable $v = 1/u$ rather than in u, due to the relation to the ASEP model that will be discussed later. Of course, the descriptions in v and u are equivalent. In the v plane the critical curve $s \to \gamma(s)$ on which zeros are located is (32)

$$v = \gamma(s) = f_D(e^{is}/4) = \frac{1}{2}\left(1-\sqrt{1-e^{is}}\right) \tag{45}$$

for $0 < s \leq 2\pi$. It is plotted in Figure 2. Inside this curve, $\psi_D(v)$ is given by (34)

$$\psi_D(v) = -\ln\left(\frac{1}{4}\left(1-(1-2v)^2\right)\right) = -\ln[v(1-v)] \tag{46}$$

while outside it, $\psi_D(v) = \ln 4$. Note that the interior of the critical curve in the v plane corresponds to the exterior of the corresponding critical curve in the u plane, and vice versa.

The phase transition point at $v_{cr} = 1/2$ is clearly visible in Figure 2. Assume that v is real. In this case, (3) and (26) entail

$$-v\psi'_D(v) = \lim_{S\to\infty} \frac{\langle N \rangle_{S,v}}{S} \tag{47}$$

where

$$\langle N \rangle_{S,v} = \frac{1}{Z_S(v)} \sum_{N=1}^{S} N Z_{S,N} v^{-N} \tag{48}$$

is the average number of excursions for a given S and v. The inverse excursion length in the thermodynamic limit is

$$m(v) = \lim_{S\to\infty} \frac{\langle N \rangle_{S,v}}{S} = \begin{cases} \frac{1-2v}{1-v} & \text{for } v \le 1/2 \\ 0 & \text{for } v > 1/2 \end{cases}. \tag{49}$$

We see that $m(v)$ is the order parameter of the phase transition between the short and long excursion phases. The phase transition is continuous, as $m(v)$ is continuous at the critical point $v_{cr} = 1/2$. Taking the next derivative we obtain a measure of fluctuations

$$-vm'(v) = \lim_{S\to\infty} \frac{\langle N^2 \rangle_{S,v} - \langle N \rangle^2_{S,v}}{S} = \begin{cases} \frac{v}{(-1+v)^2} & \text{for } v \le 1/2 \\ 0 & \text{for } v > 1/2 \end{cases} \tag{50}$$

which is discontinuous at $v_{cr} = 1/2$, so the phase transition is of the second order.

Returning to the complex plane, let us see how these critical properties are reflected in the distribution of zeros of the partition function. We are interested in the vicinity of the point $v_{cr} = 1/2$ on the complex plane. Denote $\Delta v = v - v_{cr}$ for points on the critical curve $v = \gamma(s)$ for small $s > 0$, $s \ll 1$. Equation (45) gives

$$\Delta v = e^{i3\pi/4}\sqrt{s} + o(\sqrt{s}) \tag{51}$$

The impact angle of the critical line on the real axis at this point is $3\pi/4$ in the upper half plane and $-3\pi/4$ in the lower half plane. Note that the corresponding angles in the u-plane are $\mp\pi/4$ (in the lower and upper half-planes, respectively). The calculated values $\pm 3\pi/4$ are a particular case of a general solution for the random allocation model [23] where the impact angles depend on the transition order (and the critical exponents). Impact angles in the range $(\pi/2, 3\pi/4]$ in the upper half-plane correspond to second-order phase transitions. The angle $3\pi/4$ is a boundary between the second- and third-order phase transitions.

The distribution of zeros near the critical point can be calculated from Equation (35). If we parametrize the density on the critical curve γ by the distance from the critical point, then Equation (35) takes the form $\mu(|\Delta v|)d|\Delta v| = ds/(2\pi)$ from which we deduce for $|\Delta v| \ll 1$ that $|\Delta v| \approx \sqrt{s}$ (51) and

$$\mu(|\Delta v|) = \frac{1}{2\pi} \frac{ds}{d|\Delta v|} \approx \frac{|\Delta v|}{\pi} \tag{52}$$

so the density of zero goes linearly to zero, which means that zeros are sparser near the critical point.

To conclude this section, as a side remark we note that if the Dyck walks are allowed to move into the lower half plane the corresponding generating function is given by [40]

$$\tilde{F}_D(u,z) = \frac{1}{1 - u\tilde{f}_D(z)} \tag{53}$$

where $\tilde{f}_D(z) = 2f_D(z) = 1 - \sqrt{1-4z}$. The factor reflects the fact that there are twice as many types of elementary excursions as in the previous case because each type of excursion in the upper plane has a mirrored equivalent in the lower plane, so $\tilde{w}_D(s) = 2w_D(s)$. This type of Dyck path is sometimes referred to as a Bernoulli bridge [26]. If a probabilistic approach is used directly instead of enumerating Bernoulli bridges, noting that in each step the path may go up or down with probability $1/2$ gives the corresponding probabilistic weights (39)

$$\hat{w}_D(s) = \frac{1}{2^{2s}}\tilde{w}_D(s) = \frac{2}{2^{2s}}w_D(s) = (-)^{s+1}\binom{1/2}{s} \tag{54}$$

and thus the corresponding generating function is $\hat{f}_D(z) = 1 - \sqrt{1-z}$.

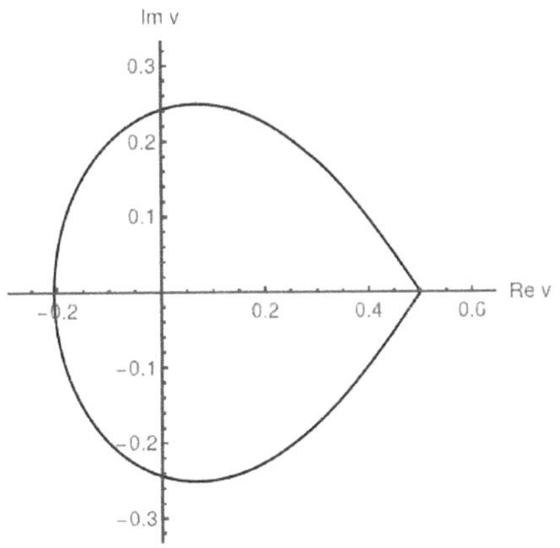

Figure 2. The analytically calculated locus of zeros for adsorbing Dyck walks in the complex v plane. The transition at $v_{cr} = 1/2$ is clearly visible, as is its second-order nature, since the impact angle of the locus of zeros is $\pm 3\pi/4$.

6. ASEPs and Lattice Paths

Blythe and Evans [4,29] were the first to observe that the Yang–Lee zeros approach to equilibrium phase transitions could also be applied to the phase transitions in the non-equilibrium steady states of the one-dimensional (totally) Asymmetric Simple Exclusion Process, or ASEP. This is defined by particles hopping from left to right on a one-dimensional lattice with open boundaries. The continuous-time variant of the model is specified by three rates.

Move	**Rate**
Particle inserted onto the left boundary site (if empty)	α
Particle removed from the right boundary site (if occupied)	β
Particle hops by one site to (an empty site on) its right	1

For the jump rate λ, the probability that a jump occurs in an infinitesimal time interval Δt is $\lambda \Delta t$. Additionally, only one particle may occupy a lattice site, see Figure 3.

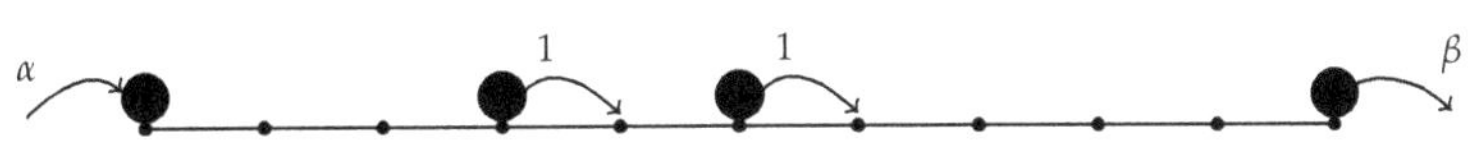

Figure 3. (Totally) Asymmetric Exclusion Process on a line with insertion rate α and removal rate β. The internal jump rate to the right is 1 and only single site occupancy is allowed.

The quantity of interest in the ASEP is Z (suggestively named to recall a partition function), which normalises the statistical weight, $w(\mathcal{C})$, of some particle configuration, $\mathcal{C}$, in the non-equilibrium steady states. This is given by

$$Z = \sum_{\mathcal{C}} w(\mathcal{C})\,, \tag{55}$$

thus the normalized probability of being in state $\mathcal{C}$ is $P(\mathcal{C}) = w(\mathcal{C})/Z$. The weights $w(\mathcal{C})$ are obtained through the stationarity condition on the transition rates $W(\mathcal{C} \to \mathcal{C}')$ which specify the model

$$\sum_{\mathcal{C}' \neq \mathcal{C}} \left[w(\mathcal{C}')W(\mathcal{C}' \to \mathcal{C}) - w(\mathcal{C})W(\mathcal{C} \to \mathcal{C}') \right] = 0\,. \tag{56}$$

$W(\mathcal{C} \to \mathcal{C}')$ is the probability of making the transition from configuration $\mathcal{C}$ to $\mathcal{C}'$ in a single timestep. Note that this is less restrictive than the detailed balance condition for equilibrium states, for which there is a term-by-term cancellation in the sum in Equation (56).

The solution of the ASEP [41] made use of a matrix product ansatz to calculate $Z_N(\alpha, \beta)$, the normalization for N sites

$$Z_N(\alpha, \beta) = \sum_{p=1}^{N} \frac{p(2N-1-p)!}{N!(N-p)!} \frac{(1/\beta)^{p+1} - (1/\alpha)^{p+1}}{(1/\beta) - (1/\alpha)}\,. \tag{57}$$

Introducing the generating function of the $Z_N(\alpha, \beta)$, or the "grand-canonical" normalization, $\mathcal{Z}(z, \alpha, \beta) = \sum_N Z_N(\alpha, \beta) z^N$ and performing the summation then gives [30]

$$\mathcal{Z}(z, \alpha, \beta) = \frac{1}{\left(1 - \frac{1}{\alpha} f_D(z)\right)\left(1 - \frac{1}{\beta} f_D(z)\right)} = F_D\left(z, \frac{1}{\alpha}\right) F_D\left(z, \frac{1}{\beta}\right) \tag{58}$$

where $f_D(z) = (1 - \sqrt{1-4z})/2$, which clearly links the ASEP normalization to the partition function of Dyck walks (38) and random allocation model. Establishing this equivalence was crucial for understanding the ASEP phase diagram. The phase diagram is shown in Figure 4. It comprises low- and high-density phases and a maximal current phase. There is a first-order transition line between the low- and high-density phases for $\alpha = \beta < 1/2$ and second-order transition lines between the low- and high-density and maximal current phases. Blythe and Evans found that this phase structure was reflected in the properties of the zeros of $Z_N(\alpha, \beta)$ in the complex α or β planes, exactly as in equilibrium models. Subsequently, it was realized that ASEP normalization (a non-equilibrium quantity) was equivalent to the partition function (an equilibrium quantity) for a class of lattice paths, so the usual Yang–Lee mechanism is in operation [4,30,42,43].

The product form of this expression shows that the grand-canonical ASEP normalization is identical to the partition function for an ensemble of pairs of non-interacting Dyck walks, with the excursion fugacities $u_1 = 1/\alpha$ and $u_2 = 1/\beta$ (or their inverses $v_1 = \alpha, v_2 = \beta$). We now mirror the discussion on obtaining the Yang–Lee zero locus from Equation (31) for the adsorbing Dyck walks in the case of ASEP. Using (58) we have

$$Z_N(\alpha, \beta) = \oint_{|z|=\epsilon} \frac{dz}{2\pi i z^{N+1}} \frac{1}{(1 - \alpha^{-1} f_D(z))(1 - \beta^{-1} f_D(z))} \tag{59}$$

Polynomials $Z_N(\alpha, \beta)$ are symmetric in α and β: $Z_N(\alpha, \beta) = Z_N(\beta, \alpha)$. We will analyze them as polynomials in α for fixed β. We are interested in the location of the phase transition for real α, and the distribution of the Yang–Lee zeros associated with this transition. The integrand (59) has a branch point singularity coming from $f_D(z) = (1 - \sqrt{1-4z})/2$ at $|z| = 1/4$, which leads to a singularity of the thermodynamic potential in the thermodynamic limit at $\alpha = 1/2$ and a pole when $\alpha \to \beta$, coming from $1/(1 - \beta^{-1} f_D(z))$, which leads to a singularity at $\alpha = \beta$. The closest singularity to the origin determines the phase so a transition occurs when two are equal, i.e., at $\alpha_{cr}(\beta) = 1/2$ for $\beta \geq 1/2$ or $\alpha_{cr}(\beta) = \beta$ for $\beta < 1/2$.

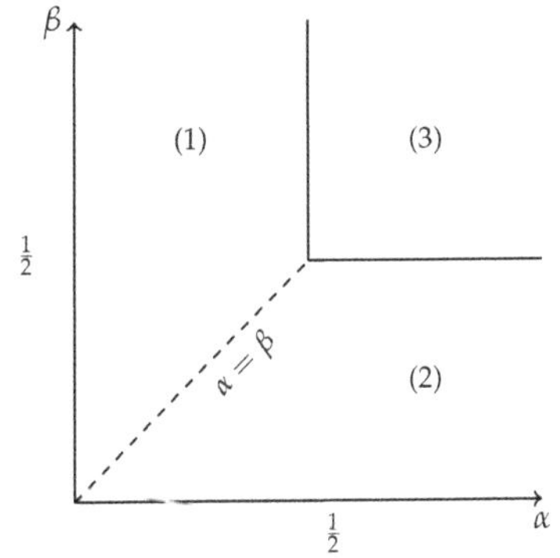

Figure 4. The ASEP phase diagram. Region (1) is a low-density phase, region (2) is a high-density phase and region (3) is the maximal current phase. The dotted transition line along $\alpha = \beta < 1/2$ is first order and the $\alpha = 1/2, \beta = 1/2$ transition lines are second order.

Taking into account the square root singularity first, the locus of zeros will be given by $\alpha = f_D(\frac{e^{is}}{4})$, for $s \in (0, 2\pi]$, when $\beta \geq 1/2$. For example, looking at $\beta = 3/4$ for the complex zeros in the variable α, we can compare the zeros calculated numerically from $Z_N(\alpha, \beta)$ in Equation (57) with this locus. We can see in Figure 5 that there is already a reasonable agreement for $N = 1000$ and as N is increased, the zeros approach the analytically determined locus further. The term $1/(1 - \beta^{-1} f_D(z))$ does not affect the limiting distribution of zeros in this case.

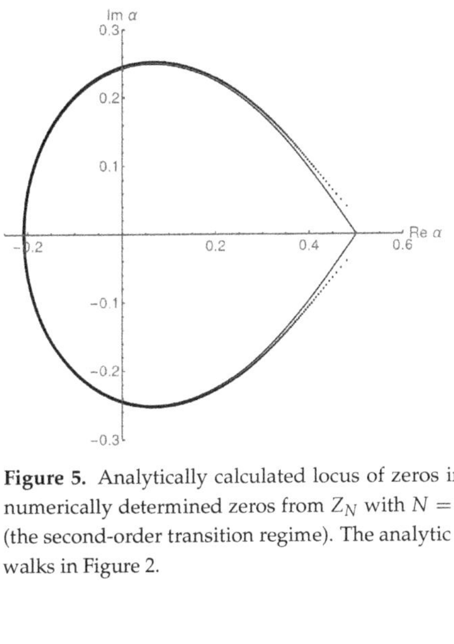

Figure 5. Analytically calculated locus of zeros in the thermodynamic limit from $\mathcal{Z}(z, \alpha, \beta)$ and numerically determined zeros from Z_N with $N = 1000$ for the ASEP normalization with $\beta = 3/4$ (the second-order transition regime). The analytic locus is identical to that for the adsorbing Dyck walks in Figure 2.

The phase transition point at $\alpha = 1/2$ appears clearly on the real axis, where the two lines of zeros meet at a right angle. Since we effectively repeat the analysis of the adsorbing Dyck walks, the impact angle of the zeros is again $\pm 3\pi/4$, as befits a second-order transition. The density of zeros vanishes at the transition point $\alpha = 1/2$ (for $\beta > 1/2$), as expected for a second-order transition (52). The locus of zeros in the complex β plane will be identical if we swap the roles of α and β.

As mentioned, for $\beta < 1/2$, the dominant singularity in the variable α is when $\alpha \to \beta$, since it comes from the pole $1/(1 - \beta^{-1} f_D(z))$. Thus, z is a solution of $\beta = f_D(z)$, which can be inverted to give $z = \beta(1-\beta) = \alpha(1-\alpha)$. The critical curve γ that gives the locus of zeros in the thermodynamic limit associated with this singularity is the f_D image of the circle of radius $\beta(1-\beta)$: $\alpha = \gamma(s) = f_D(\beta(1-\beta)e^{is})$, for $s \in (0, 2\pi]$. As an example, in Figure 6 we show the results for $\beta = 1/4$. As can be seen, the analytically determined locus and numerical zeros are already indistinguishable when $N = 1000$. In this case, the critical curve impacts the real axis at an angle $\pi/2$, which is characteristic of a first-order transition.

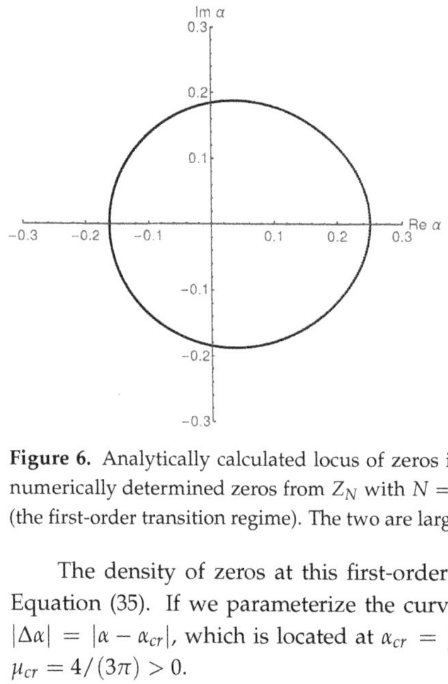

Figure 6. Analytically calculated locus of zeros in the thermodynamic limit from $\mathcal{Z}(z, \alpha, \beta)$ and numerically determined zeros from Z_N with $N = 1000$ for the ASEP normalization with $\beta = 1/4$ (the first-order transition regime). The two are largely indistinguishable.

The density of zeros at this first-order phase transition point is recovered using Equation (35). If we parameterize the curve γ by the distance from the critical point $|\Delta\alpha| = |\alpha - \alpha_{cr}|$, which is located at $\alpha_{cr} = \beta = 1/4$, the density at $\alpha_{cr} = 1/4$ is finite $\mu_{cr} = 4/(3\pi) > 0$.

7. Discussion

As well as being calculated numerically from $Z_N(\alpha, \beta)$ in [29] the analytical locus of zeros for the ASEP normalization has also been obtained previously by analytic calculations of "free energy" for the ASEP [42]

$$F = \lim_{N \to \infty} \frac{1}{N} Z_N(\alpha, \beta). \tag{60}$$

It is interesting to compare the locus of zeros for the ASEP normalization obtained in [42] with the results here (e.g., (46)). In [42] it was found that

$$\begin{aligned} F_1 = & -\ln[\alpha(1-\alpha)]\,, && \text{for } \beta > \alpha,\ \alpha < 1/2\,, && \text{region (1)} \\ F_2 = & -\ln[\beta(1-\beta)]\,, && \text{for } \alpha > \beta,\ \beta < 1/2\,, && \text{region (2)} \\ F_3 = & -\ln[1/4]\,, && \text{for } \alpha > 1/2,\ \beta > 1/2\,, && \text{region (3)} \end{aligned} \tag{61}$$

where the corresponding regions in Figure 4 are indicated. Assuming that this "free energy" derived from ASEP normalization can be treated like the free energy of an equilibrium model, which can be justified post hoc by the Dyck path equivalence, allowing a direct determination of the locus of partition function zeros since these mark the boundaries of the different phases.

Using the electrostatic analogy described in the Introduction, the loci of the zeros are determined by observing that $\Re\psi_i = \Re\psi_j$ for adjacent phases i, j. The loci of the zeros are also determined by matching the real parts of the different analytic expressions for the complex free energy in different phases in [9,10]. In the case of the ASEP, this is equivalent to

$$|\alpha(1-\alpha)| = 1/4\,, \tag{62}$$

for the second-order transition line where $F_1 = F_3$ or

$$|\beta(1-\beta)| = 1/4\,, \tag{63}$$

for the second-order transition line where $F_2 = F_3$. Across the first-order line when $\alpha = \beta < 1/2$ we have $F_1 = F_2$, so

$$|\alpha(1-\alpha)| = |\beta(1-\beta)|\,. \tag{64}$$

The locus of zeros in Figure 5 is reproduced as the f_D image of the circle $|z| = 1/4$ and it is representative of the second-order phase transition line between regions (1) and (3) in Figure 4 (or regions (2) and (3) if the roles of α and β were changed). The locus of zeros in Figure 6 is reproduced as the f_D image of the circle $|z| = \alpha(1-\alpha) = \beta(1-\beta) < 1/4$ and is representative of the first-order phase transition line between regions (1) and (2).

This "physical"/electrostatic, rather than analytical, approach to obtaining the loci of zeros also recovers the results for the random allocation model outlined in Section 4. There the electrostatic potential $\Phi(u) = \Re\psi(u)$ plays the role of the thermodynamic potential F, so matching the real parts of this for the fluid and condensed phases in Equation (34) gives $|z_0(u)| = 1$. Inverting this using Equation (33) gives the locus of zeros in the u plane found in [23], i.e., $u = 1/f(\sigma e^{is})$.

To summarize, we have obtained (analytically) the locus of the Yang–Lee zeros for various lattice path models and the ASEP normalization or, equivalently, for the partition function of an adsorbing one-transit Dyck walk. For the random allocation models, the path models and the ASEP the loci of zeros and their density are straightforwardly given by conformal mappings of a circle. The required mappings are determined from a generating function $f(z) = \sum_{s=1}^{\infty} w(s) z^s$ for random allocation models and $f_D(z) = (1 - \sqrt{1-4z})/2$ for adsorbing Dyck walks and ASEP. This follows a contour integral evaluation of the normalization/partition functions using the grand-canonical generating functions as integrands. The results obtained for the ASEP and the random allocation model are consistent with those obtained by matching the real parts of free energies (the electrostatic analogy) and with the numerically calculated zeros.

It would be interesting to explore whether obtaining the exact locus of zeros from the conformal map technique used here could be extended to models with *pair*-factorized partition functions/steady-state normalizations or urn-type models with multiple constraints.

Author Contributions: Conceptualization, Z.B. and D.A.J.; Writing—original draft, Z.B. and D.A.J. All authors have read and agreed to the published version of the manuscript.

Funding: This research received no external funding

Institutional Review Board Statement: Not applicable

Data Availability Statement: The data presented in this study are available on request from the corresponding author.

Acknowledgments: We would like to thank M. Kieburg for discussions and contributing some of the key ideas. In addition, it should be noted that Ralph Kenna contributed greatly to DAJ's understanding of partition function zeros and finite size scaling, through joint works such as [8] and elsewhere. He will be sadly missed.

Conflicts of Interest: The authors declare no conflicts of interest.

References

1. Yang, C.N.; Lee, T.D. Statistical Theory of Equations of State and Phase Transitions. I. Theory of Condensation. *Phys. Rev.* **1952**, *87*, 404–409. [CrossRef]
2. Lee, T.D.; Yang, C.N. Statistical Theory of Equations of State and Phase Transitions. II. Lattice Gas and Ising Model. *Phys. Rev.* **1952**, *87*, 410–419. [CrossRef]
3. Bena, I.; Droz, M.; Lipowski, A. Statistical Mechanics of Equilibrium and Nonequilibrium Phase Transitions. *Int. J. Mod. Phys. B* **2005**, *19*, 4269–4329. [CrossRef]
4. Blythe, R.A.; Evans, M.R. The Lee–Yang theory of equilibrium and nonequilibrium phase transitions. *Braz. J. Phys.* **2003**, *33*, 464–475. [CrossRef]
5. Abe, R. On the Zeros of the Partition Function for the Ising Model. *Prog. Theor. Phys.* **1967**, *37*, 1070–1072. [CrossRef]
6. Suzuki, M. A Theory on the Critical Behaviour of Ferromagnets. *Prog. Theor. Phys.* **1967**, *38*, 289–291. [CrossRef]
7. Fisher, M.E. *Lectures in Theoretical Physics*; Brittin, W.E., Ed.; Gordon and Breach: New York, NY, USA, 1968; Volume VII, pp. 1–46.
8. Janke, W.; Johnston, D.A.; Kenna, R. Properties of Higher-Order Phase Transitions. *Nucl. Phys. B* **2006**, *736*, 319–338. [CrossRef]
9. Biskup, M.; Borgs, C; Chayes, J.T.; Kleinwaks, L.J.; Kotecký, R. Partition Function Zeros at First-Order Phase Transitions: A General Analysis, *Commun. Math. Phys.* **2004**, *251*, 79–131. [CrossRef]
10. Biskup, M.; Borgs, C; Chayes, J.T.; Kleinwaks, L.J.; Kotecký, R. General Theory of Lee–Yang Zeros in Models with First-Order Phase Transitions, *Phys. Rev. Lett.* **2000**, *84*, 4794–4797. [CrossRef]
11. Shlosman, S.B. Unusual Analytic Properties of Some Lattice Models: Complement of Lee–Yang Theory, *Teor. Mat. Fiz.* **1986**, *69*, 273–278. [CrossRef]
12. Matveev, V.; Shrock, R. Complex-Temperature Singularities of the Susceptibility in the $d = 2$ Ising Model. I. Square Lattice. *J. Phys. A* **1995**, *28*, 1557–1573. [CrossRef]
13. Katsura, S.; Ohminami, M. Distribution of zeros of the partition function for the one dimensional Ising models. *J. Phys. A* **1972**, *5*, 95–105. [CrossRef]
14. Glumac, Z.; Uzelac, K. The partition function zeros in the one-dimensional q-state Potts model. *J. Phys. A* **1994**, *27*, 7709–7720. [CrossRef]
15. Ghulghazaryan, R.G.; Ananikian, N.S. Partition function zeros of the one-dimensional Potts model: The recursive method. *J. Phys. A* **2003**, *36*, 6297–6305. [CrossRef]
16. Krasnytska, M.; Berche, B.; Holovatch, Y.; Kenna, R. Partition function zeros for the Ising model on complete graphs and on annealed scale-free networks. *J. Phys. A* **2016**, *49*, 135001. [CrossRef]
17. Itzykson, C.; Pearson, R.; Zuber, J.B. Distribution of Zeros in Ising and Gauge Models. *Nucl. Phys. B* **1983**, *220*, 415–433. [CrossRef]
18. Baake, M.; Grimm, U.; Pisani, C. Partition Function Zeros for Aperiodic Systems. *J. Stat. Phys.* **1995**, *78*, 285–297. [CrossRef]
19. Repetowicz, P.; Grimm, U.; Schreiber, M. Planar quasiperiodic Ising models *Mater. Sci. Eng. A* **2000**, *294–296*, 638–641. [CrossRef]
20. Derrida, B. The zeroes of the partition function of the random energy model. *Physica A* **1991**, *177*, 31–37. [CrossRef]
21. Staudacher, M. The Yang-Lee edge singularity on a dynamical planar random surface. *Nucl. Phys. B* **1990**, *336*, 349–362. [CrossRef]
22. Janke, W.; Johnston, D.A.; Stathakopoulos, M. Fat Fisher Zeroes. *Nucl. Phys. B* **2001**, *614*, 494–514. [CrossRef]

23. Burda, Z.; Johnston, D.A.; Kieburg, M. Yang-Lee Zeros for Real-Space Condensation. *Phys. Rev. E* **2025**, *111*, L012101. [CrossRef]
24. Flajolet, P.; Sedgewick, R. *Analytic Combinatorics*; Cambridge University Press: Cambridge, UK, 2009.
25. Godrèche, C. Condensation for random variables conditioned by the value of their sum *J. Stat. Mech.* **2019**, 063207. [CrossRef]
26. Godrèche, C. Condensation and Extremes for a Fluctuating Number of Independent Random Variables. *J. Stat. Phys.* **2021**, *182*, 13. [CrossRef]
27. Godrèche, C.; Luck, J.M. Nonequilibrium dynamics of a simple stochastic model. *J. Phys. A* **1997**, *30*, 6245–6264. [CrossRef]
28. Corberi, F.; Sarracino, A. Probability Distributions with Singularities. *Entropy* **2019**, *21*, 312. [CrossRef] [PubMed]
29. Blythe, R.A.; Evans, M.R. Lee–Yang Zeros and Phase Transitions in Nonequilibrium Steady States. *Phys. Rev. Lett.* **2002**, *89*, 080601. [CrossRef] [PubMed]
30. Blythe, R.A.; Janke, W.; Johnston, D.A.; Kenna, R. The Grand-Canonical Asymmetric Exclusion Process and the One-Transit Walk. *J. Stat. Mech.* **2004**, P06001. [CrossRef]
31. Ritort, F. Glassiness in a Model without Energy Barriers. *Phys. Rev. Lett.* **1995**, *75*, 1190–1193. [CrossRef]
32. Bialas, P.; Burda, Z.; Johnston, D. Condensation in the Backgammon Model. *Nucl. Phys. B* **1997**, *493*, 505–516. [CrossRef]
33. Ehrenfest, P.; Ehrenfest, T. Über zwei bekannte Einwände gegen das Boltzmannsche H-Theorem. *Phys. Zeit.* **1907**, *8*, 311–314.
34. Bialas, P.; Burda, Z. Phase Transition in Fluctuating Branched Geometry. *Phys. Lett. B* **1996**, *384*, 75–80. [CrossRef]
35. Franz, S.; Ritort, F. Glassy Mean-Field Dynamics of the Backgammon Model. *J. Stat. Phys.* **1996**, *85*, 131–150. [CrossRef]
36. Bialas, P.; Burda, Z.; Johnston, D. Phase Diagram of the Mean Field Model of Simplicial Gravity. *Nucl. Phys. B* **1999**, *542*, 413–424. [CrossRef]
37. Bialas, P.; Burda, Z.; Johnston, D. Random Allocation Models in the Thermodynamic Limit. *Phys. Rev. E* **2023**, *108*, 064107. [CrossRef]
38. Bialas, P.; Burda, Z.; Johnston, D. Rényi Entropy of Zeta Urns. *Phys. Rev. E* **2023**, *108*, 064108. [CrossRef]
39. Bialas, P.; Burda, Z.; Johnston, D. Partition Function Zeros of Zeta-Urns. *Condens. Matter Phys.* **2024**, *27*, 33601. [CrossRef]
40. Janse van Rensburg, E.J. *The Statistical Mechanics of Interacting Walks, Polygons, Animals and Vesicles*; Oxford University Press: Oxford, UK, 2000; pp. 98–104.
41. Derrida, B.; Evans, M.R.; Hakim, V.; Pasquier, V. Exact Solution of a 1D Asymmetric Exclusion Model Using a Matrix Formulation. *J. Phys. A Math. Gen.* **1993**, *26*, 1493–1517. [CrossRef]
42. Blythe, R.A.; Janke, W.; Johnston, D.A.; Kenna, R. Dyck Paths, Motzkin Paths and Traffic Jams. *J. Stat. Mech.* **2004**, P10007. [CrossRef]
43. Wood, A.J.; Blythe, R.A.; Evans, M.R. Combinatorial Mappings of Exclusion Processes. *J. Phys. A Math. Theor.* **2020**, *53*, 123001. [CrossRef]

MDPI AG
Grosspeteranlage 5
4052 Basel
Switzerland
Tel.: +41 61 683 77 34

Entropy Editorial Office
E-mail: entropy@mdpi.com
www.mdpi.com/journal/entropy

www.ingramcontent.com/pod-product-compliance
Lightning Source LLC
Chambersburg PA
CBHW061843210926
51850CB00006B/954
* 9 7 8 3 7 2 5 8 5 1 4 7 8 *